Springer Series on Environmental Management

More information about this series at http://www.springer.com/series/412

Ceanothus chaparral (*Ceanothus crassifolius*) in the Trabuco Ranger District, Cleveland National Forest. Site was last burned in 1980. Photo by Richard Halsey.

Human presence in chaparral dominated ecosystems. Looking west across the Rancho Bernardo area of San Diego with Black Mountain and the Pacific Ocean in the background. Photo by Richard Halsey.

Phacelia grandiflora flowering the spring after a chaparral fire. This species is a "fire-follower", with germination stimulated by chemicals in charred wood or smoke. Photo by Richard Halsey.

Managers and researchers inspecting stream channel filled with sediment eroded from hillsides denuded by the Powerhouse Fire (2013), Angeles National Forest, Los Angeles County. Photo by Hugh Safford.

Prescribed fire in chaparral at the Bureau of Land Management's Cow Mountain Recreation Area, Mendocino County. The area is managed for a mix of uses, with a focus on recreation, including off-highway vehicle and motorcycle riding, biking, hiking, hunting, recreational shooting, horseback riding, and camping. Photo by Scott Stephens.

Chaparral restoration trials, Piru Fire (2003), Los Padres National Forest, Ventura County. Photo by Stephanie Ma.

Emma C. Underwood • Hugh D. Safford
Nicole A. Molinari • Jon E. Keeley
Editors

Valuing Chaparral

Ecological, Socio-Economic, and Management Perspectives

Springer

Editors
Emma C. Underwood
Department of Environmental Science and Policy
University of California
Davis, CA, USA

Centre for Biological Sciences
University of Southampton
Southampton, UK

Hugh D. Safford
USDA Forest Service
Pacific Southwest Region
Vallejo, CA, USA

Department of Environmental Science and Policy
University of California
Davis, CA, USA

Nicole A. Molinari
USDA Forest Service
Los Padres National Forest
Goleta, CA, USA

Jon E. Keeley
U.S. Geological Survey
Western Ecological Research Center
Three Rivers, CA, USA

Department of Ecology and Evolutionary Biology, University of California
Los Angeles, CA, USA

ISSN 0172-6161
Springer Series on Environmental Management
ISBN 978-3-319-68302-7 ISBN 978-3-319-68303-4 (eBook)
https://doi.org/10.1007/978-3-319-68303-4

Library of Congress Control Number: 2017960295

Printed on acid-free paper

This Springer imprint is published by the registered company Springer International Publishing AG part of Springer Nature.
The registered company address is: Gewerbestrasse 11, 6330 Cham, Switzerland

Foreword

California Chaparral Over the Decades: Societal Challenges and a New Vision

This book is a refreshing new analysis of the nature and functioning of the chaparral ecosystem. Its 16 chapters are devoted to the description of California chaparral, its ecological and environmental components, past and current threats to it and to society, and an in-depth history of its use. What it brings new to the table is framing these descriptions in an ecosystem service context, including intrinsic values as well as instrumental ones (values not an end in itself but ones that provide a broader contribution for human well-being, such as watershed protection for the millions of people who live in close proximity to chaparral ecosystems). Furthermore, throughout its pages the book is uniquely seasoned with contrasting perspectives of how scientists and society have viewed chaparral through the years.

Over the decades, opinions about chaparral have varied between a system to be avoided and a system to be loved. William Brewer in his journey through California in 1860–1864, as the botanist on the initial Geological Survey of the state, presented a less than joyous image of chaparral. He describes his ventures on the slopes of the San Gabriel Mountains of southern California, "… the real difficulty was the chaparral, which in places seemed absolutely impenetrable—a tangled mass of stiff, interlaced, thorny shrubs. Sometimes we broke them down … sometimes tore through, sometimes crawled on our hands and knees a long distance. At one time nearly an hour was consumed in making probably sixty or eighty rods" (Brewer 1966). This contrasts with the effervescent and rousing descriptions provided by John Muir a few decades later, where chaparral "swoops into every hollow and swells over ever ridge … in shaggy ungovernable exuberance" (Muir 1918).

In the early part of the twentieth century chaparral was recognized as a unique system to be studied. William S. Cooper, one of the pioneering ecologists in the United States, started his career at Stanford University where he taught ecology during 1914–1916 and initiated his seminal chaparral analysis. After moving to the University of Minnesota he continued his studies of the chaparral and published his in-depth monograph of the structure and functioning of chaparral (Cooper 1922). After Cooper there was a long time before more vegetation studies were pursued in chaparral. However, during this interim, the pioneering animal ecologist at the University of California,

Berkeley, Joseph Grinnell, produced his seminal paper on the niche concept based on a study of the California Thrasher in chaparral (Grinnell 1917). Incidentally, another study by Grinnell conducted in California's national parks was the first to discuss the concept of "ecological services" that were provided by animal life (Grinnell and Storer 1916). This volume synthesizes the vast storehouse of new information that has accumulated since Cooper and Grinnell emphasize the dynamics of the system and its remarkable biodiversity when viewed through successional time.

After this initial research, the 1950s witnessed a period when chaparral was viewed as a system to be converted and replaced with other vegetation types. Forty years after the Cooper study, range managers of the California Experiment Station were actively promoting the conversion of chaparral to rangelands, stating that "while not all of this land (chaparral) can be cleared profitably, large areas of it will repay the time and money spent on such a project by providing support for additional (live) stock." They used the term "worthless brush" to emphasize the value to be had by conversion (Love et al. 1952). Previously, Cooper (1922) had also noted when national forests were established in California (many of which included large swaths of chaparral) that many foresters had been conflicted between whether this brush should be replaced by something of economic value, and if not, "how can it best be made to perform its important economic function of watershed protection." This book reviews in great detail the types of ecological functions and services that chaparral can provide and exemplifies links between these services and economic valuation today.

Finally, the limiting supply of fresh water coupled with a history of fires and subsequent flooding in the southern California mountains ingrained the doctrine that chaparral was a system that needs to be managed. Its proximity to large human populations prompted the USDA Forest Service in 1933 to establish a research facility in a watershed in the San Gabriel Mountains. This facility, the San Dimas Experimental Forest, provided valuable data through time as to the best management practices for the chaparral dominated watershed. It was found that although a greater flow of water to reservoirs was enhanced by conversion of the slopes to grassland cover, there was a very unfavorable trade-off in the loss of erosion control provided by chaparral cover.

Today, the too-frequent devastating fires that are largely caused by human ignitions (as noted in this volume) result in extremely high firefighting costs to protect dwellings that are situated near or embedded within chaparral. Coupling fuel management of vegetation with improved interaction and communication with residents and city planners (for example, to regulate the siting of dwellings in vulnerable areas) can take the management of chaparral to a new level, one which embraces the human component as well.

But ultimately, this book underscores that this is a system to be treasured. The strong message is that chaparral ecosystems provide intrinsic values to society, as well as economically valued ecosystem services (carbon storage, sediment retention, water provision, and biodiversity) that can be quantified and mapped. These are important new additions to our understanding and appreciation of the chaparral ecosystem, while the many excellent chapters of this path-breaking volume update our knowledge of this system.

Stanford University, Stanford, California Harold A. Mooney

References

1. Brewer, W. 1966. Up and down California. University of California Press, Berkeley, California, USA.
2. Cooper, W. C. 1922. The broad-sclerophyll vegetation of California: An ecological study of the chaparral and its related communities. Carnegie Institute of Washington, Stanford, California, USA. Publication No. 319. 124p.
3. Grinnell, J., and T. I. Storer. 1916. Animal life as an asset of national parks. Science:375–380.
4. Grinnell, J. 1917. The niche-relationships of the California Thrasher. Auk 34:427–433.
5. Love, R. M., and B. J. Jones. 1952. Improving California brush ranges. California Agricultural Experiment Station Circular 371:4–38.
6. Muir, J. 1918. Steep trails. Houghton Mifflin, Boston, Massachusetts, USA.

Preface

In the spring of 2013, the Pacific Southwest Region of the US Forest Service convened the first Southern California Chaparral Symposium in Arcadia, California. The event focused on ecological restoration of southern California shrublands, and over 200 people attended three days of presentations, discussions, and field trips. The success of this workshop and the desire for expanded opportunities to publicly discuss management and conservation of chaparral led to a second US Forest Service-sponsored symposium titled "Understanding the Ecological Value of Chaparral Landscapes" in June, 2015, again in Arcadia. The 2015 symposium—even better attended than the 2013 event—sought to highlight the importance of chaparral landscapes, and included presentations and discussions revolving around the physical and biotic environment, human history and land use, resource management, and ecosystem services. The momentum behind the 2013 and 2015 symposia was carried forward to the 2016 Natural Areas Conference at the University of California, Davis, where three well-attended oral sessions explored current management challenges, the future of chaparral ecosystems, and chaparral ecosystem services. This book is the fruit of these seminal events, and represents a synthesis of what we know today about California's chaparral shrublands, their ecology, their management, and their contributions to human well-being. Most of the top experts in chaparral ecology, management, and conservation participated as authors or reviewers. We believe the result is essential reading for land managers, researchers, policy makers, and the general public alike.

The timing is right for this synthesis. Not only has chaparral management become a major and often controversial issue given the frequent damage to homes, infrastructure, and loss of lives from wildfire, but more generally, as indicated throughout the book, chaparral is a poorly understood and highly under-valued ecosystem. Although popular perceptions are changing—and we hope our book can catalyze an acceleration in this welcome trend—chaparral has been one of the most unappreciated of California's ecosystems. This book provides a wealth of evidence to the contrary. The carpet of chaparral that covers many low- and mid-elevation mountainsides in semi-arid parts of California is essential to retain soil cover, recharge groundwater, provide pollination of nearby crops, and provide clean drinking water for millions of households. Furthermore, the counties of California with the most chaparral are also those that support the highest terrestrial biodiversity in the contiguous 48 states. The explosion of chaparral flowers during the spring after fire attracts botanical tourists from far and wide, and chaparral covered landscapes are favorite escapes for hikers, hunters, mountain bikers, and off-road enthusiasts.

This book seeks to resurrect public and science interest in chaparral and its management. The 1970s and 1980s were decades of great academic and practical interest in chaparral. Strong ecological and evolutionary similarities among the world's Mediterranean-type climate regions (Mediterranean Basin; California US and Baja California, Mexico; southern and southwestern Australia; Cape Province of South Africa; and central coastal Chile) led to international meetings of Mediterranean ecosystem scientists (e.g., MEDECOS 1971, in Chile) and collaboratives such as the International Society for Mediterranean Ecology. Within California, the US Forest Service's Vegetation Management Alternatives for Chaparral and Related Ecosystems program in the 1980s also contributed to the flurry of attention on chaparral. However, this attention has waned over the last 25 years. Resource management budgets in the federal agencies have strongly swung towards fire management at the expense of conservation and restoration, and management focus has been redirected to more pressing and immediate needs, such as reacting to the latest disaster (flood, mudslide, fire etc.) and struggling with the day-to-day demands of managing wildlands in a rapidly growing urban environment. This book is a whole-hearted effort to realign these perceptions and priorities and engender an awareness and respect for these unique and valuable landscapes.

This book is organized into four sections. We start by placing California's chaparral into a regional and global context and highlighting its key ecological characteristics in terms of physiological attributes of the vegetation, the role of fire, and plant and animal diversity (Chaps. 1, 2 and 3). We then focus on the cultural history and Native American uses of chaparral (Chap. 4), and contrast this with perceptions and uses of chaparral in modern times (Chap. 5). Next, we highlight three key ecosystem services that chaparral landscapes provide—carbon storage and sequestration (Chap. 6), sediment erosion control and flood control (Chap. 7), and water quality and quantity (Chap. 8). These chapters on individual ecosystem services are followed by a study mapping the provision of multiple ecosystem services across the southern California landscape to identify priority areas for resource management (Chap. 9) and a detailed review of the recreational services provided by the Angeles, Los Padres, San Bernardino, and Cleveland National Forests (Chap. 10).

These details on the provision of chaparral ecosystem services are followed by a description of the current status of educational efforts related to chaparral in southern California (Chap. 11). The final chapters are devoted to the management of chaparral landscapes currently and in the future: describing threats to chaparral in terms of conversion of native shrubland to non-native grasslands (Chap. 12); ecological restoration efforts in chaparral landscapes (Chap. 13); and information on future climate projections and likely ecosystem impacts in California chaparral regions (Chap. 14). The penultimate chapter is focused on resource management in chaparral, which highlights key areas of management focus and how these interface with ecosystem services, including a series of case studies based on progressive and collaborative projects currently underway in southern California (Chap. 15). The final chapter summarizes the past, present, and future of California chaparral (Chap. 16).

By compiling hundreds of collective years of experience and knowledge of chaparral ecoystems into a single resource, this book aims to spark renewed interest and resolve for managing chaparral landscapes today and into the future. We hope that our endeavor provides a cornerstone for efforts to more sustainably manage chaparral in the face of rapid change and overwhelming threats, and a stimulus for people to teach themselves and others about the value of chaparral.

Department of Environmental Science and Policy
University of California, Davis, CA, USA
Centre for Biological Sciences
University of Southampton, UK
Emma C. Underwood

USDA Forest Service
Pacific Southwest Region, Vallejo, CA, USA
Department of Environmental Science and Policy
University of California, Davis, CA, USA
Hugh D. Safford

USDA Forest Service
Los Padres National Forest, Goleta, CA, USA
Nicole A. Molinari

U.S. Geological Survey, Western Ecological Research Center
Three Rivers, CA, USA
Department of Ecology and Evolutionary Biology
University of California, Los Angeles, CA, USA
Jon E. Keeley

Acknowledgments

First and foremost we express our greatest appreciation to the participants in the 2015 Chaparral Symposium who contributed their knowledge, experience, and valuable time to the symposium and to composing the chapters of this book. The dedication and passion with which they undertake their work is an inspiration and a critical component to increasing awareness of the value of chaparral ecosystems.

The 2015 Chaparral Symposium in Arcadia, California, was made possible by funding from the US Forest Service Pacific Southwest Region and Pacific Southwest Research Station, the US Forest Service Western Wildlands Environmental Threats Assessment Center, the California Landscape Conservation Cooperative, and The Nature Conservancy. We are extremely grateful to the Angeles National Forest for providing the Angeles Training and Conference Center and to the National Forest Foundation for logistical support of the symposium and field trip. The California Fire Science Consortium was integral to symposium planning and registration, as well as providing technical support for recording and editing presentations.

The symposium's organization, field trip, and the book itself would not have been possible without the enormous help and contributions of many individuals within the USDA Forest Service. These include Jan Beyers, Debbie Chavez, Marty Dumpis, Jen Hooper, Keith Lilley, Tracy McGuff, Mark Metcalfe, MaryBeth Najera, Sarah Sawyer, Susan Shaw, Gloria Silva, Jamie Uyehara, Katie VinZant, and Pete Wohlgemuth. We also are grateful to many others from a range of organizations spanning academia, non-profits, and federal organizations. Salient among many are Vance Russell, Edward Belden, Naomi Fraga, Stacey Frederick, Jim Quinn, and Marti Witter.

We also express our thanks to the people and organizations whose work and assistance contributed indirectly to the chapters in this book including Lorrie Flint, Alan Flint, and Bill Elliot. The education chapter involved interviews with many volunteer naturalists and organizers of naturalist programs who are essential to the mission of nature education, we especially thank Bill Howell, Chanz Boudreaux, and Fred Modern. We also thank Whitney Reynier of EcoAdapt who provided useful details of a collaborative approach to determine the vulnerability of chaparral to climate change.

Finally, we are indebted to the many colleagues who provided careful and insightful reviews of book chapters, both formal and informal: David Ackerly, Peter

Alagona, Van Butsic, Christy Brigham, Richard Cowling, Frank Davis, Stephen Fillmore, Joan Florsheim, Cynthia Fowler, Janet Franklin, Jim Grace, Jim Graham, Rick Halsey, Susan Harrison, Jeff Heys, Karen Holl, Kristie Klose, Dawn Lawson, Keith Lombardo, Wallace Meyer, Dan O'Connor, David Peterson, John Randall, Seth Riley, Cristina Sandoval, Josh Schimel, Tom Scott, Brent Sohngen, Jerre Ann Stallcup, Robert Taylor, Jan Timbrook, Kellie Uyeda, Ramon Vallejo, Kirsten Winter, Marti Witter, Peter Wohlgemuth, and Paul Zedler. We also thank Marcel Safford and Katie Nigro for assisting with editing. The time and efforts of these people has led to a book that we hope will provide a deeper understanding and appreciation of these complex, unique, and beautiful chaparral landscapes.

Contents

About the Authors

Dr. Emma C. Underwood is a research scientist with the Department of Environmental Science and Policy at the University of California, Davis and is a visiting fellow with Southampton University in the UK. A central theme of her research is the application of geospatial tools and remote sensing techniques to address biodiversity and conservation issues and inform environmental decision-making. Her research interests include conservation assessments of biodiversity, estimating conservation return on investment, evaluating ecosystem services, and mapping and predicting the distribution of invasive plant species. During the past 15 years Emma's research has spanned a variety of ecosystems including tropical forests and a global conservation assessment of Mediterranean-type climate regions. Prior to UC Davis, she worked with the World Wildlife Fund-US and has since undertaken collaborative research with The Nature Conservancy, the US Geological Survey, and the US Forest Service. Emma received her Ph.D. in Ecology from the University of California.

Dr. Hugh D. Safford is Regional Ecologist for the USDA Forest Service's Pacific Southwest Region (California, Hawaii, Pacific territories), and member of the research faculty in the Department of Environmental Science and Policy at the University of California, Davis. Hugh manages a staff of ecologists that provide expertise in vegetation, fire, and restoration ecology, climate change, inventory, and monitoring to the 18 national forests in the Pacific Southwest Region. He is manager of the Regional Research Natural Area program, Sierra Nevada region leader for the California Fire Science Delivery Consortium, and science advisory board member for a number of environmental collaboratives and NGOs. Hugh provides international technical assistance on fire, forest management, and climate change issues in partnership with the US Agency for International Development (USAID) and the International Program of the Forest Service. Hugh grew up in southwestern Montana and now splits his time between Davis and Lake Tahoe, California. He earned his Ph.D. in Ecology from the University of California in 1999.

Dr. Nicole A. Molinari is a community ecologist with a broad interest in the consequences of human induced global changes, including the effects of biological invasions, nutrient enrichment, climate change, and altered disturbance regimes on vegetation patterns. Nicole currently serves as the US Forest Service province ecologist for southern California (Angeles, Cleveland, Los Padres, and San Bernardino national forests) where she applies scientific and ecological principles to address management issues. Her focus with the US Forest Service centers on fire ecology, ecological restoration, and the response of vegetation to a changing environment. She is particularly interested in questions that couple her curiosity for the natural world with applied outcomes that can inform management and land use decisions. Nicole received her MS in Biology from Cal Poly-San Luis Obispo, and her Ph.D. in Ecology and Evolution from the University of California Santa Barbara in 2014.

Dr. Jon E. Keeley is a senior scientist with the US Geological Survey, an adjunct professor at UCLA, and a research associate at the Rancho Santa Ana Botanic Garden. He has served in Washington, D.C. as director of the ecology program for the National Science Foundation and was professor of biology at Occidental College for 20 years. He has spent sabbaticals in all of the Mediterranean-type climate regions of the world. Jon has over 350 publications in national and international scientific journals. His research on wildfires includes work on ecological and climate impacts, as well as non-native invasive plants, fire-stimulated seed germination, taxonomy of manzanitas (*Arctostaphylos*) and biochemical pathways of photosynthesis in vernal pool plants. He has received a Guggenheim Fellowship, is an Honorary Lifetime Member of the California Botanical Society, and a Fellow of the Ecological Society of America. He is senior author of a 2012 Cambridge University Press book "*Fire in Mediterranean Climate Ecosystems: Ecology, Evolution and Management.*"

Contributors

Edith B. Allen Department of Botany and Plant Sciences and Center for Conservation Biology, University of California, Riverside, CA, USA

M. Kat Anderson NRCS State Office, Davis, CA, USA

Holly M. Andrews Department of Botany and Plant Sciences, University of California, Riverside, CA, USA

Jan L. Beyers USDA Forest Service, Pacific Southwest Research Station, Riverside, CA, USA

Teresa J. Brennan U.S. Geological Survey, Western Ecological Research Center, Sequoia Field Station, Three Rivers, CA, USA

Carla M. D'Antonio Department of Ecology, Evolution, and Marine Biology and Environmental Studies, University of California, Santa Barbara, CA, USA

Jennifer R. Eberwein Department of Botany and Plant Sciences, University of California, Riverside, CA, USA

Cloé Garnache Department of Agricultural, Food and Resource Economics, Michigan State University, East Lansing, MI, USA

Rochelle Gaudette Canyoneers, San Diego Natural History Museum, San Diego, CA, USA

Richard W. Halsey California Chaparral Institute, Escondido, CA, USA

Victoria W. Halsey Ken Blanchard Companies, Escondido, CA, USA

Allan D. Hollander Department of Environmental Science and Policy, University of California, Davis, CA, USA

Patrick R. Huber Agricultural Sustainability Institute, University of California, Davis, CA, USA

G. Darrel Jenerette Department of Botany and Plant Sciences, University of California, Riverside, CA, USA

Megan K. Jennings Biology Department, San Diego State University, San Diego, CA, USA

Jon E. Keeley U.S. Geological Survey, Western Ecological Research Center, Sequoia Field Station, Three Rivers, CA, USA

Department of Ecology and Evolutionary Biology, University of California, Los Angeles, CA, USA

John B. Kim USDA Forest Service, Pacific Northwest Research Station, Corvallis, OR, USA

Keith A. Lilley County of Los Angeles, Department of Public Works, Alhambra, CA, USA

Frank Lupi Department of Agricultural, Food and Resource Economics, Michigan State University, East Lansing, MI, USA

Stephanie Ma Department of Ecology, Evolution, and Marine Biology, University of California, Santa Barbara, CA, USA

Char Miller Environmental Analysis Program, Pomona College, Claremont, CA, USA

Nicole A. Molinari USDA Forest Service, Los Padres National Forest, Goleta, CA, USA

Isaac W. Park Department of Botany and Plant Sciences, University of California, Riverside, CA, USA

Michala Phillips Department of Botany and Plant Sciences, University of California, Riverside, CA, USA

Vince H. Resh Environmental Science, Policy, and Management, Division of Organisms & Environment, University of California, Berkeley, CA, USA

Philip W. Rundel Director of the Mildred E. Mathias Botanical Garden and Department of Ecology and Evolutionary Biology, University of California, Los Angeles, CA, USA

Hugh D. Safford USDA Forest Service, Pacific Southwest Region, Vallejo, CA, USA

Department of Environmental Science and Policy, University of California, Davis, CA, USA

José J. Sánchez Pacific Southwest Research Station, USDA Forest Service, Riverside, CA, USA

Charlie Schrader-Patton RedCastle Resources, Inc., USDA Forest Service, Western Wildland Environmental Threat Assessment Center, Pineville, OR, USA

Christopher W. Solek Council for Watershed Health, Los Angeles, CA, USA

Lorie Srivastava Department of Environmental Science and Policy, University of California, Davis, CA, USA

Alexandra D. Syphard Conservation Biology Institute, Corvallis, OR, USA

Emma C. Underwood Department of Environmental Science and Policy, University of California, Davis, CA, USA

Centre for Biological Sciences, University of Southampton, UK

Kimberlyn Williams Biology Department, California State University, San Bernardino, CA, USA

Peter M. Wohlgemuth USDA Forest Service, Pacific Southwest Research Station, Riverside, CA, USA

Chapter 1
California Chaparral and Its Global Significance

Philip W. Rundel

Abstract Chaparral ecosystems represent the iconic vegetation of California, and in particular southern California, where it forms the dominant vegetation cover over broad areas of the foothills of the Coast, Transverse, and Peninsular ranges. Evergreen sclerophyll shrubs which makeup the characteristic component of chaparral communities parallel a similar dominance of this life-form in the Mediterranean Basin, central Chile, the Cape Region of South Africa, and Southwest Australia, regions of the world with a Mediterranean-type climate of warm dry summers and cool wet winters. The Mediterranean Biome comprised of these five regions are biodiversity hotspots that contain about one-sixth of the vascular plant species in the world in just 2.2% of the world's land area. Despite this global significance, these regions continue to be heavily impacted by urbanization, land-use change, climate change, and invasions by non-native species. Chaparral floras include not just the dominant woody shrubs but a diverse assemblage of annual and herbaceous perennial species, many of which have life histories linked to postfire succession. Fire is a natural component of the disturbance regime of chaparral and burns broad portions of the landscape in a coarse-grained manner, but with fine-grained differences in fuel composition and slope aspects. Short fire-return intervals of less than 10–15 years present an increasing threat to chaparral ecosystems by eliminating shrub regeneration and leading to type-conversion to non-native annual grasslands. Water availability and associated adaptive traits of drought tolerance are major factors in partitioning chaparral community composition. Nutrient availability is also important, as are, to a lesser extent, extremes of winter temperature. Although often maligned as a useless or even dangerous because of concerns over fire hazard, chaparral ecosystems provide critical ecosystem services through their roles in erosion control, hydrology, biomass sequestration, and preservation of biodiversity.

Keywords Chaparral · Conservation · Ecosystem services · Fire · Mediterranean-type shrublands · Phenology

P. W. Rundel (✉)
University of California, Los Angeles, CA, USA
e-mail: rundel@ucla.edu

E. C. Underwood et al. (eds.), *Valuing Chaparral*, Springer Series on Environmental Management, https://doi.org/10.1007/978-3-319-68303-4_1

1.1 Introduction

Chaparral communities—identified by their dense evergreen cover of woody shrubs and characteristic leathery or sclerophyllous leaf structure—form the iconic vegetation of southern California, and to a lesser extent, for the entire state. Chaparral covers much of the lower elevations of the Coast Ranges and west slope of the Sierra Nevada, as well as the Transverse and Peninsular ranges of southern California Within the political boundaries of the state, chaparral makes up over 9% of the wildland vegetation (Parker et al. 2016). Half of this cover, and the largest blocks of chaparral, are located in southern California, most notably in the foothills of the Transverse and Peninsular ranges, with the largest area in San Diego County (Fig. 1.1 and Table 1.1).

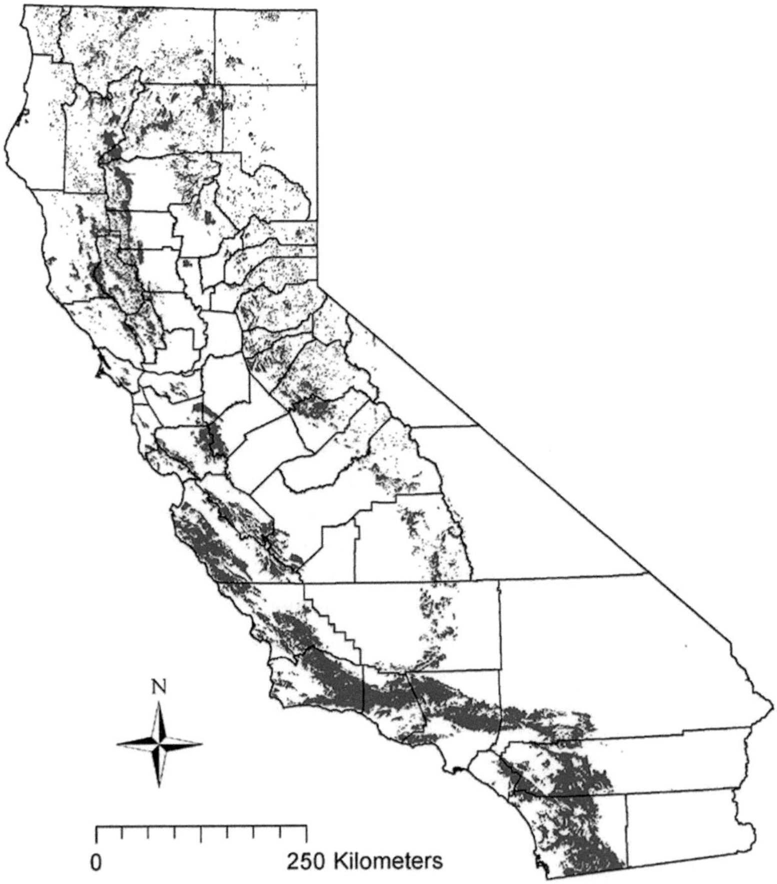

Fig. 1.1 Chaparral distribution in California. This vegetation community also extends into northwestern Baja California and a small area of southern Oregon. Map from Parker et al. (2016)

Table 1.1 Chaparral areas by California county

County	Area (ha)
San Diego	404,600
Los Angeles	223,300
Riverside	201,300
Santa Barbara	177,700
San Luis Obispo	168,400
Monterey	148,900
Ventura	131,600
San Bernardino	111,300
Sab Benito	99,400
Santa Clara	76,000
Orange	45,000
Marin	15,200
San Mateo	14,600
Santa Cruz	13,000

Based on vegetation data from Fried et al. (2004)

Evergreen sclerophyll shrublands, similar to California chaparral, form the classic Mediterranean-type climate vegetation in all five regions of the world characterized by warm dry summers and cool wet winters. This weather regime, termed a Mediterranean-type climate, results from the summer movement of subtropical high pressure cells that produce dry descending air masses capping a surface marine layer of varying levels of humidity. These conditions make summer rainfall infrequent except for irregular convective thunderstorms moving in from outside of the region. Winter conditions are influenced by the polar jet stream and associated periodic storms that bring rain and snow at higher elevations. The result is a globally unusual climate regime. Mediterranean-type climate region shrublands are termed *kwongan* in Southwest Australia, *fynbos* in the Cape Region, *maquis* or *garrigue* in the Mediterranean Basin, and *matorral* in Chile. Just as chaparral forms a significant but one of many plant communities in California, each of the five Mediterranean-type climate regions support a range of other plant communities including woodlands, grasslands, and forests.

The five Mediterranean-Type Climate (MTC) regions, including California, have attracted international interest for almost 150 years as a focal point for studies of ecosystem and evolutionary convergence. This concept of convergent evolution in plant structure and function dates back to a tacit assumption of convergence in the global vegetation classification systems developed at the end of the nineteenth century (Grisebach 1872; Drude 1890; Schimper 1903). Some of these early writings were remarkably prescient in developing concepts of ecological convergence considering the limited database available. Schimper, for example, provided a broad albeit imperfect basis for understanding the adaptive significance of structural fea-

Fig. 1.2 Chamise chaparral *Adenostoma fasciculatum* on Pine Mountain, Ventura County, California. Photo by Richard Spujt

tures in shrub species and geophytes. In describing the vegetation of the five Mediterranean-type climate regions of the world, he wrote: "…the vegetation bears essentially the same stamp, in spite of deep-seated differences in composition of the flora; it is dominated by sclerophyllous plants, and always, although to a subordinate extent, by tuberous and bulbous plants."

The concept of convergent evolution suggests that comparable climatic conditions in MTCs have selected for plants with similar functional traits, resulting in analogous vegetation types that have evolved through independent evolutionary pathways (Cody and Mooney 1978; Specht and Moll 1983; Cowling et al. 1996; Keeley et al. 2012a). The similarities that are shared in the case of convergent evolution are not the result of evolution from a common ancestor, but rather are explained as shared adaptive solutions to similar environmental pressures. Textbook examples of convergent evolution include structural traits such as the multiple origins of wings in bats and birds, the morphological and physiological adaptations to aridity seen in New World cacti and African succulent euphorbias, and the evolution of functionally similar but distinct antifreeze proteins in divergent species of Antarctica and Arctic fish. Across the world's MTC regions sclerophyllous, evergreen, deep-rooted shrubs are a defining characteristic (Fig. 1.2).

Although Mediterranean-type climate ecosystems represent widely cited examples of ecological convergence, the five regions nevertheless display striking examples of divergence as well (Cody and Mooney 1978; Rundel 2011; Rundel et al. 2016). Researchers have categorized MTC regions in terms of their histories and abiotic selective regimes to help explain patterns of convergences and divergences among them. Thus, MTC regions have been differentiated in terms of climate, e.g., amount of summer rain and reliability of winter rainfall (Cowling et al. 2005), soil nutrient status (Specht and Moll 1983), fire regime (Keeley et al. 2012a), topography (Carmel and Flather 2004), and the interactions between climate, fire and soil nutrient status (Keeley et al. 2012a; Rundel et al. 2016).

1.2 Global Significance of Mediterranean-Type Climate Regions

Mediterranean-type climate regions, exemplified by chaparral, have an important place in the global biodiversity of plant species because they harbor the world's richest extra-tropical floras (Cowling et al. 1996, 2015; Kreft and Jetz 2007; Rundel et al. 2016). While this biodiversity includes plant species in woodland, grassland and forest communities, the core of this species richness and endemism resides in the evergreen sclerophyll shrublands. Outside of Mediterranean-type climate regions, evergreen sclerophyll shrublands may be widespread but are generally unexceptional in plant species diversity.

All five MTC regions have been categorized as biodiversity hotspots, i.e., regions of global significance that are home to large numbers of species and rich in endemic taxa (Myers et al. 2000). Assuming a global sum of about 300,000 vascular plant species, MTC floras comprise about one-sixth of this total despite covering only about 2.2% of the world's land area (Cowling et al. 1996). The origin of this diversity is complex and leads to a number of important questions that are relevant in understanding global patterns of species richness. Enigmatically, this high level of species richness and associated endemism is present across all spatial scales (Cowling et al. 2015) (see Chap. 2).

However, MTC regions around the world have been and continue to be heavily impacted by human activities. In a global study of the world's biomes Sala et al. (2000) estimated the mediterranean biome will experience the greatest proportional change in biodiversity by 2100 owing to its sensitivity to a suite of drivers including land use change, climate, non-native species, and nitrogen deposition. Specifically, these threats include habitat degradation and conversion, non-native species, altered fire regimes (i.e., increased fire frequencies for chaparral) and climate change. For example, in the California-Baja California Mediterranean-type climate lowlands (<300 m or 984 ft) 20% was classified as urban (1990) compared to only 2% for lowlands in the four other MTC regions (Underwood et al. 2009). As a consequence, high levels of biodiversity combined with increasing threats have made MTC regions the focal regions for conservation activities (Rundel et al. 1998).

1.3 Evolution of Chaparral Ecosystems and Diversity

To understand chaparral today, it is important to consider the evolutionary origin of this ecosystem through geological time and the origin of its remarkable biodiversity. It was once widely held that the novel Mediterranean-type climate of California and of the other four MTC regions of the world first appeared about 4–3 Ma in the Pliocene (e.g., Suc 1984; Axelrod 1989). However, there is emerging evidence that the onset of a proto-Mediterranean-type climate occurred much earlier, at least in the mid-Miocene (Rundel et al. 2016). A key event leading to the onset of global Mediterranean-type climate regimes was the end of the Middle Miocene Climate Optimum 17–14 Ma, associated with global cooling and growth of the East Antarctic ice sheet. Atmospheric and oceanic circumpolar circulation intensified during this period, resulting in increased strength of the Hadley Cell, ocean current circulation, and seasonal movement of the subtropical high pressure centers, thereby promoting conditions favorable for a Mediterranean-climate formation. More speculatively, there may have been periods of Mediterranean-type climate formation even earlier contemporaneously with Antarctic glaciation in the Oligocene (Rundel et al. 2016). Although the global patterns of atmospheric circulation that determine this climate regime are clear, the seasonal intensity of Mediterranean-type climate has likely varied through time. Changing offshore ocean currents influence the nature of these regimes, with cold currents intensifying summer drought, and warm currents increasing summer rainfall.

The chaparral flora today includes a number of paleo-endemic taxa that predate the development of a Mediterranean-type climate regime, for example Fabaceae (*Pickeringia*), Rosaceae (*Adenostoma*), Cactaceae (*Bergerocactus*), Rutaceae (*Cneoridium*), Papaveraceae (*Dendromecon*), Anacardiaceae (*Malosma*), and Hydrangeaceae (*Carpenteria*). However, none of these genera has undergone significant speciation and collectively form only a small part of chaparral species richness. A similar pattern of floristic assembly is present in the relatively young and dynamic landscapes of the Mediterranean Basin and central Chile where a limited number of paleo-endemic sclerophyll shrubland lineages have persisted while adding little to floristic diversity. In contrast, Southwest Australia and the Cape Region, with their relatively quiet geomorphic and climatic histories through the Cenozoic, exhibit many highly diverse woody plant lineages that have ancient origins in evergreen sclerophyll shrublands on oligotrophic soils as early as the Upper Cretaceous—early Cenozoic (Lamont and He 2012).

For California, the development of the modern chaparral flora is associated with immigration and diversification from a large regional species pool. This occurred under the influence of a novel climatic seasonality and predictable crown fire regimes associated with the development of a Mediterranean-type climate regime in the Miocene. In addition to the evolution of key life history strategies to cope with fire, other ecological factors not unique to Mediterranean-type climate regions have contributed to species diversity. These include adaptations to diverse spatial patterns of climatic, topographic, and edaphic heterogeneity during the

Pliocene and Quaternary. Much of this endemic diversification has been centered on annual plants and herbaceous perennials in clades within such families as the Asteraceae, Boraginaceae, Brassicaceae, Fabaceae, Lamiaceae, Onagraceae, Polemoniaceae, and Polygonaceae. Only two genera of woody chaparral shrubs exhibit extensive diversification: *Arctostaphylos* (Ericaceae; Boykin et al. 2005) and *Ceanothus* (Rhamnaceae; Burge et al. 2011). This pattern contrasts with evolution of the Mediterranean-type climate ecosystem floras of the Cape region and Southwestern Australia where many woody plant lineages have diversified into large genera (Rundel et al. 2016).

1.4 Chaparral Vegetation Structure and Classification

Much of the rich diversity of chaparral communities is hidden by the closed canopy structure that makes individual species difficult to discern at a distance for much of the year. The most widespread and characteristic species exemplifying chaparral is chamise (*Adenostoma fasciculatum*) which often forms virtual monocultures on dry south-facing slopes or rocky areas with shallow soils (Keeley and Davis 2007, Fig. 1.3). Chamise chaparral extends over the entire range of the biome in California and south into Baja California. Less xeric north-facing slopes or those with deeper soils are often simply termed mixed chaparral with a diverse assemblage of species sharing dominance (Fig. 1.3). The most common co-dominant species are scrub oak (*Quercus berberidifolia*) and species of ceanothus (*Ceanothus* spp.) and manzanita (*Arctostaphylos* spp.), the two largest genera of chaparral shrubs, with 46 and 61 species, respectively, and many local endemics. Some coastal areas of chaparral may be dominated by a single and often locally endemic species of *Ceanothus* or *Arctostaphylos*.

Looking at the life-forms of plant species growing within chaparral, woody shrubs comprise only about 19% of the flora, with annuals and herbaceous perennials forming 35% and 39% of the flora, respectively (Table 1.2; Halsey and Keeley 2016). This breakdown of life-forms is proportionally far richer in shrub species than the flora of the entire California Floristic Province and lower in proportional richness of herbaceous perennials. Chaparral shrub species are high in endemics and contain many species listed as rare, endangered, and threatened (Fig. 1.4).

However, the rich diversity of dominant or co-dominant shrub species that may be present across even relatively small landscape gradients (Moody and Meentemeyer 2001) has led to classification systems based solely on these dominants. One of the first attempts to define vegetation units of chaparral at a broad scale was a top-down approach (Holland 1977; Holland and Keil 1989) who separated 11 different chaparral community types. Six of these types were recognized by dominant species *Adenostoma fasciculatum*, redshank (*Adenostoma sparsifolium*), *Arctostaphylos* spp., *Ceanothus* spp., *Quercus berberidifolia*, and mixed chaparral), four geographically distinct forms which they termed maritime, Channel Island, montane, and semi-desert, and one edaphic form as serpentine chaparral. Further refining a classification

Fig. 1.3 Slope impact on community structure in the Santa Monica Mountains of Los Angeles County. Photo shows mixed chaparral with dominance of hoaryleaf ceanothus (*Ceanothus crassifolius*) on north-facing slopes and coastal sage scrub on drier south-facing slopes. Photo by Noah Elhardt

Table 1.2 The life-form distribution of the chaparral flora compared to the total California flora

Life-form	California native flora	Relative %	Chaparral species	Relative %
Annual herb	1469	30	415	35
Perennial herb	2524	52	460	39
Shrub	599	12	228	19
Tree	84	2	22	2
Others	170	4	52	4
Total	4846		1177	

Adapted from Halsey and Keeley (2016)

system, R.F. Holland (1986) divided chaparral further by identifying 44 different community types. This work generally followed V.L. Holland's original terminology but with further specific refinements, as with the division of maritime chaparral into northern, central, and southern forms, and adding edaphic controls. *Ceanothus* and *Arctostaphylos* chaparral communities were further divided on the basis of dominant species. Most recently, Sawyer et al. (2009) have adopted an approach for California from the national hierarchical vegetation classification system in which chaparral alliances are defined by one or two dominant species as well as a number of associations within alliances based on semi-quantitative plot measurements. The result is more than 60 classified alliances of chaparral vegetation, and many more associa-

Fig. 1.4 Old-growth chaparral dominated by mission manzanita (*Xylococcus bicolor*) in San Diego County, California. Photo by Richard Halsey

tions. Although this system of classification has a practical value as a naming system for resource management and environmental impact studies, it provides little ecological insight into habitat conditions because of the independent distribution of individual chaparral species (Zedler 1997).

1.5 Chaparral Geography

Chaparral is widely distributed across California, with its typical occurrence in the foothills of the Coast Ranges and Sierra Nevada. The upper elevational distribution of chaparral ranges from about 800–1200 m (2625–3937 ft) in the northern Sierra Nevada and 1400–1600 m (4593–5249 ft) in the southern Sierran foothills as well as the Transverse and Peninsular ranges. Above this foothill zone, these communities are replaced by ponderosa pine forest and mixed conifer forest. At higher elevations within the lower and upper montane zones, local stands of montane chaparral occur in azonal conditions associated with shallow soils, serpentine substrates, and/or postfire successional sequences. In this latter case, these stands may have long-term persistence as a consequence of self-reinforcing, high-intensity wildfires. Montane chaparral stands may include high-elevation species of *Ceanothus* and *Arctostaphylos*, but overall the shrub flora is distinct from foothill chaparral, and indicator genera like *Adenostoma*, *Pickeringia*, *Garrya* (Garryaceae), *Rhamnus* (Rhamnaceae), and *Heteromeles* (Rosaceae) are absent.

North of San Francisco the distribution of chaparral becomes more restricted, moving inland and progressively dropping in its cover and exhibiting reduced diversity. On more mesic sites, chaparral is replaced by an evergreen sclerophyllous woodland of temperate affinity, dominated by the evergreen tanbark oak (*Notholithocarpus densiflorus*), California bay (*Umbellularia californica*), madrone (*Arbutus menziesii*), and Douglas fir (*Pseudotsuga menziesii*) along with the winter deciduous black oak (*Quercus kelloggii*). A few elements of chaparral extend into the Rogue River basin in Oregon.

Transitions from chaparral to other vegetation associations in southern California occur with abiotic changes in water availability, temperature extremes, soil type, and aspect and elevation that impact these factors. At its drier margin along the coastal areas below about 300 m (984 ft), chaparral is commonly replaced by a drought deciduous community termed sage scrub. Sage scrub species are characteristically semi-woody and lose their leaves with the onset of summer drought, and flushing new leaves after fall rains (Rundel 2007). In addition, the shallow-rooted sage scrub dominants typically do not resprout, or resprout poorly following fire, and lack soil seed pools (Rundel 2007). Many of these sage scrub species do well at higher elevations in the chaparral zone on areas of landslides or other disturbances that restrict establishment of long-lived chaparral shrubs. However, they are shaded out by larger and longer lived evergreen chaparral shrubs when facing competition from them. A similar but less species-rich sage scrub community replaces chaparral at its drier inland margin where it grades gradually into desert associations.

Within the chaparral zone itself, valley oak (*Quercus lobata*) woodlands replace chaparral in valley bottoms with deep soils in the Coast Ranges. Depending on soil depth and exposure, many of the foothill areas of the Coast Ranges exhibit mosaics of chaparral, blue oak (*Quercus douglasii*) woodland, and open non-native annual grassland. In the foothills of the Sierra Nevada, mosaics exist with *Adenostoma fasciculatum* on shallow rocky soils, *Quercus douglasii* woodland on deeper soils with fractured substrate allowing deep root penetration, and mixed evergreen woodlands co-dominated by the evergreen interior live oak (*Quercus wislizenii*), mountain mahogany (*Cercocarpus betuloides*), foothill pine (*Pinus sabiniana*), and California buckeye (*Aesculus californica*). Where there are steep transitions to Mojave or Sonoran Desert habitats, a desert-chaparral ecotone is present with a mix of shrub species from both biomes, including shrubby evergreen oaks.

To the south, chaparral continues into northwest Baja California, and with small disjunct populations of a subset of species on mountain slopes as far south on the peninsula as 28°N. In addition, elements of chaparral species and communities extend eastward from southern California across the higher elevations of the Mojave Desert into upland areas of northern Arizona (Fig. 1.5). These landscapes are most apparent on the slopes of the Mogollon Rim in the middle of Arizona (Knipe et al. 1979), a region with a bi-seasonal pattern of precipitation. Similar stands can be seen in northern Mexico in Nuevo León in a region with a summer rainfall regime and dry winters (Vankat 1989; Keeley et al. 2012a) (Fig. 1.6).

Fig. 1.5 Chaparral in Prescott National Forest, Arizona, USA. Photo by Alan Stark

Fig. 1.6 Broad generalized distribution of chaparral in North America. Adapted from Rundel and Vankat (1989)

1.6 Chaparral and Fire

Fire is a natural component of the disturbance regime of chaparral. Mature shrublands typically range from 1 to 5 m in height and form a dense closed canopy that excludes most herbaceous surface fuels. As stands age, a substantial amount of dead branch tissue is maintained in the canopy, providing a fuel structure that reinforces the crown-fire regime where combustion spreads through the canopies with relatively little surface fire. These chaparral fires burn broad portions of the landscape in a rather coarse-grained manner, although there are fine-grained differences in plant associations on different slope aspects. Natural fire-return intervals in chaparral are not well known because of the absence of records to measure fire frequency before the advent of humans in California. In any case, the natural return interval in fire regimes clearly varies greatly across the state, from as often as 30 years in areas of northern California with frequent lightning activity to 100 years or more in coastal southern California where natural sources of ignition are rare (Van de Water and Safford 2011). Historical fire records for chaparral landscapes in southern California indicate fires have increased in number since the 1930s (Safford 2007), and some locations in southern California are experiencing such frequent fire that chaparral has transitioned to weedy grassland. Almost all of these excess ignitions relate to unintentional or deliberate human activities (Syphard et al. 2007) (Fig. 1.7).

Fig. 1.7 Maritime chaparral with prescribed management burn at Fort Ord, Monterey County, California. Photo by US Army Corps of Engineers

During the first spring following fire on most sites with moderate fertility, a postfire flora of annuals and herbaceous perennials develop as the first stage of postfire recovery. Many of these are fire-following annuals with germination from soil seed pools stimulated by chemical cues within the ash of the fire. This ephemeral flora produces a significant load of fine fuels that help to reduce erosion and stabilize soil nutrient pools (Rundel and Parsons 1984). It is noteworthy that in areas where moderate fire intensities have left blackened skeletons of chaparral shrubs, this dead material plus dried ephemeral vegetation can contain significant fuel loads often in excess of 10 ton ha^{-1} only a year after fire (Keeley et al. 2012b). Non-native annual grasses can also enhance this impact. The fine fuels are easily ignited and are sufficient to carry fire in very young stands, which can be highly detrimental to the recovery of many of the prefire shrub dominants.

The postfire recovery of chaparral shrubs takes place either through resprouting from under-ground root crowns or alternatively from germination of obligate seeding species from soil seed pools (see Chap. 2). The great majority of chaparral shrubs re-establish their canopy dominance by resprouting. For many of these species there are no soil seed pools and they recruit seedlings during fire-free intervals. Such shrub species include *Quercus berberidifolia*, hollyleaf redberry (*Rhamnus ilicifolia*), lemonade berry (*Rhus integrifolia*), toyon (*Heteromeles arbutifolia*), holly-leaved cherry (*Prunus ilicifolia*), and laurel sumac (*Malosma laurina*). Seedlings of these species have a relatively high degree of shade tolerance, and exhibit a similar or greater water stress tolerance to the facultative seeders, suggesting they face strong competition for soil water in mature chaparral (Pratt et al. 2008). Other resprouters are facultative seeders which recruit seedlings in open microsites but have greater survival where drought and shade stress are reduced. Chaparral shrubs with obligate seeding strategies where parent plants are killed by fire are largely restricted to the genera *Arctostaphylos* and *Ceanothus* which exhibit both resprouting and obligate seeding strategies (Pratt et al. 2008). Consistent with the open postfire habitats where seedlings become established, seedlings are tolerant of water stress and intolerant of shade. An intermediate situation is present in *Adenostoma fasciculatum*, which is an active resprouter but additionally exhibits fire-stimulated seedling establishment from relatively short-lived soil seed pools (Stohlgren et al. 1984; Rundel et al. 1987).

Typically, chaparral shrub canopies recover and close up within 10 years following fire, after which the ephemeral flora persists in dormant soil seedbanks. Over the subsequent decade, shrub canopies expand and the ratio of live to dead fuel remains high. The relationship between stand age and live/dead ratio may be an important determinant of flammability under all but the most extreme conditions. As a consequence of structural difference in successional stages, chaparral communities go through a change from being highly vulnerable to fires during the first 5 years because of herbaceous flash fuels, then reduced susceptibility for a decade or two until dead fuels accumulate in the shrub canopies (Schoenberg et al. 2003). However, chaparral stands of any age become highly flammable under conditions of low relative humidity, drought stress, and high temperatures, as are associated with Santa Ana wind conditions in southern California.

Landscape patterns of chaparral distribution can significantly affect chaparral fire regimes (Keeley et al. 2009). Chaparral shrublands dominate a decreasing proportion of the landscape moving from south to north in California, and thus is seems unsurprising that chaparral fires tend to be the largest in the southern half of the state. Even within this region there are marked differences in fuel patterns that affect fire size (Keeley and Zedler 2009). Many of the largest fires (>50,000 ha or 123,552 acres) have occurred either in San Diego County or further north in Santa Barbara/ Ventura counties where the topography supports large contiguous east-west swaths of shrubland fuels and where strong offshore and onshore wind flows can drive fire over very long distances.

1.7 Chaparral Phenology

Most chaparral shrubs use an ecophysiological strategy based on evergreen leaves that are able to photosynthesize throughout the year in the relatively moderate climate in which they grow. They typically exhibit a broad range of optimal temperature for photosynthesis, which allows for moderate rates of carbon fixation even under winter conditions (Oechel et al. 1981; Mooney and Miller 1985). However, a general trade-off in having thick sclerophyllous leaves is that maximum rates of net assimilation are relatively low compared to those of thinner and less leathery leaf structure. During the dry summer and autumn months, as water becomes less available, most chaparral shrubs reduce their rates of carbon fixation by stomatal control, to reduce loss of water through transpiration.

While chaparral shrubs are characterized by having the functional trait of evergreen sclerophyllous leaves—in contrast to the drought deciduous leaves that characterize sage scrub dominants—leafing phenology is more complex. Evergreenness is not a simple trait but instead comes with multiple forms of leaf retention and levels of sclerophylly. The classic chaparral shrub maintains 2 years of leaves, as with most evergreen species of *Quercus, Adenostoma,* and *Arctostaphylos*, shedding the older set soon after or at the same time as new leaves are formed in spring. Many species of *Rhamnus* and *Ceanothus*, however, retain leaves for only 13–15 months, retaining only a single cohort of leaves for most of the year. At the other extreme, some chaparral shrubs such as *Heteromeles arbutifolia*, and the coastal *Malosma laurina* and *Rhus integrifolia* may retain leaves for 4–6 years (Field et al. 1983; Sharifi and Rundel unpublished data). A small but significant number of chaparral shrubs exhibit winter deciduous behavior, including chaparral ash (*Fraxinus dipetala*) and many species of *Ribes* (golden currant [*R. indecorum*] and chaparral currant [*R. malvaceum*]). Coastal areas of northwestern Baja California have chaparral-like stands of maritime scrub dominated by deciduous shrubs of lower California buckeye (*Aesculus parryi*), Baja California hop tree (*Ptelea aptera*), and *Fraxinus dipetala* (Fig. 1.8). This is a trait widely present in riparian tree species within chaparral dominated landscapes, such as willows (*Salix*), western sycamore (*Platanus racemosa*), and white alder (*Alnus rhombifolia*) where it reflects a temperate forest ancestry.

The vegetative growth of chaparral shrubs is influenced by a variety of abiotic and biotic factors, with available soil moisture, temperature, and photoperiod as the most important factors. The Mediterranean-type climate presents strong challenges as soil moisture is most available in the winter months when temperatures are lower than those optimal for growth, while favorable warm summer temperatures occur in summer when drought conditions prevail (Davis and Mooney 1986). It is not surprising, then, that the peak growing season for most shrubs is in spring, when temperatures and photoperiod rise and soils are still moist (Mooney et al. 1977).

Flowering phenology in chaparral shrubs peaks in this same spring season for many species but is highly variable depending on a variety of factors including the phylogenetic lineages of the species. A number of species flower in winter or very early spring from preformed buds set in the previous growing season on mature stems. This form of flowering can be best seen in species of *Ceanothus*, *Arctostaphylos*, and *Ribes*. At the other extreme are species such as *Adenostoma sparsifolium* that flower in mid-summer.

1.8 Water Availability and Drought Tolerance

Because a protracted summer dry season of 4–6 months is characteristic of chaparral habitats, morphological traits of rooting architecture and ecophysiological traits of water use efficiency and drought tolerance are important for survival of chaparral shrubs and for the establishment of seedlings (Mooney 1989). In addition, much of the range of chaparral experiences high summer temperatures and solar irradiance that far exceeds the plant's ability to use this energy for photosynthesis. Water availability is to a major degree a function of soil depth, but slope aspect, substrate geology, and local hydrology may also have major influences on its availability to chaparral shrubs. While it seems intuitive that arid south-facing chaparral slopes would experience the longest periods of soil moisture stress, this is commonly not the case. The lower leaf area index and low rates of transpiration from drought tolerant shrub species compared to more mesic north-facing slopes means that soil moisture stress may occur earlier on these north-facing slopes (Ng and Miller 1980). Areas along the coast influenced by the upwelling and the cold California Current have some mitigation of the summer dry season due to fog occurrence which may provide additional moisture but also conditions which reduce transpiration (Vasey et al. 2012). Along the central California coast and in areas of northwestern Baja California these conditions may promote the development of a distinctive maritime chaparral. However, the southern California coast largely lacks regular fog and provides few opportunities for the development of such communities.

Chaparral shrubs have been categorized along a continuum by the degree of water availability that they experience during the summer dry season, as measured as the minimum seasonal water potential (Davis and Mooney 1986; Bhaskar and

Fig. 1.8 Maritime chaparral with coastal sage scrub at Torrey Pines State Park, San Diego County, California. Photo by User Nauticashades

Ackerly 2006). At one end of the continuum are shallow-rooted species that experience low water potentials as surface soils dry. Good examples of this morphology is exemplified by *Ceanothus* species in the subgenus *Cerastes* and shrubby species of drought deciduous *Salvia* such as black sage (*S. mellifera*) (Thomas and Davis 1989; Jacobsen et al. 2007). These drought tolerators typically exhibit hydraulic traits in their xylem system that restrict the formation of embolisms and allow them to survive extremely low water potentials (Venturas et al. 2016). At the other end of the continuum are deep-rooted shrubs that avoid water stress by tapping subsurface pools of water. This group is exemplified by Anacardiaceae of subtropical ancestry including *Malosma laurina*, sugar bush (*Rhus ovata*), and *Rhus integrifolia* (Thomas and Davis 1989; Jacobsen et al. 2007). Most chaparral shrubs experience water potentials between these two extremes of drought tolerators and avoiders, generally suggesting intermediate rooting depths.

Species tolerant of low water potentials typically exhibit a suite of traits that allow them to maintain a broader range of physiological function at more negative water potentials. Key among these is greater resistance to water stress-induced xylem cavitation, caused by air bubbles pulled into xylem conduits where they embolism (Kolb and Davis 1994; Davis et al. 1998, 2002). Species that experience more negative minimum seasonal water potentials have greater cavitation resistance (Davis et al. 1998, 1999a, b; Jacobsen et al. 2007; Pratt et al. 2007). As a group, evergreen chaparral shrubs typically have greater cavitation resistance in their stems compared to deciduous shrubs that occur in the chaparral community. As might be expected, greater cavitation resistance is correlated with increased survival of drought in chaparral seedlings (Pratt et al. 2008).

1.9 Temperature Limitations

Although extremes of winter temperatures across the range of chaparral distribution are relatively mild by temperate standards, these regions can intermittently experience periods of very low winter temperatures. Extremes of winter lows may reach −8 °C to −12 °C every few years, even in the Coast Ranges of southern California. Some chaparral shrubs are susceptible to freezing injury caused by xylem embolism, and this sensitivity may limit their distribution (Langan et al. 1997; Davis et al. 1999a, b). The best examples of cold tolerance as a limiting factor in the distribution of chaparral shrubs can be seen in taxa of tropical ancestry, as with members of the Anacardiaceae such as *Malosma laurina* and *Rhus integrifolia*, which are restricted to coastal foothill areas of southern California and Baja California. Low temperature tolerance has been shown to vary among chaparral species as well as between adult shrubs and seedlings in the ability of their leaves to acclimate (Boorse et al. 1998).

1.10 Nutrient Availability

In addition to water availability, soil nutrients may also be an important limiting factor for growth of chaparral shrubs. Young and relatively skeletal soils that characterize much of the chaparral region of California are often low in nitrogen, leading to adaptive strategies to minimize loss of nitrogen and other nutrients as leaves senesce. Indeed, the evergreen habitat is often associated with plant species growing on low nutrient soils, and has been widely suggested as an adaptation to increase the efficiency of nutrient utilization (Rundel 1982). The dynamics of nutrient cycling has been studied in some detail in *Adenostoma fasciculatum*, a species that serves as a model for other chaparral shrubs. Seasonal changes in the nitrogen and phosphorus content of leaf tissues indicate that the plant takes up nutrients during the winter rainy season prior to the initiation of above-ground vegetative growth (Mooney and Rundel 1979). This seasonality of uptake allows a sustainable conservation of key nutrients that would otherwise be lost as decomposition and leaching occur during the wet winter season, with the evergreen leaves providing a sink for nutrient retention during these periods without above-ground growth.

Periodic fire in chaparral ecosystems has an important impact on nutrient cycling with a temporal cycle of change which involves the initial loss of nutrients that were previously held in the above-ground biomass, litter, and surface soils (Christensen and Muller 1975; DeBano and Conrad 1978; Rundel and Parsons 1980). This fire-induced change in the relative distribution and abundance of nutrients is of significance to the entire biotic community. For example, the relative availability of nutrients in the soil determines plant growth, while foliar nutrient contents determine the suitability and attractiveness of the foliage as browse for grazing animals. Burning produces very profound effects on nutrient cycles by rapidly mineralizing above-ground biomass and litter into ash. Changes in available forms of nitrogen and increased microbial activity following fire provide important means of promoting favorable nutrient conditions for new growth. A significant initial increase in soil concentrations of ammonium and organic nitrogen is well documented in the first weeks following fire (Christensen 1973). The ammonium is quickly mineralized, resulting in the commonly observed high nitrate concentrations in recently burned areas. These high nitrate levels, together with increases in phosphorous, organic nutrients, and selected mineral elements during the first 18 months following fire (Christensen and Muller 1975), create a highly favorable condition for a postfire flush of herbaceous species as well as shrub growth during the first few years following a chaparral fire.

Hot chaparral fires volatilize significant amounts of nitrogen from above-ground biomass, litter and surface soils, and this nitrogen is lost to the atmosphere. Studies reviewed by Rundel and Vankat (1989) found a fire loss of 119–241 kg ha^{-1} of nitrogen from *Adenostoma* stands, amounting to as much as 7% or more of total system nitrogen. Further losses of nitrogen on the order of 8–15 kg ha^{-1} can occur through erosion and runoff (DeBano and Conrad 1978). The rapid establishment of postfire annual species forms the major biomass pool in the first spring and often second year after fire, and plays an important ecosystem role in sequestering nutrients that might otherwise be loss through erosion and leaching (Rundel and Parsons 1984). Chronosequence studies of chamise chaparral have shown that these substantial losses of nitrogen are replaced within 5–10 years after fire through a variety of inputs. The primary source of this nitrogen comes from the legume subshrub deerweed (*Acmispon glaber*, formerly known as *Lotus scoparius*). This species widely germinates in large numbers from soil seedbanks following chaparral fires and fixes 10–15 kg ha^{-1} $year^{-1}$ (Nilsen and Schlesinger 1981). A second source of nitrogen input comes from dry deposition associated with atmospheric aerosols. This input has been estimated to be about 1–2 kg ha^{-1} $year^{-1}$ in pristine areas of chaparral (Schlesinger and Hasey 1980) and up to 15 times this amount in polluted air masses associated with chaparral stands in the foothills of the Transverse Ranges (Riggan et al. 1985).

A number of woody chaparral shrubs have the ability to form symbiotic associations with nitrogen-fixing bacteria, but the significance of inputs of symbiotic nitrogen fixation for mature chaparral stands is not well established. The genus *Ceanothus* is well known for its potential to fix significant amounts of atmospheric nitrogen in moist forest environments in the western United States, but does not appear to have a significant impact on soil nitrogen pools in chaparral (Pratt et al. 1997). Another widespread chaparral shrub with symbiotic nitrogen fixation is *Cercocarpus betuloides*. Stands of *Cercocarpus* in the Great Basin actively fix nitrogen (Lepper and Fleschner 1977). Nitrogen fixation is also known to occur in chaparral pea

(*Pickeringia montana*), the only native woody legume shrub present in chaparral, and in mountain misery (*Chamaebatia foliolosa*) and southern mountain misery (*C. australis*) (Rundel et al. 1981).

1.11 Ecosystem Services Provided by Chaparral

Chaparral has often been described with such words as useless, dense, and impenetrable, and is often maligned as dangerous because of its flammability (see Chaps. 5 and 12). Such misconceptions too often lead to irrational public policy that promotes destructive land management practices to eliminate chaparral through broad scale removal of native shrublands through burning, mastication, and herbicide treatments. However, both historically and continuing today, chaparral ecosystems provide a variety of significant ecosystem services (see Chapters 2, 3, 4, 6, 7, 8, 9 and 11). Ecosystem services describe the ways that ecosystems directly or indirectly provide a positive benefit to people. Such services can be categorized as regulating (e.g. climate amelioration, flood control), provisioning (e.g., food, fuel, fresh water), supporting (e.g. nutrient cycling and carbon sequestration), and cultural (e.g., aesthetic, educational, recreation) (Millennium Ecosystem Assessment 2005).

The growth of chaparral cover on steep hillsides helps to reduce flooding, erosion, and mudslides that can occur during winter rains (Gabet and Dunne 2002) (see Chap. 7). This service is especially apparent after chaparral crowns have been removed by intense crown fire, and heavy winter storms cause costly and lethal mudslides (Ren et al. 2011). A second regulating service comes with the energy balance as chaparral absorbs sunlight and transpires water, thereby helping to regulate temperature during the hot summer months compared to highly urbanized areas that experience the "heat island effect" (LaDochy et al. 2007).

Provisioning services of chaparral center on filtration of water, which helps to maintain fresh drinking water in aquifers and reduce eutrophication in the ocean and reservoirs that receive runoff (see Chap. 8). This is important in areas of southern California where nitrogen deposition from air pollution is high and nitrate is prone to leach into groundwater and collect in downstream bodies of water. Watersheds of the San Bernardino and San Gabriel mountains northeast of Los Angeles exhibit some of the highest levels of nitrogen pollution in the United States (Fenn and Poth 1999). Areas that were formerly chaparral and now converted to grassland have been shown to be less effective at filtering water and yield greater nitrate runoff (Riggan et al. 1985). Moreover, chaparral ecosystems also provide critical food and habitat resources for a diversity of native animal species and help to stabilize trophic chains.

Supporting services provided by chaparral include a significant role in carbon sequestration with stands of chaparral (see Chap. 6). Mature stands of chaparral can support 40–80 tons ha^{-1} or more of above-ground biomass (Rundel and Vankat 1989). Because chaparral stands continue to maintain high rates of productivity with age, even old stands remain significant carbon sinks (Luo et al. 2007). These large amounts of carbon biomass have led to suggestions that chaparral could be harvested on a sustainable basis to provide biofuel for the generation of electricity (Riggan and Dunn 1981). Experimental type-conversion of chaparral slopes to grassland to increase water yield has also been attempted in the past (Hill and Rice 1963; Meixner and Wohlgemuth 2003). The results indeed decreased transpirational water loss from the deeply-rooted chaparral canopies, but came at the expense of landscape instability, reduction in water quality, reduction in temperature regulation associated with canopy energy balance, and loss of wildlife value.

There has been an increasing realization in recent years of the significance that chaparral communities play in providing pollination services for adjacent agricultural developments. The pollination services provided by native bees are associated with the amount of nearby natural habitat where these bees reside (Kremen et al. 2004).

Although less easy to quantify, chaparral habitats provide significant cultural services in the role they play for outdoor recreation (see Chap. 10). These activities are particularly significant in and around the large urban areas of southern California where millions of visitors enjoy hiking, biking, horseback riding, and camping. Many families, including many from minority communities in inner cities, flock to picnic areas of local parks and reserves on weekends and holidays.

1.12 Chaparral Conservation in an Era of Global Change

Despite the significant role that chaparral shrublands play in providing ecosystem services and as hotspots of biodiversity, informed management of California chaparral ecosystems has often been neglected (see Chap. 15). Nevertheless, chaparral ecosystems remain disproportionately vulnerable to major global threats to sustainability and biodiversity. These threats can be best mitigated when local, state, and federal agencies coordinate their activities to utilize the best available science and adaptive management practices.

Across the state of California there are over 6,000,000 ha (~15,000,000 acres) of shrublands, accounting for almost 15% of the landscape, but management authority is split between multiple federal, state, and local agencies. The largest areas are managed by the USDA Forest Service, with 40%, and the Bureau of Land Management with 15% of the area (Table 1.3). About one-third (31%) of shrubland is under private ownership. All other organizations, such as the National Park Service and California Department of Parks and Recreation, each manage <2.5% of the shrubland in the state.

Global change models predict that the climate of California will be increasingly warmer and, at least for southern California, drier in the coming decades (Hayhoe et al. 2004; Neelin et al. 2013) (see Chap. 14). These changes will impact chaparral ecosystems in a number of ways, with complex interactions between temperature means and extremes, precipitation amounts and seasonality, and local soil moisture storage capacities. It has been suggested that increasing CO_2 associated with global change may increase water-use efficiency and alter patterns of fuel moisture in ways

Table 1.3 Ownership of shrubland in the state of California. Shrubland data are derived from the California Department of Forestry and Fire Protection's Fire and Resource Assessment Program vegetation data (FRAP 2015) and the California Protected Areas Database (version 2014a)

	Area (ha)	Area (acres)	Percent (%)
USDA Forest Service	2,458,600	6,075,309	40.45
Private	1,887,200	4,663,354	31.05
Bureau of Land Management	952,400	2,353,422	15.67
National Park Service	150,000	370,657	2.47
Department of Defense	147,900	365,467	2.43
Local Government	142,800	352,865	2.35
CA Dept. of Parks and Recreation	95,500	235,985	1.57
Bureau of Indian Affairs	72,800	179,892	1.2
CA Dept. of Fish and Wildlife	59,800	147,768	0.98
Non-Profit Conservancies and Trusts	57,300	141,591	0.94
Other State Lands	27,200	67,212	0.45
US Fish and Wildlife Service	10,200	25,205	0.17
Other Federal Lands	7900	19,521	0.13
Bureau of Reclamation	6400	15,815	0.1
CA Dept. of Forestry and Fire Protection	1500	3707	0.03

that potentially could offset increasing fire hazard due to warmer temperatures (Oechel et al. 1995). However there are secondary effects that may negate this advantage. Increased atmospheric CO_2 under future climates will also stimulate biomass accumulation and influence high-intensity fires. Increased CO_2 may well have unexplored and unanticipated impacts on soil microbial communities with associated changes in the dynamics of litter decomposition and nutrient cycling, with cascading effects across food webs (Oechel et al. 1995).

A significant indirect impact of climate change on chaparral may well come through changes in fire regime. Chaparral stands are generally not resilient to fire-return intervals less than about 10–15 years (Keeley et al. 2012b), and the increased number and frequency of anthropogenic ignitions in southern California have already led to major areas of type-conversion from chaparral to non-native annual grassland (see Chaps. 12 and 13). Once converted, an alternate stable state may be reached where ignitions can occur almost any time of the year because of the fine grass fuels. This said, land-use changes from urbanization and agricultural development over the coming decades may well play as important or more important a role as climate change in the conservation of chaparral and related shrubland ecosystems (Riordan and Rundel 2014).

Historically the primary management focus on chaparral, particularly in southern California, has been on management of fuels and fire hazard, with little emphasis on the sustainability of chaparral ecosystems and the associated ecosystem services provided. In simple terms, chaparral has been widely ignored by federal and state management agencies as an uninteresting but flammable landscape that produces threats to the built environment of California. One example of this lack of interest can be seen in the history of management plans for the national forests in southern California. Unlike other units of the national forest system, the Angeles, Cleveland, San Bernardino, and Los Padres national forests were originally set aside to protect watershed values rather than manage timber resources. Although important communities of conifers are present, chaparral ecosystems characterize a major component of their landscape. The dominance of chaparral over timber makes these four national forests distinct from any other forest units in the federal system. However, for a variety of historical and cultural reasons that fail to value chaparral like a commodity, land managers have neither given adequate attention to chaparral as an important natural resource nor appreciated its ecological and ecosystem value. As a result, chaparral has been treated more as a fuel problem than a native plant community worthy of preservation, and chaparral management plans have largely ignored sustainability and ecosystem services, and have centered instead on approaches to fuel reduction.

Today there is an increasing understanding at many government levels that chaparral ecosystems provide critical ecosystem services, most directly through their role in erosion control, hydrology, biomass sequestration, and preservation of biodiversity. These functions will increase in significance in the future under conditions of reduced precipitation and warmer temperatures. The presence of chaparral communities at or near the expanding boundaries of suburban development leads to inevitable conflicts between the impacts of chaparral wildfire and the protection of human life and structures. Such conflicts will continue without informed regional and local policies for planning and land use development.

References

Axelrod, D. I. 1989. Age and origin of chaparral. Pages 7-19 *in* S. C. Keeley, editor. The California chaparral: paradigms reexamined. Natural History Museum of Los Angeles County, Los Angeles, California, USA.

Bhaskar, R., and D. D. Ackerly. 2006. Ecological relevance of minimum seasonal water potentials. Physiologia Plantarum 127:353-359.

Boorse, G. C., F. W. Ewers, and S. D. Davis. 1998. Response of chaparral shrubs to below-freezing temperatures: acclimation, ecotypes, seedlings vs. adults. American Journal of Botany 85:1224-1230.

Boykin, L. M., M. C. Vasey, V. T. Parker, and R. Patterson. 2005. Two lineages of *Arctostaphylos* (Ericaceae) identified using the internal transcribed spacer (ITS) region of the nuclear genome. Madroño 52:139-147.

Burge, D. O., D. M. Erwin, M. B. Islam, J. Kellermann, S. W. Kembel, D. H. Wilken, and P. S. Manos. 2011. Diversification of *Ceanothus* (Rhamnaceae) in the California Floristic Province. International Journal of Plant Sciences 172:1137-1164.

Carmel, Y., and C. H. Flather. 2004. Comparing landscape scale vegetation dynamics following recent disturbance in climatically similar sites in California and the Mediterranean Basin. Landscape Ecology 19:573-590.

Christensen, N. L. 1973. Fire and the nitrogen cycle in California chaparral. Science 181:66-68.

Christensen, N. L., and C. H. Muller. 1975. Effects of fire on factors controlling plant growth in *Adenostoma* chaparral. Ecological Monographs 45:29-55.

Cody, M. L., and H. A. Mooney. 1978. Convergence versus nonconvergence in Mediterranean-climate ecosystems. Annual Review of Ecology and Systematics 9:265-321.

Cowling, R. M., F. Ojeda, B. B. Lamont, P. W. Rundel, and R. Lechmere-Oertel. 2005. Rainfall reliability, a neglected factor in explaining convergence and divergence of plant traits in fire-prone Mediterranean-climate ecosystems. Global Ecology and Biogeography 14:509-519.

Cowling, R. M., A. J. Potts, P. L. Bradshaw, J. Colville, M. Arianoutsou, S. Ferrier, F. Forest, N. M. Fyllas, S. D. Hopper, F. Ojeda, S. Proches, R. J. Smith, P. W. Rundel, E. Vassilakis, and B. R. Zutta. 2015. Variation in plant diversity in Mediterranean-climate ecosystems: the role of climatic and topographical stability. Journal of Biogeography 42:552-564.

Cowling, R. M., P. W. Rundel, B. B. Lamont, M. K. Arroyo, and M. Arianoutsou. 1996. Plant diversity in Mediterranean-climate regions. Trends in Ecology and Evolution 11:362-366.

Davis, S. D., F. W. Ewers, J. S. Sperry, K. A. Portwood, M. C. Crocker, and G. C. Adams. 2002. Shoot dieback during prolonged drought in Ceanothus (Rhamnaceae) chaparral of California: a possible case of hydraulic failure. American Journal of Botany 89:820-828.

Davis, S. D., F. W. Ewers, J. Wood, J. J. Reeves, and K. J. Kolb. 1999a. Differential susceptibility to xylem cavitation among three pairs of *Ceanothus* species in the Transverse Mountain Ranges of southern California. Ecoscience 6:180-186.

Davis, S. D., K. J. Kolb, and K. P. Barton. 1998. Ecophysiological processes and demographic patterns in the structuring of California chaparral. Pages 297-310 in P. W. Rundel, G. Montenegro, and F. Jaksic, editors. Landscape disturbance and biodiversity in Mediterranean-type ecosystems. Springer-Verlag, New York, USA.

Davis, S. D., and H. A. Mooney. 1986. Water use patterns of four co-occurring chaparral shrubs. Oecologia 70:172-177.

Davis, S. D., J. S. Sperry, and U. G. Hacke. 1999b. The relationship between xylem conduit diameter and cavitation caused by freezing. American Journal of Botany 86:1367-1372.

DeBano, L. F., and C. E. Conrad. 1978. The effect of fire on nutrients in a chaparral ecosystem. Ecology 59:489-497.

Drude, O. 1890. Handbuch der Pflanzengeographie. Verlag von J. Engelhorn, Stuttgart, Germany.

Fenn, M. E., and M. A. Poth. 1999. Temporal and spatial trends in streamwater nitrate concentrations in the San Bernardino Mountains, southern California. Journal of Environmental Quality 28:822-836.

Field, C., J. Merino, and H. A. Mooney. 1983. Compromises between water-use efficiency and nitrogen-use efficiency in five species of California evergreens. Oecologia 60:384-389.

Fried, J. S., C. L. Bosinger, and D. Beardsley. 2004. Chaparral in southern and central coastal California in the mid-1990's: area, ownership, condition, and change. Resource Bulletin PNW-RB-240. USDA Forest Service, Pacific Northwest Research Station, Portland, Oregon, USA.

Gabet, E. J., and T. Dunne. 2002. Landslides on coastal sage-scrub and grassland hillslopes in a severe El Niño winter: the effects of vegetation conversion on sediment delivery. Geological Society of America Bulletin 114:983-990.

Grisebach, A. 1872. Die Vegetation der Erde nach Ihrer Klimatischen Anordnung. W. Engelmann, Leipzig, Germany.

Halsey, R. W. and J. E. Keeley. 2016. Conservation issues: California chaparral. Reference Module in Earth Systems and Environmental Sciences. https://doi.org/10.1016/B978-0-12-409548-9.09584-1

Hayhoe, K., D. Cayan, C. B. Field, P. C. Frumhoff, E. P. Maurer, N. L. Miller, S.C. Moser, S.H. Schneider, K. N. Cahill, E. E. Cleland, L. Dale, R. Drapek, R. M. Hanemann, L. S. Kalkstein, J. Lenihan, C. K. Lunch, R. P. Neilson, S. C. Sheridan, J. H. Verville 2004. Emissions pathways, climate change, and impacts on California. Proceedings of the National Academy of Sciences 101:12422-12427

Hill, L. W., and R. M. Rice. 1963. Converting from brush to grass increases water yield in southern Californi Journal of Range Management 16:300-305

Holland, R. F. 1986. Preliminary descriptions of the terrestrial natural communities of California. State of California, The Resources Agency, Non-game Heritage Program, Department of Fish and Game, Sacramento, California, USA.

Holland, V. L. 1977. Major plant communities of California. Pages 3-41 *in* D. R. Walters, M. McLeod, A. G. Meyer, D. Rible, R. O. Baker, and L. Farwell, editors. Native plants: a viable option. Symposium. California Native Plant Society Special Publication 3, Sacramento, California, USA.
Holland, V. L., and D. J. Keil. 1989. California vegetation. El Corral Publications, San Luis Obispo, California, USA.
Jacobsen, A. L., R. B. Pratt, F. W. Ewers, and S. D. Davis. 2007. Cavitation resistance among twenty-six chaparral species of southern California. Ecological Monographs 77:99-115.
Keeley, J. E., W. J. Bond, R. A. Bradstock, J. G. Pausas, P. W. Rundel. 2012a. Fire in Mediterranean ecosystems: ecology, evolution and management. Cambridge University Press, Cambridge, UK.
Keeley, J. E., and F. W. Davis. 2007. Chaparral. Pages 339-366 *in* M. G. Barbour, T. Keeler-Wolf, and A. A. Schoenherr, editors. Terrestrial vegetation of California. University of California Press, Berkeley, California, USA.
Keeley, J. E., C. J. Fotheringham, and P. W. Rundel. 2012b. Postfire chaparral regeneration under Mediterranean and non-Mediterranean climates. Madroño 59:109-127.
Keeley, J. E., H. Safford, C. J. Fotheringham, J. Franklin, and M. Moritz. 2009. The 2007 southern California wildfires: lessons in complexity. Journal of Forestry 107:287-296.
Keeley, J. E., and P. H. Zedler. 2009. Large, high-intensity fire events in southern California shrublands: debunking the fine-grain age patch model. Ecological Applications 19:69-94.
Kolb, K. J., and S. D. Davis. 1994. Drought tolerance and xylem embolism in co-occurring species of coastal sage and chaparral. Ecology 75:648-659.
Knipe, O., C. Pase, and R. Carmichael. 1979. Plants of the Arizona chaparral. General Technical Report RM-GTR-54. USDA Forest Service, Rocky Mountain Forest and Range Experiment Station, Fort Collins, Colorado, USA.
Kreft, H., and W. Jetz. 2007. Global patterns and determinants of vascular plant diversity. Proceedings of the National Academy of Sciences 104:5925-5930.
Kremen, C., N. M. Williams, R. L. Bugg, J. P. Fay, and R. W. Thorp. 2004. The area requirements of an ecosystem service: Crop pollination by native bee communities in California. Ecology Letters 7:1109-1119.
LaDochy, S., R. Medina, and W. Patzert. 2007. Recent California climate variability: spatial and temporal patterns in temperature trends. Climate Research 33:159-169.
Lamont, B. B., and T. He. 2012. Fire-adapted Gondwanan angiosperm floras evolved in the Cretaceous. BMC Evolutionary Biology 12:223.
Langan, S. J., F. W. Ewers, and S. D. Davis. 1997. Differential susceptibility to xylem embolism caused by freezing and water stress in two species of chaparral shrubs. Plant, Cell and Environment 20:425-437.
Lepper, M. G., and M. Fleschner. 1977. Nitrogen fixation by *Cercocarpus ledifolius* (Rosaceae) in pioneer habitats. Oecologia 27:333-338.
Luo, H., W. C. Oechel, S. J. Hastings, R. Zulueta, Y. Qian, and H. Kwon. 2007. Mature semiarid chaparral ecosystems can be a significant sink for atmospheric carbon dioxide. Global Change Biology 13:386-396.
Meixner, T., and P. M. Wohlgemuth. 2003. Climate variability, fire, vegetation recovery, and watershed hydrology. Pages 651-656 *in* K. G. Renard, S. A. McElroy, W. J. Gburek, E. H. Canfield, and R. L. Scott, editors. Proceedings of the First Interagency Conference on Research in the Watersheds, Benson, Arizona, October 27-30, 2003. US Department of Agriculture, Agricultural Research Service, USA.
Millennium Ecosystem Assessment. 2005. Ecosystems and human well-being. World Resources Institute, Washington, D.C., USA.
Moody, A., and R. K. Meentemeyer. 2001. Environmental factors influencing spatial patterns of woody plant diversity in chaparral, Santa Ynez Mountains, California. Journal of Vegetation Science 12:41-52.

Mooney, H. A. 1989. Chaparral physiological ecology—paradigms revisited. Pages 85-90 *in* S. C. Keeley, editor. The California chaparral: paradigms reexamined. Natural History Museum of Los Angeles County, Los Angeles, California, USA.

Mooney, H. A., J. Kummerow, A. W. Johnson, D. J. Parsons, S. Keeley, A. Hoffmann, R. I. Hays, J. Giliberto, and C. Chu. 1977. Convergent evolution in the consumer organisms of the Mediterranean Chile and California. Pages 85-143 in H. A. Mooney, editor. Convergent evolution in Chile and California: Mediterranean climate ecosystems. Dowden, Hutchinson, and Ross, Stroudsberg, Pennsylvania, USA.

Mooney, H. A., and P. C. Miller. 1985. Chaparral. Pages 213-231 *in* B. F. Chabot and H. A. Mooney, editors. Physiological ecology of North American plant communities. Chapman and Hall, New York, New York, USA.

Mooney, H. A., and P. W. Rundel. 1979. Nutrient relations of the evergreen shrub, *Adenostoma fasciculatum*, in the California chaparral. Botanical Gazette 140:109-113.

Myers N., R. A. Mittermeier, C. G. Mittermeier, G.A.B. da Fonseca, and J. Kent. 2000. Biodiversity hotspots for conservation priorities. Nature 403:853–858.

Ng, E., and P. C. Miller. 1980. Soil moisture relations in the southern California chaparral. Ecology 61:98-107.

Neelin, J. D., B. Langenbrunner, J. S. Meyerson, A. Hall, and N. Berg. 2013. California winter precipitation change under global warming in the coupled model intercomparison project phase 5 ensemble. Journal of Climate 26:6238-6256.

Nilsen, E. T., and W. H. Schlesinger. 1981. The influence of the Mediterranean climate on nutrients, particularly nitrogen in an even-aged stand of the summer deciduous, fire response shrub *Lotus scoparius*. Oecologia 50:217-224.

Oechel, W. C., S. J. Hastings, G. L. Vourlitis, M. A. Jenkins, and C. L. Hinkson. 1995. Direct effects of elevated CO_2 in chaparral and Mediterranean-type ecosystems. Pages 58-75 *in* J. L. Moreno, and W. C. Oechel, editors. Global change and Mediterranean-type ecosystems. Springer-Verlag, New York, New York, USA.

Oechel, W. C., W. T. Lawrence, J. Mustafa, and J. Martinez. 1981. Energy and carbon acquisition. Pages 151-183 *in* P. C. Miller, editor. Resource use by chaparral and matorral. Springer-Verlag, New York, New York, USA.

Parker, V. T., R. B. Pratt, and J. E. Keeley. 2016. Chaparral. Pages 479-508 *in* H. Mooney, and E. Zavaleta, editors. Ecosystems of California—a source book. University of California Press, Berkeley, USA.

Pratt, R. B., A. L. Jacobsen, K. A. Golgotiu, J. S. Sperry, F. W. Ewers, and S. D. Davis. 2007. Life history type and water stress tolerance in nine California chaparral species (Rhamnaceae). Ecological Monographs 77:239-253.

Pratt, R. B., A. L. Jacobsen, R. Mohla, F. W. Ewers, and S. D. Davis. 2008. Linkage between water stress tolerance and life history type in seedlings of nine chaparral species (Rhamnaceae). Journal of Ecology 96:1252-1265.

Pratt, S. D., A. S. Konopka, M. A. Murry, F. W. Ewers, and S. D. Davis. 1997. Influence of soil moisture on the nodulation of post fire seedlings of *Ceanothus* spp. growing in the Santa Monica Mountains of southern California. Physiologia Plantarum 99:673-679.

Ren, D., R. Fu, L. M. Leslie, and R. E. Dickinson. 2011. Modeling the mudslide aftermath of the 2007 southern California wildfires. Natural Hazards 57:327-343.

Riggan, P. J., and P. H. Dunn. 1981. Harvesting chaparral biomass for energy—an environmental assessment. Pages 22-26 *in* C. E. Conrad, and W. C. Oechel, editors. Proceedings of the Symposium on Dynamics and Management of Mediterranean-Type Ecosystems, San Diego, California, June 22–26, 1981. General Technical Report PSW-GTR-58. USDA Forest Service, Pacific Southwest Forest and Range Experiment Station, Berkeley, California, USA.

Riggan, P. J., S. Franklin, and J. A. Brass. 1985. Fire and chaparral management at the chaparral/urban interface. Fremontia 14:28-30.

Riordan, E. C., and P. W. Rundel. 2014. Land use compounds habitat losses under projected climate change in a threatened California ecosystem. PLoS ONE 9:e86487.

Rundel, P. W. 1982. Nitrogen use efficiency in Mediterranean-climate shrubs of California and Chile. Oecologia 55:409-413.
Rundel, P. W. 2007. Sage scrub. Pages 208-228 *in* M. G. Barbour, T. Keeler-Wolf, and A. A. Schoenherr, editors. Terrestrial vegetation of California. University of California Press, Berkeley, California, USA.
Rundel, P. W. 2011. Convergence and divergence in Mediterranean-climate ecosystems: what we can learn by comparing similar places. Pages 93-108 *in* M. Price, and I. Billick, editors. The ecology of place. University of Chicago Press, Chicago, Illinois, USA.
Rundel, P. W., M. K. Arroyo, R. M. Cowling, J. E. Keeley, B. B. Lamont, and P. Vargas. 2016. Mediterranean biomes: evolution of the floras, vegetation, and climate regime. Annual Review of Ecology, Evolution and Systematics 47: 383-407.
Rundel, P. W., G. A. Baker, D. J. Parsons, and T. J. Stohlgren. 1987. Post-fire demography of resprouting and seedling establishment by *Adenostoma fasciculatum* in the California chaparral. Pages 575-596 *in* J. Tenhunen, F. M. Catarino, O. L. Lange, and W. C. Oechel, editors. Plant response to stress - functional analysis in Mediterranean ecosystems. Springer-Verlag, Heidelberg, Germany.
Rundel, P. W., G. Montenegro, and F. Jaksic, editors. 1998. Landscape disturbance and biodiversity in Mediterranean-type ecosystems. Springer-Verlag, Berlin, Germany.
Rundel, P. W., and D. J. Parsons. 1980. Nutrient changes in two chaparral shrubs along a fire-induced age gradient. American Journal of Botany 67:51-58.
Rundel, P. W., D. J. Parsons, and G. A. Baker. 1981. Productivity and nutritional responses of *Chamaebatia foliolosa* (Rosaceae) to seasonal burning. Pages 191-196 *in* N. Margaris, and H. A. Mooney, editors. Components of productivity of Mediterranean-climate regions. Dr. W. Junk Publishers, The Hague, Netherlands.
Rundel, P. W., and D. J. Parsons. 1984. Post-fire uptake of nutrients by diverse ephemeral herbs in chamise chaparral. Oecologia 61:285-288.
Rundel, P. W., and J. L. Vankat. 1989. Chaparral communities and ecosystems. Pages 127-139 *in* S. Keeley, editor. The California chaparral; paradigms reexamined. Los Angeles County Museum of Natural History, Los Angeles, California, USA.
Safford, H. D. 2007. Man and fire in southern California: doing the math. Fremontia 35:25-29.
Sala, O. E., F. S. Chapin, J. J. Armesto, E. Berlow, J. Bloomfield, R. Dirzo, E. Huber-Sanwald, L. F. Huenneke, R. B. Jackson, A. Kinzig, R. Leemans, D. M. Lodge, H. A. Mooney, M. Oesterheld, N. L. Poff, M. T. Sykes, B. H. Walker, M. Walker, and D. H. Wall. 2000. Global biodiversity scenarios for the year 2100. Science 287:1770-1774.
Sawyer, J. O., T. Keeler-Wolf, and J. M. Evens. 2009. A manual of California vegetation. California Native Plant Society Press, Sacramento, California, USA.
Schimper, A. F. W. 1903. Plant-geography upon a physiological basis. Clarendon Press, Oxford, UK.
Schlesinger, W. H., and M. M. Hasey. 1980. The nutrient content of precipitation, dry fallout, and intercepted aerosols in the chaparral of southern California. American Midland Naturalist 103:114-122.
Schoenberg, F. P., R. Peng, Z. Huang, and P. W. Rundel. 2003. Detection of non-linearities in the dependence of burn area on fuel age and climatic variables. International Journal of Wildland Fire 12:1-6.
Specht, R. L., and E. J. Moll. 1983. Mediterranean-type heathlands and sclerophyllous shrublands of the world: an overview. Pages 41-65 *in* F. J. Kruger, D. T. Mitchell, J. U. M. Jarvis, editors. Mediterranean-type ecosystems: the role of nutrients. Springer-Verlag Heidelberg, Germany.
Stohlgren, T. J., D. J. Parsons, and P. W. Rundel. 1984. Population structure of *Adenostoma fasciculatum* in mature stands of chamise chaparral in the southern Sierra Nevada, California. Oecologia 64:87-91.
Suc J.-P. 1984. Origin and evolution of the Mediterranean vegetation and climate of Europe. Nature 307:429-432.

Syphard, A. D., V. C. Radeloff, J. E. Keeley, T. J. Hawbaker, M. K. Clayton, S. I. Stewart, and R. B. Hammer. 2007. Human influence on California fire regimes. Ecological Applications 17:1388-1402.

Thomas, C. M., and S. D. Davis. 1989. Recovery patterns of three chaparral shrub species after wildfire. Oecologia 80:309-320.

Underwood, E. C., J. H. Viers, K. R. Klausmeyer, R. L. Cox, and M. R. Shaw. 2009. Threats and biodiversity in the Mediterranean biome. Diversity and Distributions 15:188-197.

Van de Water, K. M., and H. D. Safford. 2011. A summary of fire frequency estimates for California vegetation before Euroamerican settlement. Fire Ecology 7:26-58.

Vankat, J. 1989. Water stress in chaparral shrubs in summer rain versus summer drought climates: whither the Mediterranean type climate paradigm. Pages 117-124 *in* S. C. Keeley, editor. The California chaparral: paradigms reexamined. Natural History Museum of Los Angeles County, Los Angeles, California, USA.

Vasey, M. C., M. E. Loik, and V. T. Parker. 2012. Influence of summer marine fog and low cloud stratus on water relations of evergreen woody shrubs (*Arctostaphylos*: Ericaceae) in the chaparral of central California. Oecologia 170:325-337.

Venturas, M. D., E. D. MacKinnon, H. L. Dario, A. L. Jacobsen, R. B. Pratt, and S. D. Davis. 2016. Chaparral shrub hydraulic traits, size, and life history types relate to species mortality during California's historic drought of 2014. PLoS ONE 11:e0159145.

Zedler, P. H. 1997. Review: A manual of California vegetation by John O. Sawyer and Todd Keeler-Wolf. Madroño 44:214-219.

Chapter 2
Drivers of Chaparral Plant Diversity

Jon E. Keeley

Abstract Chaparral diversity has marked spatial and temporal variation. Evolutionary diversity at the genetic, specific, and lineage level contribute to a very diverse flora. Ecological diversity is evident in life histories that comprise a range of physiological and morphological strategies for dealing with drought, and demographic patterns centered around different seedling recruitment strategies. Community or alpha diversity varies markedly through time. Mature chaparral ranges from monotypic stands of chamise (*Adenostoma fasciculatum*) to mixed chaparral often with up to a dozen shrub species. The understory contributes relatively little other than a few diminutive annuals and occasional herbaceous perennial resprouts. However, after fire, diversity increases dramatically and is often dominated by annuals that arise from a dormant seedbank with significant contribution of geophytes resprouting and flowering from dormant bulbs and corms. This flora has very diverse life histories, with some present only a year or two and then existing as a dormant seedbank or bulbs until the next fire. Others may persist much longer, often in gaps in the shrub canopy. Postfire dominance-diversity patterns fit a geometric model as most communities are dominated by a few species and the bulk of the flora comprise subordinates that occupy specific microhabitats. Postfire community assembly is a result of competitive interactions and environmental filtering effects. Beta diversity plays a role in community assembly for as heterogeneity of communities in the landscape increases, the potential species pool for a community increases. Gamma diversity is particularly high because species turnover across latitudinal and elevational gradients is high. The role of diversity in conferring community resilience is complex and a function of the life history of shrub dominants and the historical patterns of fires. Under some circumstances low diversity may be more resilient than high diversity, for example under high fire frequency monotypic stands of *Adenostoma fasciculatum* may resist change better than diverse stands that include obligate seeding shrubs sensitive to short interval fires. Postfire annuals also are sensitive to short interval fires as these disturbances enhance the invasion by more competitive non-native grasses. Expected increases in anthropogenic ignitions due to population growth are the biggest threat to biodiversity in chaparral.

J. E. Keeley (✉)
U.S. Geological Survey, Three Rivers, CA, USA

University of California, Los Angeles, CA, USA
e-mail: jon_keeley@usgs.gov

E. C. Underwood et al. (eds.), *Valuing Chaparral*, Springer Series on Environmental Management, https://doi.org/10.1007/978-3-319-68303-4_2

Keywords Alpha diversity · Beta diversity · Dominance diversity · Ecological diversity · Gamma diversity · Genetic diversity · Postfire diversity

2.1 Introduction

In ecology there is a marked interest in issues of species diversity. In terms of conservation management, diversity of taxa is used as a measure of success based on the notion that more is better. However, there are sometimes more important goals than the maximum number of species at a single point in time, for example, long-term sustainability of species is a matter of concern in communities subjected to frequent disturbance. Diversity is widely thought to have synergistic effects on community stability and on ecosystem productivity (Diamond 1975). In addition, diversity is considered to be an important feature of ecosystem resilience to disturbance (Oliver et al. 2015). In chaparral, resilience is challenged by regular high intensity fires and rapid recovery is facilitated by diverse regeneration strategies.

This chapter will focus on describing patterns of diversity in California chaparral (Fig. 2.1) and the role of ecosystem processes such as periodic fires in driving diversity patterns. It will explore the possible insights into chaparral structure and function by examining diversity at different spatial scales. Two aspects are considered: (1) ecological diversity and (2) evolutionary diversity. This artificial division has

Fig. 2.1 Landscape mosaic of chaparral illustrating pure chamise (*Adenostoma fasciculatum*) chaparral on the south-facing slopes (on the right) and more diverse mixed chaparral on north-facing slopes (on the left). This landscape mosaic includes sage scrub or soft chaparral (lower right), annual grassland, and oak woodland (bottom third). Photo by Jon Keeley

Fig. 2.2 Postfire chaparral in the first spring after fire illustrating resprouting shrubs, shrub seedling recruitment, annuals, and resprouting herbaceous perennials. Photo by Teresa Brennan

some heuristic value, but it needs to be kept in mind that ecological and evolutionary processes are inextricably intertwined.

Ecological diversity is particularly evident in strategies for surviving summer drought and in regeneration modes after fire. Chaparral shrubs have been categorized by the degree of water stress they experience during the dry season (Parker et al. 2016), and there is a close linkage between mode of drought tolerance and options for postfire regeneration (Keeley 1998), and this has contributed to the diversity of life histories. This diversity increases dramatically after fire, in part because the number of species present is an order of magnitude greater than prior to fire. Many of the postfire species are successional, though they are often dominants in the closely aligned sage scrub or 'soft chaparral' vegetation (Keeley et al. 2005a). Postfire floras comprise annuals, herbaceous perennials, suffrutescents, and woody plants, but even within these life-forms there is a great variety of growth forms, e.g., prostrate, erect, vine-like, diminutive, or robust (Fig. 2.2).

Such diversity may contribute to higher species diversity (Fig. 2.3), perhaps by exploiting different resources.

These life-forms represent a variety of life histories exhibiting very different regeneration characteristics.

Evolutionary diversity is evident in chaparral by divergence at different taxonomic scales. For example, mixed chaparral (Fig. 2.4) typically comprises plant lineages represented by an assortment of plant families. Most investigations have

Fig. 2.3 Relationship between growth form and species richness in postfire chaparral (from Keeley et al. 2005b)

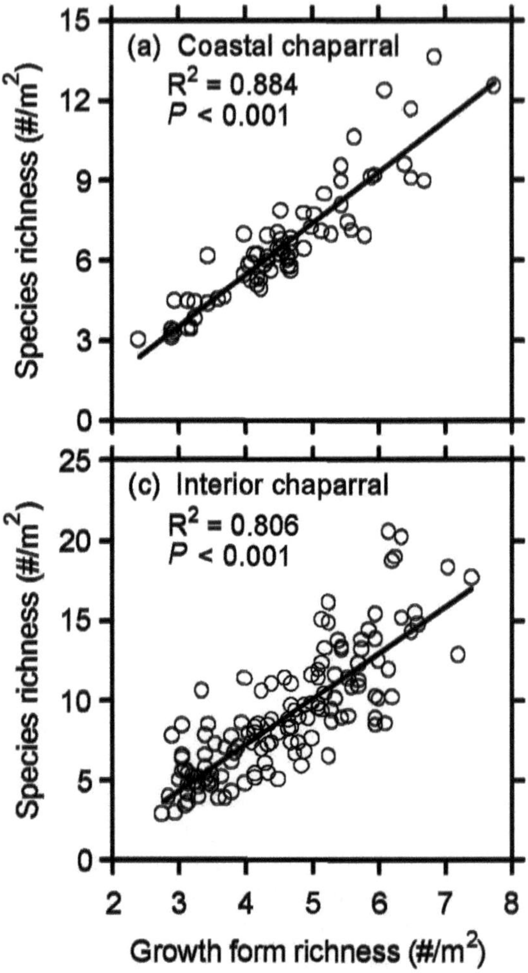

focused on the species assemblages that make up chaparral communities. However, finer scales of diversity are evident even within species, for example, subspecific variation represented by different morphs in the burl forming Eastwood's manzanita (*Arctostaphylos glandulosa*) (Ericaceae, Fig. 2.4). Populations of this shrub often comprise a mixture of clones with different foliage coloration, a pattern perpetuated by its resprouting habit after fire.

Fig. 2.4 Southern California mixed chaparral with Eastwood's manzanita (*Arctostaphylos glandulosa*) (Ericaceae), chamise (*Adenostoma fasciculatum*) (Roseaceae), sugar bush (*Rhus ovata*) (Anacardiaceae), mission manzanita (*Xylococcus bicolor*) (Ericaceae), and southern mountain misery (*Chamaebatia australis*) (Roseaceae), among others. Note the white glaucous-leaved and green morphs of *A. glandulosa*. Photo by Jon Keeley

2.2 Ecological Diversity

An important driver of diversity in the dominant shrubs is soil drought. Species are distributed along a continuum of soil moisture; at one end are water stress tolerators that experience low water potentials during the drought, and at the other end are those that avoid water stress (Keeley 2000; Parker et al. 2016). The former often have shallow roots and physiological mechanisms for tolerating drought, whereas avoiders have deep roots that have greater access to water during droughts (Davis et al. 1998; Pratt et al. 2007; Jacobsen et al. 2008). Avoiders and tolerators sort out along the aridity gradient, being differently distributed at various geographical scales based on factors such as slope aspect, elevation, and latitude. Tolerators and avoiders are extremes in a continuum of physiological and morphological traits selected for in response to soil aridity.

There is a close linkage between mode of drought tolerance and options for postfire regeneration and this is largely tied to constraints on seedling recruitment (Keeley 1998; Pratt et al. 2010; Keeley et al. 2012a). Species able to tolerate drought have the capacity for successful seedling recruitment after fire whereas avoiders, which depend on deep roots for surviving summer drought, have seedlings that are

Table 2.1 Life-form spectra for chaparral plant species (from Keeley and Davis 2007)

Life-form	Chaparral species	Chaparral only species	% Chaparral species
Annuals	415	28	35.3
Annual or perennial	23	1	1.9
Herbaceous perennial	460	22	39.1
Suffrutescent	11	0	0.9
Woody	279	46	23.7
Total	1177	110	

less successful in the postfire environment. As a consequence there is a close link between the avoider strategy and postfire resprouting as a means of regeneration. Also linked to this is the fact that these taxa generally occupy more mesic north-facing and higher elevation sites, where vigorous resprouting dominates postfire conditions, reducing safe sites for seedling recruitment. Drought tolerators generally occupy the more arid south-facing slopes and lower elevations. Thus, as landscape heterogeneity produces heterogenous conditions with respect to soil drought, this contributes to diversity in shrub species.

Chaparral comprises various life-forms (Table 2.1). Annuals compose over a third of the species, herbaceous perennials slightly more, and woody species nearly a quarter of the flora. Peak diversity is typically evident in the first and second springs after fire and diversity declines in subsequent years, although with slight pulses of increased diversity in high rainfall years during early seral stages (Keeley et al. 2006). In other ecosystems, colonization from outside the burned area is important for community recovery, but in chaparral this is not a factor, as regeneration is largely an endogenous process. Roughly 90% of the plant cover in 5 year old postfire chaparral comprises species present from resprouts and dormant seedbanks in the first year after fire (Keeley et al. 2005a, 2006).

In mature chaparral a few annuals occur regularly in the understory. These are sparsely distributed and diminutive species, including natives such as field parsley (*Aphanes occidentalis*) (Rosaceae), wild celery (*Apiastrum angustifolium*) (Apiaceae), dichondra (*Dichondra* spp.) (Convolvulaceae), annual bedstraw (*Galium aparine*) (Rubiaceae), small venus looking-glass (*Triodanis biflora*) (Campanulaceae), among others, and non-natives including daggerleaf cottonrose (*Logfia* [anagram of *Filago* to which it was formerly assigned] *gallica*) (Asteraceae) (nomenclature according to Baldwin et al. 2012). These species are generalists found throughout the state of California.

However, in the first year or two after fire, annuals are usually the dominant life-form, comprising 50% or more of all species on a site, differing in the extent to which recruitment is tied to fire (Keeley et al. 2005a). A group of taxa known as pyro-endemics are entirely fire-dependent and are unknown in unburned chaparral, either in the understory or openings in the vegetation. There are at least a dozen or more species we can assign as being rather strict pyro-endemics and they are best represented in the waterleaf family Hydrophyllaceae (in Baldwin et al. 2012 it is considered a subfamily of the borage family Boraginaceae, but recent work suggests it should be given coordinate recognition with that family), including whispering bells (*Emmenanthe penduliflora*), shortlobe phacelia (*Phacelia brachyloba*) as well as several other species of scorpionweed (*Phacelia*). Other pyro-endemics include sticky gilia (*Allophyllum glutinosum*) and *A. gilioides* (Polemoniaceae), San Diego wild cabbage (*Caulanthus heterophyllus*) and California mustard (*C. lasiophyllus*) (Brassicaceae), San Luis blazingstar (*Mentzelia micrantha*) (Loasaceae), fire poppy (*Papaver californicum*) (Papaveraceae), and multinerved catchfly (*Silene coniflora*) (Caryophyllaceae) (Keeley and Davis 2007). These taxa are abundant in the first spring after fire and often in the second postfire year, however they are generally absent in subsequent years, although seeds remain dormant in the soil until the next fire. How long the seeds survive is unknown but these species do recruit after fires in century-old stands (Keeley et al. 2005c), suggesting considerable seed longevity.

Other annuals are opportunistic species in that they greatly increase after fire but often persist at low levels in inter-stitial spaces between shrubs in mature chaparral. These include taxa such as pussy paws (*Calyptridium monandrum*) (Portulacaceae), prickly-nut crypthanta (*Cryptantha muricata*) (Boraginaceae), blue-thimble flower (*Gilia capitata*) phlox (Polemoniaceae), toadflax *Nuttallanthus* (formerly *Linaria canadensis*) (Plantaginaceae), *Phacelia cicutaria* (Hydrophyllaceae), *chia* (*Salvia columbariae*) (Lamiaceae), and many others (Keeley and Davis 2007).

Herbaceous perennials are mostly all present as bulbs, corms, or rhizomes in the understory of chaparral and produce above-ground growth occasionally, although seldom flowering. This life history is not closely tied to fire in that their natural cycle involves dieback in late spring, prior to the fire season, and then emergence with winter rain. The primary difference after fire is that due to the higher light, temperature, moisture, and nutrient conditions a much greater proportion of the bulbs resprout and produce more vigorous vegetative growth (e.g., Fig. 2.5). Also, flowering increases dramatically in the first postfire season. Indeed, it has been shown for one common species, common star lily (*Toxicoscordion* [formerly *Zigadenus*] *fremontii*) (Melanthiaceae), that prior to fire only a few percent of the population flowered in the understory but this increased to nearly 100% in the first growing season after fire, declining rapidly in subsequent years (Tyler and Borchert 2002). This pattern is evident in other geophytes, e.g., mariposa lily (*Calochortus* spp.). For most herbaceous perennials seedling recruitment is generally nil in the first season after fire due to the lack of a dormant soil-seedbank. However, seeds produced by the first year resprouts germinate in subsequent years and thus recruitment is fire-dependent on a narrow window of opportunity prior to shrub canopy recovery. This is a common pattern evident in a number of herbaceous perennials and also some subshrubs and further adds to the diversity of regeneration strategies in chaparral (Keeley et al. 2012a).

Suffrutescents are sometimes classified as subshrubs and other times as herbaceous perennials as these are plants that die back each year, but remain woody near the base. Included here are diverse taxa such as golden eardrops (*Ehrendorferia* [formerly *Dicentra*] *chrysantha*) (Papaveraceae), deerweed (*Acmispon glaber*) (formerly *Lotus scoparius*), and peak bush-rose (*Helianthemum scoparium*) (Cistaceae).

Fig. 2.5 Postfire resprouts of the bulb-forming star lily (*Toxicoscordion* [formerly *Zigadenus*] *fremontii*) (Melanthiaceae) after fire. Photo by Jon Keeley

These taxa do not resprout after fire but regenerate from soil-stored seedbanks where dormancy is broken by smoke or heat shock.

In terms of the shrub dominants, the majority of genera regenerate after fire by resprouting from basal or under-ground parts. A few of these (e.g., *Adenostoma fasciculatum* (Rosaceae), manzanita (*Arctostaphylos* spp.) (Ericaceae), ceanothus (*Ceanothus* spp.) (Rhamnaceae), mission manzanita (*Xylococcus bicolor*) (Ericaceae), and possibly tassel bush (*Garrya* spp.) (Garryaceae) resprout from a lignotuber at the base of the stem produced as a normal part of development (Keeley et al. 2012a). Other resprouters regenerate from unspecialized under-ground vegetation structures.

Many shrub species produce seeds that remain dormant in the soil for decades until stimulated to germinate after fire (primarily *A. fasciculatum,* and most all manzanita [*Arctostaphylos*] and ceanothus [*Ceanothus*] species). About two-thirds of the taxa in the latter two genera lack the capacity to resprout and regenerate after fire entirely from soil-stored seeds (obligate seeders), and the remaining taxa, and *A. fasciculatum,* regenerate from resprouts and seedling recruitment (facultative seeders). The majority of the shrub genera in chaparral regenerate after fire strictly from resprouting (obligate resprouters). In these taxa seedling recruitment is generally restricted to the shrub understory in older stands (Keeley 1992a).

Further diversity in species that recruit after fire is observed in the germination triggers that break seed dormancy. The majority of pyro-endemic annuals exhibit "smoke"-stimulated germination, where chemicals volatilized in smoke or leached from charred wood trigger germination (Keeley and Pausas 2016). Alternatively, however, some species with hard water-impermeable seed coats (e.g., Fabaceae) are triggered by heat shock (Keeley 1991).

Further diversity is evident in herbaceous perennials, in which most species produce seeds that are not dormant at dispersal and form transient seedbanks. Since flowering is often restricted to the immediate postfire year, recruitment is limited to early successional years. Essentially all obligate resprouting (ones with no postfire seedling recruitment) shrub species produce transient seedbanks and recruitment is generally restricted to older stands of chaparral (Keeley 1992a). Many annuals, suffrutescents, and shrubs have polymorphic seedbanks where a portion of the seeds lack dormancy and germinate in years of normal or above normal rainfall and another portion require a fire stimulus (Keeley and Fotheringham 2000). Some species exhibit ecotypic variation in that they are pyro-endemics in chaparral, but are distributed in other vegetation types where they are not tied to fire. For example, *Emmenanthe penduliflora* (Hydrophyllaceae) is a strict pyro-endemic in chaparral, but also occurs in desert scrub where it is not fire-dependent. Arroyo lupine (*Lupinus succulentus*) (Fabaceae) is much the same, being only found after fire in chaparral but occurring in associated grasslands where it does not exhibit a strict dependence on fire. It is likely that these patterns reflect genetic diversity in seed dormancy-breaking traits, however, environmentally induced differences during seed development could explain this as well (Keeley 1991).

2.3 Evolutionary Diversity

According to one estimate, the chaparral flora in California includes 1177 species (Halsey and Keeley 2016). Species are generally the focus of diversity studies although lineages (e.g., families) are often important for phylogenetic corrections when examining the ecological basis for trait evolution (e.g., Godoy et al. 2014). The term species diversity has been variously described in the ecological literature, oftentimes measured with indices such as the Shannon-Weiner Index or Simpson's Index that weight species by their dominance or evenness in the community. However, there are good reasons for limiting the term diversity to species density, i.e., species per unit area (Hurlbert 1971), often termed species richness.

Diversity takes on unique characteristics at different spatial scales and common metrics include *alpha* (α), *beta* (β), and *gamma* (γ), diversities. Whittaker (1960) first introduced these terms to describe different scales of diversity; α-diversity being the richness in species of a particular stand of a community, β-diversity the mixture of communities that occur in a complex environmental setting, and γ-diversity the number of species across a number of community samples.

Typically community or α-diversity is measured at the scale of a tenth hectare (0.25 acre) or larger (Fig. 2.6), although for comparative purposes between communities and for plotting species area curves, smaller spatial scales are sometimes

Fig. 2.6 Landscape mosaic with separate patches of chamise chaparral, each of which represents α-diversity and different patches representing γ-diversity, and the landscape mosaic of different plant communities, which represents β-diversity. Photo by Jon Keeley

used, from point diversity at 1 m^2 and upwards. β-Diversity has not always been applied consistently (Tuomisto 2010), but it generally captures the diversity contributed by diverse communities (e.g., chaparral, coastal sage scrub, grassland) juxtaposed in a complex environmental setting. At a landscape scale with a mosaic of different plant communities juxtaposed with one another, β-diversity is a measure of landscape diversity. In Whittaker's formulation γ-diversity was the diversity over broad regions, but others such as Cody (1986) used it in a more restrictive context, to measure diversity within a single habitat type over a broad region. Cody's approach allows one to examine species turnover in chaparral communities along a gradient (e.g., elevational or edaphic) and this approach has been followed by others studying diversity patterns in Mediterranean-type climate ecosystems (e.g., Simmons and Cowling 1996).

2.3.1 Factors Driving Community or α-Diversity

Although species richness is highest in early postfire years, there is marked spatial variation between chaparral communities, determined by complex temporal and spatial effects. The vast majority of taxa are *satellite* species, being found in a very

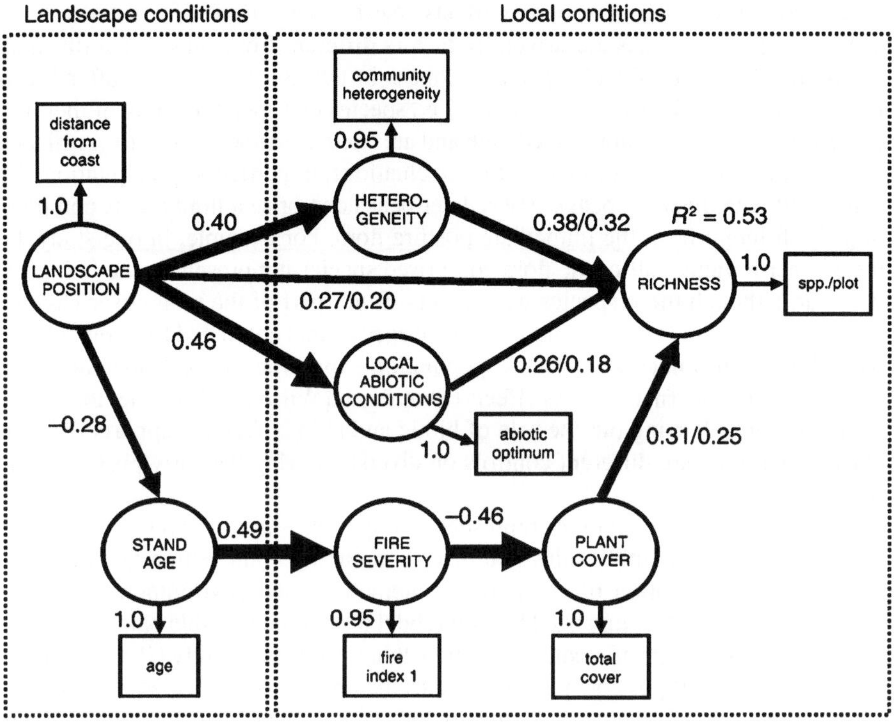

Fig. 2.7 Structural equation model of direct and indirect factors affecting species richness in postfire chaparral (from Grace and Keeley 2006)

small fraction of microsites within a community as well as within a small fraction of sites across the landscape (Keeley et al. 2005b). This could result in metapopulation dynamics if species had poorly developed dispersal ability, but that is not generally the case in these taxa and thus it seems likely that in most cases, species are restricted to a small fraction of the landscape due to microhabitat specialization.

It is hypothesized that postfire diversity is a function of at least four factors (Keeley et al. 2005a): (1) event characteristics, such as fire severity or subsequent precipitation, (2) historical effects, e.g., short intervals between fires may threaten recovery of obligate seeding species requiring sufficient time to accumulate a soil-stored seedbank, (3) inter-specific regulatory factors, such as successional changes in woody cover correlated with decreases in herb cover, and (4) environmental filters, reflected in landscape patterns such as differences between coastal and interior sites.

It has been found that both local conditions and landscape position were required to produce a model that would explain more than 50% of site to site variation in postfire chaparral diversity (Fig. 2.7). Studies have shown that peak richness occurs where plant abundance is moderately high, within-site heterogeneity is high, rock cover is high, and soils are low in nitrogen and high in sand (Grace and Keeley 2006).

Temporal variability in resources affects year-to-year variation in diversity, however, early postfire peaks are driven by factors different from subsequent diversity peaks (Keeley et al. 2005a). Species richness in the first year or two after fire is influenced by life history specialization, i.e., species that spend the bulk of their life cycle as a dormant soil-stored seedbank and are cued by smoke or heat to germinate. However, in subsequent years resource fluctuations, in particular precipitation, are associated with diversity peaks. These later peaks comprise a flora that is compositionally different from the immediate postfire flora. For example, in one study, by the 5th year roughly half of the flora comprised species not present in the first postfire year, although these species were a very minor part of the total cover (Keeley et al. 2005a, 2006). However, these were not species that colonized from outside the burned area, but rather were due to population expansion of local populations in other parts of the burn, i.e., mass effects (Shmida and Wilson 1985) rather than colonization events. Parsing out the role of biotic and abiotic factors supports the conclusion that there are different controls on diversity during the early postfire years (Grace et al. 2012).

Postfire seedling populations represent an important stage of community assembly and there are assembly rules that drive community composition and these are related to a combination of competitive interactions and environmental filtering effects. Competition is suggested by null model tests of all possible pairs of species in a large number of postfire sites, that show there was significantly ($P < 0.001$) less co-occurrence of all possible pairs of species than predicted if species were distributed randomly (Keeley and van Mantgem 2008). One mechanism could be different nitrogen use strategies responding to the fine scale pattern of nutrient distribution in postfire chaparral (Rice 1993). For example, some pyro-endemic annuals have a preference for ammonium nitrogen over nitrate, thereby being able to take advantage of high levels of the ammonium form in postfire environments, whereas more opportunistic species favor nitrate (Swift 1991). In addition, postfire communities comprise both nitrogen-fixing and non-nitrogen fixing species (Guo 2001), which may further contribute to postfire diversity.

Environmental filters operate at different scales. For example, at the site scale, different patterns of soil heating during fire can drive assembly patterns in the postfire environment (Odion and Davis 2000), while at a broader scale, proximity to the coast is a driver of diversity patterns (Keeley et al. 2005b). Additionally, environmental filters may change over time. During early succession there are often multiple cohorts of seedling recruitment and it is apparent that different filters are at work in early versus later succession (Table 2.2). Another factor determining how species subdivide resources is drought stress. Postfire seeding species tend to occur on more xeric sites and seedlings of these taxa are physiologically better able to deal with drought stress (Keeley 1998; Meentemeyer et al. 2001). Obligate resprouting shrubs dominate on more mesic slopes and their drought sensitive seedlings tend to be most successful later in succession in the understory of the mature chaparral. It is likely these shrubs have been selected to avoid postfire seedling recruitment not just because of the more arid conditions on postfire sites but to avoid competition with the vigorous resprouting of the parent plants on mesic slopes. Even within species

Table 2.2 Regression analysis relating the proportion (total 5 year postfire seedling density) for each seedling guild to environmental factors for soft chaparral; ns $P > 0.05$, *$P < 0.05$, **$P < 0.01$, ***$P < 0.001$, $n = 48$ sites (from Keeley and van Mantgem 2008)

	r-Value				
Seedling cohort	Coast	Insolation	Sand	Soil P	Fire severity
First-year pulse	+0.44**	+0.31*	+0.48***	−0.35*	+0.38**
Second-year pulse	−0.51***	ns	+0.32*	ns	ns
Multi-year recruits	ns	−0.36*	−0.46*	+0.42**	−0.38**

there appear to be site preference differences between seeding and resprouting subspecies in the genus *Arctostaphylos*, likely driven by conditions that favor postfire seedling recruitment on more open sites (Keeley et al. 2016).

Does diversity matter in terms of community resilience? While it is appealing to consider that as diversity increases there are more options for recovery in the face of ecosystem perturbations, however, this may not always be the case. For example, monotypic stands of chamise chaparral might be expected to be highly resilient to frequent fires because this species couples both resprouting and seedling recruitment. Plus, because resprouts flower in the first postfire year, the seedbank accumulates rapidly. In contrast, more diverse mixed chaparral communities comprise a mixture of regeneration strategies, some of which include obligate seeding species that may require a decade or more without fire to accumulate a seedbank sufficient to withstand a repeat fire (e.g., Zedler et al. 1983). For example, in a study of chaparral burned twice in four years, the obligate seeding *Ceanothus tomentosus* was extirpated from the site but the facultative seeder *Adenostoma fasciculatum* after the second burn only suffered a 40% reduction in seedling recruitment over what was observed after the first fire (data from Keeley and Brennan 2012).

On the other hand, diverse communities may be more resilient to longer fire-return intervals since they will likely include shrub species that can reach arborescent proportions and recruit seedlings in the absence of fire (Keeley 1992a). Thus, to understand the extent to which diversity contributes to ecosystem resilience, the issue needs to be evaluated in the context of life history characteristics and fire regimes. In short, there is little reason to believe a simple relationship between diversity and resilience exists in chaparral.

2.3.2 *Landscape β-Diversity*

Due to a combination of topographic, edaphic, and disturbance patterns chaparral is often juxtaposed with a variety of other plant communities, e.g., annual grassland, sage scrub, woodlands, and riparian forests (Wells 1962). Thus, β-diversity is often high on chaparral landscapes. Despite this fine-grained pattern of communities, fires tend to burn in a coarse-grained pattern and seldom stops at the boundaries between different communities. Fire has rather different impacts on diversity of these communities as woodlands and riparian communities recover almost entirely from resprouting by the woody species and have relatively little increase in diversity due to seedling recruitment of dormant seeds (Keeley and Syphard 2018).

β-Diversity patterns may have marked effects on α-diversity patterns in chaparral, for example, impacts on non-native species invasion. Generally non-native species do not fare well in competition with intact chaparral but often dominate in open grasslands and oak savannas. These species recover after fire in grasslands from seed germination and often present a formidable invasion risk for adjacent burned chaparral. Other ways β-diversity may affect α-diversity is when chaparral is juxtaposed with more mesic vegetation, and fires may burn through chaparral and then die down in adjacent communities. This creates refugia for small mammals that may feed in adjacent burned chaparral, impacting seedling recruitment and altering post-fire diversity (van Mantgem et al. 2015).

2.3.3 Community Turnover or γ-Diversity

As defined by Cody (1986), γ-diversity measures the changes in chaparral composition from one site to another, which contributes to broad scale diversity patterns throughout the range. Statewide there are marked differences in chaparral diversity in different regions, with the highest diversity in coastal central California (Fig. 2.8). Climate variables are generally the strongest predictors of plant diversity and their role differs with life-form (Richerson and Lum 1980). Harrison et al. (2006) found plant diversity was highest in the northern part of the state and attributed this to the observation that the state is very arid and rainfall (and thus productivity) are less

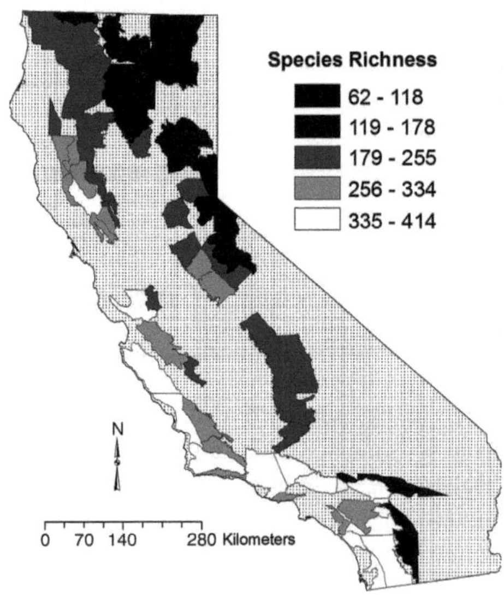

Fig. 2.8 Regional variation in species richness in chaparral (from Keeley and Davis 2007)

limiting at higher latitudes in the state; they concluded that while precipitation was important in determining regional diversity patterns, temperature was not. This contrasts with what is observed for chaparral (Fig. 2.8), suggesting that different plant communities may have very different drivers of diversity.

This issue of high chaparral diversity in the central coastal region (Fig. 2.8) was addressed previously where it was suggested that the region had several characteristics in common with the highly diverse South African fynbos (Keeley 1992b). In this analysis it was concluded that temperature was important, in particular the less stressful extreme high summer temperatures reduced the elimination of taxa in coastal regions, but was perhaps a factor in reducing diversity in the interior foothills of the Sierra Nevada, and chaparral sites further north. This role of a more benign climate also appears to favor persistence of rare species (Harrison et al. 2008). Another factor affecting diversity in the central coast chaparral is the Pleistocene marine terraces with low nutrient sandy substrates that contribute to reduced productivity, which may reduce the competitive exclusion of many species, much like what is thought to be an important driver of fynbos diversity (Simmons and Cowling 1996).

Composition varies markedly across the state, comprising over 40 different communities or more than 70 alliances (Halsey and Keeley 2016). The composition of chaparral turns over in association with distance; for example, the southern-most county is 42% dissimilar to chaparral in other parts of southern California and 73% dissimilar to counties in the northern part of the state (Keeley and Davis 2007), a general pattern seen in the vascular plant flora (e.g., Burge et al. 2016). The two largest genera *Arctostaphylos* and *Ceanothus* (Rhamnaceae) exhibit remarkable latitudinal species turn-over in chaparral. Cody (1986) estimated a local turnover rate of 50% of the species in these genera within a distance of 100–300 km (62–186 miles) along a latitudinal gradient in the state and a similar elevational turn-over has been described by Zedler (1995). This potentially could impact the capacity for adjusting to future climate changes (see Chap. 14). With different taxa turning over along a latitudinal gradient it may increase seed sources for natural migrations.

This landscape turnover is likely driven by changes in climate but fire regimes likewise may play a role. The southern part of the state has very high fire frequency but further north fires are less frequent (Safford and Van de Water 2014). As discussed earlier, different life histories are favored by different fire frequencies. Species turnover in chaparral communities is of course also impacted by substrate diversity as unusual substrates such as serpentine typically select for different chaparral dominants (Safford et al. 2005).

Chaparral communities have substantial turnover along other gradients as well. On coastal sites is a unique set of shrub species known as maritime chaparral that is largely restricted to sites with summer fog (Vasey et al. 2014). Climate parameters play a significant role in diversity patterns and are responsible for substantial turnover in species composition along the gradient from coast to interior valleys and foothills. Life histories also change, for example a greater incidence of postfire obligate seeding species occur in the interior than at the coast.

Table 2.3 (a) Top-ranking habitats of California's rare plants according to CNPS lists (Skinner and Pavlik 1994). Observed values contrasted with those expected based on the amount of land area occupied by each habitat, and (b) distribution of rare plant life-forms in chaparral, expected base on proportion of life-forms from postfire chaparral studies (from Keeley et al. 2005a, b, c)

(a)

		Taxa	
Rank	Habitat	Observed	Expected
2	Lower coniferous forests	359	294
3	Cismontane woodland	311	362
4	Valley/foothill grassland	247	431
5	Coastal scrub	211	132

(b)

	Annual	Herbaceous perennial	Suffrutescent	Subshrub/Shrub
Observed	43	79	17	63
Expected	131	27	16	28

A similar pattern is evident along elevational gradients where obligate seeding species dominate in the lower elevations and are progressively replaced by resprouting species at higher elevations (Keeley et al. 2012a). At the highest elevations is montane chaparral that is often inter-spersed on severe substrates with coniferous forests. Compositionally it has relatively little overlap with shrubland chaparral, and is typically dominated by obligate resprouting shrubs such as huckleberry oak (*Quercus vaccinfolia*) (Fagaceae), giant chinquapin (*Chrysolepis chrysophylla*) (Fagaceae), bitter cherry (*Prunus emarginata*) (Rosaceae), or facultative seeders such as green leaved manzanita (*Arctostaphylos patula*) (Ericaceae), and mountain whitethorn (*Ceanothus cordulatus*) (Rhamnaceae).

2.3.4 Endemic and Rare Species

California has a flora rich in rare and endemic plant species (Harrison 2013). However, in chaparral, despite more than 1100 plant species, only about 10% are restricted to chaparral and most of these are shrub dominants. These endemics are largely postfire seeders including *Adenostoma, Arctostaphylos* and *Ceanothus* species.

Rare species are of particular interest but comprise a diverse collection of taxa. Some are widely distributed geographically but locally very rare, whereas some may be frequent within a region but geographically highly restricted (Rabinowitz 1981). Rare species in chaparral comprise both types. Relative to other vegetation types, chaparral has significantly more rare species than would be expected based

Table 2.4 Typical species richness reported for 1 m² and 1000 m² scales for selected MTC region sites (from Keeley and Fotheringham 2003)

	Species richness	
	1 m²	1000 m²
California		
Mature chaparral	2	24
Postfire chaparral (coast)	7	46
Postfire chaparral (interior)	7	53
Arizona		
Postfire chaparral	8	88
Mediterranean Basin		
Mature Eastern maquis	–	27
Disturbed Eastern maquis	–	130
Western Australia		
Mallee	7	49
Heathland	13	67

on area alone (Table 2.3). It is noteworthy that rarity is strongly associated with life-form. Chaparral annuals are rather depauperate in rare species whereas herbaceous perennials and shrubs have substantially more than to be expected based on number of species in the community.

2.4 Comparisons with Other Mediterranean-Type Climate Shrublands

Chaparral is one of the dominant vegetation types within California's Mediterranean-type climate (MTC) region and there is something to be learned by comparisons with shrublands that dominate the other four MTC regions of the world (elaborated in more detail in Chap. 1). At the scale of communities, chaparral exhibits similar species richness as other MTC communities (Keeley and Fotheringham 2003). One of the primary differences is peak diversity in chaparral is tied to postfire conditions, whereas this is not the case with some MTC communities such as South African fynbos or Western Australian heathland, where diversity does not decline markedly between burned and mature stands (e.g., Table 2.4).

It is of interest that North American chaparral in a non-MTC region such as southeastern Arizona has a substantially higher diversity than California chaparral (Keeley et al. 2012b). Also, eastern Mediterranean Basin disturbed maquis has extraordinarily high community diversity. In both cases these extraordinarily high diversities are theorized to be driven by biogeographical patterns from the mixing of floras from different regions (Keeley et al. 2012a), which would support the idea that there are not strong ecological constraints on diversity (Harmon and Harrison 2015). However, this only holds up where closed canopy chaparral is prevented

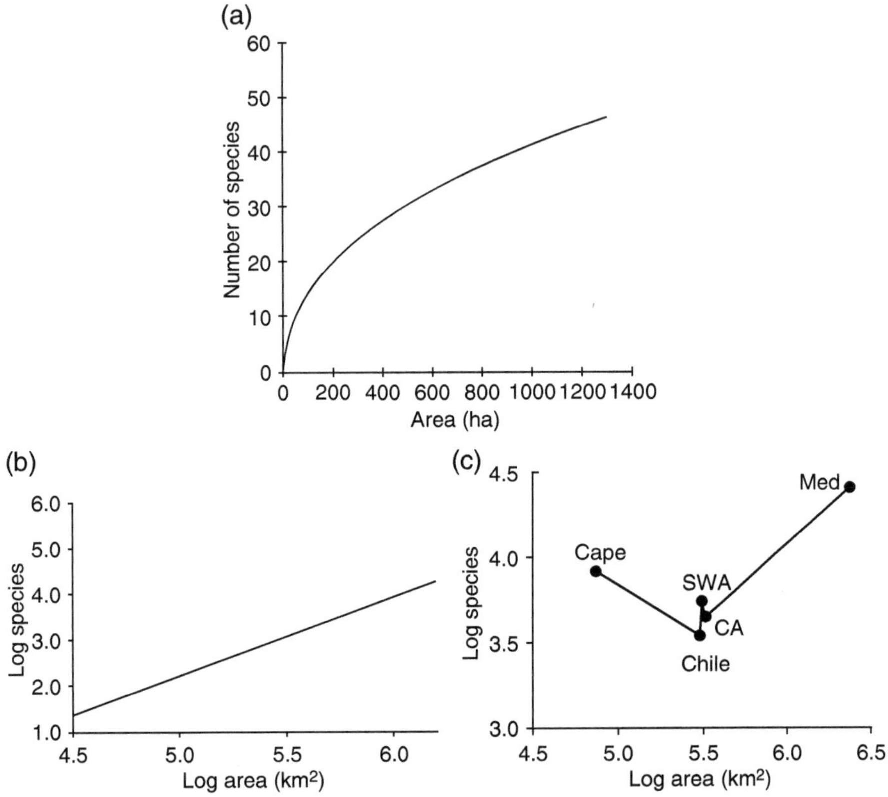

Fig. 2.9 Species-area relationships: (**a**) Idealized species-area relationship on an arithmetic scale, (**b**) loglog scale, and (**c**) observed species-area relationship for the five MTC regions. *Med* Mediterranean Basin, *CA* California, *Chile* central Chile, *Cape* Cape Region of South Africa, *SWA* Southwest Australia (from Keeley et al. 2012a, b)

from developing, either through soil water deficits in Arizona or disturbance in the eastern Mediterranean Basin (Table 2.4).

At a global scale, the five MTC regions differ markedly in plant diversity (see Chap. 1) but they also differ greatly in the land area covered. One of the commonly held generalizations about species diversity is that it increases with area (Fig. 2.9a, b). This should not be surprising since with increasing area the probability of encountering new species increases (Fig. 2.9a, b), what has been termed one of the few "laws" in ecology (Lomolino 2001). However, this species-area relationship only approaches the status of a "law" when describing patterns in nested samples (Dunn and Loehl 1988). That is, subsamples of different size taken from within the boundaries of larger area samples. When samples are not nested, as with the comparison of MTC regions (Fig. 2.9c), there is a glaring lack of fit to an idealized species area relationship (Fig. 2.9b), pointing up some of the important differences in diversity between these MTC regions. As discussed by Rundel (see Chap. 1), these differences arise from subtle variations in climate, not so subtle variations in geol-

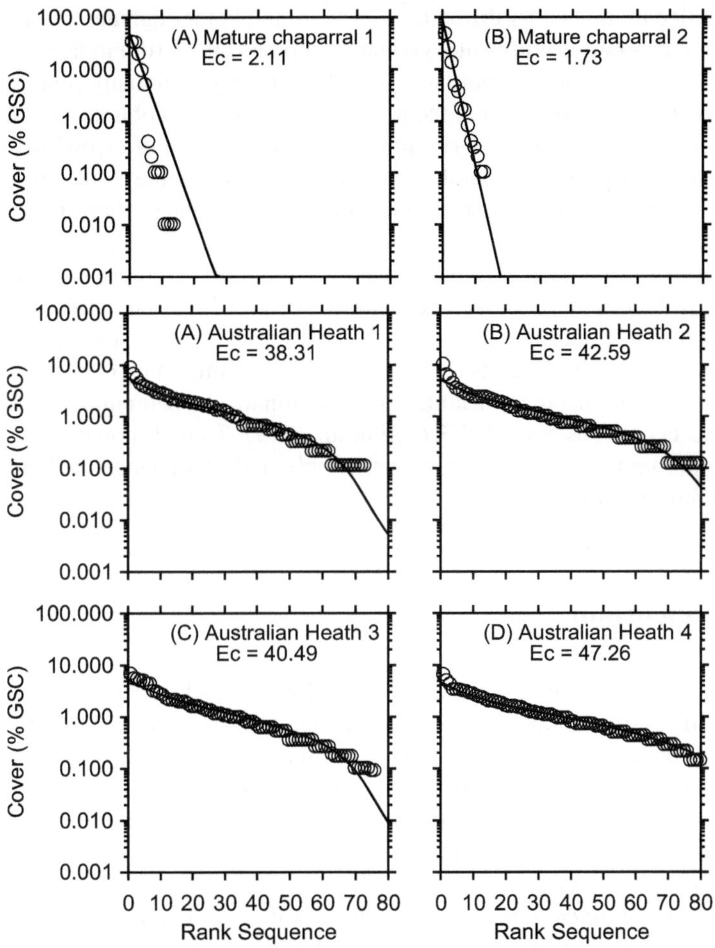

Fig. 2.10 Dominance-diversity curves based on percentage ground surface covered for: (**a, b**) two 2 year old postfire chaparral sites, (**c**) 30 year old mature chaparral, (**d**) 90 year old mature chaparral, (**e, f**) two mature Western Australian heathland sites. Ec = Whittaker's equitability index (from Keeley and Fotheringham 2003)

ogy, and to their interaction with fire, as well as to phylogenetic and biogeographic histories (Keeley et al. 2012a).

Comparisons of species-area curves across broad regions may reveal important characteristics about global scale diversity patterns (Rosenzweig 1995). At the scale of plant communities differences in species-area curves have been attributed to sampling artifacts (Rosenzweig 1995), but Keeley and Fotheringham (2003) contended they may provide insights into factors driving community assembly. For example, following fire, species-area curves are very different between California chaparral and Western Australian heathland; chaparral fits a power model and heathland exhibits a better fit to an exponential model (Keeley et al. 2012a). It is hypothesized that these different model fits are the result of different dominance-diversity relationships tied to life history differences (Keeley and Fotheringham 2003). Chaparral dominance-diversity patterns fit a geometric model (Fig. 2.10a) indicating diversity is controlled by the strong dominance of a few species with many subordinate species. This arises because most of the postfire flora comprises obligate and facultative seeders and a small number of vigorous resprouters, which often dominate postfire cover. Although postfire endemics are usually abundant the majority of annuals occur in relatively small numbers and much of this annual diversity is relatively transient.

In contrast, Australian heathland species are distributed more equitably, with a better fit to MacArthur's broken stick model (Fig. 2.10b). The implication of this model is that populations reach a stable equilibrium without the development of dominance by any one species (Whittaker 1972). An important contributor to these patterns is the fact that these heathlands are dominated by perennials, and more than three-quarters are resprouters (George et al. 1979). Thus, following fire there are fewer opportunities for community reassembly and less chance of dominance by one or a few species.

2.5 Conclusion

Chaparral diversity is most markedly affected by fire and response to fire is a function of life-form and mode of regeneration. Diversity is controlled by event-dependent factors such as fire severity of subsequent precipitation, timing of repeat burns, and rate of canopy recovery as well as environmental filters. The role of diversity in conferring community resilience is complex and a function of the life history of shrub dominants and the historical patterns of fires. Different fire regimes may favor different levels of diversity. For example, low diversity monotypic stands of *Adenostoma fasciculatum* may be highly resilient to frequent fires because of its facultative seeder life history that combines seeding and resprouting. Often such stands are far more resilient than high diversity mixed stands that include obligate seeding shrubs, which are often extirpated when fire intervals are too short. Future global changes that are thought to most strongly impact chaparral is primarily human population growth, and this is likely to translate in more frequent fires, which we expect to greatly threaten biodiversity in chaparral.

References

Baldwin, B. G., D. H. Golman, D. J. Keil, R. Patterson, T. J. Rosatti, and D. H. Wilken, editors. 2012. The Jepson manual. Vascular plants of California. University of California Press, Los Angeles, California, USA.

Burge, D. O., J. H. Thorne, S. P. Harrison, B. C. O'Brien, J. P. Rebman, J. R. Shevock, E. R. Alverson, L. K. Hardison, J. D. Rodriguez, S. A. Junak, T. A. Oberbauer, H. Rieman, S. E.

Vanderplank, and T. Barry. 2016. Plant diversity and endemism in the California Floristic Province. Madroño 63:3-206.

Cody, M. L. 1986. Diversity, rarity and conservation in Mediterranean-climate regions. Pages 122-152 *in* M. E. Soulé, editor. Conservation biology. The science of scarcity and diversity. Sinauer Associates, Sunderland, Massachusetts, USA.

Davis, S. D., K. J. Kolb, and K. P. Barton. 1998. Ecophysiological processes and demographic patterns in the structuring of California chaparral. Pages 297-310 *in* P. W. Rundel, G. Montenegro, and F. M. Jaksic, editors. Landscape disturbance and biodiversity in Mediterranean-type ecosystems. Springer-Verlag, New York, USA.

Diamond, J. M. 1975. Assembly of species communities. Pages 342-444 *in* M. L. Cody and J. M. Diamond, editors. Ecology and evolution of communities. Harvard University Press, Cambridge, Massachusetts, USA.

Dunn, C.P., and C. Loehl. 1988. Species-area parameter estimation: testing the null model of lack of relationship. Global Ecology and Biogeography 9:59-74.

George, A. S., A. J. M. Hopkins, and N. G. Marchant. 1979. The heathlands of Western Australia. Pages 211–230 *in* R. L. Specht, editor. Ecosystems of the world 9A. Heathlands and related shrublands: descriptive studies. Elsevier Scientific, New York, USA.

Godoy, O., N. J. B. Kraft, and J. M. Levine. 2014. Phylogenetic relatedness and the determinants of competitive outcomes. Ecology Letters 17:836-844.

Grace, J. B., and J. E. Keeley. 2006. A structural equation model analysis of postfire plant diversity in California shrublands. Ecological Applications 16:503-514.

Grace, J. B., J. E. Keeley, D. Johnson, and K. A. Bollen. 2012. Structural equation modeling and the analysis of long-term monitoring data. Pages 325-359 *in* R. A. Gitzen, J. J. Millspaugh, A. B. Cooper, and D. S. Licht, editors. Design and analysis of long-term ecological monitoring studies. Cambridge University Press, Cambridge, UK.

Guo, Q. 2001. Early post-fire succession in California chaparral: changes in diversity, density, cover and biomass. Ecological Research 16:471-485.

Halsey, R. W., and J. E. Keeley. 2016. Conservation issues: California chaparral. Reference Module in Earth Systems and Environmental Sciences. https://doi.org/10.1016/B978-0-12-409548-9.09584-1.

Harmon, L. J., and S. Harrison. 2015. Species diversity is dynamic and unbounded at local and continental scales. American Naturalist 185:584-593.

Harrison, S. P. 2013. Plant and animal endemism in California. University of California Press, Los Angeles, California, USA.

Harrison, S., H. D. Safford, J. B. Grace, J. H. Viers, and K. F. Davies. 2006. Regional and local species richness in an insular environment: serpentine plants in California. Ecological Monographs 76:41-56.

Harrison, S., J. H. Viers, J. H. Thorne, and J. B. Grace. 2008. Favorable environments and the persistence of naturally rare species. Conservation Letters 1:65-74.

Hurlbert, S.H. 1971. The nonconcept of species diversity: a critique and alternative parameters. Ecology 52:577-586.

Jacobsen, A. L., R. B. Pratt, S. D. Davis, and F. W. Ewers. 2008. Comparative community physiology: nonconvergence in water relations among three semi-arid shrub communities. New Phytologist 180:100-113.

Keeley, J. E. 1991. Seed germination and life history syndromes in the California chaparral. Botanical Review 57:81-116.

Keeley, J. E. 1992a. Recruitment of seedlings and vegetative sprouts in unburned chaparral. Ecology 73:1194-1208.

Keeley, J. E. 1992b. A Californian's view of fynbos. Pages 372-388 *in* R. M. Cowling, editor. The ecology of fynbos: nutrients, fire and diversity. Oxford University Press, Oxford, UK.

Keeley, J. E. 1998. Coupling demography, physiology and evolution in chaparral shrubs. Pages 257-264 *in* P. W. Rundel, G. Montenegro, and F. M. Jaksic editors. Landscape degradation and biodiversity in Mediterranean-type ecosystems. Springer, New York, New York, USA.

Keeley, J. E. 2000. Chaparral. Pages 203-253 *in* M. G. Barbour and W. D. Billings, editors. North American terrestrial vegetation. Second edition. Cambridge University Press, Cambridge, UK.

Keeley, J. E., W. J. Bond, R. A. Bradstock, J. G. Pausas, and W. Rundel. 2012a. Fire in Mediterranean climate ecosystems: ecology, evolution and management. Cambridge University Press, Cambridge, UK.

Keeley, J. E. and T. J. Brennan. 2012. Fire driven alien invasion in a fire-adapted ecosystem. Oecologia 169:1043-1052.

Keeley, J. E., and F. W. Davis. 2007. Chaparral. Pages 339-366 *in* M. G. Barbour, T. Keeler-Wolf, and A. A. Schoenherr, editors. Terrestrial vegetation of California. Third edition. University of California Press, Los Angeles, California, USA

Keeley, J. E., and C. J. Fotheringham. 2000. Role of fire in regeneration from seed. Pages 311-330 *in* M. Fenner, editor. Seeds: the ecology of regeneration in plant communities. Second edition. CAB International, Wallingford, UK.

Keeley, J. E., and C. J. Fotheringham. 2003. Species area relationships in Mediterranean-climate plant communities. Journal of Biogeography 30:1629-1657.

Keeley, J. E., C. J. Fotheringham, and M. Baer-Keeley. 2005a. Determinants of postfire recovery and succession in Mediterranean-climate shrublands of California. Ecological Applications 15:1515-1534.

Keeley, J. E., C. J. Fotheringham, and M. Baer-Keeley. 2005b. Factors affecting plant diversity during postfire recovery and succession of Mediterranean-climate shrublands in California, USA. Diversity and Distributions 11:535-537.

Keeley, J. E., C. J. Fotheringham, and M. Baer-Keeley. 2006. Demographic patterns of postfire regeneration in Mediterranean-climate shrublands of California. Ecological Monographs 76:235-255.

Keeley, J. E., C. J. Fotheringham, and P. W. Rundel. 2012b. Postfire regeneration of chaparral under Mediterranean and non-Mediterranean climates. Madroño 59:109-127.

Keeley, J. E., V. T. Parker and M. C. Vasey. 2016. Resprouting and seeding hypotheses: a test of the gap-dependent model using resprouting and obligate seeding subspecies of *Arctostaphylos*. Plant Ecology 217:743-750.

Keeley, J. E., and J. G. Pausas. 2016. Evolution of 'smoke' induced seed germination in pyroendemic plants. South African Journal of Botany. http://dx.doi.org/10.1016/j.sajb.2016.07.012.

Keeley, J. E., A. H. Pfaff, and H. D. Safford. 2005c. Fire suppression impacts on postfire recovery of Sierra Nevada chaparral shrublands. International Journal of Wildland Fire 14:255-265.

Keeley, J. E., and A. D. Syphard. 2018. South Coast Bioregion. Pages 319-351 *in* J. W. van Wagtendonk, N. G. Sugihara, S. L. Stephens, A. E. Thode, K. E. Shaffer, and J. A. Fites-Kaufman, editors. Fire in California's ecosystems. Second edition. University of California Press, Berkeley, California, USA.

Keeley, J. E., and P. J. van Mantgem. 2008. Community ecology of seedlings. Pages 255-273 *in* M. A. Leak, V. T. Parker, and R. L. Simpson, editors. Seedling ecology and evolution. Cambridge University Press, Cambridge, UK.

Lomolino, M. V. 2001. The species-area relationship: new challenges for an old pattern. Progress in Physical Geography 25:1-21.

Meentemeyer, R. K., A. Moody, and J. Franklin. 2001. Landscape-scale patterns of shrub-species abundance in California chaparral. Plant Ecology 156:19-41.

Odion, D. C., and F. W. Davis. 2000. Fire, soil heating, and the formation of vegetation patterns in chaparral. Ecological Monographs 70:149-169.

Oliver, T. H., M. S. Heard, N. J. B. Isaac, D. B. Roy, D. Procter, F. Eigenbrod, R. Freckleton, A. Hector, C. D. L. Orme, O. L. Petchey, V. Proenca, D. Raffaelli, K. B. Suttle, G. M. Mace, B. Martin-Lopez, B. A. Woodcock, and J. M. Bullock. 2015. Biodiversity and resilience of ecosystem functions. Trends in Ecology and Evolution 30:673-684.

Parker, V. T., R. B. Pratt, and J. E. Keeley. 2016. Chaparral. Pages 479-507 *in* H. Mooney, and E. Zavaleta, editors. Ecosystems of California. University of California Press, Berkeley, California, USA.

Pratt, R. B., A. L. Jacobsen, K. A. Golgotiu, J. S. Sperry, F. W. Ewers, and S. D. Davis. 2007. Life history type and water stress tolerance in nine California chaparral species (Rhamnaceae). Ecological Monographs 77:239-253.

Pratt, R. B., G. B. North, A. L. Jacobsen, F. W. Ewers, and S. D. Davis. 2010. Xylem root and shoot hydraulics is linked to life history type in chaparral seedlings. Functional Ecology 24:70-81.

Rabinowitz, D. 1981. Seven forms of rarity. Pages 205-217 *in* H. Synge, editor. The biological aspects of rare plant conservation. John Wiley and Sons, New York, New York, USA.

Rice, S. K. 1993. Vegetation establishment in post-fire *Adenostoma* chaparral in relation to fine-scale pattern in fire intensity and soil nutrients. Journal of Vegetation Science 43:115-124.

Richerson, P. J., and K. L. Lum. 1980. Patterns of plant species diversity in California: relation to weather and topography. American Naturalist 116:504-536.

Rosenzweig, M. I. 1995. Species diversity in space and time. Cambridge University Press, Cambridge, UK.

Safford, H. D., and K. M. Van de Water. 2014. Using fire return interval departure (FRID) analysis to map spatial and temporal changes in fire frequency on National Forest lands in California. Research Paper PSW-RP-266. USDA Forest Service, Pacific Southwest Research Station, Albany, California, USA.

Safford, H. D., J. H. Viers, and S. P. Harrison. 2005. Serpentine endemism in the California flora: a database of serpentine affinity. Madroño 52:222-257.

Shmida, A., and M. V. Wilson. 1985. Biological determinants of species diversity. Journal of Biogeography 12:1-20.

Simmons, M. T., and R. M. Cowling. 1996. Why is the Cape Peninsula so rich in plant species? An analysis of the independent diversity components. Biodiversity and Conservation 5:551-573.

Skinner, M. W. and B. M. Pavlik. 1994. California native plant society's inventory of rare and endangered vascular plants of California. Fifth edition. California Native Plant Society, Sacramento, California, USA.

Swift, C. 1991. Nitrogen utilization strategies of post-fire annual species in the chaparral. University of California, Los Angeles, California, USA.

Tuomisto, H. 2010. A diversity of beta diversities: straightening up a concept gone awry. Part 1. Defining beta diversity as a function of alpha and gamma diversity. Ecography 33:2-22.

Tyler, C., and M. Borchert. 2002. Reproduction and growth of the chaparral geophyte, *Zigadenus fremontii* (Liliaceae), in relation to fire. Plant Ecology 165:11-20.

van Mantgem, E. F., J. E. Keeley, and M. Witter. 2015. Faunal responses to fire in California chaparral and sage scrub in California, USA. Fire Ecology 11:128-148.

Vasey, M. C., V. T. Parker, K. D. Holl, M. E. Loik, and S. Hiatt. 2014. Maritime climate influence on chaparral composition and diversity in the coast range of central California. Ecology and Evolution 4: 3362-3674.

Wells, P. V. 1962. Vegetation in relation to geological substratum and fire in the San Luis Obispo quadrangle, California. Ecological Monographs 32:79-103.

Whittaker, R.H. 1960. Vegetation of the Siskiyou Mountains, Oregon and California. Ecological Monographs 30:279-338.

Whittaker, R. H. 1972. Evolution and measurement of species diversity. Taxon 21:213-251.

Zedler, P. H. 1995. Fire frequency in southern California shrublands: biological effects and management options. Pages 101-112 *in* J. E. Keeley and T. Scott, editors. Brushfires in California: ecology and resource management. International Association of Wildland Fire, Fairfield, Washington, USA.

Zedler, P. H., C. R. Gautier, and G. S. McMaster. 1983. Vegetation change in response to extreme events: the effect of a short interval between fires in California chaparral and coastal scrub. Ecology 64:809-818.

Chapter 3
Faunal Diversity in Chaparral Ecosystems

Megan K. Jennings

Abstract Chaparral ecosystems, although known primarily for their rich plant diversity, are also home to a suite of diverse and unique animals, helping to make California's Floristic Province one of the biodiversity hotspots in the world. The air, land, and freshwater habitats in chaparral ecosystems host nearly 400 species of vertebrate fauna and an unknown number of invertebrates. These species play important roles in maintaining ecological function which also provide ecosystem services for the large human populations that live in and amidst California's chaparral. However, these human populations pose a threat to the faunal biodiversity of chaparral, putting these ecosystems at risk from loss of habitat, fragmentation, and disturbance. Preserving the integrity of chaparral systems for their human and wildlife inhabitants requires an understanding of the diversity of the species occurring in chaparral and the role they play in how ecosystems function.

Keywords Amphibian · Animal · Bird · Chaparral · Conservation · Ecosystem services · Faunal · Fish · Invertebrate · Mammal · Reptile · Wildlife

3.1 Introduction

The shrubland ecosystems of Mediterranean-type climates, such as chaparral, are some of the most biologically diverse ecosystems of the world (Myers et al. 2000). Chaparral provides habitat for a broad range of animal species and also exhibits a high degree of rarity and endemism in both plant and animal species due to unique abiotic and biotic features. The mild climate and predominantly coastal locale that are key characteristics of chaparral have also made these areas attractive to human development, which has resulted in broad scale habitat degradation, fragmentation, and loss, threatening many native chaparral animals. Forty species within the five classes of vertebrates, as well as invertebrates in the phylum Arthropoda, in California's chaparral are listed under the federal Endangered Species Act. Here, I

M. K. Jennings (✉)
San Diego State University, San Diego, CA, USA
e-mail: mjennings@mail.sdsu.edu

E. C. Underwood et al. (eds.), *Valuing Chaparral*, Springer Series on Environmental Management, https://doi.org/10.1007/978-3-319-68303-4_3

provide an overview of faunal diversity and the critical role these species play in chaparral ecosystems. This will include examples of species, communities, and populations that exhibit distinctive behaviors or roles in their habitats and unique relationships among species. This chapter examines the threats and stressors chaparral fauna face including habitat loss, threats from non-native species, and increasingly frequent fires in chaparral. This chapter provides a review of adaptations to these threats whereby the same process, such as fire, can be a threat to some species and a boon to others, depending on diet and habitat niche breadth.

As described in Chap. 2, biodiversity can be categorized and quantified in a range of ways (i.e., genetics, species, populations, ecosystems, and functional diversity) and across multiple spatial scales (i.e., alpha, beta, and gamma diversity). By any combination of metric and scale, Mediterranean-type climate (MTC) regions in general, and California's chaparral ecosystems specifically, are some of the most biodiverse in the world outside of the tropics (see Chap. 1). The faunal diversity within these ecosystems is no exception. Arthropods and the five groups of vertebrates are well represented across the range of chaparral ecosystem types, and many species are known only from the region (i.e., the California Floristic Province), or even from specific microhabitats within chaparral. California's Floristic Province, where chaparral is the second most common vegetation type after coniferous forest, is one of the top 25 biodiversity hotspots in the world, areas that cover 1.4% of the land area on the planet and contain 44% of all plant species and 35% of vertebrate species (Myers et al. 2000). The region is home to more than 580 vertebrate species, with more than 12% of those being endemic (Myers et al. 2000), and at least 26,000 described invertebrates (Kimsey 1996). The same climatic, topographic, and geologic diversity that supports the rich diversity of plants in the region, along with the plant diversity itself, creates a broad range of niches in which these animals thrive.

3.2 Origins of Faunal Diversity in Chaparral Ecosystems

Whereas plant diversity within MTC regions like chaparral is notably characterized by a vast array of rare and endemic species (see Chaps. 1 and 2), the degree of endemism in chaparral's fauna cannot compare (Hobbs et al. 1995; Blondel and Aronson 1995; Davis and Rutherford 1995). Although the local climates, topography, and geology of the MTC region of California are the primary drivers of plant diversity (see Chap. 2), the diversity of animals in chaparral does not follow these same patterns, nor does it appear to be related to the rise of rich and unique plant associations during the climatic shifts of the late Miocene (see Chap. 3, Calsbeek et al. 2003). It has been proposed that the origin of the broad and unique range of animal species is more likely a function of biogeography than a result of adaptations to the MTC or the associated plant diversity (Blondel and Aronson 1995; Davis and Rutherford 1995; Hobbs et al. 1995). The patterns of molecular evolution and diversification within the California Floristic Province indicate that faunal diversity and endemism may have arisen with the geologic change that occurred between 2 and 5 million years ago when mountains began to form and aridification of the region began (Calsbeek et al. 2003). These geologic shifts and associated climatic changes likely created both physical and physiological barriers affecting species' ranges and movement. Further studies of chaparral animal origins (LaPointe and Rissler 2005; Vandergast et al. 2008) and endemism (Harrison 2013) have provided evidence to support this proposed mechanism for faunal diversity in chaparral. It has been demonstrated that overall diversity and genetic divergence in wildlife tend to be higher in ecotones (along steep environmental gradients) and along suture zones (joining along fault zones, Vandergast et al. 2008). These evolutionary hotspots may have been the source of speciation and divergence in the past, and could provide adaptive capacity for climatic and landscape shifts in the future.

Both hard chaparral (dominated by chamise [*Adenostoma* spp.], ceanothus [*Ceanothus* spp.], mountain mahogany [*Cercocarpus betuloides*], manzanita [*Arctostaphylos* spp.], scrub oak [*Quercus berberidifolia*], and toyon [*Heteromeles arbutifolia*]) and soft chaparral (or coastal sage scrub, dominated by California sagebrush [*Artemesia californica*] and laurel sumac [*Malosma laurina*] associations) dominate the south and central coast regions of California. The region is also characterized by varied and rugged topography, diverse geology, and a range of climatic conditions. In particular, the southern part of the state that is largely dominated by shrublands exhibits broad diversity among plants and animals at varying levels of taxonomic classification (Stebbins and Major 1965; Myers et al. 2000; Rissler et al. 2006). Although chaparral is widespread across California, its distribution is patchy in the central and northern parts of the state and lacks connectivity. For reptiles, this pattern of distribution may have resulted in limited speciation and endemism whereas for the other vertebrate taxa, it may have promoted speciation via isolation. From the Transverse range to the Mexican border, southern California supports the largest and most contiguous patches of chaparral and is where faunal endemism is higher than elsewhere in the state, particularly among mammals (Quinn 1990) and reptiles (Harrison 2013).

3.3 Current Faunal Diversity in Chaparral Ecosystems

Among states nationwide, California ranks fifth in vertebrate richness (804 species) and first in number of endemic vertebrates at 62 (Stein et al. 2000). California has the highest richness of mammals, the fourth highest bird and reptile species richness, and fifth greatest amphibian richness (Stein et al. 2000; Harrison 2013). There is a paucity of information on the exact richness and diversity of wildlife in chaparral, but general estimates can be gleaned from several sources. Of the 712 "regularly-occurring" mammals, birds, reptiles, and amphibians mapped and modeled in the California Wildlife Habitat Relationships System (CWHR, version 9.0, CDFW 2013), 388 species occur, at least in part, in hard or soft chaparral habitats, 374 of which are native. This includes 30 species of amphibians, 172 birds, 114 mammals, and 58 reptiles (Table 3.1). Although the scale of these data may not accurately

Table 3.1 Taxonomic diversity of native fauna in chaparral ecosystems

	Total species in chaparral	Federal ESA	State ESA	State SSC	BLM/USFS sensitive	Total species with listing status	Percent listed
Amphibian	30	2	2	10	11	12	40.0%
Bird	172	6	6	28	16	36	20.9%
Mammal	114	13	9	31	26	44	38.6%
Reptile	58	5	5	15	17	23	39.7%
Invertebrate	Unknown	8	0	N/A	11	18	Unknown
Fish	9[a]	6	8	3	3	8	88.9%
Total	374	40	30	87	84	141	37.7%

Data were derived from the California Wildlife Habitats Relationship System (CDFW 2013) and are summarized by total species richness, as well as listing status under the Federal Endangered Species Act (ESA), State ESA, or as State Species of Special Concern (SSC), Bureau of Land Management (BLM) Sensitive, and US Forest Service (USFS) Sensitive. The total of species with some listing status, as well as the percent listed are also included

[a]Denotes species likely found in chaparral-adjacent streams

capture all the species that may occasionally occur in chaparral habitats (especially migratory birds), from this estimate, it is likely that over half of the faunal biodiversity assessed under this program utilize chaparral to some degree. Classifying which native fish species exist in streams surrounded by chaparral or that are adjacent to chaparral is more challenging. Of the 116 native fish species in California, only a handful are described as occurring in streams within or adjacent to chaparral, as stream morphology and water conditions are more pertinent characteristics for defining and describing fish habitat than surrounding vegetation associations. Geospatial data on vertebrate species distribution in the state (CDFW 2010) reveal that in ecoregions dominated by chaparral vegetation types, richness among mammals, reptiles, and birds is higher than elsewhere in California (Fig. 3.1).

More conservative estimates of wildlife diversity in chaparral, likely representing only regular inhabitants of these ecosystems, are much lower. Quinn and Keeley (2006) estimated that chaparral ecosystems host approximately 50 species of mammals, more than 50 year-round resident bird species with another 50 that are seasonal/migratory, at least 24 species of reptiles, few amphibians, and many invertebrates, with notable diversity among hymenopterans, lepidopterans, and coleopterans. The combination of rare and restricted populations and widespread habitat loss and fragmentation in the region have resulted in a large number of 'at-risk' faunal species deserving of conservation efforts. Many of these species are now protected under the federal Endangered Species Act and California Endangered Species Act (Table 3.1).

Arthropods are the most diverse but the least well-studied and documented fauna of chaparral. Estimates of statewide insect diversity alone have ranged from 30,000–35,000 (Powell and Hogue 1979) to more than 100,000 (Schierenbeck 2014), a number that does not include arachnids like spiders, mites, ticks, and scorpions. Though chaparral contains only a portion of this total diversity of insects as well as other invertebrates, it has high levels of invertebrate diversity and endemism (Powell and Hogue 1979), in part because plant species diversity is high. These invertebrates

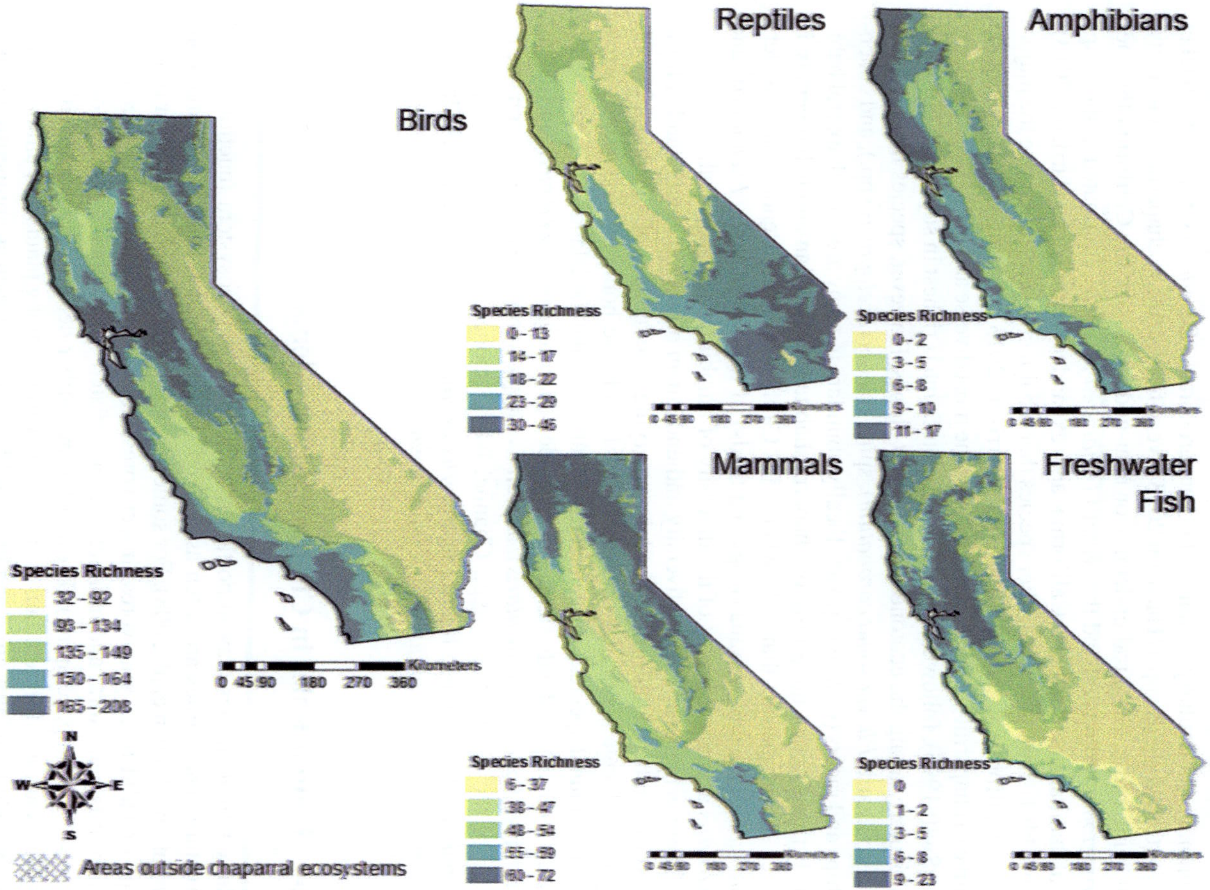

Fig. 3.1 Species richness by faunal taxonomic group in subecoregions of California dominated by chaparral

likely encompass the most common species and undoubtedly include many endemics, but we know very little about those species as they are proportionally understudied globally as well as locally.

For mammals, much of the diversity and endemism in chaparral is among small mammals, particularly rodents like kangaroo rats (*Dipodomys* spp.) and pocket mice (*Chaetodipus* spp.) (Harrison 2013). Very few birds use chaparral exclusively, but some of the chaparral endemics or near-endemics, like the California thrasher (*Toxostoma redivivum*) and the wrentit (*Chamaea fasciata*), originated in the chaparral habitats of southern California and expanded northward (Harrison 2013). These birds, among several other species that inhabit chaparral, comprise the majority of California's endemic avifauna (Harrison 2013). Among the herpetofauna native to chaparral systems, the lizards are the most diverse and well-studied (Keeley and Swift 1995), followed by snakes. Herpetofauna is richest in southern California, particularly along the ecotones of the desert slopes. However, species like the coast horned lizard (*Phrynosoma coronatum*), spiny lizards (*Sceloporus* spp.), and coachwhips and racers (*Coluber* sp.) are locally abundant in many areas of chaparral habitat. Perhaps due to the aridity of chaparral habitats, amphibian and fish diversity is lower in chaparral than elsewhere in California. The slender salamanders (*Batrachoseps*) and anurans (frogs and toads) of the genera *Rana, Pseudacris,* and *Anaxyrus* account for most of the diversity of amphibians in chaparral.

There is highly localized diversity and endemism on the Channel Islands off the coast of southern California, which host varying assemblages of vegetation including coastal sage scrub and chaparral. The islands are home to a range of narrow endemic faunal species, with chaparral inhabitants like the Island fox (*Urocyon littoralis*), Channel Islands slender salamander (*Batrachoseps pacificus*), and Island scrub jay (*Aphelocoma insularis*), as well as unique subspecies that occur on one or more of the eight islands (e.g., Channel Island spotted skunk [*Spilogale gracilis amphiala*], Island fence lizard [*Sceloporus occidentalis becki*], and San Clemente loggerhead shrike [*Lanius ludovicianus mearnsi*]).

3.4 Faunal Roles in Chaparral Ecosystems

Many faunal species play important roles in chaparral ecosystem functioning and structure. These roles are often associated with activities related to resource consumption (Whelan et al. 2008). For some species, this is tied directly to their trophic type or guild (e.g., granivores, insectivores, nectarivores, Table 3.2), whereas for others it is behavioral and more closely associated with resource acquisition (e.g., digging, gleaning, scraping). These roles are further defined by inter-specific relationships among different faunal species as well as among plants and animals. This can include both intra-guild and inter-guild interactions such as competition or predation, interactions among predator and prey or herbivore and plant populations, as well as how animals relate to the abiotic environment.

Some of the roles and interactions among faunal species and their habitats in chaparral ecosystems also provide a benefit to people. These services can be viewed

Table 3.2 Faunal roles in chaparral ecosystems categorized by trophic type or guild and resources acquisition behavior. The faunal trait or behavior is listed as well as description of the role served and the category of ecosystem service provided

Category	Faunal trait/behavior	Description of role	Category of ecosystem service
Trophic type	Carnivore	Affect prey populations, maintain trophic interactions	Regulating
	Invertivore/Insectivore	Affect prey populations, maintain trophic interactions	Regulating
	Herbivore	Affect plant populations	Regulating
	Granivore	Disperse seeds	Supporting
	Frugivore	Disperse seeds	Supporting
	Nectarivore	Pollination	Supporting
	Detritivore	Facilitate decomposition, nutrient cycling	Regulating
	Scavenger	Facilitate decomposition, nutrient cycling	Regulating
	Coprophagous	Facilitate decomposition, nutrient cycling	Regulating
Behavior	Digging	Facilitate decomposition, nutrient cycling	Regulating
	Sifting through litter	Facilitate decomposition, nutrient cycling	Regulating

as ecosystem services, or the value people enjoy from ecosystems (Millennium Ecosystem Assessment 2003), described in more detail in Chap. 9. Ecosystem services are generally defined by four categories: supporting, regulating, provisioning, and cultural (Millennium Ecosystem Assessment 2003). The supporting and regulating services provided by wildlife largely involve their roles in ecological functioning described above, which may provide indirect benefits to humans (Table 3.2). However, provisioning and cultural services are more directly beneficial to people (Fig. 3.2). Wildlife provides agricultural and consumptive services directly for humans by, for example, pollinating crops and being the target of hunting and fishing. Finally, the cultural benefit from wildlife includes consumptive (e.g., hunting and fishing) and non-consumptive (e.g., hiking, bird-watching) recreation as well as the overall aesthetic value of wildlife merely existing (see Chap. 10).

3.4.1 Pollination and Seed Dispersal

Plant reproduction, growth, composition, and distribution in chaparral have been linked to wildlife activities in several ways. Granivores in chaparral, such as kangaroo rats and harvester ants (*Pogonomyrmex* spp.), are known to consume seeds of many native chaparral plants and may play a role in seed dispersal. For example, harvester ants collect seeds, often from buckwheat (*Eriogonum* spp.), remove them

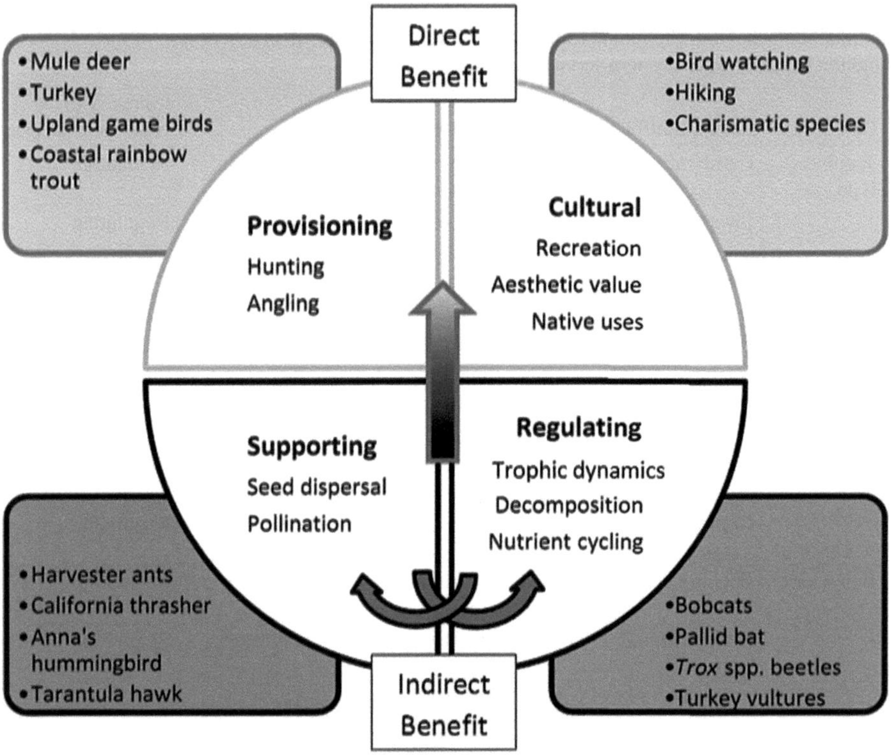

Fig. 3.2 Ecosystem services provided by wildlife in chaparral ecosystems. The types of services are described in the inner segments and examples of the species that provide these services in chaparral are in the outer boxes. Dark gray denotes services contributing to overall ecosystem functioning that indirectly benefit humans. Light gray identifies services that directly benefit humans. Gray arrows indicate feedback where supporting and regulating services each enhance the other as well as the cultural and provisioning services

from the husk outside their colony, and leave the chaff on the outside of their nest (Quinn and Keeley 2006). Despite how commonly this behavior is observed, recent research has documented native harvester ants (*P. rugosus*) preferentially selecting the seeds of non-native plants over those of natives, but with no major effects of this preference observed in the seedbank (Briggs and Redak 2016). This suggests that the native harvester ants may actually play a limited role in seed dispersal. However, the food preferences and activities of other granivores may affect the distribution of plants and seedbank composition (Parker 2015). In an assessment of 159 woody plant species in San Diego County, animals, particularly birds, were identified as the most probable mechanism of seed dispersal for 64 (40%) of those plant species (Zedler 1995). These faunal seed dispersers, especially birds with large home ranges or dispersal capabilities, have the potential to play a critical role in facilitating plant migration and range shifts, for example, in response to climate change.

The range of pollinators in chaparral appears to be as diverse as the plants that comprise these vegetation communities. Insects are likely the most important, and certainly the most diverse, pollinators of chaparral plants (Fulton and Carpenter 1979). Across a range of scrub dominated sites in California, 648 insect taxa were documented, the most common of which were bees (Force 1990; Cody et al. 1977). This diversity may be supported by a high degree of pollinator specialization that has been documented in chaparral systems, with notably more specialization and greater diversity of pollinators in chaparral compared to coastal sage scrub types (Cody et al. 1977). Despite a high degree of specialization among the pollinators, woody and herbaceous chaparral plants do not exhibit this same type of specialization for their pollinators. In fact, the majority of plants are pollinated by at least two species, and nearly half are pollinated by three or more species (Moldenke 1976). The exception to this are entire genera of chaparral plants that have flower morphology specifically evolved for pollination by sphynx moths (10 genera) and hummingbirds (39 genera, Moldenke 1976). For example, nectarivores in chaparral such as Anna's hummingbirds (*Calypte anna*) serve as pollinators for many native species, including currant or gooseberry plants (*Ribes* spp.), monkey flowers (*Mimulus* spp.), and penstemons (*Penstemon* spp. or *Keckiella* spp.). They also play a role in the pollination of many ornamental species.

Some species change their diets or behaviors with different life stages or seasonally, with their roles in ecosystem functioning changing accordingly. Many butterflies fit this description, with larvae feeding on foliage while the adults feed on the nectar of flowers, pollinating them in the process. The rare Hermes copper butterfly (*Lycaena hermes*), endemic to southern California and Baja California, Mexico, lays eggs exclusively on its host plant, spiny redberry (*Rhamnus crocea*), where the larvae then hatch and feed. The adult butterfly, on the other hand, almost exclusively feeds on the nectar of the small flowers of California buckwheat (*Eriogonum fasciculatum*), serving as a pollinator of this common chaparral shrub. This butterfly is threatened by habitat loss from urbanization and wildfires (Marschalek and Klein 2010). The tarantula hawk (*Pepsis* sp.), one of the largest species of Hymenopteran, is one of the most readily recognizable insects in chaparral owing to its bright, aposematic coloring—orange wings paired with a metallic blue body—meant to warn potential predators of its painful sting. The tarantula hawk performs different roles in the ecosystem during different stages of its life cycle. While the adults feed on nectar and pollinate native plants, the larvae consume animal matter. The female tarantula hawk will sting and paralyze the tarantula spider with its powerful venom, drag the large spider to her den or back into the spider's den, then proceed to lay a single egg on the abdomen of the spider. When the larvae emerges, it will consume the spider.

3.4.2 *Inter-specific Interactions and Trophic Dynamics*

Chaparral has few large herbivores like mule deer (*Odocoileus hemionus*), but browsing by the species does have direct effects on growth, reproduction, and survival of individual plants (Côté et al. 2004). This is particularly true in postfire landscapes where fresh forage is an attractive food source for deer (Lawrence 1966). The degree to which these individual level effects scale up to populations or the community in chaparral ecosystems is unclear. There is evidence that deer play a larger role in affecting plant composition and community dynamics in forested environments (Côté et al. 2004), but this has not been studied in chaparral. There are also many species of smaller herbivores such as the woodrats (*Neotoma* spp.) that inhabit chaparral, consuming seeds, shoots, and green vegetation of many plants and using twigs and other materials for their conspicuous nests. Even more abundant are plant-eating insects that range from leaf miners to leaf gallers, sap-suckers, and chewing insects (Parker et al. 2016). Again, there is little information about the ecosystem level impacts of herbivory on the chaparral plant community, but it is possible that herbivory may indirectly affect plant competition and interactions.

In aquatic systems found in chaparral, herbivores include the arroyo chub (*Gila orcuttii*), a fish that primarily eats algae (Greenfield and Deckert 1973), the Baja California treefrog (*Pseudacris hypochondriaca*), an algae grazer, and Pacific pond turtle (*Actinemys marmorata*). However, the treefrog and pond turtle switch diets, exhibiting carnivorous behavior as adults. In the adult life stage, the adult treefrog eats primarily flying insects, and pond turtles may feed on invertebrates, like crayfish and larvae of aquatic insects, as well as frogs and fish. California newts (*Taricha torosa*) are predators of insects, among other things, and consume a wide variety of prey items in terrestrial and aquatic systems (see Box 3.1). All of these species play a role in maintaining trophic structure and interactions in the aquatic systems found within chaparral habitats.

Insectivores are arguably some of the most important animals in terms of the ecological role they play in chaparral as well as many other ecological communities. Members of each vertebrate group feed on and may affect insect populations that are pests to humans and agricultural areas as well as those species that pose a threat to native plants and wildlife. This includes insect vectors of diseases like Lyme disease (see Box 3.2) or West-Nile virus, and non-native species like the goldspotted oak borer (*Agrilus auroguttatus*) that has affected native oak populations in San Diego County and the polyphagous shot hole borer (*Euwallaecea* sp.) that kills a variety of agricultural and native trees throughout southern California. Although there is scant evidence of regulation of insect populations by wildlife, the introduction and spread of these non-native insects that, in the absence of native predators, have become pests in native and agricultural systems suggests this could be an important role animals play in chaparral ecosystems. Among the chaparral insectivores are native bats such as the low-flying, ground-foraging pallid bat (*Antrozous pallidus*) and the high-flying moth specialist, the spotted bat (*Euderma maculatum*). There are also many insectivorous species of birds in chaparral. For example, the California thrasher is a year-round resident of shrubland habitats that feeds on insects from the ground, using its long, curved bill to sift through leaf litter to locate prey. This disturbance of the litter layer may also facilitate decomposition of dead plant matter. The thrasher switches its diet in the dry season, foraging on berries, fruits, and seeds of native shrubs like *Ribes* spp., toyon (*Heteromeles arbutifolia*), and buckthorns (*Rhamnus* spp.), and distributing seeds of those species (Cody 2012).

Carnivores also affect prey populations, playing an important role in the trophic dynamics of the food web. For example, mountain lions (*Puma concolor*), coyotes (*Canis latrans*), and occasionally bobcats (*Lynx rufus*) prey on mule deer and have

Box 3.1 Species Highlight—California Newt (*Taricha torosa*)

The California newt (*Taricha torosa*) is a voracious predator in both aquatic and terrestrial habitats within chaparral ecosystems. Their bright orange aposematic coloration, meant to warn potential predators, is coupled with toxic skin secretions containing tetrodotoxin (Fig. 3.3). Adults are thought to be very long-lived, on the order of 20 years or more (Jennings and Hayes 1994), during which time they can have huge predatory effects in streams.

In terrestrial environments, adult newts will feed on small invertebrates such as worms, snails, slugs, and sowbugs, although a small nestling bird was once found in the stomach of one newt (Hanson et al. 1994). When they return to the aquatic environment to reproduce, they feed on amphibian eggs and larvae (Stebbins 1972; Hanson et al. 1994), including newt larvae and newt eggs (Kats et al. 1992). When feeding on the ground, newts project a sticky tongue to capture prey, whereas in the aquatic environment, they open their mouths and suck in prey items (Findeis and Bemis 1990). The larvae of the species eat small aquatic invertebrates, decomposing organic matter, as well as other newt larvae. The only native predators of California newts are species of garter snake (*Thamnophis* sp., Jennings and Hayes 1994). However, invasive aquatic species such as mosquitofish (*Gambusia affinis*) and crayfish (*Procambarus clarkia*) pose a threat to the California newt as they consume newt larvae and eggs and may extirpate the species from invaded stream reaches (Gamradt and Kats 1996).

Fig. 3.3 Breeding ball of California newts in San Mateo Canyon Wilderness, Riverside County, California. Photo by Megan K. Jennings

Box 3.2 Species Highlight—Western Fence Lizard (*Sceloporus occidentalis*) and Lyme Disease

The western fence lizard (*Sceloporus occidentalis*) is one of the most widespread and abundant species of lizard in chaparral (Fig. 3.4). In areas where they co-occur with western black-legged ticks (*Ixodes pacificus*), the most common carrier of the Lyme disease spirochete (*Borrelia burgdorferi*) in California, they are a common food source for the nymphal ticks (Lane and Loye 1989). Despite the fact that fence lizards are an important host species for ticks, studies in the 1980s and 1990s found that the lizards were not infected with the spirochete that causes Lyme disease (Lane and Loye 1989) and in fact, fewer adult ticks carried the spirochete compared to nymphs (Lane and Quistad 1998). Experimental studies that exposed tick nymphs to lizard sera resulted in spirochete death, leading researchers to conclude that western fence lizard blood contains a borreliacidal factor that kills the spirochetes (Lane and Quistad 1998), reducing the risk of Lyme disease to humans. Further study found a similar spirochete response with exposure to southern alligator lizard (*Elgaria multicarinata*) sera (Kuo et al. 2000). Interestingly, after the experimental removal of western fence lizards where incidence of spirochetes was expected to increase, there was instead a large decline in the population of infected ticks, suggesting the nymphs were not able to switch prey species, resulting in a decrease in the disease risk (Swei et al. 2011).

Fig. 3.4 Western fence lizard in San Bernardino County, California. Photo by Megan K. Jennings

been shown to affect survival rates, primarily of fawns, but also females in other regions of the west (Forrester and Wittmer 2013). There has been comparatively little focus on mule deer in chaparral. However, recent research found that predation of mule deer by mountain lions was highest in chaparral vegetation types in the Santa Monica Mountains and Simi Hills (Benson et al. 2016). In this urbanized region, both deer and mountain lions may benefit from the cover thick chaparral vegetation provides. Smaller predators also play a role in chaparral trophic dynamics. Bobcats, coyotes, and gray foxes (*Urocyon cinereoargenteus*) consume rabbits (*Sylvilagus* spp. and *Lepus californicus*), woodrats, and ground squirrels (*Otospermophilus beecheyi*), as well as many other species, opportunistically. Although these species are all sensitive to urbanization and habitat fragmentation (Crooks 2002), as habitat generalists with varying degrees of diet breadth, they have been able to persist in chaparral with habitat patches that are large enough or closely connected. Coyotes, the most widespread of these three mammalian predators, also have the broadest diet and greatest degree of urban association (Riley et al. 2003). Although the gray fox has similar diet breadth to the coyote, they avoid interactions with coyotes both spatially and temporally to escape competition and predation pressure (Fedriani et al. 2000). Bobcats, a strict carnivore, exhibits a greater degree of avoidance of urban areas in chaparral systems (Riley et al. 2003).

Birds and snakes also play important roles as predators in chaparral. The greater roadrunner (*Geococcyx californianus*) preys on lizards and snakes on the ground and is one of the only known predators of the tarantula hawk. Red-tailed hawks (*Buteo jamaicensis*) and sharp-shinned hawks (*Accipiter striatus*) are aerial predators, with the former preying primarily on small mammals and the latter specializing in birds. The four species of venomous rattlesnake that occur in California's chaparral—the speckled (*Crotalus mitchellii*), the southern Pacific (*C. oreganus helleri*), northern Pacific (*C. oreganus oreganus*), and red diamond (*C. ruber*)—primarily consume small mammals. In contrast, California kingsnakes (*Lampropeltis getula*) are a constricting species that preys on other snakes, most notably, rattlesnakes (*Crotalus* sp.).

3.4.3 Nutrient Cycling

Many insects are decomposers that return nutrients to the system from various sources. For example, termites (*Reticulitermes hesperus*, *Heterotermes aureus,* and *Coptotermes formosanus*) decompose wood, and *Trox* spp. beetles consume the scat of larger species, particularly mammals, breaking down fecal matter and returning those nutrients into the soil. Turkey vultures (*Cathartes aura*) and corvids (*Corvus* spp., *Aphelocoma* spp.) are common scavengers in chaparral, along with the much rarer California condor (*Gymnogyps californianus*), which all consume the carcasses of dead animals. These species, among other birds and mammals, facilitate decomposition along with a host of insects. Another form of nutrient cycling occurs through soil disturbance. Fossorial animals like the pocket gopher (*Thomomys bottae*) mine soil, tilling it up to the surface as they create networks of under-ground tunnels. Given that they also eat plant material, this freshly tilled soil can make for suitable plant nursery habitat (Cox 1986; Cox et al. 1995).

In freshwater streams surrounded by chaparral, fish like the southern steelhead trout (*Onchorynchus mykiss*) feed on a range of insects and other invertebrates, drifting organisms, and even other fish, but their most important role is in nutrient cycling. In southern California, water temperatures and flows are highly variable in the low-elevation portions of creeks inhabited by *O. mykiss*, while the upper reaches of these streams, surrounded by chaparral, provide the deeper canyons and cooler waters needed to sustain the species (Moyle et al. 2015). Within these streams the species also helps maintain biodiversity and trophic structure in the food web. Both the resident inland trout (*Onchoryhnchus mykiss irideus*), which spends its entire life in these streams, as well as the anadromous steelhead that migrates to the ocean, breed in these freshwater streams. Although they are iteroparous and do not die after a single spawning, their eggs and fry provide food resources for many other species and provide nutrient input back into the system (Childress et al. 2014).

3.4.4 Cultural and Provisioning Services

In addition to their ecological importance, chaparral fauna also have cultural and aesthetic value. The major metropolitan areas in southern California that are surrounded by coastal sage scrub and chaparral are home to more than 20 million people, many of whom recreate in the nearby preserves, parks, and national forests. Whether these recreationists enter these natural landscapes to watch native birds, hunt, or even opportunistically to catch a glimpse of some of the fauna in these chaparral ecosystems, many appreciate that there is value in the mere existence of these animals. Certainly the more charismatic a species, the more appreciation it is likely to receive as a chaparral inhabitant. Although some people may never have the opportunity to see certain animals, they still assign value to the knowledge that these animals exist in the hillsides blanketed in chaparral.

Consumptive uses of chaparral fauna, in the form of hunting and fishing, serve as both cultural and provisioning services. A number of chaparral's wildlife species targeted by hunters for food and trophies include mule deer, turkeys (*Meleagris gallopavo*), and other upland game birds. Gray fox, and until recently bobcats, were also hunted and trapped for their fur. Although much of the freshwater fishing opportunities in streams adjacent to chaparral are for non-native fish species, the native coastal rainbow trout is a prized target among anglers.

3.5 Threats to Chaparral Biodiversity and Ecosystem Services

Globally, Mediterranean-type climate regions are at high risk of habitat and biodiversity loss from climate change and continued development—more so than any other biome (Sala et al. 2000; Underwood et al. 2009). This is most pronounced in California, which has the highest population density and percent urban area of five MTC regions (Underwood et al. 2009). In fact, the California Floristic Province has already lost over 75% of the original vegetation due to increased development, associated infrastructure, and shifts in land use in California's four large metropolitan areas—Los Angeles, San Francisco, Sacramento, and San Diego—which has resulted in loss and fragmentation of the surrounding chaparral (Brooks et al. 2002). Additional stressors such as non-native species, too frequent wildfire, and climate change put wildlife within chaparral at further risk. With the percent of urbanization and population density increasing in California's MTC region more than 10% in the last decade of the twentieth century, the rate of habitat loss (Underwood et al. 2009) and concomitant species extinctions (Stein et al. 2000; Harrison 2013) pose a threat to ecosystem services, many of which are still undiscovered or not well understood.

The greater the degree of interface between natural and human environments, the greater the risk of spillover, or edge effects (Lovejoy et al. 1986), from the human dominated landscape. In chaparral, the edge effects that may affect wildlife include non-native species, exposure to environmental pollutants and toxicants like anticoagulant rodenticides (Riley et al. 2007), unauthorized roads and trails, dumping, and trampling of vegetation (Sauvajot et al. 1998), altered wildfire regimes (see Chap. 12), and a changing climate (see Chap. 14). Moreover, roadways and other infrastructure (e.g., powerlines, water and sewer lines) can hamper wildlife movement, acting as a barrier to movement and gene flow (Riley et al. 2006; Delaney et al. 2010; Lee et al. 2012), a source of mortality (Vickers et al. 2015), and affecting wildlife distribution and reproduction. Thus, there can be long-term repercussions at the population and community level that lead to reduced species richness in urban fragments (Soulé et al. 1988; Bolger et al. 1997a, b; Sauvajot et al. 1998).

Low-density or exurban development in the foothills has resulted in an expansive wildlife-urban interface (WUI) between chaparral dominated wildlands and homes, and thus higher fire risk in these areas (Syphard et al. 2007, 2009). Although much of the native chaparral fauna have adapted to fire as part of their environment, more fires on the landscape and more frequent fires can affect postfire re-colonization rates and distributions of populations or entire species that experience mortality or flee from fires (Diffendorfer et al. 2012; Schuette et al. 2014). With shorter fire-return intervals in chaparral, vegetation-type conversion is also a risk. Non-native plants like annual grasses (e.g., *Bromus* spp., *Avena* spp.) emerge after fires can alter the overall composition and suitability of habitat for many species, or can affect the availability of individual habitat elements for foraging and cover (Keeley et al. 2005). In California's MTC region, establishment and expansion of non-native species is

Box 3.3 Species Highlight: Argentine Ant (*Linepithema humile*)

The Argentine ant (*Linepithema humile*) is a non-native ant species that has become widespread in California. This prolific species competes with native ants, and has been shown to reduce overall invertebrate diversity of invaded sites, particularly for springtails (Collembola), spiders (Arachnida), and flies (Diptera, Human and Gordon 1997). In addition to reducing invertebrate diversity, Argentine ants are also known to disrupt two unique roles native harvester ants play in chaparral. One is as a specialized dispersers of seeds of the native tree or bush poppy (*Dendromecon rigida*), and the other is as the primary food source for the horned lizard (*Phrynosoma* sp.). The tree poppy is a myrmecochorous plant, meaning its seeds are designed specifically to attract ants as its primary dispersal mechanism. However, when Argentine ants displace native harvester ants, the dispersal of the tree poppy decreases substantially and has been associated with increased seed predation by birds (Carney et al. 2003). Another consequence of Argentine ant displacement of native harvester ants is reduction in the primary prey resource for horned lizards (Suarez et al. 2000) (Fig. 3.5). Already threatened by habitat loss and fragmentation, horned lizards, an ant specialist, continue to decline in habitat fragments where Argentine ants have invaded. Furthermore, in invaded sites where no native ant species remain, horned lizards also exhibit reduced growth rates (Suarez and Case 2002).

Fig. 3.5 Horned lizard in Hauser Canyon San Diego County, California. Photo by Megan K. Jennings

the second most common threat, after habitat loss, to species surrounded by anthropogenic development (MacDonald et al. 1988). In addition to non-native plants, non-native animals are also a threat to native chaparral animals. Non-native insects like Argentine ants (*Linepithema humile*, see Box 3.3) can compete with native species, affect the food sources of native species, or damage and kill vegetation. Feral pigs are another non-native animal that occur in chaparral. This species is widespread and prolific and can increase the risk of establishment and spread of non-native plant species, damage sensitive soils and aquatic habitats, and competes with—and even preys on—native wildlife.

Climate disruption is expected to greatly affect California's chaparral ecosystems and their services. Climate models for the western US predict that temperatures will increase overall and there will be a general trend toward increasing aridity and drought events (Hannah et al. 2002; Westerling et al. 2003), both of which are likely to extend fire seasons and increase fire frequency (Swetnam and Betancourt 1998; Brown et al. 2004). The potential effects of these changing environmental conditions on wildlife may include shifts in habitat components such as vegetation type and cover, more restricted water sources, and changes in the food web and trophic structure of communities as some species become extirpated or go extinct. Temperature shifts may drive migration upslope to cooler climates (Hughes 2000) or westward to areas with greater marine influence and lower temperatures. Phenological mismatches—such as when the timing of insect emergence and flowering are out of sync—could mean the loss of key pollinators for rare and endemic species, disrupting reproduction of plant species, and affecting the persistence of both plants and animals. Species with narrow niches or in certain age classes (Russell et al. 2012), such as juveniles, may be more susceptible to mortality caused by environmental stochasticity. Maintaining regional biodiversity and the ecological functioning of our natural landscapes in the face of these climate-induced shifts is one of the central and burgeoning issues facing land managers today.

3.6 Resilience and Adaptation in Chaparral Fauna

How individual species and populations respond to environmental disturbances and shifts depends on both intrinsic factors like biology and life history traits as well as extrinsic factors such as exposure to disturbances or habitat loss and fragmentation. The intrinsic characteristics can be separated into sensitivity and adaptive capacity (Foden et al. 2013), whereas exposure may depend on the geographical area considered, the range of the species, or the populations considered. Life history traits relating to sensitivity include high trophic level, slow life history, large home ranges (Purvis et al. 2000), habitat specialization, narrow environmental tolerances, and rarity, while the ability to respond through adaptation may be hampered by poor dispersal potential and low evolutionary potential (Foden et al. 2013).

Generally, species' responses to disturbances and climatic shifts are defined by their adaptive capacity, such that species with a greater evolutionary potential or plasticity of ecological responses will have a higher capacity to adapt to environmental changes (Williams et al. 2008). Though difficult to measure directly, adaptive capacity can often be measured indirectly through evaluation of dispersal ability and habitat specialization, which may indicate sensitivity to threats and stressors (Williams et al. 2008; Glick et al. 2011; Gardali et al. 2012). This is currently a focal topic with respect to climate change, but plays a role in how species respond to stochastic events such as wildfires, disease, and floods, but also to permanent changes like habitat loss and fragmentation. The availability of refugia, species' mobility, and the breadth of habitat and diet preferences all affect a species' ability to behaviorally adapt to disturbances such as wildfire in chaparral. During a fire, many small mammals, especially those preferring open habitats like the deer mouse (*Peromyscus maniculatus*) and kangaroo rats, can seek shelter under ground and emerge after the fire, quickly re-colonizing the burned area (van Mantgem et al. 2015). In contrast, the woodrats dwell in rock piles or in stick nests that burn rapidly. These species often take much longer to re-colonize burned areas (Diffendorfer et al. 2012). More mobile species of birds and large mammals (e.g., coyotes, mule deer) often flee the flames and return when food and cover resources are adequate (van Mantgem et al. 2015).

Species with small population sizes, disconnected habitats, and low genetic variability or gene flow are more vulnerable and less resilient to stochastic events such as fire or disease outbreaks. Many of the threatened and endangered species in chaparral in Table 3.1 face these threats as populations decline in response to habitat loss. The coastal sage scrub specialist, California gnatcatcher (*Polioptila californica*), has suffered significant population declines from habitat loss and fragmentation. Now faced with increasing fire frequency that results in degraded conditions after fire, the gnatcatcher is at risk of further declines (Winchell and Doherty 2014), despite substantial investment in conservation efforts for the species. Mountain lions in southern California have also suffered from habitat fragmentation and a disconnected landscape, resulting in limited gene flow among populations and inbreeding (Ernest et al. 2014). Furthermore, if shortened fire-return intervals convert shrublands to grasslands (Jennings et al. 2016), the persistence of the species within the region is in question.

3.7 Conservation of Fauna in Chaparral

The sheer number of at-risk species in chaparral ecosystems (Table 3.1), many with impairments to adapt to change and disturbance, begs the question of how to preserve not only individual species, but the functioning of the ecosystem as a whole. There are often trade-offs between species-specific management goals and those targeting ecosystem function and equivalency. When faced with these management conundrums, considering ecosystem services and the specific functions provided by chaparral fauna may offer useful context. In particular, evaluating functional redundancy (Lawton and Brown 1993) or replaceability, (i.e., whether the functional niche of one species could be filled by another species), is one option for prioritizing conservation needs. Although there is evidence that functional diversity is correlated with species richness and an overall loss of species may reduce ecosystem function (Flynn et al. 2009), some animals play unique roles that cannot be replicated by other species. Considering which species and roles are critical to ecological functioning or ecosystem processes may help guide conservation decision-making or prioritization in chaparral ecosystems.

Determining the ecological value of faunal species could be approached by focusing attention on keystone species, ecosystem engineers, or strong interactors. For example, meso-predator release can result in trophic cascades (Soulé et al. 1988). In southern California's scrub covered urban habitat fragments, the loss of an apex chaparral predator like coyote was thought to have released meso-predators like gray fox and domestic cats (*Felis catus*) from predation or competition, which was correlated with decreased diversity and abundance of native scrub-breeding birds (Soulé et al. 1988). Although documenting the ecological importance of the functional roles animals play in their ecosystems can be difficult, identifying the most critical of these roles or services is an approach to conservation and management decision-making that may protect ecosystem function and viability. For example, pollination of native and agricultural species is arguably one of the most economically valuable ecological services at risk (Kremen and Chaplin-Kramer 2007). From sharp declines of insect pollinators associated with pesticide use to the reduction of bird and mammal pollinators as a function of habitat loss and fragmentation, many pollinators have either been lost or are at risk (Allen-Wardell et al. 1998) without a formal recognition of the extent of their ecological role. Therefore, in certain areas, conservation investment and action might focus on identifying the pollinators of dominant or rare plant species to inform a strategy for protecting those pollinators to ensure long-term ecosystem viability.

The primary method by which conservation action is taken is through habitat preservation and management. This requires identifying the locations to be preserved, and determining the appropriate reserve design and management actions. In some areas of California's chaparral, ecological networks of preserves have been assembled, partially in response to the state's Natural Community Conservation Planning Act. Many of these networks, however, focus attention on the large number of state and federally protected species in these ecosystems. To ensure chaparral ecosystems continue to function and support a rich diversity of flora and fauna, more work needs to be done to improve resilience and adaptive capacity to the threats and stressors in California's chaparral while also mitigating those threats and stressors. This can be done by focusing attention not only on species-specific preservation, but on maintaining functional diversity at the community or ecosystem level. This will also require conservation efforts to focus on preserving or re-establishing connections between the increasingly fragmented patches of habitat in chaparral ecosystems.

Efforts to develop proactive, adaptive planning for landscape linkages under climate and land use change are increasingly being employed for conservation planning in other regions of the western US (e.g., Penrod et al. 2012; Nuñez et al. 2013). Landscape connectivity allows for movement among patches of suitable habitat, reduces the chance of extinction for small populations (Brown and Kodric-Brown

1977), and maintains gene flow in patchy landscapes (Noss 1987). Over longer time scales, and in the face of changing environmental conditions, connectivity will prove critical for facilitating range shifts in response to landscape changes caused by changing climate and altered disturbance regimes (Hannah et al. 2002; Heller and Zavaleta 2009). By identifying and protecting climate refugia, and creating habitat connectivity through habitat linkages or corridors and wildlife bridges and crossings, we can move toward protecting wildlife diversity as well as chaparral ecosystems and functions.

References

Allen-Wardell, A. G., P. Bernhardt, R. Bitner, A. Burquez, S. Buchmann, J. Cane, P. A. Cox, V. Dalton, P. Feinsinger, M. Ingram, D. Inouye, C. E. Jones, K. Kennedy, P. Kevan, H. Koopowitz, R. Medellin, S. Medellin-Morales, and G. P. Nabhan. 1998. The potential consequences of pollinator declines on the conservation of biodiversity and stability of food crop yields. Conservation Biology 12:8-17.

Benson, J. F., J. A. Sikich, and S. P. D. Riley. 2016. Individual and population level resource selection patterns of mountain lions preying on mule deer along an urban-wildland gradient. PloS One 11:e0158006.

Blondel, J., and J. Aronson. 1995. Biodiversity and ecosystem function in the Mediterranean basin: human and non-human determinants. Pages 43-119 *in* G. W. Davis, and D. M. Richardson, editors. Mediterranean-type ecosystem: functions of biodiversity. Springer, Heidleberg, Germany.

Bolger, D. T., A. C. Alberts, R. M. Sauvajot, P. Potenza, C. McCalvin, D. Tran, S. Mazzoni, and M. E. Soulé. 1997b. Response of rodents to habitat fragmentation in coastal southern California. Ecological Applications 7:552-563.

Bolger, D. T., T. A. Scott, and J. T. Rotenberry. 1997a. Breeding bird abundance in an urbanizing landscape in coastal southern California. Conservation Biology 11:406-421.

Briggs, C. M., and R. A. Redak. 2016. Seed selection by the Harvester Ant *Pogonomyrmex rugosus* (Hymenoptera: Formicidae) in coastal sage scrub: interactions with invasive plant species. Environmental Entomology 45:983-990.

Brooks, T. M., R. A. Mittermeier, C. G. Mittermeier, G.A. da Fonseca, A. B. Rylands, W. R. Konstant, P. Flick, J. Pilgrim, S. Oldfield, G. Magin, and C. Hilton-Taylor. 2002. Habitat loss and extinction in the hotspots of biodiversity. Conservation Biology 16:909-923.

Brown, T. J., B. L. Hall, and A. L. Westerling. 2004. The impact of twenty-first century climate change on wildland fire danger in the western United States: an applications perspective. Climatic Change 62:365-388.

Brown, J. H., and A. Kodric-Brown. 1977. Turnover rates in insular biogeography: effect of immigration on extinction. Ecology 58:445-449.

CDFW [California Department of Fish and Wildlife]. 2010. ACE-II: Areas of Conservation Emphasis Map. Extent: California. https://map.dfg.ca.gov/ace/

CDFW [California Department of Fish and Wildlife]. 2013. California interagency wildlife task group, CWHR Version 9.0 personal computer program. Sacramento, California, USA.

Calsbeek, R., J. N. Thompson, and J. E. Richardson. 2003. Patterns of molecular evolution and diversification in a biodiversity hotspot: the California Floristic Province. Molecular Ecology 12:1021-1029.

Carney, S. E., M. B. Byerley, and D. A. Holway. 2003. Invasive Argentine ants (*Linepithema humile*) do not replace native ants as seed dispersers of *Dendromecon rigida* (Papaveraceae) in California, USA. Oecologia, 135:576-582.

Childress, E. S., J. D. Allan, and P. B. McIntyre. 2014. Nutrient subsidies from iteroparous fish migrations can enhance stream productivity. Ecosystems 17:522-534.

Cody, M. L. 2012. California Thrasher (*Toxostoma redivivum*). *In* A. Poole, editor. The Birds of North America. Ithaca, New York, USA: Cornell Laboratory of Ornithology. https://birdsna.org/Species-Account/bna/species/calthr

Cody, M. L., E. R. Fuentes, W. Glanz, J. H. Hunt, and A. R. Moldenke. 1977. Convergent evolution in the consumer organisms of the Mediterranean Chile and California. Pages 144-192 *in* H. A. Mooney, editor. Convergent evolution in Chile and California: Mediterranean climate ecosystems. Dowden, Hutchinson, and Ross, Stroudsberg, Pennsylvania, USA.

Côté, S. D., T. P. Rooney, J. P. Tremblay, C. Dussault, and D. M. Waller. 2004. Ecological impacts of deer overabundance. Annual Review of Ecology, Evolution, and Systematics 35:113-147.

Cox, G. W. 1986. Mima mounds as an indicator of the presettlement grassland-chaparral boundary in San Diego County, California. American Midland Naturalist 16:64-77.

Cox, G. W., L. C. Contreras, and A. V. Milewski. 1995. Role of fossorial animals in community structure and energetics of Pacific Mediterranean ecosystems. Pages 383-398 *in* M. T. K. Arroyo, P. H. Zedler, and M. D. Fox, editors. Ecology and biogeography of Mediterranean ecosystems in Chile, California, and Australia. Springer, New York, New York, USA.

Crooks, K. R. 2002. Relative sensitivities of mammalian carnivores to habitat fragmentation. Conservation Biology 16:488-502.

Davis, G. W., and M. C. Rutherford. 1995. Function in Mediterranean-type ecosystems: clarifying the role of diversity. Pages 151-183 *in* G. W. Davis, and D. M. Richardson, editors. Mediterranean-type ecosystem: functions of biodiversity. Springer, Heidelberg, Germany.

Delaney, K. S., S. P. D. Riley, and R. N. Fisher. 2010. A rapid, strong, and convergent genetic response to urban habitat fragmentation in four divergent and widespread vertebrates. PLoS One 5:e12767.

Diffendorfer, J., G. M. Fleming, S. Tremor, W. Spencer, and J. L. Beyers. 2012 The role of fire severity, distance from fire perimeter and vegetation on postfire recovery of small-mammal communities in chaparral. International Journal of Wildland Fire 21:436-448.

Ernest, H. B., T. W. Vickers, S. A. Morrison, M. R. Buchalski, and W. M. Boyce. 2014. Fractured genetic connectivity threatens a southern California puma (*Puma concolor*) population. PLoS One 9:e107985.

Fedriani, J. M., T. K. Fuller, R. M. Sauvajot, and E. C. York. 2000. Competition and intraguild predation among three sympatric carnivores. Oecologia 125:258-270.

Findeis, E. K., and W. E. Bemis. 1990. Functional morphology of tongue projection in *Taricha torosa* (Urodela: Salamandridae). Zoological Journal of the Linnean Society 99:129–157.

Flynn, D. F. B., M. Gogol-Prokurat, T. Nogeire, N. Molinari, B. T. Richers, B. B. Lin, N. Simpson, M. M. Mayfield, and F. DeClerck. 2009. Loss of functional diversity under land use intensification across multiple taxa. Ecology Letters 12:22-33.

Foden, W. B., S. H. Butchart, S. N. Stuart, J. C. Vié, H. R. Akçakaya, A. Angulo, L. M. DeVantier, A. Gutsche, E. Turak, L. Cao, and S. D. Donner. 2013. Identifying the world's most climate change vulnerable species: a systematic trait-based assessment of all birds, amphibians and corals. PLoS One 8:e65427.

Force, D. C. 1990. Ecology of insects in California chaparral. Research Paper PSW-RP-201. USDA Forest Service, Pacific Southwest Research Station, Albany, California, USA.

Forrester, T. D. and H. U. Wittmer. 2013. A review of the population dynamics of mule deer and black-tailed deer *Odocoileus hemionus* in North America. Mammal Review 43:292-308.

Fulton, R. E., and F. L. Carpenter. 1979. Pollination, reproduction, and fire in California *Arctostaphylos*. Oecologia 38:147-157.

Gamradt, S. C. and L. B. Kats. 1996. Effect of introduced crayfish and mosquitofish on California newts. Conservation Biology 10:1155-1162.

Gardali, T., N. E. Seavy, R. T. DiGaudio, and L. A. Comrack. 2012. A climate change vulnerability assessment of California's at-risk birds. PLoS ONE 7:e29507.

Glick, P., B. A. Stein, and N. A. Edelson. 2011. Scanning the conservation horizon: a guide to climate change vulnerability assessment. National Wildlife Federation, Washington, D.C., USA.

Greenfield, D. W. and Deckert, G. D. 1973. Introgressive hybridization between *Gila orcutti* and *Hesperoleucus symmetricus* (Pisces: Cyprinidae) in the Cuyama River Basin, California: II. Ecological aspects. Copeia 1973:417-427.

Hannah, L., G. F. Midgley, and D. Millar. 2002. Climate change-induced conservation strategies. Global Ecology and Biogeography 11:485-495.

Hanson, K., J. Snyder, and L. Kats. 1994. Natural history notes: *Taricha torosa* (California newt). Diet. Herpetological Review 25:62.

Harrison, S. P. 2013. Plant and animal endemism in California. University of California Press, Berkeley, California, USA.

Heller, N. E., and E. S. Zavaleta. 2009. Biodiversity management in the face of climate change: a review of 22 years of recommendations. Biological Conservation 142:14-32.

Hobbs, R. J., R. H. Groves, S. D. Hopper, R. J. Lambeck, B. B. Lamont, S. Lavorel, A. R. Main, J. D. Majer, and D. A. Saunders. 1995. Function of biodiversity in the Mediterranean-type ecosystems of southwestern Australia. Pages 233-284 *in* G. W. Davis and D. M. Richardson, editors. Mediterranean-type ecosystem: functions of biodiversity. Springer, Heidelberg, Germany.

Hughes, L., 2000. Biological consequences of global warming: is the signal already apparent? Trends in Ecology & Evolution 15:56-61.

Human, K. G. and D. M. Gordon. 1997. Effects of Argentine ants on invertebrate biodiversity in northern California. Conservation Biology 11:1242-1248.

Jennings, M. K., R. L. Lewison, T. W. Vickers, and W. M. Boyce. 2016. Puma response to the effects of fire and urbanization. Journal of Wildlife Management 80:221-234.

Jennings, M. R. and M. P. Hayes. 1994. Amphibian and reptile species of special concern in California. California Department of Fish and Game, Inland Fisheries Division, Rancho Cordova, California, USA.

Kats, L. B., S. A. Elliott, and J. Curt-ens. 1992. Intraspecific oophagy in stream-breeding California newts (*Taricha torosa*). Herpetological Review 23:7-8.

Keeley, J. E., C. J. Fotheringham, and M. Baer-Keeley. 2005. Determinants of postfire recovery and succession in Mediterranean-climate shrublands of California. Ecological Applications 15:1515-1534.

Keeley, J. E., and C. C. Swift. 1995. Biodiversity and ecosystem functioning in Mediterranean-climate California. Pages 121-183 *in* G. W. Davis and D. M. Richardson, editors. Mediterranean-type ecosystem: functions of biodiversity. Springer, Heidelberg, Germany.

Kimsey, L. S. 1996. Status of terrestrial insects. Pages 735-742 *in* Sierra Nevada ecosystem project: final report to Congress. Volume II: assessments and scientific basis for management options. Centers for Water and Wildlands Resources, University of California, Davis, California, USA.

Kremen, C. and R. Chaplin-Kramer. 2007. Insects as providers of ecosystem services: crop pollination and pest control. Pages 349-382 *in* A. J. A. Stewart, T. R. New, and O. T. Lewis, editors. Insect conservation biology: proceedings of the Royal Entomological Society's 23rd Symposium, Brighton, England, September 12-14, 2005. CABI Publishing, Wallingford, UK.

Kuo, M. M., R. S. Lane, and P. C. Giclas. 2000. A comparative study of mammalian and reptilian alternative pathway of complement-mediated killing of the Lyme disease spirochete (*Borrelia burgdorferi*). Journal of Parasitology 86:1223-1228.

Lane, R. S. and J. E. Loye. 1989. Lyme disease in California: interrelationship of *Ixodes pacificus* (Acari: Ixodidae), the western fence lizard (*Sceloporus occidentalis*), and *Borrelia burgdorferi*. Journal of Medical Entomology 26:272-278.

Lane, R. S., and G. B. Quistad. 1998. Borreliacidal factor in the blood of the western fence lizard (*Sceloporus occidentalis*). Journal of Parasitology 84:29-34.

LaPointe, F. J. and L. J. Rissler. 2005. Congruence, consensus, and the comparative phylogeography of codistributed species in California. American Naturalist 166:290-299.

Lawrence, G. E. 1966. Ecology of vertebrate animals in relation to chaparral fire in the Sierra Nevada Foothills. Ecology 47:278-291

Lawton, J. H., and V. K. Brown. 1993. Redundancy in ecosystems. Pages 255-270 *in* H. A. Mooney, editor. Biodiversity and ecosystem function. Springer-Verlag, Berlin, Germany.

Lee, J. S., E. W. Ruell, E. E. Boydston, L. M. Lyren, R. S. Alonso, J. L. Troyer, K. R. Crooks, and S. VandeWoude. 2012. Gene flow and pathogen transmission among bobcats (*Lynx rufus*) in a fragmented urban landscape. Molecular Ecology 21:1617-1631.

Lovejoy, T. E., R. O. Bierregaard Jr, A. B. Rylands, J. R. Malcolm, C. E. Quintela, L. H. Harper, K. S. Brown Jr, A. H. Powell, G. V. N. Powell, H. O. R. Schubart, and M. B. Hays. 1986. Edge and other effects of isolation on Amazon forest fragments. Pages 257-285 *in* M. E. Soulé, editor. Conservation biology: the science of scarcity and diversity. Sinauer Associates, Sunderland, UK.

MacDonald, I. A. W., D. M. Graber, S. DeBenedetti, R. H. Groves, and E. R. Fuentes. 1988. Introduced species in nature reserves in Mediterranean-type climatic regions of the world. Biological Conservation 44:37-66.

Marschalek, D. A., and M.W. Klein Sr. 2010. Distribution, ecology, and conservation of Hermes copper (Lycaenidae: *Lycaena* [Hermelycaena] *hermes*). Journal of Insect Conservation 14:721-730.

Millennium Ecosystem Assessment. 2003. Ecosystems and their services. Pages 49-70 *in* Millennium Ecosystem Assessment. Ecosystems and human well-being: a framework for assessment. Island Press, Washington, D.C., USA.

Moldenke, A. R. 1976. California pollination ecology and vegetation types. Phytologia 34:305-361.

Moyle, P. B., R. M. Quiñones, J. V. E. Katz, and J. Weaver. 2015. Fish species of special concern in California. California Department of Fish and Wildlife, Sacramento, California, USA. https://www.wildlife.ca.gov/Conservation/Fishes/Special-Concern. Accessed 15 May 2016

Myers, N., R. A. Mittermeier, C. G. Mittermeier, G.A.B. da Fonseca, and J. Kent. 2000. Biodiversity hotspots for conservation priorities. Nature 403:853-858.

Noss, R. F. 1987. Corridors in real landscapes: a reply to Simberloff and Cox. Conservation Biology 1:159-164.

Nuñez, T. A., J. J. Lawler, B. H. McRae, D. J. Pierce, M. B. Krosby, D. M. Kavanagh, P. H. Singleton, and J. J. Tewskbury. 2013. Connectivity planning to address climate change. Conservation Biology 27:407-416.

Parker, V. T. 2015. Seed bank divergence between *Arctostaphylos* Adans. (Ericaceae) and *Ceanothus* L. (Rhamnaceae) suggests different seed predator interactions. Ecologia Mediterranea 41:5-13.

Parker, V. T., R. B. Pratt, and J. E. Keeley. 2016. Chapter 24: chaparral. Pages 479-508 *in* H. A. Mooney and E. Zavaleta, editors. Ecosystems of California. University of California Press, Berkeley, California, USA.

Penrod, K., P. Beier, E. Garding, and C. Cabañero. 2012. A linkage network for the California deserts. Produced for the Bureau of Land Management and The Wildlands Conservancy. Science and Collaboration for Connected Wildlands, Fair Oaks, California, USA.

Powell, J. A. and C. A. Hogue. 1979. California insects. University California Press, Berkeley and Los Angeles, California, USA.

Purvis, A., J. L. Gittleman, G. Cowlishaw, and G. M. Mace. 2000. Predicting extinction risk in declining species. Proceedings of the Royal Society of London B: Biological Sciences 267:1947-1952.

Quinn, R. D. 1990. Habitat preferences and distribution of mammals in California chaparral. Research Paper PSW-RP-202. USDA Forest Service, Pacific Southwest Research Station, Berkeley, California, USA.

Quinn, R. D., and S. C. Keeley. 2006. Introduction to California chaparral, Volume 90. University of California Press, Berkeley, California, USA.

Riley, S. P. D., C. Bromley, R. H. Poppenga, F. A. Uzal, L. Whited, and R. M. Sauvajot. 2007. Anticoagulant exposure and notoedric mange in bobcats and mountain lions in urban southern California. Journal of Wildlife Management 71:1874-1884.

Riley, S. P. D., J. P. Pollinger, R. M. Sauvajot, E. C. York, C. Bromley, R. K. Fuller, and R. K. Wayne. 2006. FAST-TRACK: a southern California freeway is a physical and social barrier to gene flow in carnivores. Molecular Ecology 15:1733-1741.

Riley, S. P. D., R. M. Sauvajot, T. K. Fuller, E. C. York, D. A. Kamradt, C. Bromley, and R. K. Wayne. 2003. Effects of urbanization and habitat fragmentation on bobcats and coyotes in southern California. Conservation Biology 17:566-576.

Rissler, L. J., R. J. Hijmans, C. H. Graham, C. Moritz, and D. B. Wake. 2006. Phylogeographic lineages and species comparisons in conservation analyses: a case study of California herpetofauna. American Naturalist 167:655-666.

Russell B. D., C. D. G. Harley, T. Wernberg, N. Mieszkowska, S. Widdicombe, J. M. Hall-Spencer, and S. D. Connell. 2012. Predicting ecosystem shifts requires new approaches that integrate the effects of climate change across entire systems. Biology Letters 8:164-166.

Sala, O. E., F. S. Chapin, J. J. Armesto, E. Berlow, J. Bloomfield, R. Dirzo, E. Huber-Sanwald, L. F. Huenneke, R. B. Jackson, A. Kinzig, and R. Leemans. 2000. Global biodiversity scenarios for the year 2100. Science 287:1770-1774.

Sauvajot, R. M., M. Buechner, D. A. Kamradt, C. M. Schonewald. 1998. Patterns of human disturbance and response by small mammals and birds in chaparral near urban development. Urban Ecosystems 2:279-297.

Schierenbeck, K. A. 2014. Phylogeography of California: an introduction. University of California Press, Berkeley, California, USA.

Schuette, P. A., J. E. Diffendorfer, D. H. Deutschman, S. Tremor, and W. Spencer. 2014. Carnivore distributions across chaparral habitats exposed to wildfire and rural housing in southern California. International Journal of Wildland Fire 23:591-600.

Soulé, M. E., D. T. Bolger, A. C. Alberts, J. Wrights, M. Sorice, and S. Hill. 1988. Reconstructed dynamics of rapid extinctions of chaparral-requiring birds in urban habitat islands. Conservation Biology 2:75-92.

Stebbins, R. C. 1972. Amphibians and reptiles of California. California Natural History Guides (31). University of California Press, Berkeley and Los Angeles, California, USA and London, UK.

Stebbins, G. L., and J. Major. 1965. Endemism and speciation in the California flora. Ecological Monographs 35:1-35.

Stein, B. A., L. S. Kutner, and J. S. Adams. 2000. Precious heritage: the status of biodiversity in the United States. Oxford University Press, Oxford, UK.

Suarez, A. V. and T. J. Case. 2002. Bottom-up effects on persistence of a specialist predator: ant invasions and horned lizards. Ecological Applications 12:291-298.

Suarez, A. V., J. Q. Richmond, and T. J. Case. 2000. Prey selection in horned lizards following the invasion of Argentine ants in southern California. Ecological Applications 10:711-725.

Swei, A., R. S. Ostfeld, R. S. Lane, and C. J. Briggs. 2011. Impact of the experimental removal of lizards on Lyme disease risk. Proceedings of the Royal Society of London B: Biological Sciences 278:2970-2978.

Swetnam, T. W., and J. L. Betancourt. 1998. Mesoscale disturbance and ecological response to decadal climatic variability in the American Southwest. Journal of Climate 11:3128-3147.

Syphard, A. D., V. C. Radeloff, J. E. Keeley, T. J. Hawbaker, M. K. Clayton, S. I. Stewart, and R. B. Hammer. 2007. Human influence on California fire regimes. Ecological Applications 17:1388-1402.

Syphard, A. D., V. C. Radeloff, T. J. Hawbaker, and S. I. Stewart. 2009. Conservation threats due to human-caused increases in fire frequency in Mediterranean-climate ecosystems. Conservation Biology 23:758-769.

Vickers, T. W., J. N. Sanchez, C. K. Johnson, S. A. Morrison, R. Botta, T. Smith, B. S. Cohen, P. R. Huber, H. B. Ernest, and W. M. Boyce. 2015. Survival and mortality of pumas (*Puma concolor*) in a fragmented, urbanizing landscape. PLoS ONE 10:e0131490.

Westerling, A. L., A. Gershunov, T. J. Brown, D. R. Cayan, and M. D. Dettinger. 2003. Climate and wildfire in the western United States. Bulletin of the American Meteorological Society 84:595-604.

Whelan, C. J., D. G. Wenny, and R. J. Marquis. 2008. Ecosystem services provided by birds. Annals of the New York Academy of Sciences 1134:25-60.

Williams, S. E., L. P. Shoo, J. L. Isaac, A. A. Hoffmann, and G. Langham. 2008. Towards an integrated framework for assessing the vulnerability of species to climate change. PLoS Biology 6:2621-2626.

Winchell, C. S. and P. F. Doherty Jr. 2014. Effects of habitat quality and wildfire on occupancy dynamics of Coastal California Gnatcatcher (*Polioptila californica californica*). The Condor116:538-545.

Underwood, E. C., J. H. Viers, K. R. Klausmeyer, R. L. Cox, and M. R. Shaw. 2009. Threats and biodiversity in the Mediterranean biome. Diversity Distributions 15:188-197.

Vandergast, A. G., A. J. Bohonak, S. A. Hathaway, J. Boys, and R. N. Fisher. 2008. Are hotspots of evolutionary potential adequately protected in southern California? Biological Conservation 141:1648-1664.

van Mantgem, E. F., J. E. Keeley, and M. Witter. 2015. Faunal responses to fire in chaparral and sage scrub in California, USA. Fire Ecology 11:128-148.

Zedler, P. H. 1995. Plant life history and dynamic specialization in the chaparral/coastal sage shrub flora in southern California. Pages 89-115 *in* M. T. K. Arroyo, P. H. Zedler, and M. D. Fox, editors. Ecology and biogeography of Mediterranean ecosystems in Chile, California and Australia, Springer-Verlag, New York, New York, USA.

Chapter 4
Native Peoples' Relationship to the California Chaparral

M. Kat Anderson and Jon E. Keeley

Abstract Ethnographic interviews and historical literature reviews provide evidence that for many tribes of California, chaparral plant communities were a rich source of food, medicines, and technologies and that they supplemented natural fires with deliberate burning of chaparral to maximize its ability to produce useful products. Many of the most important chaparral plant species used in the food and material culture have strong adaptations to fire. Particularly useful were many annual and perennial herbs, which proliferate after fire from seed and bulb banks, shrub resprouts that made superb cordage and basketry material, as well as animals that were more readily caught in postfire environments. The reasons for burning in chaparral are grouped into seven ecological categories, each relying on a known response to fire of the chaparral community. The authors posit that tribes employed intentional burning to maintain chaparral in different ages and size classes to meet diverse food and material needs, tracking the change in plant and animal abundance and diversity, and shifts in shrub architecture and habitat structure during the recovery of the chaparral community. Areas were burned in ways designed to create a mosaic of open grassland and recently burned, young and mature stands of chaparral with different combinations of species and densities. This management conferred on chaparral plant communities a degree of spatial, structural, successional, and biotic diversity that exceeded what would have been the case in the absence of human intervention. These impacts are still evident on contemporary landscapes.

Keywords California Indians · Chaparral · Ethnobiology · Fire ecology · Fire management · Indian burning · Native Americans · Tribes · Type-conversion · Vegetation patterns

---

M. K. Anderson (✉)
NRCS State Office, Davis, CA, USA
e-mail: mkanderson@ucdavis.edu

J. E. Keeley
U.S. Geological Survey, Three Rivers, CA, USA

University of California, Los Angeles, CA, USA

E. C. Underwood et al. (eds.), *Valuing Chaparral*, Springer Series on Environmental Management, https://doi.org/10.1007/978-3-319-68303-4_4

4.1 Introduction

Although plant communities we would recognize as chaparral have been present in western North America for at least 10 million years (Keeley et al. 2012), it is impossible to fully understand the ecology, composition, and distribution of this vegetation without reference to how it has been impacted, at the very tail end of this long history, by humans. People of mostly western European extraction have indeed heavily influenced California chaparral during the past 200 years by clearing chaparral for agriculture, mining, rangeland, and residential and industrial development, by altering fire regimes, and by changing the global climate, but chaparral in California was shaped by human activities in ways equally profound, though less evident, well before the Gold Rush or the establishment of the missions in Spanish colonial times. These vegetation-altering activities were carried out, of course, by the Native peoples of California, who, not long after their arrival ~12,000 years ago, developed a relationship with chaparral based on both using its resources extensively and carefully managing them.

For many of the tribes of California, chaparral plant communities were a rich source of food, medicines, and materials for baskets, ceremonial items, clothing, living structures, tools, and other items of technology. Exploiting these resources over long periods of time, the Indians who lived in and near chaparral zones learned through observation, trial-and-error experimentation, and accidental discovery that many of the chaparral-based resources they found most valuable could be sustained by judicious harvest methods and enhanced, in quality or quantity, by certain kinds of manipulation. For example, replanting the bulblets of the checker lily and wild onion bulbs harvested for food could insure the availability of bulbs in subsequent years. Saving the seeds of red maids (*Calandrinia ciliata*) and scattering them where they had not grown before could expand the populations and thus the harvests of these food plants. Pruning flannelbush (*Fremontodendron californicum*) in a certain way would rejuvenate older shrubs and encourage the growth of shoots ideal for cordage manufacture.

These management techniques were all important, but by far the most consequential was the use of fire. Done properly, burning areas of chaparral could yield multiple and long-lasting beneficial effects by creating diverse landscape mosaics of chaparral and herbaceous communities. Tribes discovered that if you burned the chaparral at the right time of year and at the right frequency, you could simultaneously open up the habitat for ease of movement and hunting, increase its ability to support deer and small mammals, increase the productivity of the plants used for food and medicine, encourage shrubs to produce shoots ideal for basketry, and reduce the vulnerability of villages to out-of-control, lightning-ignited summer wildfires (Timbrook et al. 1982; Lewis 1993; Shipek 1993; Keeley 2002; Anderson and Rosenthal 2015).

Native peoples in California applied fire and other management techniques to chaparral to achieve specific, relatively short-term, and human-centered results, but over the many thousands of years during which this management regime was practiced (and undoubtedly refined), its consequences reached deeply into the foundation of what defines chaparral: the distribution, interactions, and genomes of its constituent species. We will likely never know the exact extent to which Indian management of chaparral in the pre-historic era altered the distribution and ecology of the chaparral plant community, but the evidence indicates that it almost surely did. The implications for present-day management of chaparral are considerable. Our goal in this chapter is to present the foundational knowledge and evidence that need be taken into account when considering how management and use of chaparral by the first peoples of the state could inform today's management of chaparral dominated lands.

4.2 Native Peoples' Use of the Chaparral Community

Colonization of what is now California by Europeans completely disrupted Native peoples' lifeways, including their harvesting of plants and animals from the environment (Anderson 2005a). Infectious diseases brought by the early explorers and Spanish missionaries spread through villages in wave after wave beginning in the 1600s, decimating many populations, even among tribes living far from Spanish influence (Preston 1996, 2002). Using methods of both persuasion and coercion, the padres brought surviving Indians to the missions to work the fields, make adobe blocks, process livestock skins for leather, and perform the many other tasks that kept the missions' economies running (Costo and Costo 1987). As a result of these processes, the traditional subsistence and management practices of the tribes in the chaparral zones of the central and southern Coast Ranges dwindled in many areas by the end of the eighteenth century (Castello 1978). In the chaparral zones of the Sierra Nevada foothills and northern California, tribes kept traditional practices going into the nineteenth century despite demographic disruption, but then the massive influx of whites that came during and after the Gold Rush drastically curtailed them (Mason 1881; Holliday 1999). Within a few decades, appropriation of land, violent repression, genocide, disease, and legal strictures had severely limited Native harvesting of chaparral resources in the foothills and northern regions and supplanted Native management with neglect and sometimes wanton destruction (Anderson 2005a; Madley 2016).

By the early twentieth century, when trained scientists began serious study of California's vegetation and anthropologists began to document elements of Native cultures, the reciprocal relationship that California Indians had established with chaparral was only a memory in many Native families. The extensive and regular use of fire as a management tool in chaparral had waned in most regions. Few Indians existed entirely apart from the emerging industrial economy, and what traditional practices remained were isolated remnants of the former lifeways.

Fortunately, we know how Native peoples in California used resources from the chaparral community in the times pre-EuroAmerican colonization, through evidence in varying forms, ranging from observations recorded more than 200 years ago to pollen grains buried in lake sediments. Archaeological sites yield clues to the importance of chaparral in the form of charred fruits, seeds, and bulbs from cooking hearths, and bone, antler, and hide remains of terrestrial mammals. In museums and other repositories, plant foods, baskets, and other artifacts, collected by anthropologists and others before the vast curtailment of traditional practices, tell us what plants and animals had importance in indigenous economies, as do the written accounts of missionaries and early settlers. Remarkably, some Native people today still practice gathering and hunting traditions and remember what their grandparents told them about former indigenous burning and other chaparral management techniques, and they hold knowledge passed along by their ancestors about the plants and animals of the land. In addition, the anthropologists, ethnographers, and naturalists who studied the partially intact Native cultures that still existed around the turn of the twentieth century were tenacious and thorough in their search for authentic examples of the former lifeways and left us a rich record of documentation. These forms of evidence, combined with data from growth rings, pollen deposits, soil, and charcoal deposits, allow us to form a rough picture of the role of plants and animals of the chaparral in the pre-historic cultures of California.

As we explore the different plant and animal species of the chaparral that helped support tribes' subsistence economies and supplied much of their physical cultures, it is important to keep in mind that the chaparral plant communities that existed in California pre-EuroAmerican contact were somewhat different from what we see today. Why this would be so is the subject of discussion later in this chapter. For now, it suffices to recognize that the pre-EuroAmerican contact chaparral was, at least in some locations, more spatially heterogeneous than what exists today. Its different successional stages formed mosaics of shrublands mixed with open grasslands and woodlands. This structural complexity, with its greater vegetational diversity and ecotonal area, translated into greater floristic and habitat diversity and an abundance of different niches for wildlife (Verner and Boss 1980). Simply put, the chaparral in much of pre-EuroAmerican contact California was species-rich in a way that a contemporary person struggling through a sea of chamise (*Adenostoma fasciculatum*) or contemplating a hillside of seemingly pure buckbrush (*Ceanothus cuneatus*) would find difficult to appreciate.

In pre-historic times the heterogenous chaparral zones of California supported populations of mule deer, grizzly and black bears, mountain lions, various smaller mammals, and numerous species of birds, reptiles, and insects, many of which were taking advantage of this fire-prone landscape (White et al. 1980; Quinn 1990). Mule deer (*Odocoileus hemionus*) fed on the herbaceous plants at the edges of brush land and browsed the young, tender sprouts of recently burned chaparral shrubs such as, mountain mahogany (*Cercocarpus betuloides*) (plant nomenclature according to Baldwin et al. 2012), Utah service-berry (*Amelanchier utahensis*), and chamise (*Adenostoma fasiculatum* var. *fasciculatum*). Bears (*Ursus americanus* and *U. arctos*) feasted on the berry bushes along chaparral borders, rooted up bulbs from dense Mariposa lily (*Calochortus* spp.) patches, and denned in dense thickets of chaparral. Mourning doves (*Zenaida macroura*) fed on the jet-black seeds of *Calandrinia ciliata* growing in fire-created openings and tortoiseshell butterfly larvae (*Nymphalis californica*) partook of young *Ceanothus cuneatus* seedlings. Small mammals such as voles, moles, ground squirrels, and rabbits hid in the shrub cover and became meals for various raptors. The high diversity and abundance of insects fed larger predatory reptiles, birds, and mammals.

Fig. 4.1 Estefana Salazar, Tubatulabal, with tobacco (*Nicotiana attenuata*) leaves gathered from a patch in Weldon, Kern County, California. *Nicotiana attenuata* was widely used by tribes for ceremonial and medicinal purposes, and clearing areas and enhancing *N. attenuata* patches was one of the most commonly recorded reasons for burning in chaparral. Museum number 15-10544, courtesy of the Phoebe A. Hearst Museum of Anthropology, and the Regents of the University of California. Photo by E.W. Voegelin, July 1932

Chaparral and its ecotonal margins supported a diversity of shrubs, herbs, and grasses that as a group ranked among the most useful and most needed of food and non-food plants. Annuals such as *Calandrinia ciliata*, chia (*Salvia columbariae*), and perennial grasses such as blue wild-rye (*Elymus glaucus*) produced edible seeds. Shrubs and perennial grasses such as sumac, also known as sourberry (*Rhus aromatica*), *Ceanothus cuneatus*, deerbrush (*Ceanothus integerimus*), redbud (*Cercis occidentalis*), and deer grass (*Muhlenbergia rigens*) yielded prized basketry materials. A few special shrubs and perennials like cascara sagrada (*Frangula purshiana*), yerba santa (*Eriodictyon californicum*), California lomatium (*Lomatium californicum)*, and jimsonweed (*Datura wrightii*), and annual herbs such as tobacco (*Nicotiana attenuata* and *N. quadrivalvis*), provided materials for medical and ceremonial use (Fig. 4.1) (Goddard 1903; Lake 1982). Many species that today might be difficult to find growing in a chaparral community were more abundant and more widespread. The native people of California made wide use of these and other chaparral plants and animals, as we detail in the following pages.

4.3 Useful Chaparral Plants

The indigenous people of California were fortunate to have a great diversity of plants to draw from in the chaparral. More than 400 species that make up the chaparral plant community are known to have been used in some manner by at least one tribe. Together, these plants furnished an essential portion of the subsistence economy of tribes that had access to chaparral dominated lands.

One reason why the number of chaparral plants documented as useful is so high is that Native people could find uses for just about any plant. If it did not produce fruit, seeds, shoots, or below-ground parts that could be eaten or used for medicine, then its stems or roots might have been fibrous enough to be used for cordage or basketry, or perhaps they contained compounds that could be extracted as dyes. If none of these uses seemed to be possible, then the stems or branches could at least be used for structures like summer houses or hunting blinds, or the branches and trunks could be burned as cooking or heating fuel.

David Prescott Barrows (1967), an anthropologist, took note of this ability to see and find uses in plants. "There are few plants in Indian country," he wrote, "that have not been experimented with by its native inhabitants." At the time, he was speaking of the Cahuilla, with whom he had conducted field work in the 1890s, but his statement would apply to all of the tribes of California. The Cahuilla's use of so many plants from the chaparral and other plant communities for such an enormous variety of purposes shows "how diligent and acute…the investigation for useful things has been" (Barrows 1967).

Native people gathered native plants from every type of chaparral, from the lower elevation chaparral on the slopes of the coastal mountains to the montane chaparral higher up and in the Sierra Nevada. In the chamise chaparral, the Luiseño cut the branches of *Adenostoma fasciculatum* to form the rough coiled structures of the acorn granary and used the young shoots to form the foreshafts of arrows (Sparkman 1908; Beemer 1980). In the mesic scrub oak chaparral on north-facing slopes, the Cahuilla plucked acorns of scrub oak (*Quercus berberifolia*) for food (Bean and Saubel 1972). On the lower slopes of the San Gabriel and San Bernardino Mountains, the Tongva broke off the carbohydrate-rich young flowering stalks of chaparral yucca (*Hesperoyucca whipplei*) for baking in an earth oven and collected the leaves of ephedra (*Ephedra* spp.) for tea (Johnston 1962). The Kawaiisu gathered the edible seeds of the dominant bush poppy (*Dendromecon rigida*) in the desert chaparral of the Coast Ranges bordering the San Joaquin Valley (Zigmond 1981). In the red shanks (*Adenostoma sparsifolium*) chaparral of southern California the Kumeyaay harvested the wood and roots of *Adenostoma sparsifolium* for firewood (Hinton 1975). In the higher-altitude montane chaparral of the Cascades, the Wintu collected the tasty fruits of *Amelanchier utahensis* (Du Bois 1935).

Of all the native chaparral plants, perhaps the many species of manzanita (*Arctostaphylos* spp.) are the most emblematic of the chaparral's cultural importance. Indians used the leaves, bark, and berries of *Arctostaphylos* spp. for various medicines and transformed its bark and wood into arrow straighteners, awl handles,

Fig. 4.2 Mollie Cheepo, North Fork Mono, pounding the berries of manzanita (*Arctostaphylos* spp.) in a bedrock mortar, a major step in processing the fruits for cider. Tasineu Village, North Fork. Museum Number 15-6227, courtesy of the Phoebe A. Hearst Museum of Anthropology and the Regents of the University of California

digging sticks, canes, brooms, reels for string, containers, firewood, house construction, pipes, mush paddles, and ear- and nose-piercing sticks. Because the wood of *Arctostaphylos* spp. burned so hot and made excellent coals, the Mono, Coast Miwuk and other tribes preferred it for heating rocks for cooking acorn mush and parching seeds and other foods (Collier and Thalman 1991; Anderson 2009). A vitamin-rich cider-like drink unique to the tribes of California was made from the fruits of *Arctostaphylos* spp. (Fig. 4.2). So important was this drink that it typically accompanied foods such as venison and acorn mush or soup, and it was mixed with foods such as ground *Salvia columbariae* seeds and yellowjacket larvae, and moistened other foods such as wildflower seeds that were crushed and pressed into balls. "[I]n color and flavor," wrote ethnographer C. Hart Merriam in 1902, this beverage was "like the very best apple cider…cooling and delicious" (Merriam 1902). Since any particular locality supported only a few of the 14 species of *Arctostaphylos* utilized in California chaparral,[1] each tribe had its own local cider reflecting that particular *terroir* (Anderson and House 2012). The berries of *Arctostaphylos* spp. are one of the most common paleoethnobotanical remains in archaeological sites located in chaparral habitat (Wohlgemuth 2004), indicating the plants' importance to tribes.

[1] *Arctostaphylos canescens, A. glandulosa, A. glauca, A. manzanita, A. mewukka, A. myrtifolia, A. nevadensis, A. parryana, A. patula, A. pumila, A. pungens, A. tomentosa, A. uva-ursi,* and *A. viscida.*

4.3.1 Plants for Food and Medicine

Ethnographers working with California Indian tribes in the late 1800s and early 1900s noted their uniformly sweet breath, teeth with no decay, clear eyes, and the rarity of obesity (Hudson 1901a; Powers 1976). Good health was partially a result of a very diverse diet of plants, animals, fish, shellfish, red meat, and mushrooms, and for many tribes much of this diet came from chaparral. The variety of edible chaparral plants in the traditional diet is impressive and includes four plant-part categories: potato-like under-ground stems (called bulbs, corms, and tubers); seeds, grains and nuts, leaves; stems and flowers for greens; and the fleshy fruits (the pomes, drupes, and berries).

Edible chaparral plants were important parts of the maintenance of social relations, being offered as gifts and offerings. The berries of *Arctostaphylos* spp., for example, were brought as a present to a Foothill Yokuts mother's ceremony after childbirth and were scattered as offerings during Nisenan healing ceremonies (Hudson 1902; Gayton 1948a). Today various foods and drinks made from chaparral plants, such yerba buena tea, sumac berries, ephedra tea, Sierra mint tea, manzanita cider, and gray pine nuts, are given to friends and guests at major social events and offered as snacks and refreshments at Indian homes. They are extremely important in maintaining ethnic identity (Anderson 2005a).

Annual cycles of food-gathering were closely tied to the phenology of chaparral plants. Particularly important were the plants with edible under-ground storage organs, called *geophytes* by botanists and ecologists and *Indian potatoes* or *root foods* by Native Americans. Indians dug these bulbs, corms, and tubers for food with digging sticks in disturbed, open ground of chaparral and they were eaten raw or boiled, baked in an earth oven, or roasted in coals. Perhaps the most widely dug geophytes in the chaparral were soap plants (*Chlorogalum pomeridianum*), Gairdner's yampah (*Perideridia gairdneri*), and blue dicks (*Dichelostemma capitatum*). The tremendous diversity of geophtyes available for harvest is illustrated by the large number of species utilized in the *Brodiaea* complex. The corms of at least five kinds of brodiaeas (harvest brodiaeae [*Brodiaea elegans*] subsp. *elegans*, California brodiaeae [*B. californica*], crown brodiaeae [*B. coronaria*], dwarf brodiaeae [*B. minor*], and Kaweah brodiaeae [*B. insignis*]), all species of *Dichelostemma* (ookow [*D. congestum*], roundtooth snakelily [*D. multiflorum*], and twining snakelily [*D. volubile*]), and four species of *Triteleia* (largeflower triteleia [*Triteleia grandiflora*], white brodiaeae [*T. hyacinthine*], pretty face [*T. ixioides*], and common triteleia [*T. laxa*]) were dug and eaten (Dixon 1905; Brubaker 1926; Barrett and Gifford 1933; McMillin 1956; Duncan 1964; Powers 1976; Latta 1977; Eastwood n.d.; Howell n.d.; Hudson n.d.). To the north, important chaparral root foods included pussy ears (*Calochortus tolmiei*) bulbs, yellow fritillary (*Fritillaria pudica*) bulbs, and hairy brackenfern (*Pteridium aquilinum* var. *pubescens*) rhizomes (Fowler 1986; Goddard 1903; McMillin 1956; Knudtson 1977). In southern California common goldenstar (*Bloomeria crocea*), *Calochortus concolor*, and Palmer's mariposa lily (*Calochortus palmeri* var. *palmeri*) were valued (Sparkman 1908; Voegelin 1938; Bean and Saubel 1972). The contribution of these and other root crops to the subsistence economies of tribes was substantial (Anderson and Lake 2016).

In addition to gathering the carbohydrate-rich geophytes growing in early successional chaparral, Native Americans gathered many plant foods that grew above-ground: sprouts, young stems, fleshy fruits, and seeds. Seasonally abundant, easily harvested in large quantities, rich in vitamins and phytonutrients, and easily dried or prepared for storage and trading, the above-ground plant foods from chaparral plants were an essential dietary component for many tribes.

While many of the above-ground plant foods gathered in chaparral could be found in other vegetation types, chaparral and openings in chaparral were favorite gathering spots. These habitats supported a great variety of edible species and produced abundant and predictable crops. Green vegetables gathered from chaparral included the sweet raw shoots of whitehead mule-ears (*Wyethia helenioides*), the tender tops of horseweed (*Erigeron canadensis*), the stems of different kinds of thistles (*Cirsium* spp.), the young leaves of docks (*Rumex* spp.), phacelias (*Phacelia* spp.), fiddlenecks (*Amsinckia* spp.), and buckwheats (*Eriogonum* spp.), mature stems of common cowparsnip (*Heracleum maximum*), and young fiddleheads of bracken ferns (*Pteridium*). The young flower stalks and basal portions of mature stalks of the *Hesperoyucca whipplei* were eaten after being roasted in a pit oven with hot stones (Voegelin 1938; Gayton 1948a).

Another important source of food, particularly in areas around villages, was patches of chaparral shrubs with edible fruits. Many different kinds of fruits, most of them berries, were dried and soaked in water to re-constitute them before being eaten. Dried berries were also made into cakes and fruit leathers. The chaparral shrubs producing edible berries are numerous: lemonade berry (*Rhus integrifolia*), sumac, or *Rhus aromatica*, sugar bush (*Rhus ovata*), *Arctostaphylos* spp., blue elderberry (*Sambucus mexicana*), bitter cherry (*Prunus emarginata*), western chokecherry (*Prunus virginiana* var. *demissa*), Sierra plum (*Prunus subcordata*), holly-leaf cherry (*Prunus ilicifolia*), spiny redberry (*Rhamnus crocea*), gooseberry (*Ribes amarum*, *Ribes menziesii*, *Ribes quercetorum*, and other spp.), *Amelanchier utahensis* and other *Amelanchier* spp., *Heteromeles arbutifolia*, California blackberry (*Rubus ursinus*), barberry (*Mahonia* spp.), snowberry (*Symphoricarpos albus* var. *laevigatus*), rose (*Rosa* spp.), whitebark raspberry (*Rubus leucodermis*), California wild grape (*Vitis californica*), California juniper (*Juniperus californica*), western juniper (*Juniperus occidentalis*), nightshade (*Solanum xanti*), ninebark (*Physocarpus capitatus*), and mission manzanita (*Xylococcus bicolor*) (Timbrook 2007; Anderson and Rosenthal 2015). Some of these fruits are still gathered today and eaten raw or made into jams, jellies, cakes, and pies.

Many kinds of sun-loving wildflowers including *Salvia columbariae*, *Calandrinia ciliata*, California compassplant (*Wyethia angustifolia*), common madia (*Madia elegans*), valley popcorn flower (*Plagiobothrys canescens*), western buttercup (*Ranunculus occidentalis*), farewell-to-spring (*Clarkia* spp.), and whitestem blazingstar (*Mentzelia albicaulis*) were harvested for their edible seeds in the open patches of chaparral (Barrett and Gifford 1933; Voegelin 1938; Anderson et al. 2012). The seeds of forbs and grasses were important enough to be stored in their own granaries or baskets separate from acorns (Barrett and Gifford 1933). Seeds were roasted, baked into bread, or boiled into mush or soup. A favored seed collection method was to beat the inflorescences of wildflowers and grasses with a seed beater, a shallow basket with a handle, so that the seeds would fall into a wider-mouthed basket or burden basket.

In open areas in and around chaparral grew many kinds of wildflowers, as well as various native grasses that were valued for their edible grains. These included California brome *(Bromus carinatus)*, *Elymus glaucus*, and slender hair grass (*Deschampsia elongata*) (Duncan 1964; Powers 1976; Bunnell 1980; Anderson et al. 2012). Seeds of subshrubs and shrubs such as white sage (*Salvia apiana*), black sage (*Salvia mellifera*), *Ceanothus cuneatus* and deer brush (*Ceanothus* spp.) were important too (Dixon 1905; Miller 1928; Anderson et al. 2012; Hudson n.d.). The Serrano gathered the seeds of *Salvia apiana* and ate them raw and also peeled and ate the new stems (Lerch 2002). The Chumash gathered the seeds of at least two kinds of salvias: *Salvia columbariae* and *S. carduacea* (Timbrook 2007). The Mono mixed the seeds of *Ceanothus cuneatus* with the pupae of the California tortoise-shell butterfly, and the Konkow soaked *Ceanothus* spp. seeds and then scorched them to remove the bitterness. They were pounded into flour, sifted in a winnowing basket, and then eaten plain or mixed with acorn mush, meat, or other foods (Duncan 1964). The nuts of various kinds of trees that occur in chaparral were gathered for food as well. The most important were the oaks such as canyon live oak (*Quercus chrysolepis*), blue oak (*Q. douglasii*), coastal scrub oak (*Q. dumosa*), Engelmann oak (*Q. engelmannii*), Oregon white oak (*Q. garryana* var. *semota*), turbinella oak (*Q. turbinella*), and interior live oak (*Q. wislizeni*). Also important were California buckeye (*Aesculus californica*), California bay (*Umbellularia californica*), and gray pine (*Pinus sabiniana*) (Fig. 4.3) (Beals 1933; Bean and Saubel 1972; Clark 1987; Anderson 1988).

Fig. 4.3 Elizabeth Enos, Nisenan Maidu, peeling California buckeye (*Aesculus californica*) nuts for food. Museum number 25-5034, courtesy of the Phoebe A. Hearst Museum of Anthropology and the Regents of the University of California. Photo taken by Samuel A. Barrett, 1983

California tribes living in chaparral not only gathered food plants but also plants with medicinal properties. Indigenous pharmacopeias contained hundreds of plant species that could be collected in the chaparral plant community. Each plant or plant part was designated for very specific treatments such as curing fevers, reducing internal pains, tonifying vital organs, alleviating colds and coughs, or healing stomach troubles. Many medicinal plants were taken internally in some form: the Shasta chewed the roots of chaparral clematis (*Clematis lasiantha*) for colds (Holt 1946), tribes up and down California boiled the leaves of *Eriodictyon* spp. and made a tea used as a cough medicine (Anderson 2016), the Kawaiisu made an infusion of the leaves of pallid silk tassel bush (*Garrya flavescens*) and drank it to cure stomach aches (Zigmond 1981).

A wide variety of chaparral herbs were also made into salves or poultices and applied topically for skin, muscle, and eye problems and some were collected for baths to treat aching and sore muscles, arthritis, bruises, cuts, sores, wounds, and sore eyes. The Konkow applied the milky juice of purple milkweed (*Asclepias cordifolia*) externally for warts and other skin problems such as skin cancer (Duncan 1964). The Nisenan mashed the root of narrow petal wakerobin (*Trillium angustipetalum*), cooked it, and applied it as a poultice for toothaches, stiff necks, and sore throats, and in powdered form it was applied to small cuts (Duncan 1964). The Paiute and Shoshoni made a decoction of the plant Scouler's St. John's wort (*Hypericum scouleri*) and used it to bathe aching feet (Train et al. 1941). The Southern Sierra Miwuk put mugwort (*Artemisia douglasiana*) leaves in bathwater to sooth aching bones (Anderson 1988).

The Kawaiisu, Sierra Miwuk, and Luiseño, each separated by other tribal territories and speaking mutually unintelligible languages, all used different species of a chaparral spurge (*Chamaesyce*)[2] applied as a poultice to draw out the poison of rattlesnake bites and reduce the swelling (Sparkman 1908; Barrett and Gifford 1933; Zigmond 1981). That three different tribes would have discovered this plant's effectiveness for snakebite, attests to the proficiency of Indian healers and their exhaustive experimentation with chaparral plants.

Many chaparral plants were so effective in treating ailments that they were adopted by the Franciscan missionaries as they came into contact with the Indians. The padres gave several of these plants names that included the words *sagrada* (sacred) and *santa* (holy or saint). Later, some of these chaparral plants were used by American doctors and became part of American medical care. Their official listing in the US Pharmacopeia by the medical establishment[3]—and the present-day use of a few, such as *Frangula purshiana*, attests to their efficacy (Voegel 1970).

[2] *Chamaesyce albomarginata* for the Kawaiisu; *Chamaesyce ocellata* and *C. serphyllifolia* for the Sierra Miwok; and *Chamaesyce polycarpa* var. *polycarpa* for the Luiseño.

[3] In the first US Pharmacopeia issued in 1820, almost half of the substances were native plants used by American Indians.

Fig. 4.4 Justin Farmer, Ipai, splitting a managed branch of sumac (*Rhus aromatica*) into three pieces to be used as sewing strands or weft in baskets. Young growth is what weavers prefer as it is flexible, long, without side-branching and no insects or diseases. This desirable shrub architecture and quality are created through deliberate Indian burning and pruning in chaparral areas. Photo by Kat Anderson

4.3.2 Plants as Raw Material for Technologies

If the chaparral plant community could be considered a grocery store and pharmacy because of the abundance of its edible and medicinal plants, then it was also the Indians' hardware store, home building center, tobacco shop, and craft supply store. Like aboriginal people anywhere, the people of California made a variety of items and structures necessary for human survival and comfort, including cordage, baskets, mats, shelters, clothing, weapons, and tools, and for many tribes much of the raw material for these items came from chaparral (Figs. 4.4 and 4.5). The diverse properties of the wood and fiber from chaparral plants and shrubs are what made the chaparral such an important source of raw material for technologies. The exceptional hardness of *Cercocarpus betuloides* wood, for example, made it ideal for spears and digging sticks, similarly because the wood of *Arctostaphylos* spp. had "the same hardness all the way through" (Gayton 1948a) it was perfect for making pipes.

This section addresses five primary cagories: basketry, cordage, clothing, utensils, tools, weapons, and structures and fencing. Baskets, which served many different functions, were among the most important items of technology for all tribes in

Fig. 4.5 Justin Farmer, Ipai, pointing out the tiny knot, also called a "dimple" that forms where the lateral branch grows from the main stem on older-growth sumac (*Rhus aromatica*) making it unsuitable for basketry material. Photo by Kat Anderson

California. They held drinking water, stone-boiled foods, trapped fish, transported commodities, winnowed seeds, and stored household items. Many parts from chaparral plants were used to make baskets: leaves of *Hesperoyucca whipplei*, branches and roots of *Pinus sabinianas*, flower stalks of *Muhlenbergia rigens*, stems of rushes, and young shoots of *Cercis occidentalis*, *Rhus aromatica*, *Ceanothus* spp., *Ceanothus cuneatus*, *Vitis californica*, bush penstemon (*Keckiella breviflora*), wild mock orange (*Philadelphus lewisii*), pink honeysuckle (*Lonicera hispidula*), ocean spray (*Holodiscus discolor*), and *Fremontodendron californicum* (Merriam 1902; Goddard 1903; Merrill 1923; Brubaker 1926; Voegelin 1938; Goodrich et al. 1980; Bates 1982; Hedges and Beresford 1986).

Thousands of young shoots of chaparral shrubs were gathered by each weaving family every year to make burden baskets, seedbeaters, cradleboards, winnowers, sifting baskets, rough work baskets, and many other kinds (Barrett and Gifford 1933; Duncan 1964). Such large quantities of basketry materials were needed that a typical Indian house might be filled with bundles of straight sticks of *Ceanothus* spp., *Cercis occidentalis*, *Ceanothus cuneatus*, *Rhus aromatica*, and *Prunus emarginata*, with coils of *Pinus sabiniana* root hanging from ceilings, hung on walls, or stored in baskets (Barrett and Gifford 1933; Neely 1971; Chesnut 1974; Heizer 1978; Bates 1982; Bethel et al. 1984; Anderson 1988). The Southern Sierra Miwuk considered *Ceanothus cuneatus* to be the "strongest basket material" and used it to make cooking pots, drying baskets to hold acorns and fruit, and bath tubs (Hudson 1901a; Merriam 1955). Basketweaver Norma Turner Behill, Mono/Dumna (pers. comm. 2006) described the extensive use of this chaparral shrub, "The whole rods are used for winnowing baskets. The two year old ones are split for lacing and go around seed beaters, the tops of the burden baskets and the tops of baby baskets. They use those little fine sticks to make those little fine baskets".

Chaparral plants were an important source of the fibers used to make cordage. The bast fibers of dogbane (*Apocynum* spp.) were employed in constructing fences for deer and rabbit drives and also for fishnets and fishing line (Duncan 1964). The stem bast fibers of *Asclepias cordifolia*, showy milkweed (*Asclepias speciosa*), and narrow-leaf milkweed (*A. fascicularis*) were used in many parts of California and made into string for bows, sling-shots, belts, carrying straps, net bags, hairnets, fishing lines and nets, and for lashing mush-stirrers and binding dwelling frames and posts (Barrett and Gifford 1933; Gayton 1948a; Duncan 1964; Eastwood n.d.). Clothing, regalia, and jewelry were sewn with milkweed thread. *Fremontodendron californicum* branches were a major tying material in the central foothills of California and throughout southern California, they were split into thin strips to make ties for house frames, rafter bindings, thatch bindings, and lashings for acorn granaries. In addition the outer bark of the branches was stripped off for use in making tumplines and fiber rope for ferrying things across streams (Barrett and Gifford 1933; Voegelin 1938; Gayton 1948a; Bethel et al. 1984; Hudson n.d.). In southern California, the fibers extracted from the leaves of Mohave yucca (*Yucca schidigera*) were used for bowstrings, netting, strings for shell money, ropes, mats, and coiled rope soles for sandals (Bean and Saubel 1972). Cordage was particularly important to tribes because of its central role in the technology used for hunting and fishing. Knowledge of string-making and its application in a great variety of hunting equipment, such as nets, snares, and spears, enabled tribes to tap the rich animal resources of the chaparral.

Most tribes wore clothing including accessories like earrings, belts, bracelets, and purses that derived in part from chaparral plants. The Sierra Miwuk wore sleeveless buckskin clothing belted with *Adenostoma fasciculatum* (Hudson 1901a). Similarly, the deerskins worn by Nisenan women were often sewn with *Adenostoma fasciculatum* (Hudson 1902). The Maidu made rod armor for warfare using *Cercocarpus betuloides* sticks (Kroeber 1976). Indian children in different tribes were swaddled with the shredded bark of *Fremontodendron californicum* or the stems of soft rush (*Juncus effusus*) (Hudson n.d.). Ohlone women wore skirts and aprons woven with the leaves of *Eriodictyon* spp. (Bocek 1984). Hupa women wore aprons under their skirts consisting of many long strands of knobcone pine (*Pinus attenuata*), nut shells strung on twine, and adorned themselves with necklaces made with the black fruits of oval-leaved viburnum (*Viburnum ellipticum*), and hair ties made of the sprigs of yerba buena (*Clinopodium douglasii*) (Goddard 1903).

A great variety of utilitarian items such as utensils for cooking and eating, tools for gathering, hunting, construction, and materials processing, and weapons for defense, were made from the wood, leaves, and bark of various chaparral shrubs and trees. Indians combed their hair with *Chlorogalum pomeridianum* brushes or with combs made of small sticks of *Cercocarpus betuloides* tied with milkweed string (Barrett and Gifford 1933; Voegelin 1938). The needles of gray and other pines scattered in chaparral were used for thatching, bedding, and floor covering (Neely 1971). *Muhlenbergia rigens* leaves were sat on while playing gambling games and game pieces consisted of chaparral plants (Goddard 1903; Anderson 1994). Pipes for pleasure smoking were made of *Sambucus* spp., *Arctostaphylos* spp., or *Cercocarpus betuloides*. Canes made of California-nutmeg, silk tassel bush (*Garrya elliptica*), or *Cercocarpus betuloides* aided walking (Hudson 1902; Goddard 1903; Gifford 1932). Kitchens were stocked with soup paddles and stirrers made of *Ceanothus cuneatus*, *Arctostaphylos* spp., *Fremontodendron californicum*, *Quercus douglasii*, or *Pinus sabiniana*, rocks were fetched from the fire with tongs made of *Quercus douglasii*, *Cercocarpus betuloides*, or *Fremontodendron californicum*. Floors were swept with brooms of *Cercocarpus betuloides* or *Ceanothus* spp. twigs, and axes for chopping firewood had *Ceanothus cuneatus* handles (Hudson 1902; Barrett and Gifford 1933; Voegelin 1938; Gayton 1948b; Bethel et al. 1984; Goode 1992).

Tools for gathering plant parts such as knocking sticks to knock down acorns and retrieve firewood from tree canopies, digging sticks for prying bulbs and corms from the earth, and sickles to cut grasses and forbs, were frequently made of chaparral plants. The twigs and leaves of naked buckwheat (*Eriogonum nudum*) formed brushes to clear the ground under *Arctostaphylos* spp. bushes before knocking off berries, and *Cercocarpus betuloides* brushes tied with *Fremontodendron californicum* fiber were used to sweep under oaks before knocking the acorns (Barrett and Gifford 1933, Ruby Pomona, North Fork Mono, pers. comm. 2006). Two essential pieces of equipment that formed fire-making kits—drills and hearth plates—were often made of *Sambucus* spp., *Aesculus* spp., *Arctostaphylos* spp., or desert almond (*Prunus fasciculata*). A Sierra Miwuk hunter in the early 1800s carried a foxskin quiver full of arrows made from *Sambucus* spp. with *Adenostoma fasciculatum* foreshafts, each arrow fitted with the feathers of a red-tailed hawk (*Buteo jamaicensis*) or roadrunner (*Geococcyx californianus*) (Barrett and Gifford 1933; Clark 1987; Hudson n.d.). On the north coast, Yuki women and children would pry mussels off rocks with a fire-hardened stick of *Garrya elliptica* (Merriam 1967).

Indian homes were well supplied with a wide assortment of substances made from chaparral plants that aided in running the household. *Chlorogalum pomeridianum* bulbs and pitch from *Pinus sabiniana*, Coulter pine (*Pinus coulteri*), or Torrey pine (*Pinus torreyana*) were used to make adhesives, while acorns and the wet inner bark of white alder were used to make dyes for coloring basketry material, bows, and ceremonial paraphernalia. Indian men and women washed their bodies and hair with the mashed bulbs of *Chlorogalum pomeridianum* or the pulverized roots of California goosefoot (*Chenopodium californicum*) and applied the leaves of *Rhus aromatica* as an underarm deodorant (Barrett and Gifford 1933; Voegelin 1938; Melba Beecher, Mono, pers. comm. 2008). Sprigs of *Umbellularia californica* were lit on fire, purifying the air in homes (Duncan 1964).

All manner of structures such as shade ramadas, acorn granaries, dwellings, and sweathouses, were built with chaparral plants. Branches and trunks of *Quercus wislizeni* formed platforms for acorn granaries, frames of houses, and roof beams of sweat houses (Gayton 1948a, R. Pomona and N. Turner Behill, Anderson unpublished field notes 2006). Mono lean-tos were made of willow, *Umbellularia californica*, and *Cercocarpus betuloides* with roofs of *Vitis californica* (Gladys McKinney,

Fig. 4.6 Chaparral brush was used extensively for different types of structures and fencing. One example is the building of shade structures such as this brush structure designed to shield North Fork Mono women from the sun as they pounded acorn and other foods in mortar holes in a granite outcrop, North Fork, Madera County, California. Museum Number 15-6221, courtesy of the Phoebe A. Hearst Museum of Anthropology, UC Berkeley

Mono, pers. comm. 1992). The Tubatulabal made deer hunting blinds of *Quercus berberifolia* (Voegelin 1938). Shade ramadas made of *Ceanothus cuneatus*, *Adenostoma fasciculatum*, *Quercus berberifolia*, and *Umbellularia californica* were built over grinding rocks to shield women from the sun as they pounded acorns into flour (Fig. 4.6) (Voegelin 1938; Bethel et al. 1984; Theodoratus et al. 1985; Anderson unpublished field notes 2006). The Sierra Miwuk made acorn granaries that were covered with *Ceanothus* spp. (Tadd 1988). Golden-fleece (*Ericameria arborescens*) was also used to make Mono granaries and the cross-sticks were of *Cercocarpus betuloides* (Clines 1997 unpublished field notes; Anderson 2009).

Deerweed (*Acmispon glaber*) was an important thatch material for Ohlone and Chukchansi structures and Mono acorn granaries (Bocek 1984; Hudson n.d.; N. Turner Behill, Anderson unpublished field notes 2004). The Chukchansi used Spanish clover as a thatch for structures and Foothill Yokuts used brush of an unidentified species for thatching dwellings (Gayton 1948a; Hudson n.d.).

Drift fences were built of *Adenostoma fasciculatum* or unidentified brush from chaparral in the Sierra foothills. The Mono built a drift fence of brush in the shape of a "V," with each wing about a kilometer long. About 30 people would drive mountain quail (*Oreortyx pictus*) to the fence, and the birds would run along it looking for openings in which the Mono had put snares equipped with nooses made of milkweed. The Sierra Miwuk made a drift fence of *Adenostoma fasciculatum* 1 km

(0.6 mile) long in thick brush for hunting quail. Nets for hunting rabbits were 274–366 m (300–400 yards) and held in place by forked *Adenostoma fasciculatum* sticks (Barrett and Gifford 1933; Aginsky 1943).

4.4 Useful Animals of the Chaparral

The chaparral of California occupied a central place as hunting terrain for birds, reptiles, and both large and small mammals (Wallace 1978; Anderson 2009). Black-tailed deer, black bears, grizzly bears, and mountain lions were often hunted in chaparral or the grassland at its margins (Loud 1918; Anderson 2009). Hunters might stalk them with bows and arrows or set up different types of snares, deadfalls, or pitfalls to capture the animals.

California Indians were particularly fond of eating venison and hunted deer with bows and arrows or spears. In addition to meat, these animals provided many kinds of products: skins for clothing, tallow for paint, medicine for dressing wounds, antlers for glue and pressure flaking tools, the liver and blood for arrow poison, bones for fish hooks, fish spear tips, and basketry awls, sinew for bows, and brains for tanning hides (Gifford 1932; Barrett and Gifford 1933; Voegelin 1938; Gayton 1948a, b; Latta 1977). Bears hibernating in dense chaparral were roused and killed by Native American hunters to provide food, skins for blankets, capes, and sandals (Gifford 1932; Barrett and Gifford 1933; Gayton 1948a, b; Kroeber 1976; Librado 1979). Some tribes, such as the Sierra Miwuk, ate the meat of grizzly bears and used the claws as a charm in ceremonies (Barrett and Gifford 1933; Hudson n.d.). Mountain lions (*Puma concolor*) were valued for their meat, their skins were made into clothing, blankets, and quivers and their bones were used in gambling games (Barrett and Gifford 1933; Voegelin 1938; Gayton 1948a; Latta 1977).

All of these large mammals relied strongly on the chaparral and the ecotones between chaparral and oak woodland and chaparral and grassland, as habitat. Deer found many of their favored foods in the young chaparral vegetation, black bears and grizzly bears often used the chaparral for denning, and mountain lions prowled chaparral for prey.

A number of other, smaller mammals that live in the chaparral were valued by many tribes. The fur of gray fox (*Urocyon cinereoargenteus*) was used for quivers, breech clouts, cloaks, and blankets, and the meat was eaten (Hudson 1901a; Goddard 1903; Gayton 1948a, b; Merriam 1967; Latta 1977; Librado 1979). The Sierra Miwuk ate the meat of coyotes (*Canis latrans*) and used their skins for pillows, quivers, and blankets (Barrett and Gifford 1933; Voegelin 1938). The Yokiah Pomo made bags for carrying *Nicotiana* spp. out of skins of long-tailed weasels (*Mustela frenata*), which used chaparral of the Coast Ranges as habitat (Merriam 1955). Audubon's cottontail (*Sylvilagus audubonii*), brush rabbits (*Sylvilagus bachmani*), and black-tailed jack rabbits (*Lepus californicus*), all denizens of chaparral, were hunted and trapped to provide food, skins for blankets and clothing, bones for whistles, and toes and claws for charm necklaces (Barrett and Gifford 1933; Drucker

1937; Voegelin 1938; Gayton 1948a, b; Merriam 1967; Librado 1979). Foothill Yokuts, Mono, and Tubatulabal people ate the meat of dusky-footed woodrats (*Neotoma fuscipe*) (Voegelin 1938; Gayton 1948a, b) and the Foothill Yokuts ate the meat of Botta's pocket gophers (*Thomomys bottae*) (Gayton 1948a; Kroeber 1976) and California ground squirrels (*Spermophilus beecheyi*) (Gayton 1948a, b; Kroeber 1976; Theodoratus and Parsons 1980). The Mono ate the meat of American badgers (*Taxidea taxus*) and the Sierra Miwuk used badger skins for quivers in ceremonial dances (Barrett and Gifford 1933; Gayton 1948a).

A number of birds that frequent chaparral were valued by tribes for many uses. Common raven (*Corvus corax*) feathers were used to make cloaks and skirts for ceremonies, doctor's outfits, and headdresses (Hudson 1901a; Barrett and Gifford 1933; Gayton 1948a). The feathers of yellow-hammers formed the headbands of Wappo dancers (Beard 1979). Feathers from greater roadrunners (*Geococcyx californianus*) were used in Sierra Miwuk arrows and Mono headdresses (Hudson 1901b; Barrett and Gifford 1933; Gayton 1948a). The Sierra Miwuk ate the meat of mourning doves and snared large numbers of California quail (*Callipepla californica*) so that they could use their feathers to decorate basketry (Barrett and Gifford 1933; Gayton 1948a; Spier 1978). All tribes ate quail meat, the Miwuk hunted quail during migration and killed enough birds to store them over the winter (Barrett and Gifford 1933). The red-tailed hawk (*Buteo jamaicensis*) was hunted for its valuable feathers, which were placed on arrows, sewn into ceremonial clothing, and assembled into fans for fanning fires (Barrett and Gifford 1933; Gayton 1948a; Latta 1977). The Sierra Miwuk used turkey vulture (*Cathartes aura*) feathers to make head ornaments, cloaks, skirts, and dance sticks, the Foothill Yokuts used them to make fire fans for hunting and for fanning coals under foods being cooked or roasted (Barrett and Gifford 1933; Gayton 1948a). Golden eagle (*Aquila chrysaetos*) down and feathers were used in ceremonial regalia such as belts, plume sticks, and dance skirts (Barrett and Gifford 1933; Voegelin 1938; Gayton 1948b; Hudson n.d.). The feathers fanned fires and were used in war arrows, and eagle bones were used in fish spears and panpipes (Hudson 1901b; Gayton 1948a). Foothill Yokuts groups used golden eagles for a variety of purposes: the feathers were important ceremonially, the leg made a container, leg bones were fashioned into whistles, and the tallow served as a salve and binder for face paint (Gayton 1948b).

Many reptiles that frequent chaparral were useful (Basey and Sinclear 1980). The western rattlesnake (*Crotalus viridis*) provided medicine and food for the Sierra Miwuk, a rheumatism remedy for the Chumash, and arrow poison for the Tongva (Barrett and Gifford 1933; Johnston 1962; Librado 1979). Gopher snake (*Pituophis melanoleucus*) meat was eaten by both the Sierra Miwuk and Chukchansi (Barrett and Gifford 1933; Gayton 1948a). The Wintu removed the bright red bellyskin of a red-bellied snake that inhabits chaparral (probably ring-necked snake [*Diadophis punctatus*]) and wrapped it in a spiral around their bows as a decoration (Merriam 1955).

Insects that live part or all of their life-cycles in chaparral formed an important part of California Indian economies by serving as food. Some of the most important insects used for food were June or rain beetles (*Pleocoma fimbriata, P. hoppingi, and P. tulerensis*) (Fig. 4.7), tortoiseshell butterflies, sphinx moths (*Hyles lineata*), and

Fig. 4.7 From left to right: male and female rain beetles (*Pleocoma tulerensis*) an important food source of tribes in the central and southern Sierra Nevada that live their entire life cycle in chaparral. Specimens in the Bohart Museum of Entomology, UC Davis collections. Photo by Kat Anderson 2009

red-legged grasshoppers (*Melanoplus femurrubrum*) (Hovore 1979, 1998; Anderson 2005b). Insects formed an ideal supplemental food source because they are high in protein, could be dried and cached over the winter in large quantities, and were considered a delicacy (Barrett and Gifford 1933; Bean and Saubel 1972; Beard 1979).

Insects were especially attractive as a food source because many are relatively easy to capture during at least one of their life history stages: the larval and pupal life stages move slowly or not at all, and the adults of certain species concentrate in large numbers in cyclical events called "outbreaks" (Powell 1972). The adults of some species can be strategically "herded" to a destination by many people sweeping the ground with branches or by setting a fire to direct their course for capture.

Tribes also utilized chaparral insects for ceremonies. One of the most important in this regard was the ceanothus silk moth (*Hyalophora euryalus*) (Peigler 1994; Collins 2011). The inner lining of each moth's cocoon (which protects the pupal life stage) was removed and some small pebbles, sand, or seeds were placed inside the hard outer shell. Several of these cocoons were tied to stick handles to make loud rattles. These accompanied various kinds of ceremonies such as the Bear Dance and the Shamans' Contest and were used in curing illness and pain and to prevent snakebites (Dixon 1905; Gayton 1948a; Kroeber 1976). The chaparral host plants for this moth include *Ceanothus* spp., *Arctostaphylos* spp., *Cercocarpus betuloides*, *Rhamnus crocea*, and California coffee berry (*Frangula californica*) (Tuskes et al. 1996).

4.5 Native Management of Chaparral

The emergent qualities of pre-EuroAmerican chaparral plant communities that made them so important to indigenous subsistence economies and cultures, i.e., their biodiversity, productivity, and abundance, were not merely products of natural ecological processes. Native people deliberately manipulated chaparral to enhance these qualities. They did so because chaparral contained and supported so many useful species, but the converse was in a sense true as well. Chaparral supported an abundance of many useful species *because* of Native management. Through creation of landscape mosaics of chaparral and herbaceous communities, this management conferred a degree of spatial, structural, successional, and biotic diversity that exceeded what would have been the case in the absence of human intervention (Anderson and Rosenthal 2015).

As we noted earlier in the chapter, fire was by far the most important management tool used in chaparral plant communities. Its impacts were spatially widespread, multi-faceted, quickly manifested, ecologically consequential, and probably long-lasting. Further, fire had a disproportionately beneficial effect on the chaparral plant species most useful to Native people. In part, this was due to the adaptations of useful chaparral species to predictable disturbance by fire. Regular burning allowed them to enhance reproduction, reduce competition from other plants, and maintain them in a state of high growth and production postfire. Because of fire's pre-eminent status for Native land managers and because fire is the factor of greatest managerial relevance today, it will receive the most attention in the following discussion of Native management of chaparral.

It is well established by various lines of evidence that Native people did indeed burn chaparral plant communities with regularity and conscious intent. Nevertheless, it is helpful to examine this evidence in depth, because in addition to establishing the use of fire in indigenous management, it informs how, when, why, and where fire was used in California chaparral.

4.5.1 The California Landscape Was Pyrogenic at Euro-American Contact: Archaeological Evidence

We know that California has been peopled for many millennia. Archeological evidence suggests that humans began to occupy certain areas of California more than 12,000 years ago (Erlandson et al. 2011; Rosenthal and Fitzgerald 2012). Shortly after the demise of the North American mega-fauna, evidence for human occupation in California is widespread (Rosenthal and Fitzgerald 2012). The density of pre-Columbian people in California was, as is the case with contemporary populations, many times greater than that of most other parts of the West (Krech 1999). The native population in California is estimated to have been over 300,000 (Cook 1978), much greater than, for example, the 25,000 estimated to have lived in Montana (Baker 2002). Indians were widely dispersed along the California coast and throughout the coastal foothills and valleys, averaging 1–3 persons per km^2 (247 acres) Cook 1951). The regions that are today the best agricultural areas correlate with very high Indian population densities at Euro-American contact. For example, in Courtland, south of Sacramento, there were an estimated 6.4 people per km^2 at contact (Anderson and Wohlgemuth 2012). In the Santa Barbara region the Chumash achieved a density of 7.7 people per km^2 (Milliken 2006; John Johnson pers. comm. 2008). These densities are much higher than those in other areas considered densely populated in pre-historic times, such as the area where the Kongkandji lived in Australia (1.9 people per km^2) and where the Puyallup lived in Washington (also 1.9 persons per km^2). Some California tribes are believed to have achieved, at the time of Euro-American contact, the greatest population densities of any Native group in North America (Ubelaker 2006), and perhaps any hunter-gatherer group on earth (Kelly 1995).

Archaeological evidence suggests that virtually every part of the California landscape was inhabited, at least part of the year, including the Mojave Desert and alpine Sierra Nevada (Jones and Klar 2007). Permanent settlements were typically established in well-watered valleys and along upland rivers and streams with between 10 and 250 individuals (Kroeber 1976; Heizer 1978). In these politically autonomous lineages, families lived in closely spaced houses on cleared sites. Land use was intensive from the valley bottom up to the crest on each side of the drainage (Shipek 1993). Food resources varied from year-to-year in accordance with rainfall and thus families also maintained resource-extraction zones at scattered locations beyond their home valleys but typically within a half-day walk. This "home range" was jealously guarded (Beals and Hester 1974), and inter-tribal conflicts often occurred in accordance with resource fluctuations (McCorkle 1978). Many violent deaths during the late Holocene have been interpreted as resulting from population pressures (Fiedel 1992) and most accounts of warfare list resources as the primary reason for conflict (James and Graziani 1991). It appears that people were living at or near the carrying-capacity of their local environments (Baumhoff 1981).

By the early Holocene, broad-spectrum economies based on the exploitation of large and small mammals, fish, birds, shellfish, and nuts and seeds were widely established throughout California (Rosenthal and Fitzgerald 2012). One archaeological signature of a wide diet breadth is the common occurrence of plant processing tools. Handstones and millingslabs along with a more general set of pounding, chopping, and scraping tools are found at most Early Holocene sites throughout a broad range of environmental settings in western California (Rosenthal and Fitzgerald 2012).

While paleontological, fire scar, and archeological studies suggest that Indians were burning the vegetation to some extent as early as many millennia ago (Stephens et al. 2007; Klimaszewski-Patterson et al. 2015), evidence of technological innovations, greater diet breadth, and increasing complexity of social organization in the late pre-historic sequence makes archaeologists suspect that fire was increasingly used in California as a vegetation management tool, as part of an overall strategy for economic intensification (Hammett 1991; Cuthrell 2013; Lightfoot et al. 2013a, b). Fire scar studies along north-coastal California indicate late Holocene fire regimes with fire-return intervals at a frequency much greater than what would have been possible from lightning-strike ignitions alone (Stephens and Fry 2005; Stuart and Stephens 2006). Abrams and Nowacki (2008) propose that by the time Euro-Americans arrived in the New World, many of the landscapes they encountered were pyrogenic—products of both human- and lightning-caused fires.

4.5.2 *Native People Managed Chaparral with Fire: Historical and Anthropological Evidence*

Numerous early European explorers took note of the mosaic of grasslands, shrublands, and woodlands they passed through or observed from just offshore, correctly inferring a human influence over its inviting pattern. The coast north of Cape Mendocino featured, according to Vancouver, "a great variety of hills and dales, agreeably interspersed with wood-land, and clear spots, as if in a state of cultivation" (Lamb 1984). Explorers, missionaries, and early white settlers also directly witnessed Indian burning in many California landscapes, either learning from the Indians or surmising that the purpose was to clear the brush (Bolton 1927). "In all of New California from Fronteras northward," Spanish explorer José Longinos Martinez reported in 1792, "the gentiles have the custom of burning the brush" (Simpson 1938).

From the early 1900s to the early 1960s, a string of anthropologists under the tutelage of Dr. Alfred Kroeber and other faculty of the University of California, Berkeley conducted field research with many tribes, recording the widespread practice of Indian burning to reduce the brush, either in current practice or in tribal memory. For example, anthropologist Llewellyn Loud (1918) noted that the grasslands within Wiyot territory in northwestern California were kept open and free of brush with Indian burning, and wrote that the Wiyot's use of fire was of "incalculable value" in encouraging the grasslands to produce not only "vegetable products, but also...game". Anthropologist Omer Stewart's field notes from 1935 have many entries from multiple Pomo consultants who spoke of deliberately setting fires in California's Coast Ranges and valleys for such purposes as encouraging clovers, fostering wildflowers with edible bulbs and seeds, eliminating brush, enhancing grass, and driving game and grasshoppers. Anthropologist Julian Steward (1938) recorded of the Owens Valley Paiute that "The brush in basins in the hills near the winter villages was burned and *Mentzelia* and *Chenopodium* seeds were broadcast. There is no question that this practice was native."

A second wave of anthropologists, ecologists, and ethnobiological researchers conducted interviews in the late 1970s into the 2000s with Native elders who still retained specialized local forms of knowledge about burning practices. They found that in addition to having observed fires being set by their parents or grandparents, often in chaparral, some elders had even burned patches of vegetation on the sly up until recent times. With long-term ties to the ecology of the places they have lived, these indigenous consultants have made a significant contribution to regional fire histories (Knudtson 1977; Peri et al. 1982; Shipek 1981, 1993; Heffner 1984; Anderson 2005a; Lake 2007).

While the majority of the ethnographic evidence of burning in chaparral is from central and northern California, tribes in southern California used many of the same chaparral plants for the same purposes. For example, *Rhus aromatica* for basketry, *Nicotiana* spp. for ceremonies, *Eriodictyon* spp. for medicine, *Acmispon glaber* for thatch, *Salvia columbariae*, *Calandrinia ciliata*, *Dichelostemma* spp., and

Calochortus spp. for food, and *Muhlenbergia rigens* for basketry. It is difficult to imagine these fire-adapted plants yielding sufficient quantities of useful products in the absence of regular burning.

4.5.3 Native Burning Shaped the Distribution of Chaparral on the Landscape: Biogeographic Evidence

Some researchers have concluded that the practice of burning chaparral and coastal sage scrub repeatedly to affect localized type-conversion to grassland and to maintain grass/shrub mosaics was widespread (Knowles 1953; Baumhoff 1981; Anderson 1994; Anderson and Moratto 1996) (Fig. 4.8).

The contemporary pattern throughout the central and southern Coast Ranges of California is a mosaic of chaparral, sage scrub, grassland, and oak woodland. While the boundaries of these vegetations may seem timeless, ecological analyses have concluded that disturbance has played a prominent role in their formation. Specifically, some researchers believe these patterns may have been initiated by Native Americans and perpetuated by Spanish/Mexican and American settlers (Keeley 2002). In general, the vegetation patterns are consistent with the hypothesis that Native Americans utilized high fire-frequency to drive type-conversion from woody shrublands/woodlands to herbaceous associations.

Fig. 4.8 Chaparral burned at short intervals thins out chaparral and increases herbaceous growth, which increases flammability and is conducive to repeat burns before the woody vegetation has had time to recover (photo by Jon Keeley)

Fig. 4.9 Contemporary landscape mosaic of grasslands and chaparral in central Coast Ranges of California. Cooper (1922) hypothesized this pattern derived from repeated burning in the valleys and a diminishing influence of high fire frequency further into the range. Image from Google Earth

Wells (1962) examined the substrate and slope aspect characteristics associated with grassland, shrubland, and woodland vegetation in the San Luis Obispo Quadrangle of the central coast. Grasslands, all of which were dominated by non-native annuals, were well represented on at least half a dozen different substrates and these same substrata also supported abundant woody vegetation. Indeed, he commonly found grassland and shrubland or woodland juxtaposed side-by-side on the same soil type. He hypothesized that some time after humans entered California in the late Pleistocene, their intentional acceleration of fire frequencies initiated a long process of type-conversion of ligneous (woody) associations to herbaceous communities. These conclusions are supported by many other studies, as summarized in Keeley (2002).

Cooper (1922) made some profound observations about vegetation patterns in the Coast Ranges of California (e.g., Fig. 4.9) and ascribed historical Indian burning as the key determinant.

It is worth quoting him at length:

> "[M]ountains are controlled by chaparral and the plains by grasses. The character of the transition zone between the types is as follows: The first hills are as a rule entirely grass covered, though even on these, and occasionally out upon the valley-floor, are patches of chaparral. These show absolutely no correlation with altitude, slope-exposure, or soil type. Their edges are sharp and the shrubs are uniformly developed throughout. They are obviously remnants…. Penetrating farther into the mountain mass, the chaparral patches become more and more numerous…. In short, everywhere near the valleys and plains the hills are grass, while in the depths of the ranges they are covered with scrub. The larger the extent of

the mountain mass the greater is the central area of chaparral. Conversely, a small isolated area of hills, though of considerable altitude, may have none. This arrangement is so nearly universal where chaparral and grassland meet that specific examples are hardly necessary."

Cooper (1922) concluded that the mechanism driving these patterns was fire. If fire occurred with great frequency, it favored grassland at the expense of the chaparral, and yearly burning would inevitably destroy the brush completely and prevent invasion by it. "The patchy transition between grassland and chaparral is also explained," he wrote, "for fires started in the valleys, where most of the Indian population lived." These Indian-set fires, Cooper surmised, "would spread into the surrounding ranges in various directions and to varying distances. Certain areas would escape, and these would be larger and more mountain systems, where paucity of population would reduce the starting of fires to a minimum" (Cooper 1922).

"The most convincing proofs of former control of present-day grassland by chaparral," Cooper wrote, "are the frequent remnants [of chaparral]." These remnants are "sharply limited patches in the midst of other vegetation, in which *Adenostoma* is usually most prominent." Summarizing the results of using his method, which Cooper claimed "has in some cases been corroborated by historical testimony", he wrote that "it has been possible to demonstrate that dense chaparral once covered extensive areas which are now grassland."

Other early observers noted similar patterns in the northern Coast Ranges (Sterling 1904). In the southern Tehachapi Mountains, Bauer (1930) noted the following: "In the grassland the islands of shrub growth, with sharp boundaries and uniform vegetative composition, indicate a more or less remote fire or fires… It is reported that in aboriginal days the natives intentionally burned the rank herbaceous vegetation yearly."

Working in San Diego County, Dodge (1975) concluded that localities described in the diaries of early Spanish explorers as grasslands are today covered by shrublands, presumably due to the exclusion of Indian burning. Other researchers have reached the same conclusion: Native Americans maintained the southern California landscape in a mixture of grassland and shrubland through repeated burning (Aschmann 1959). Timbrook et al. (1982) came to a similar conclusion about the impact of Chumash Indian burning in the Santa Barbara region.

At the northern end of the central coastal region around the San Francisco Bay there are numerous reports of relatively recent shrub re-establishment into grasslands following the elimination of grazing and burning due to the incorporation of these areas into parks and reserves (McBride and Heady 1968). More vigorous suppression of natural fires is often invoked to explain shrub "invasion," but since natural lightning fires are rare in the region (Keeley 2005), it seems more probable that invasion is due to enhanced prevention of anthropogenic fires. It is likely that much of the grassland in this area originated with Native American burning, as this region was densely populated with over 2000 inhabitants spread across as many as 100 village sites (Cook 1951).

The idea that chaparral and other shrubland vegetation has spread into areas formerly maintained as grassland by Indian burning has gained support in recent decades with the discovery of numerous former Indian habitation sites in areas of dense chaparral. Obscured by the chaparral for more than a century, these sites have been revealed by wildfires burning through the thick chaparral. The 1995 Saddle Fire in Sequoia National Park, the 2003 Cedar Fire in Cuyamaca Rancho State Park, the 2013 Springs Fire in Point Mugu State Park, and the 2016 Scherpa Fire in Santa Barbara, along with others, have all revealed various kinds of archaeological sites, including roasting pits for yucca, milling stations, shell middens, and shell scatters (Nathan Stephenson pers. comm. 1998, Schneider 2009, Barbara Tejada pers. comm. 2016). These sites, located in all types of chaparral terrain, were certainly not established by native people when the immediate surroundings were covered with chaparral vegetation. Tribes must have burned off the chaparral to make suitable conditions for daily activities and living. However, it is likely that patches or tracts of chaparral were maintained nearby, because this vegetational diversity would have maximized the availability of important resources. Once the Native peoples left the sites and regular burning ceased, the nearby chaparral re-invaded.

How much of the vegetation physiognomy was altered by Native American burning? A starting point would be to look at the current distribution of grasslands in the 10 coastal counties from Monterey southward. Today they cover almost two million hectares (~five million acres) or 25% of the landscape. They are dominated almost entirely by non-native annuals, and less than 1% of this grassland landscape has significant patches of native perennial bunchgrass (Huenneke 1989). If we accept the conclusions of Cooper (1922), Wells (1962), and Hamilton (1997) that the origin of much of these non-native grasslands lies in anthropogenic type-conversion from shrubland/woodland, and assume minimal expansion of grassland since EuroAmerican settlement, then perhaps one quarter of the indigenous landscape was altered by fire-driven type-conversion of shrublands and chaparral.

4.5.4 What Native People Desired to Accomplish by Burning Chaparral

We know from the evidence summarized above that Native people began using fire many thousands of years ago to shape the landscape to their advantage not long after their arrival in what is now California. By the time Euro-Americans first arrived, the Natives had been burning chaparral for many centuries, if not longer, in order to intensify their exploitation of its resources. It is useful, then, to explore in greater detail exactly how burning accomplished this overall goal. We can do so by looking at specific resources—the useful chaparral plants and animals described earlier in this chapter—and examining how burning made them more useful by increasing their quantity or enhancing their quality.

Tribes in California were very much aware of the different things fire could do, depending on where, how, and when it was used, and they used this knowledge to achieve specific objectives. The ethnographic and historical literature is full of examples of Indians describing their reasons for burning. For instance, how burning a particular hillside every few years was necessary for maintaining the health of a patch of *Cercis occidentalis*, for example, or that the *Salvia columbariae* would decline if that valley over there was not burned every few years or so. Such evidence indicates that over the many centuries during which they exploited the abundant biotic resources of the chaparral, the indigenous people learned a great deal about the biological needs of the plant and animal species on which they depended, enough to allow them to manage each species with burning and other methods so that it would be available and flourish.

If we put the goals of burning that relate to management of plant and animal resources together with those connected to more general goals, such as "to keep down the brush," we come up with seven distinct categories (Anderson and Rosenthal 2015). In using fire, Native people sought to: (1) enhance the growth and production of plants with edible above-ground parts (seeds, greens, and berries), (2) enhance the growth and production of plants with edible below-ground parts (corms, bulbs, tubers, and rhizomes), (3) promote the growth of basket-weaving and cordage materials, (4) maintain in optimal condition the habitats used frequently by game birds and mammals, (5) control pathogens and insect predators of valued plants, (6) increase water resources, and (7) keep areas open to improve accessibility and reduce the chance of catastrophic fire.

Although a single fire might achieve several different objectives at once, and some of the objectives were overlapping in the sense that realizing one necessarily meant realizing another, the evidence indicates that Native people had different and distinct desirable outcomes in mind when they set fires. Below, each of these seven objectives is discussed in turn.

1. Enhance the growth and production of plants with edible above-ground parts (seeds, greens, and berries)

   Certain chaparral lands were cleared by burning the shrubs right down to the ground. With frequent enough burning these areas were type-converted and managed for patches of herbaceous plants used for foods and medicines. For example, many of the edible seed resources used by Native Americans were annuals that were abundant for a short period after fire. These included blow-wives (*Achyrachaena mollis*), *Salvia columbariae*, farewell-to-springs (*Clarkia biloba*, *C. purpurea* subsp. *viminea*, *C. unguiculata*, *C. rhomboidea*, *C. williamsonii*), *Calandrinia ciliata*, and tarweeds (*Centromadia fitchii*, *Madia elegans*, *M. gracilis*, *M. sativa*).
2. Enhance the growth and production of plants with edible below-ground parts (corms, bulbs, tubers, and rhizomes)

   As good carbohydrate sources that could be stored for long periods of time, "root foods" such as *Dichelostemma* spp., *Calochortus* spp., wild onions (*Allium* spp.), *Chlorogalum* spp., sanicles (*Sanicula* spp.), and *Perideridia* spp., were critical food resources. Burning of the chaparral served these plants well for the same reasons it benefited annuals with edible seeds: it created the open habitat they needed, reduced competition, released nutrients, and encouraged vigorous growth (Anderson and Lake 2016). In addition to using fire where these plants grew, the Native people also harvested the under-ground parts in a way that actively promoted the plants' asexual reproduction and enlarged their populations. The roots or bulbs were dug up with sticks, which loosened the soil. Only the largest and most mature parts were removed for consumption, immature plants, parts of roots and rhizomes, and the asexual propagules of bulbs were left in the soil and often replanted with care, sometimes outside of the established population so that the patch would grow in size. Small trees, shrubs, and other plants that might compete with the food plants were pulled up and removed.

3. Promote the growth of basket-weaving and cordage materials

   Islands of chaparral within grassland and the diverse ecotones between chaparral tracts and grasslands contained *Cercis occidentalis*, *Rhus aromatica*, *Ceanothus integerimus*, *Fremontodendron californicum*, *Prunus virginiana var. demissa*, *Prunus emarginata*, and other shrubs that were managed specifically as sources of wattling material. Fire and pruning encouraged the plants to produce straight and flexible shoot growth, which was used for basketry, cordage, drying racks, fish weirs, housing materials, tools, household utensils, digging sticks, and many other items (Anderson and Rosenthal 2015). Patches of *Muhlenbergia rigens* in chaparral were burned in the fall or winter every 2–5 years to remove dead material and increase flower stalks for the foundations of coiled baskets (Anderson 1996).

4. Maintain in optimal condition the habitats used frequently by game birds and mammals

   Using fire to benefit game animals was always a major part of Native land management, because these animals were so important to Native economies and cultures. Unlike the management of plant-based resources, however, the management of deer, bear, mountain lion, rabbits, quail, mourning doves, and other animals was not direct. Native people aimed at keeping the populations of these animals at optimal sizes and in optimal health by managing the habitats and vegetation on which the animals depended for food, bedding, and cover and by keeping the overall landscape open enough to facilitate their ability to move and migrate. Fire, of course, was the most powerful tool for doing this.

   When Indians set fires to the chaparral lands to manage habitat for game birds and mammals, they actually accomplished three distinct but overlapping goals at the same time. They wanted to make hunting easier by reducing the vegetation that might hide animals or deter the pursuit of wounded prey, they wanted to make the habitats more attractive to the game animals so that they would congregate there more frequently, and they wanted to maximize the amount of food available to the animals so that their populations could be as large as possible.

   The hypothesis that Native Americans utilized fire to open up dense shrublands to increase deer and other animal resources is well supported by contemporary game management practices (Lawrence 1966). We know from studies of deer management that undisturbed stands of chaparral are nearly impenetrable and the new growth in older stands is commonly produced out of reach of deer. Indeed, it was this observation by wildlife managers that led to the widely popular myth that old chaparral becomes senescent (Keeley 1992). Immediately after fire the food available for deer from shrubs increases 40 fold or more (Hendricks 1968) and the majority of species comprising the temporary postfire flora are also important food resources for deer (Cronemiller and Bartholomew 1950). Herds increase several-fold in postfire environments, although the effect is short-lived, as the vegetation closes in after about 5 years (Biswell 1961). Repeated burning produces a mosaic of grassland and shrub patches, which is ideal habitat, and results in a permanent three- to five-fold increase in deer herds (Taber 1956). Other important resources such as California quail, brush rabbits, and mourning doves increase several fold in open brush and grassland mosaics compared to undisturbed chaparral (Biswell et al. 1952). Jack rabbits, which completely avoid dense shrublands, will expand into these the chaparral-grassland mosaics created by burning. Opening up these shrublands would have been crucial to Native American exploitation of these animal resources because approaching prey undetected would have been unlikely in undisturbed shrublands, and lack of maneuverability would have prevented the use of bows and arrows or the boomerang-like throwing stick (McCawley 1996).

5. Control pathogens and insect predators of valued plants

   Many culturally significant plants that occur in chaparral are susceptible to insects and diseases and if attacked are rendered useless for basketry, medicines, cordage and other items (Sinclair et al. 1987). Although native people were not aware of pathogenic microorganisms as such, and may not have completely understood the life cycles of insect "pests," they did know that fire was useful for limiting the damage to valued plants caused by insects and diseases. The ethnographic literature contains many examples of elders claiming that fire or smoke was good for "getting rid of" pests. Modern research corroborates such claims. For example, burning is thought to be an effective control for a pathogen called *Passalora* that blackens the leaves, pods, and stems of showy milkweed, an important plant for cordage, food, and medicine, and for the native pathogen called black knot (*Apiosporina morbosa*) on *Prunus virginiana* var. *demissa*, a plant used for food and basketry (David Rizzo, pers. comm. 2013).

6. Increase water resources

   Vegetation is known to affect the flow, quantity, and recharge of groundwater through various mechanisms. Mature chaparral vegetation, with its deep roots and large leaf surface area, appears to remove more water from the soil through transpirational losses than the types of herbaceous vegetation that would have replaced chaparral in the presence of Indian burning. Native people observed that springs and seeps were more productive when there was less brush. When patches of chaparral were burned off, the water table rose and water sources flowed more strongly or reappeared. This was an important effect of fire because the availability of water affected landscape scale interactions, including where human villages might be located and the migration patterns of large mammals.

   Ethnographic research has found that Central Sierra Miwuk elders remember the connection between available water and burning of brush. "The Indians [Central Sierra Miwuk] used to keep the brush burned off," Miwuk elder Louis Williams told anthropologist Gary Maniery in 1980, "thus making the springs useful and productive" (Maniery 1987). Similarly, North Fork Mono elder Rosalie Bethel (pers. comm. 1991) remembered that "burning brush helped to save water." Experimental research on small chaparral watersheds in Arizona and California show a marked increase in the flow of springs and streams when brush is converted to grassland with controlled burns (Biswell and Schultz 1958; DeBano 1983; Biswell 1989).

7. Keep areas open to improve accessibility and reduce the chance of catastrophic fires

Chaparral was so important to Native people for hunting, harvesting plant food, gathering basketry and cordage materials, and so on that they often chose to live in close proximity to it. Maintaining this kind of close relationship required management focused on the spatial attributes of the vegetation. Native people needed to be able to move through chaparral to hunt and to access its resources, and areas of dense brush could not be located too close to villages because of the danger of out-of-control wildfire. Therefore, chaparral was often burned for the general purposes of keeping it open, maintaining a network of trails, and eliminating the brush that might carry a catastrophic fire. In the ethnographic literature, there are many examples of native informants speaking about burning for the purpose of keeping the landscape open and preventing fires that might burn trees or villages.

North Fork Mono elder Rosalie Bethel explained this rationale for burning to one of the authors in 1991:

> "I'm going by what the elders told me happened in the 1800s. Burning was in the fall of the year when the plants were all dried up when it was going to rain. They'd burn areas when they would see it's in need. If the brush was too high and too brushy it gets out of control. If the shrubs got two to four feet in height it would be time to burn. They'd burn every two years. Both men and women would set the fires. The flames wouldn't get very high. It wouldn't burn the trees, only the shrubs. They burned around the camping grounds where they lived and around where they gathered. They also cleared pathways between camps. They burned in the valleys and foothills" (pers. comm. 1991).

Based on his ethnographic work, Duncan (1964) described what the northern Sierra Nevada foothills may have looked like before the Gold Rush:

> "There was considerably less chaparral and underbrush, due to the Maidu practice of burning off the areas near where they lived each fall and winter. They preferred an open, grassy, oak savannah habitat for several reasons. Open country is much easier to travel in than country with thick underbrush as it is easier to find game and harder for enemies to sneak up on a camp. More bulbs and greens grow in such an environment, and it is easier to gather acorns on bare ground."

4.6 Impacts of Native Use of Fire on Chaparral

While there were a multitude of motivations for Native people to use fire and plenty of evidence for its use, there is some debate on the impact of Indian fire management practices on the distribution of chaparral vegetation. Stewart (1956) was convinced that fires set by Indians were of the utmost importance in determining many landscape patterns throughout the western hemisphere, and many researchers today share his perspective and assert its relevance for understanding chaparral in

California. Indeed, there are reasons to believe that these shifts in fire regimes initiated by the earliest inhabitants in North America were transformative events that greatly altered landscape patterns (Pinter et al. 2011). Representing another side of the debate, Vale (2000) contended that lightning-ignited fires were so frequent in western North America that whatever additional burning was done by Native peoples altered vegetation in only limited areas. This view is supported by Jones and Hadick (2016). What this difference in perspective tells us is that some researchers remain unconvinced that Native people, as opposed to the "natural" force of lightning strikes, could have controlled the fire regime in chaparral.

So let us review what is known about human-versus lightning-caused fire in California. In the absence of human influence, the natural fire regime in California varied both spatially and temporally. The Coast Ranges were ignition-limited and experienced century-long fire-free periods, in contrast to interior montane landscapes where annual lightning ignitions generated more frequent and more predictable fires (Keeley and Safford 2016) (Fig. 4.10). In the central coastal region, modeling studies conclude that fire-return intervals from just natural lightning ignitions were substantially longer before Native Americans arrived on the scene compared to after (Greenlee and Langenheim 1990). Fossil pollen from the central Coast Ranges has also been interpreted as providing evidence for burning by Indians (Mensing 1998; Anderson et al. 2015). Other circumstantial evidence of Native American influence is from charcoal deposition studies that show that the frequency of large fires in the front range of the Santa Ynez Mountains of Santa Barbara County prior to EuroAmerican colonization was similar to the contemporary period (Mensing et al. 1999). Today humans are responsible for the vast majority of ignitions in this region (Keeley and Syphard 2018), suggesting that Native Americans likewise were a dominant source of ignition in pre-history.

So, on the chaparral dominated landscapes of central and southern California at least, natural ignitions were few and far between. Further, much of this was a densely populated landscape, far denser than the average Native American density

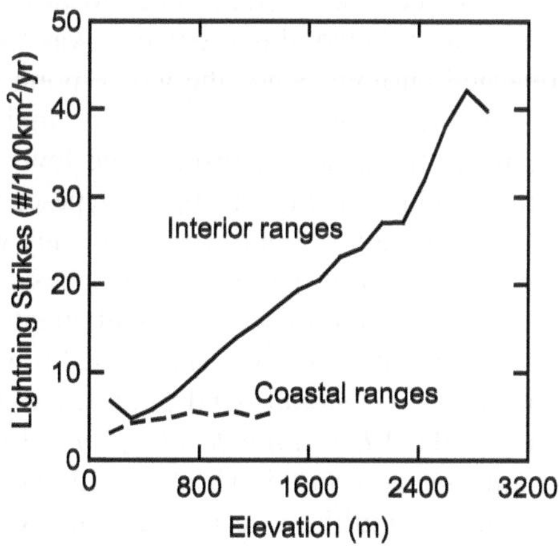

Fig. 4.10 Distribution of lightning strikes in interior versus coastal ranges of California illustrating the ignition-limited characteristic of this landscape dominated by Native peoples (from Keeley 2006)

across western North America. It is very unlikely that California tribes relied solely upon natural fires to generate the postfire herbaceous resources they needed to support this dense population because natural fires in the Coast Ranges occur at long intervals, perhaps only once or twice a century (Keeley and Syphard 2018). Thus, we are confident in arguing that Indian burning significantly decreased the fire-return interval relative to the "natural" background interval, altering chaparral shrublands and associated vegetation. At the same time, we acknowledge that impacts may have varied locally depending on population density, topography, elevation, species composition, tribal culture, access to non-chaparral based food sources like inter-tidal invertebrates, and other factors.

So exactly how did Indian burning alter the chaparral dominated landscapes of California and affect the distribution of chaparral vegetation? In considering this question, it is important to keep in mind that Native people wielded fire with intention, in order to realize specific objectives, and guided their use of this powerful tool with in-depth knowledge about how fire affected plant growth. The many objectives that Native people sought to realize from burning chaparral dominated landscapes indicate that what they wanted to achieve, in terms of overall landscape physiognomy, was a mosaic of open, herbaceous dominated plant communities interspersed with large and abundant patches of woody chaparral vegetation. This type of heterogeneous landscape, with its structural and ecological diversity and large amount of ecotonal boundary, would have maximized productivity and biotic diversity and satisfied native requirements for accessibility and habitability at the same time. This could be achieved only with the skilled use of fire.

Repeated burning by Indians would maintain the herbaceous elements in the area and diminish the capacity of the woody cover to close in, thus placing the vegetation on a trajectory that favored persistence of a strong herbaceous component. Continued burning would produce a new quasi-equilibrium, where shrub re-colonization was slowed by weak seed dispersal or poor seedling establishment in grasslands (Keeley and Brennan 2012). As a consequence, once the stand of chaparral was opened up, less frequent burning would have been needed to preclude shrub recolonization. Thus, Horne's (1981) contention that "annual burning of shrublands" did not occur is almost certainly correct: once localized type-conversion to herbaceous associations was effected, this vegetation was likely stable for a decade or longer without repeated burning. Since the whole point was to create a vegetational mosaic containing significant woody elements, Indian land managers would have wanted to keep the frequency of disturbance low enough to avoid eliminating shrubs altogether and producing a complete type-conversion to grassland.

Careful calibration of fire frequency would have allowed native people to create the optimal mixture of herbaceous and woody elements *and* control the species composition of the woody elements. Frequent fires (e.g., more than one per decade) would extirpate chaparral shrubs that recruit entirely by seed (e.g., many species of *Ceanothus* and *Arctostaphylos*), and thin out facultatively seeding shrubs like *Adenostoma fasciculatum* (Keeley and Syphard 2018). Under such fire frequency, resprouting shrubs would persist as islands in a matrix of herbaceous vegetation and these resprouters include some important Native American food and basketry resources: *Prunus ilicifolia*, *Heteromeles arbutifolia*, *Sambucus* spp., *Ribes* spp., *Rhus aromatica*, other species of *Arctostaphylos*, and *Quercus berberifolia*. There is also value added to this scenario in that these resources are far more accessible when present in isolated island remnants, plus fruit production increases following such stand thinning due to reduced plant competition for soil water resources (Keeley and Keeley 1988). Burning removed dead biomass and encouraged maximal growth of fruit-bearing canes and branches (Anderson and Rosenthal 2015).

The widespread existence of shrub islands and vegetational mosaics is substantiated in Pomo testimony related to Omer Stewart (unpublished field notes, 1935): "When John was a boy the hills were bare from brush—all bald. The brush was much less thick and *Arctostaphylos* spp. was limited to a few spots. The areas where productive brush was located was protected from yearly fires which burned grass and other brush. The grass fires did not bother the big trees. Small trees were burned at time in the hills. This was used for wood. Each fall the whole country was burned. This made the grass grow better."

If humans had not found a way to migrate from Asia to North America and California had remained unpeopled, the Euro-Americans sailing along the coast in the late sixteenth and early seventeenth centuries would have glimpsed a landscape very different from what they actually encountered. Instead of a land "agreeably interspersed with wood-land, and clear spots, as if in a state of cultivation," they would have looked upon hillsides covered in brush, uninviting and difficult to penetrate. Making trails through this dense shrubland, they would have found a less diverse flora and probably less wildlife. Fortunately for these explorers, California was peopled, and the people had worked for perhaps millennia shaping the landscape, especially its chaparral and allied vegetation, into something more productive and diverse than nature alone could accomplish. Although centuries have passed since Indians actively managed the chaparral and other plant communities with fire, we still see the impacts of that early land management.

4.7 A Future for Indigenous Burning?

There is increasing interest among resource managers in incorporating traditional ecological knowledge into land management practices, and there are notable cases where it has played a significant role in understanding contemporary issues. As just one example, the very lethal 1993 Four Corners "Navajo Flu" outbreak was a total mystery to scientists until local medicine men reported that it had been observed several times in the twentieth century and was associated with high rainfall followed by a population explosion of mice. Here was a case where the combination of traditional ecological knowledge and contemporary scientific investigation had a synergistic effect on bringing to light the very serious health issue known as hanta virus, a lethal virus transmitted through mice feces. Undoubtedly there is much to be learned from a fuller understanding of traditional ecological knowledge.

With respect to fire, there is a growing interest in restoring indigenous peoples' fire management practices to forests, savannas, and other landscapes throughout the world (Trauernicht et al. 2015). In California there is a strong case to be made for this in many forested landscapes in the Sierra Nevada and northern California where fire suppression has greatly altered natural fire regimes (Keeley and Safford 2016). On these landscapes not only have traditional burning practices been eliminated but natural lightning-ignited fires have been suppressed (though not always successfully). As a consequence abnormal accumulations of living and dead fuels have made these ecosystems extremely vulnerable to high-intensity crown fires that cause high rates of tree mortality.

Western scientists and resource managers are increasingly recognizing that indigenous burning in various vegetation types contributed not only to community livelihood, but also to many conservation values such as landscape heterogeneity and resiliency (USDA Forest Service 2012). The outcomes that indigenous people were aiming for when burning chaparral, such as increased water flow, enhanced wildlife habitat, and the maintenance of many kinds of flowering plants and animals, are congruent and dovetail with the values that public land agencies, non-profit organizations, and private landowners wish to preserve and enhance through wildland management. As a result, Indian burning is being emulated by some non-Indian land managers. In Whiskeytown National Recreation Area, in chaparral areas most likely traditionally managed with fire by Wintu cultural groups, prescribed burning and brush thinning favors open diverse understories, stimulating the germination and growth of long dormant bulb and seedbanks (Jennifer Gibson, pers. comm. 2016).

For over 20 years, the staff biologists of Occidental Arts and Ecology Center (OAEC), a 28 ha (70 acre) Wildland Preserve in western Sonoma County, have been stewarding 2.8 ha (7 acres) of coastal prairie using guidelines derived from horticultural and traditional practices (Dolman 2016). They are saving the seeds of native bulbs, wildflowers, and grasses, reintroducing frequent low-intensity fire to keep coyote brush and other chaparral species from encroaching, and broadcasting the collected seeds into recently burned areas before major winter rains. The results are markedly heightened patches of native perennial bunchgrasses such as *Elymus glaucus*, California oatgrass (*Danthonia californica*), and purple needlegrass (*Stipa pulchra*), and wildflowers that include indigenous food sources such as multiple species of *Perideridia* spp., *Dichelostemma* spp., *Triteleia* spp., *Brodiaea* spp., and yellow mariposa lily (*Calochortus luteus*) (Dolman 2016).

Fire-based management informed by knowledge of pre-historic practices is also being carried out by Native people themselves. Some tribal elders and indigenous resource managers still retain detailed knowledge of how, why, and when to apply fire to the land. Members of the Amah Mutsun Land Trust, a tribally owned trust, in partnership with Pinnacles National Park, are bringing back onto their traditional lands the practices of burning of *Muhlenbergia rigens* to heighten flower stalk production and burning to keep chaparral from engulfing bunchgrass colonies. Don Hankins, Plains Miwuk, with other Konkow practitioners, and staff and students have been burning since 2010 in oak-chaparral communities in Big Chico Creek Ecological Reserve in Butte County to increase native grass dominance and culturally significant plants, benefit oaks, maintain landscape patchiness and representation of various ecological states, similar to what might have been done under the traditional management of the Konkow (Don Hankins, pers. comm. 2016). In northern California, the Karuk tribe and Orleans/Somes Bar Fire Safe Council have conducted fuel treatments over the past 15 years on the vegetation of Offeld Mountain, which includes chaparral, setting the stage for bringing back the ceremonial use of fire on the mountain. The US Forest Service is working with the Karuk and local communities around Happy Camp to restore controlled burns to high-elevation ridge systems to create landscape scale fuelbreaks.

Southern California chaparral represents a very different situation and one in which restoring traditional fire practices on any significant scale would not improve fire hazard and instead would likely cause ecological damage. The primary reason is that indigenous burning in the region has been replaced by even more anthropogenic burning than Indians ever did.

Some would argue that the problem with today's large catastrophic fires in southern California is the result of not using traditional fire management practices, which would prevent fires from spreading due to a mosaic of different aged fuels. The primary basis for this belief is the idea that large fires in this region are the result of highly successful fire suppression that has resulted in abnormal fuel accumulation (Minnich 1983). However, it is apparent that over the last century on this landscape fire suppression has failed to exclude fire and the region has had an abnormally high fire frequency (Safford and van de Water 2014). So much so that regionally there is no evidence of anomalously high fuel accumulation and that fuel age and continuity have very little control on fire size (Keeley et al. 1999). Rather it has been shown that large fire events are the result of extreme droughts, high temperatures, and high winds (Keeley and Zedler 2009).

Some advocates of restoring Indian burning maintain that we should restore those early landscapes that type-converted shrublands to a mosaic of shrubs, grass, and herbs because of its inherent cultural value. However, these cultural landscapes were comprised of native shrubs and native herbs, but today the herbaceous flora is dominated by non-native species. Repeated burning of chaparral is invaded by these non-native species, greatly diminishing the resource value, and is contrary to conservation goals of maintaining native vegetation. In addition, such type-conversion increases the highly flammable flashy fuels and results in increased ignitions and fire spread into more hazardous chaparral fuels (see Chap. 12).

References

Aginsky, B. W. 1943. Cultural element distributions. XXIV: Central Sierra. University of California Anthropological Records 8(4).

Abrams, M. D., and G. J. Nowacki. 2008. Native Americans as active and passive promoters of mast and fruit trees in the eastern USA. The Holocene 18:1123-1137.

Anderson, M. K. 1988. Southern Sierra Miwok plant resource use and management of the Yosemite region. Thesis. Wildland Resource Science, Department of Forestry and Resource Management, University of California, Berkeley, California, USA.

Anderson, M. K. 1994. Prehistoric anthropogenic wildland burning by hunter-gatherer societies in the temperate regions: a net source, sink, or neutral to the global carbon budget? Chemosphere 29:913-934.
Anderson, M. K. 1996. The ethnobotany of deer grass, *Muhlenbergia rigens* (Poaceae): its uses and fire management by California Indian tribes. Economic Botany 50:409-22.
Anderson, M. K. 2005a. Tending the wild: Native American knowledge and the management of California's natural resources. University of California Press, Berkeley, California, USA.
Anderson, M. K. 2005b. Identification of the fungi, insects, and plant parts of the Charles P. Wilcomb Collection. Final unpublished report to the Oakland Museum, Oakland, California, USA.
Anderson, M. K. 2009. Historical Native American resource use, harvesting, and management of California's oak communities. Final unpublished report to the University of California Integrated Hardwood Range Management Program, Berkeley, California, USA.
Anderson, M.K. 2016. The original medicinal plant gatherers and conservationists, Journal of Medicinal Plant Conservation Spring:26-29.
Anderson, M. K., J. Effenberger, D. Joley, and D. J. Lionakis Meyers. 2012. Edible seeds and grains of California Tribes and the Klamath Tribe of Oregon. Unpublished report on file in the Phoebe Apperson Hearst Museum of Anthropology collections, University of California Berkeley, USA.
Anderson, R. S., A. Ejarque, J. Rice, S. J. Smith, C. G. Lebow. 2015. Historic and Holocene environmental change in the San Antonio Creek Basin, mid-coastal California. Quaternary Research 83:273-286.
Anderson, M. K., and J. House. 2012. California's ancient cornucopia: a story of abundance, diversity, and Indigenous stewardship. Wise Traditions 13:30-37.
Anderson, M. K., and F. K. Lake. 2016. Beauty, bounty, and biodiversity: the story of California Indians' relationship with edible native geophytes. Special Issue on California Geophytes. Fremontia: Journal of the California Native Plant Society 44:44-51.
Anderson, M. K., and M. J. Moratto. 1996. Native American land-use practices and ecological impacts. Pages 187-206 *in* SNEP Team, editors. Sierra Nevada ecosystem project: Final report to Congress, Vol. 2, Assessments and scientific basis for management options. University of California, Centers for Water and Wildland Resources, Davis, California, USA.
Anderson, M. K., and J. Rosenthal. 2015. An ethnobiological approach to reconstructing indigenous fire regimes in the foothill chaparral of the western Sierra Nevada. Journal of Ethnobiology 35:4-36.
Anderson, M. K., and E. Wohlgemuth. 2012. California Indian proto-agriculture: its characterization and legacy. Pages 190-224 *in* P. Gepts, T. R. Famula, R. L. Bettinger, S. B. Brush, A. B. Damania, P. E. McGuire, and C. O. Qualset, editors. Biodiversity in agriculture: Domestication, evolution, and sustainability. Cambridge University Press, Cambridge, UK.
Aschmann, H. 1959. The evolution of a wild landscape and its persistence in southern California. Annals of the Association of American Geographers 49:34-56.
Baker, W. L. 2002. Indians and fire in the US Rocky Mountains: the wilderness hypothesis renewed. Pages 41-76 *in* T. R. Vale, editor. Fire, native peoples, and the natural landscape. Island Press, Covelo, California, USA.
Baldwin, B. G., D. H. Goldman, D. J. Keil, R. Patterson, T. J. Rosatti, and D. H. Wilken. 2012. The Jepson manual: vascular plants of California. Second edition. University of California Press, Berkeley, California, USA.
Barrett, S. A., and E. W. Gifford. 1933. Miwok material culture. Bulletin of the Public Museum of the City of Milwaukee 2:117-376.
Barrows, D. P. 1967. Ethnobotany of the Coahuilla Indians. Malki Museum Press, Morongo Indian Reservation, Banning, California, USA.
Basey, H. E., and D. A. Sinclear. 1980. Amphibians and reptiles. Pages 13–74 *in* J. Verner and A. S. Boss, technical coordinators. California wildlife and their habitats: western Sierra Nevada. General Technical Report PSW-GTR-37. USDA Forest Service, Pacific Southwest Research Station, Berkeley, California, USA.

Bates, C. D. 1982. Coiled basketry of the Sierra Miwok. *San Diego Museum Papers* No. 15. San Diego Museum of Man, San Diego, California, USA.
Bauer, H. L. 1930. On the flora of the Tehachapi Mountains, California. Bulletin of the Southern California Academy of Sciences 29:96-99.
Baumhoff, M. A. 1981. The carrying capacity of huntergatherers. Pages 77-87 *in* S. Koyama and D. H. Thomas, editors. Affluent foragers. Pacific coasts east and west. Senri Ethnological Studies no. 9. National Museum of Ethnology, Senri Expo Park, Suita, Osaka, Japan.
Beals, R. L. 1933. Ethnology of the Nisenan. University of California Publications in American Archaeology and Ethnology 31:335-414.
Beals, R. L., and J. A. Hester, Jr. 1974. California Indians I. Indian land use and occupancy in California, Vol. I. Garland Publishing Inc., New York, New York, USA.
Bean, L. J., and K. S. Saubel. 1972. Temalpakh: Cahuilla Indian knowledge and usage of plants., Malki Museum Press, Morongo Indian ReservationBanning, California, USA
Beard, Y. S. 1979. The Wappo: a report. Malki Museum Press, Morongo Indian Reservation. Banning, California, USA.
Beemer, E. 1980. My Luiseño neighbors: excerpts from a journal kept in Pauma Valley Northern San Diego County, 1934 to 1974. Acoma Books, Ramona, California, USA.
Bethel, R., P. V. Kroskrity, C. Loether, and G. A. Reinhardt. 1984. A practical dictionary of Western Mono, The American Indian Studies Center, University of California, Los Angeles, California, USA.
Biswell, H. H. 1961. Manipulation of chamise brush for deer range improvement. California Fish and Game 47:125-144.
Biswell, H. H. 1989. Prescribed burning in California wildlands vegetation management. University of California Press, Berkeley, California, USA.
Biswell, H. H., and A. M. Schultz. 1958. Effects of vegetation removal on spring flow. California Fish and Game 44:211-230.
Biswell, H. H., R. D. Taber. D. W. Hedrick, and A. M. Schultz. 1952. Management of chamise brushlands for game in the north coast range of California. California Fish and Game 38:453-484.
Bocek, B. R. 1984. Ethnobotany of Costanoan Indians, California, based on collections by John P. Harrington. Economic Botany 38:240-255.
Bolton, H. E. 1927. Fray Juan Crespi, missionary explorer on the Pacific Coast, 1769-1774. University of California Press, Berkeley, California, USA.
Brubaker, F. 1926. Plants used by Yosemite Indians. Yosemite Nature Notes 5:73-79.
Bunnell, L. H. 1980. Discovery of the Yosemite, and the Indian War of 1851. F. Revell, Chicago, Illinois, USA.
Castello, E. D. 1978. The Impact of Euro-American exploration and settlement. Pages 99-127 *in* R. F. Heizer, editor. Handbook of North American Indians. Vol 8: California, Smithsonian Institution, Washington D.C., USA.
Chesnut, V. K. 1974. Plants used by the Indians of Mendocino County California. Mendocino. Reprint of 1902 edition. Pages 295-422 *in* Contributions from the US National Herbarium Government Printing Office 1900-1902, volume 7. Mendocino County Historical Society, Fort Bragg, California, USA.
Clark, G. 1987. Indians of the Yosemite Valley and vicinity: their history, customs and traditions. Reprint of 1904 edition. Diablo Books, Walnut Creek, California, USA.
Collier, M. E. T., and S. B. Thalman. 1991. Interview with Tom Smith and Maria Copy: Isabel Kelly's ethnographic notes on the Coast Miwok Indians of Marin and Southern Sonoma Counties, California. Mapom Occasional Papers Number 6, Miwok Archeological Preserve of Marin, San Rafael, California, USA.
Collins, M. M. 2011. Cocoons: reflections on their unappreciated natural history. News of the Lepidopterists' Society 53:39-43.
Cook, S. F. 1951. The aboriginal population of upper California. Pages 66-83 *in* R. F. Heizer and M. A. Whipple, editors. The California Indians: a source book. University of California Press, Los Angeles, California, USA.

Cook, S. F. 1978. Historical demography. Pages 91-98 *in* R. F. Heizer, editor. Handbook of North American Indians, Vol. 8: California. Smithsonian Institution, Washington, D.C., USA.

Cooper, W. S. 1922. The broad-sclerophyll vegetation of California. An ecological study of the chaparral and its related communities. Publication No. 319. Carnegie Institution of Washington, Washington D.C., USA.

Costo, R., and J. H. Costo. 1987. The missions of California: a legacy of genocide. Indian Historian Press. San Francisco, California, USA.

Cronemiller, F. P., and P. S. Bartholomew. 1950. The California mule deer in chaparral forests. California Fish and Game 36:343-365.

Cuthrell, R. Q. 2013. Archaeobotanical evidence for Indigenous burning practices and foodways at CA-SMA-113. California Archaeology 5:265-290.

DeBano, L. F. 1983. Multiresource management of mixed conifer and chaparral watersheds. Forestry Sciences Laboratory, Arizona State University, Tempe, Arizona, USA.

Dixon, R. B. 1905. The Northern Maidu. Bulletin of the American Museum of Natural History 17:119-346.

Dodge, J. M. 1975. Vegetational changes associated with land use and fire history in San Diego County. Dissertation. University of California, Riverside, California, USA.

Dolman, B. 2016. Mending the wild at the Occidental Arts and Ecology Center. Fremontia 44:52-56.

Drucker, P. 1937. The Tolowa and their southwest Oregon Kin. University of California Publications in American Archeology and Ethnology 36:221-300.

Du Bois, C. 1935. Wintu ethnography. University of California Publications in American Archeology and Ethnology 36:1-148.

Duncan III, J. W. 1964. Maidu ethnobotany. Unpublished Thesis. Department of Anthropology, California State University, Sacramento, California, USA.

Eastwood, A. n.d. Some plants of Tulare County, with the Indian names and uses. Unpublished manuscript at the Grace Hudson Museum and Sun House, Ukiah, California, USA.

Erlandson, J. M., T. Rick., T. J. Braje, M. Casperson, B. Culleton, B. Fulfrost, T. Garcia, D. A. Guthrie, N. Jew, D. J. Kennett, M. L. Moss, L. Reeder, C. Skinner, J. Watts, L. Willis. 2011. Paleoindian seafaring, maritime technologies, and coastal foraging on California's Channel Islands. Science 311:1181-1185.

Fiedel, S. J. 1992. Prehistory of the Americas, Second edition.Cambridge University Press, Cambridge, UK.

Fowler, C. 1986. Subsistence. Pages 64-97 *in* W. L. D'Azevedo, editor. Handbook of North American Indians. Vol. 11: Great Basin. Smithsonian Institution, Washington D.C., USA.

Gayton, A. H. 1948a. Yokuts and Western Mono ethnography II: Northern Foothill Yokuts and Western Mono. Anthropological Records 10:143-302.

Gayton, A. H. 1948b. Yokuts and Western Mono ethnography I: Tulare Lake, Southern Valley, and Central Foothill Yokuts, Anthropological Records 10:1-140.

Gifford, E. W. 1932. The Northfork Mono. University of California Publications in American Archeology and Ethnology 31:15-65.

Goddard, P. E. 1903. Life and culture of the Hupa. University of California Publications in American Archaeology and Ethnology 1:1-88.

Goode, R. W. 1992. Cultural traditions endangered. Unpublished report on file at the US Forest Service, Sierra National Forest, Sierra, California, USA.

Goodrich, J., C. Lawson, and V. P. Lawson. 1980. Kashaya Pomo plants. American Indian Monograph Series, No. 2. American Indian Studies Center, Los Angeles, California, USA.

Greenlee, J. M., and J. H. Langenheim. 1990. Historic fire regimes and their relation to vegetation patterns in the Monterey Bay area of California. American Midland Naturalist 124:239-253.

Hamilton, J. G. 1997. Changing perceptions of pre-European grasslands in California. Madroño 44:311-333.

Hammett, J. E. 1991. The ecology of sedentary societies without agriculture: Paleoethnobotanical indicators from Native California. Dissertation. University of North Carolina, Chapel Hill, North Carolina, USA.

Hedges, K., and C. Beresford. 1986. Santa Ysabel ethnobotany. San Diego Museum of Man Ethnic Technology Notes No. 20. San Diego Museum of Man, San Diego, California, USA.

Heffner, K. 1984. Following the smoke: contemporary plant procurement by the Indians of Northwest California. Unpublished document on file at the US Forest Service, Six Rivers National Forest, Eureka, California, USA.

Heizer, R. F., editor. 1978. Handbook of North American Indians, vol. 8: California. Smithsonian Institution, Washington D.C., USA.

Hendricks, J. H. 1968. Control burning for deer management in California. Proceedings of the 8th Tall Timbers Fire Ecology Conference 8:219-233.

Hinton, L. 1975. Notes on La Huerta Diegueño ethnobotany. Journal of California Anthropology 2:214-222.

Holliday J. S. 1999. Rush for riches: gold fever and the making of California. University of California Press, Berkeley, California and the Oakland Museum, Oakland, California, USA.

Holt, C. 1946. Shasta ethnography. University of California Anthropological Records 3:299-349.

Horne, S. P. 1981. The Inland Chumash: ethnography, ethnohistory, and archeology. Dissertation. University of California, Santa Barbara, California, USA.

Hovore, F. T. 1979. Rain beetles. Small things wet and wonderful. Terra 17:10-14.

Hovore, F. T. 1998. Pleocoma. Email to Cheryl B. Barr. Essig Museum of Entomology, University of California, Berkeley, March 3.

Howell, J. n.d. A partial list of plants found on nature trail, Indian Grinding Rock State Historic Park. Unpublished manuscript, Indian Grinding Rock State Historic Park Archives, Volcano, California, USA.

Hudson, J. 1901a. Field notebook. G.H.M. Acc. No. 20,004. Collection of Grace Hudson Museum and Sun House, Ukiah, California, USA.

Hudson, J. 1901b. Field notebook. G.H.M. Acc. No. 20,002. Collection of Grace Hudson Museum and Sun House, Ukiah, California, USA.

Hudson, J. 1902. Field notebook. G.H.M. Acc. No. 20,011. Collection of Grace Hudson Museum and Sun House, Ukiah, California, USA.

Hudson, J. n.d. Vocabulary. Tribes of Moquelumnian stock. G.H.M. Acc. No. 20,169a. Collection of Grace Hudson Museum and Sun House, Ukiah, California, USA.

Huenneke, L. F. 1989. Distribution and regional patterns of Californian grasslands. Pages 1-12 in L. F. Huenneke, and H. A. Mooney, editors. Grassland structure and function: California annual grassland. Kluwer Academic Publishers, Dordrecht, Netherlands.

James, S. R., and S. Graziani. 1991. California Indian warfare. Pages 50-109 *in* L. J. Bean and S. V. Vane, editors. Ethnology of the Alta California Indians I. Precontact. Garland Publishing Inc., New York, New York, USA.

Johnston, B. E. 1962. California's Gabrielino Indians. Sowthwest Museum, Los Angeles, California, USA.

Jones, T. L., and K. Hadick. 2016. Indigenous California. Pages 169-184 *in* H. Mooney and E. Zavaleta, editors. Ecosystems of California. University of California Press, Los Angeles, California, USA.

Jones T. L., and Klar K. A. 2007. California prehistory: colonization, culture, and complexity. Altamira Press, Lanham, Maryland, USA.

Keeley, J. E. 1992. Demographic structure of California chaparral in the long-term absence of fire. Journal of Vegetation Science 3:79-90.

Keeley, J. E. 2002. Native American impacts on fire regimes in California coastal ranges. Journal of Biogeography 29:303-320

Keeley, J. E. 2005. Fire history of the San Francisco East Bay region and implications for landscape patterns. International Journal of Wildland Fire 14:285-296.

Keeley, J. E. 2006. South Coast Bioregion, Pages 350–390 *in* N. G. Sugihara, J. W. van Wagtendonk, K. E. Shaffer, J. Fites-Kaufman, and A. E. Thode, editors. Fire in California's ecosystems. University of California Press, Berkeley, Calfornia, USA.

Keeley, J. E., and T. J. Brennan. 2012. Fire driven alien invasion in a fire-adapted ecosystem. Oecologia 169:1043-1052.

Keeley, J. E., W. J. Bond, R. A. Bradstock, J. G. Pausas, and P. W. Rundel. 2012. Fire in Mediterranean ecosystems: ecology, evolution and management. Cambridge University Press, Cambridge, Massehusetts, USA.

Keeley, J. E., C. J. Fotheringham, and M. Morais. 1999. Reexamining fire suppression impacts on brushland fire regimes. Science 284:1829-1832.

Keeley, J. E., and S. C. Keeley. 1988. Temporal and spatial variation in fruit production by California chaparral shrubs. Pages 457-463 *in* F. di Castri, C. Floret, S. Rambal, and J. Roy, editors. Time scales and water stress. Proceedings of the 5th International Conference on Mediterranean Ecosystems (MEDECOS V). International Union of Biological Sciences, Paris, France.

Keeley, J. E., and H. D. Safford. 2016. Fire as an ecosystem process. Pages 27-45 *in* H. Mooney and E. Zavaleta, editors. Ecosystems of California. University of California Press, Oakland, California, USA.

Keeley, J. E., and A. D. Syphard. 2018. South Coast Bioregion. Pages 319-351 *in* J. W. van Wagtendonk, N. G. Sugihara, S. L. Stephens, A. E. Thode, K. E. Shaffer, and J. A. Fites-Kaufman, editors. Fire in California's ecosystems. Second edition. University of California Press, Berkeley, California, USA.

Keeley, J. E., and P. H. Zedler. 2009. Large, high intensity fire events in southern California shrublands: debunking the fine-grained age-patch model. Ecological Applications 19:69-94.

Kelly, R. L. 1995. The foraging spectrum: diversity in hunter-gatherer lifeways. Smithsonian Institution Press, Washington, D.C., USA.

Klimaszewski-Patterson, A., S. Mensing, and L. Gassaway. 2015. Potential non-climate forest structure change in the southern Sierra Nevada range using paleoenvironmental and archaeological proxy data. Pacific Climate Workshop 2015, March 8–11, 2015. Asilomar, California, USA.

Knowles, C. 1953. Vegetation burning by California Indians as shown by early records. University of California, Berkeley, Forestry Library, Pamphlet 16, Fire Vol. 28.

Knudtson, P. M. 1977. The Wintun Indians of California and their neighbors. Naturegraph, Happy Camp, California, USA.

Krech, S. I. 1999. The Ecological Indian. Myth and history. W.W. Norton and Company. New York, New York, USA.

Kroeber, A. L. 1976. Handbook of the Indians of California. Bureau of American Ethnology Bulletin 78, reprint of book published in 1925. Washington D.C. and Dover, New York, USA.

Lake, F. K. 2007. Traditional ecological knowledge to develop and maintain fire regimes in northwestern California, Klamath-Siskiyou bioregion: management and restoration of culturally significant habitats. Dissertation. Environmental Sciences, Oregon State University, Corvallis, Oregon, USA.

Lake, Jr., R. G. 1982. Chilula: people from the ancient redwoods. University Press of America, Arcata, California, USA.

Lamb, W. K., editor. 1984. A voyage of discovery to the north Pacific Ocean and round the world, 1791-1795; With an introduction and appendices. Hakluyt Society, London, UK.

Latta, F. F. 1977. Handbook of Yokuts Indians. Bear State Books, Santa Cruz, California, USA.

Lawrence, G. E. 1966. Ecology of vertebrate animals in relation to chaparral fire in the Sierra Nevada foothills. Ecology 47:278-291.

Lerch, M. K. 2002. Ethnobotanical resources of the Arrowhead east/west tunnel study area, San Bernardino National Forest, San Bernardino County, California. Final Unpublished Technical Report 00-52 prepared by Statistical Research Inc., Redlands, California for the USDA Forest Service, San Bernardino National Forest, San Bernardino, California, USA.

Lewis, H. T. 1993. Patterns of Indian Burning in California: ecology and ethnohistory. Pages 55-116 *in* T. C. Blackburn and M. K. Anderson, editors. Before the wilderness: environmental management by Native Californians. Ballena Press, Menlo Park, California, USA.

Librado, F. 1979. Breath of the sun: life in early California as told by a Chumash Indian, Fernando Librado to John P. Harrington. *In* T. Hudson, editor. Malki Museum Press and Ventura County Historical Society, Banning, California, USA.

Lightfoot, K. G., R. Q. Cuthrell, C. Boone, R. Byrne, A. B. Chavez, L. Collins, A. Cowart, R. E. Evett, P. V. A. Fine, D. Gifford-Gonzalez, M. G. Hylkema, V. Lopez, T. M. Misiewicz, and

R. E. B. Reid. 2013a. Antrhopogenic burning on the central California coast in late Holocene and early historical times: findings, implications, and future directions, California Archaeology 5:371-390.

Lightfoot, K. G., R. Q. Cuthrell, C. J. Striplen, and M. G. Hylkema. 2013b. Rethinking the study of landscape management practices among hunter-gatherers in North America. American Antiquity 78:285-301.

Loud, L. L. 1918. Ethnogeography and archaeology of the Wiyot Territory. University of California Publications in American Archeology and Ethnology 14:221-436.

Madley, B. 2016. An American genocide: the United States and the California Indian Catastrophe. Yale University Press, New Haven, Connecticut, USA.

Maniery, J. G. 1987. Six Mile and Murphys Rancherias: a study of two Central Sierra Miwok village sites. San Diego Museum Papers No. 22.

Mason, J. D. 1881. History of Amador County, California, with illustrations and biographical sketches of its prominent men and pioneers. Thompson and West, Pacific Press Publishing House, Oakland, California, USA.

McBride, J. R., and H.F. Heady. 1968. Invasion of grassland by *Baccharis pilularis* DC. Journal of Range Management 21:106-108.

McCawley, W. 1996. The First Angelinos. Ballena Press, Novato, California, USA.

McCorkle, T. 1978. Intergroup conflict. Pages 694-700 *in* R. F. Heizer, editor. Handbook of North American Indians. Volume 8: California. Smithsonian Institution, Washington D.C., USA.

McMillin, J. H. 1956. The Aboriginal human ecology of the Mountain Meadows area in southwestern Lassen County, California. Unpublished Thesis, Department of Anthropology, Sacramento State University, Sacramento, California, USA.

Mensing, S. A. 1998. 560 years of vegetation change in the region of Santa Barbara, California. Madroño 45:1–11.

Mensing, S. A., J. Michaelsen, and R. Byrne. 1999. A 560-year record of Santa Ana fires reconstructed from charcoal deposited in the Santa Barbara Basin, California. Quaternary Research 51:295-305.

Merriam, C. H. 1902. Unpublished field notes: Mariposa area. September 18-19:205-217. Library of Congress, Washington D.C., USA.

Merriam, C. H. 1955. Studies of California Indians. University of California Press, Berkeley, California, USA.

Merriam, C. H. 1967. Ethnographic notes on California Indian Tribes III: ethnological notes on Central California Indian Tribes. Pages 257-448 *in* R. F. Heizer, editor. Reports of the University of California Publications in American Archaeology and Ethnology 68, Part III. University of California Archaeological Research Facility, Department of Anthropology, Berkeley, California, USA.

Merrill, R. E. 1923. Plants used in basketry by the California Indians. University of California Publications in American Archaeology and Ethnology 20:215-242.

Miller, F. E. 1928. The medicinal plants of Yosemite National Park. Unpublished manuscript. Yosemite Research Library, Yosemite National Park, California, USA.

Milliken, R. 2006. The central California Ethnographic Community Distribution Model, version 2.0, with special attention to the San Francisco Bay Area. Cultural Resources Inventory of Caltrans District 4, Rural Conventional Highways. Report prepared for Caltrans District 4, Oakland, California, USA.

Minnich, R. A. 1983. Fire mosaics in southern California and northern Baja California. Science 219:1287-1294.

Neely, W. 1971. Miwok uses of Yosemite plants in food, medicine, baskets, weapons, and construction. Unpublished manuscript, Yosemite Research Library, Yosemite National Park, California, USA.

Peigler, R. 1994. Non-sericulture uses of moth cocoons in diverse cultures. Proceedings of the Museum of Natural History Series 3:1-20.

Peri, D. W., S. M. Patterson, and J. L. Goodrich. 1982. Ethnobotanical mitigation Warm Springs Dam—Lake Sonoma, California. *In* E. Hill and R.N. Lerner, editors. Unpublished report of Elgar Hill, Environmental Analysis and Planning, Penngrove, California, USA.

Pinter, N., S. Fiedel, and J. E. Keeley. 2011. Fire and vegetation shifts at the vanguard of Paleoindian migrations. Quaternary Science Reviews 30:269-272.

Powell, J. A. 1972. Population expansions and mass movements of *Nymphalis californica* (Nymphalidae). Journal of the Lepidopterists' Society 26:226-228.

Powers, S. 1976. Tribes of California. Reprint of book published in 1877 contributions to North American ethnology, III. Department of the Interior, US Geographical and Geological Survey of the Rocky Mountain Region. University of California Press, Berkeley, California, USA.

Preston, W. L. 1996. Serpent in the Garden of Eden: dispersal of foreign diseases into pre-Mission California. Journal of California and Great Basin Anthropology 18:2-37.

Preston, W. L. 2002. Portents of plague from California's Protohistoric Period. Ethnohistory 49:69-121.

Quinn, R. D. 1990. Habitat preferences and distribution of mammals in California chaparral. Research Paper PSW-RP-202. USDA Forest Service, Pacific Southwest Research Station, Berkeley, California, USA.

Rosenthal, J. S., and R. T. Fitzgerald. 2012. The paleo-archaic transition in western California. Pages 67-103 *in* C. B. Bousman and B. J. Vierra, editors. From the Pleistocene to the Holocene: Human organization and cultural transformations in prehistoric North America. Texas A & M University Press, College Station, Texas, USA.

Safford, H. D., and K. M. van de Water. 2014. Using fire return interval departure (FRID) analysis to map spatial and temporal changes in fire frequency on national forest lands in California. Research Paper PSW-RP-266. USDA Forest Service, Pacific Southwest Research Station, Albany, California, USA.

Shipek, F. C. 1981. A Native American adaptation to drought: the Kumeyaay as seen in the San Diego Mission records 1770-1798. Ethnohistory 28:295-312.

Shipek, F. 1993. Kumeyaay plant husbandry: fire, water, and erosion management systems. Pages 379-388 *in* T.C. Blackburn and K. Anderson, editors. Before the wilderness. Environmental management by Native Californians. Ballena Press, Menlo Park, California, USA.

Simpson, L. B. 1938. California in 1792. The Expedition of Jose Longinos Martinez. Henry E. Huntington Library and Art Gallery. San Marino, California, USA.

Sinclair, W. A., H. H. Lyon, and W. T. Johnson. 1987. Diseases of trees and shrubs. Cornell University Press, Ithaca, New York, USA.

Schneider, J. S. 2009. Archaeological documentation and data recovery at the Stacked Stone Site (CA-SDI-17666): a site incorporating stone architecture. Cuyamaca Rancho State Park California. Cultural Resource Management Project #418 (Winter 2008-09) Final Report.

Sparkman, P. S. 1908. The culture of the Luiseño Indians. University of California Publications in American Archaeology and Ethnology 8:187-234.

Spier, R. F. G. 1978. Foothill Yokuts. Pages 471-484 *in* R. F. Heizer, editor, Handbook of North American Indians, Vol. 8: California. Smithsonian Institution, Washington D.C., USA.

Stephens, S. L., and D. L. Fry. 2005. Fire history in coast redwood stands in the Northeastern Santa Cruz Mountains, California. Fire Ecology 1:2-19.

Stephens, S. L., R. E. Martin, and N. E. Clinton. 2007. Prehistoric fire area and emissions from California's forests, woodlands, shrublands, and grasslands. Forest Ecology and Management 251:205-216.

Sterling, E. A. 1904. Fire notes on the coast ranges of Monterey County, California. Bureau of Forestry, Sacramento, California, USA.

Steward, J. H. 1938. Basin-Plateau Aboriginal sociopolitical groups. Smithsonian Bureau of American Ethnology Bulletin 120. US Government Printing Office, Washington D.C., USA.

Stewart, O. C. 1956. Fire as the first great force employed by man. Pages 115-133 *in* W. L. Thomas Jr., editor. Man's role in changing the face of the earth. University of Chicago Press, Chicago, Illinois, USA.

Stuart, J., and S. L. Stephens. 2006. North coast California bioregions. Pages 147-169 *in* N. G. Sugihara, J. W. van Wagtendonk, K. E. Shaffer, J. Fites-Kaufman, and A. E. Thode, editors. Fire in California's ecosystems, University of California Press, Berkeley, California, USA.

Taber, R. D. 1956. Deer nutrition and population dynamics in the North Coast Range of California. North American Wildlife Conference Transactions 21:159-172.

Tadd, B. 1988. Miwok: one Miwok's view of native food preparations and the medicinal uses of plants. Three Forests Interpretive Association, Stanislaus, Sierra and Sequoia National Forests, California, USA.
Theodoratus, D. J., C. M. Blount, H. McCarthy, and N. H. Evans. 1985. Haas-Kings river hydroelectric project phase II, ethnographic investigations. Final report to Pacific Gas and Electric Company, San Francisco, California and SMUD-NCPA-Southern Cities, Sacramento, California, USA.
Theodoratus, D. J., and M. Parsons. 1980. Miwok ethnohistory. Manuscript on fiel at the Calaveras County Museum and Archives, San Andreas, California, USA.
Timbrook, J. 2007. Chumash ethnobotany: plant knowledge among the Chumash people of southern California. Santa Barbara Museum of Natural History, Santa Barbara and Heyday Books, Berkeley, California, USA.
Timbrook, J., J. R. Johnson, and D. D. Earle. 1982. Vegetation burning by the Chumash. Journal of California and Great Basin Anthropology 4:163-186.
Train, P., J. R. Henrichs, and W. A. Archer. 1941. Medicinal uses of plants by Indian tribes of Nevada. Contributions toward a flora of Nevada No. 33. USDA The Division of Plant Exploration and Introduction, Bureau of Plant Industry, Washington D.C., USA.
Trauernicht, C., B. W. Brook, B. P. Murphy, G. J. Williamson, and D. M. J. S. Bowman. 2015. Local and global pyrogeographic evidence that indigenous fire management creates pyrodiversity. Ecology and Evolution 5:1908-1918.
Tuskes, P. M., J. P. Tuttle, and M. M. Collins. 1996. The wild silk moths of North America: A natural history of the Saturniidae of the United States and Canada. Cornell University Press, Ithaca, New York, USA.
Ubelaker, D. H. 2006. Skeletal biology and population size: introduction. Pages 492-496 *in* D. H. Ubelaker, and W. C. Sturtevant, editors. Handbook of North American Indians, volume 3: environment, origins, and population. Smithsonian Institution, Washington, D.C., USA.
USDA Forest Service. 2012. The Okanogan-Wenatchee National Forest restoration strategy: Adaptive ecosystem management to restore landscape resiliency. Available: http://www.fs.usda.gov/InternetFSE_DOCUMENTS/stelprdb5340103.pdf
Vale, T. R. 2000. Pre-Columbian North America: pristine or humanized – or both? Ecological Restoration 18:2–3.
Verner, J., and A. S. Boss, technical coordinators. 1980. California wildlife and their habitats: Western Sierra Nevada. General Technical Report PSW-GTR-37. USDA Forest Service, Pacific Southwest Forest and Range Experiment Station, Berkeley, California, USA.
Vogel, V. J. 1970. American Indian medicine. University of Oklahoma Press, Norman, Oklahoma, USA.
Voegelin, E. W. 1938. Tubatulabal ethnography. Anthropological Records 2:1-84.
Wallace, E. 1978. Sexual status and role differences. Pages 683-689 *in* R. Heizer, editor. Handbook of North American Indians. Vol. 8: California. Smithsonian Institution, Washington D.C., USA.
Wells, P. V. 1962. Vegetation in relation to geological substratum and fire in the San Luis Obispo quadrangle, California. Ecological Monographs 32:79-103.
White, M., R. H. Barrett, A. S. Boss, T. F. Newman, T. J. Rahn, and D. F. Williams. 1980. Mammals. Pages 321-424 *in* J. Verner and A.S. Boss, technical coordinators. California wildlife and their habitats: western Sierra Nevada. Technical Coordinators, General Technical Report PSW-GTR-37. USDA Forest Service, Pacific Southwest Forest and Range Experiment Station, Berkeley, California, USA.
Wohlgemuth, E. 2004. The course of plant food intensification in native central California. Dissertation, Department of Anthropology, University of California, Davis, California, USA.
Zigmond, M. L.1981. Kawaiisu ethnobotany. University of Utah Press, Salt Lake City, Utah, USA.

Chapter 5
Essential Landscape: An Environmental History of Chaparral Ecosystems in California

Char Miller

Abstract Although chaparral is the most extensive, native plant habitat in California, it is not well understood in terms of the biodiversity it contains and the ecosystem services it provides. This was not always the case, and this chapter will explore what those inhabiting this region—native people, Spanish missionaries, US resource extractors, and conservationists—have known about this dynamic habitat, how their knowledge has changed over time, and why. Drawing on the insights of anthropology, archaeology, history and cultural studies, the chapter documents the evolving and reciprocal relationship between humans and the chaparral ecosystems that for millennia have sustained them in California.

Keywords Abbott Kinney · Biodiversity · Chaparral ecosystem · Chumash · Conservation · John Muir · LA County Fire Department · San Bernardino Mountains · San Gabriel Mountains · Sierra Mountains · Southern California · Spanish Missions · Stuart Flintham · Tongva · US Forest Service

5.1 Introduction

Chaparral is a key ecosystem in California. In southern California, it constitutes upwards of 80% of the San Gabriel Mountains and serves as the verdant backdrop to the Los Angeles region. The San Bernardino, Santa Ana, and Santa Monica Mountains are similarly clothed. Chaparral's dominant presence can also be measured in acreage per county: San Diego contains more than one million acres of chaparral and Los Angeles more than 550,000 (222,577 ha). Riverside has nearly 500,000 acres (202,343 ha) and San Bernardino has an estimated 276,000 acres (111,693 ha), Orange a modest 111,000 acres (44,920 ha), and Ventura has more than 326,000 acres (131,927 ha) (Fried et al. 2004). Chaparral also has a significant presence in the California's central Coast Ranges, with Santa Barbara County home

C. Miller (✉)
Pomona College, Claremont, CA, USA
e-mail: char.miller@pomona.edu

E. C. Underwood et al. (eds.), *Valuing Chaparral*, Springer Series on Environmental Management, https://doi.org/10.1007/978-3-319-68303-4_5

to roughly 440,000 acres (178,062 ha) of chaparral, San Benito contains close to 250,000 acres (101,171 ha), and Santa Clara 188,000 acres (76,081 ha). The plant community also ranges north to southern Oregon and is a critical habitat in the western and eastern flanks of the Sierra. Although large swaths of chaparral have been cleared away since the mid-twentieth century as a result of urban growth, suburban sprawl, and exurban development, chaparral remains a highly visible presence in the region's heavily visited canyons, foothills, and mountains. Its visibility does not necessarily mean, however, that this ecosystem is well understood in terms of its biodiversity or the diverse ecological services it provides (see Chaps. 2, 3, 4, 6, 7, 8, 9, and 10).

This lack of knowledge was not always the case. The native peoples of southern California made (and continue to make) extensive use of chaparral landscapes for food, medicine, as well as ritual and ceremonial needs, modifying and adapting to these rugged terrains. To secure those ends, they also used fire to manage these ecosystems. Their integrative approach to the management of chaparral, along with their communities' dependence on its many services, was challenged with the arrival of Euro-Americans in the eighteenth and nineteenth centuries. The successive waves of colonizers, and the economies they introduced, depended on a more extractive approach to the resources these mountainous regions contained, whether flora, fauna, or mineral. Yet by the late nineteenth century, some of these new Californians began to argue against unregulated resource extraction, believing that rampant exploitation generated considerable environmental damage, with upstream despoliation having troubling consequences downstream. Others saw the San Gabriels, San Bernardinos, and the Sierra mountains—among other high ground in California—as the source of a new wildland aesthetic that reinforced the concept that these towering terrains contained important values that needed regulation and protection. Conservationist Abbott Kinney and naturalist John Muir were among those aware of the significant values and resources that chaparral ecosystems supplied. Their insights, coupled with the boom in recreation, an acceleration in flood control concerns, and an increase in agricultural demand for water, resulted in the creation of the San Gabriel Timberland Reserve (1892). This represented the first designated public land reserve in California (and one of the first in the nation), others in southern California, the Sierra, and the Coast Ranges were designated shortly thereafter (Godfrey 2005).

Although subsequent generations of land managers often devalued native ecosystems, as suggested by afforestation initiatives and grassland conversion projects that local and federal agencies mounted, this devaluation may be undergoing a reevaluation. For example, chaparral restoration efforts in the aftermath of the 2009 Station Fire in the San Gabriel Mountains (see Chap. 15), and the establishment of the San Gabriel Mountains National Monument (2014), the Berryessa Snow Mountain National Monument (2015) in the state's northern Coast Ranges, and the Sand to Snow National Monument (2016) covering portions of the San Jacinto Mountains, may be signaling a return to a more complex appreciation of chaparral for its biodiversity, ecosystem services, and social importance.

5.2 Indigenous People and Chaparral Ecosystems

Native people have inhabited chaparral ecosystems in what is now California for more than 12,000 years (see Chap. 4). They have made extensive use of this landscape's many resources. Utilizing the region's waters for drinking and cooking purposes, they also fished in these same snow-fed creeks, streams, and rivers that found their sources in the Sierra, the Coast Ranges, and in the transverse mountains that frame southern California. The chaparral dominated canyons and foothills through which these watercourses flowed have offered a variety of roots, bulbs, seeds, bark, grasses, as well as animals, for which they have foraged and hunted across the seasons, some for food, others for ritual practices.

The Chumash, for example, actively managed their territory, which encompassed the Channel Islands and coastal regions from present day Los Angeles north to Santa Barbara, and east into the Santa Ynez Mountains. They harvested shellfish in the Santa Ynez, Ventura, and Santa Clarita watersheds. In oak groves and grasslands they gathered acorns and other foodstuffs, fashioned arrows from oak and manzanita in the chaparral biome, and used fire to stimulate the production of certain fire-followers such as chia, increase browse for deer and smaller mammals, and provide shelter for quail. Like their contemporaries throughout southern California, they also built their housing and other structures with willow and thatch harvested locally (Gamble 2011).

Making similar choices about how to manage chaparral environments were the Tongvan people who inhabited the inland valleys of the Los Angeles region west and south to the Pacific coast. Other groups also included the Luiseño in and around Orange County, the Cahuilla and Serrano whose territories included the San Bernardino and San Gabriel Mountains, the Kumeyaay in San Diego, and the Kitanemuk, Tejon, and Yokuts in the high ground that rises above the San Joaquin Valley (Patterson 2014). Over the millennia, their collective, and at times collaborative, management of chaparral habitat was so extensive that the First People in California had "ample time to affect the evolutionary course of plant species and plant communities" (Anderson 2005). Put another way, these "traditional management systems have influenced the size, extent, pattern, structure, and composition of the flora and fauna within a multitude of vegetation types throughout the state." So much so, that when the first Europeans sailed along the coast or trekked through interior valleys beginning in the seventeenth century, what they extoled as pristine terrain was arguably "a carefully tended 'garden' that was a result of thousands of years of selective harvesting, tilling, pruning, burning, sowing, weeding, and transplanting" (Anderson 2005).

5.3 Chaparral as a Landscape of Resistance

This lengthy process of integration, adaptation, and manipulation produced what Panich and Schneider (2014) call "indigenous landscapes" across what is now the US southwest. Well-tended, these habitats proved critical to the ways in which

"Native people actively negotiated Spanish colonialism on their own terms" (Panich and Schneider 2014). As the Spanish pushed north into the region they called Alta California, for example, Franciscan missionaries and Spanish soldiers initially encountered stout resistance. Signs of this confrontation have often escaped the notice of those archaeologists and historians who have focused their investigations on the mission sites alone (Deitler et al. 2016). What these scholars have failed to account for is how the broader landscape in which the mission Indians lived offered a refuge of sorts. Panich and Schneider (2014) challenge the long-held argument that Native people unilaterally capitulated before European firepower, that they "entered the orbit of the missions never again to fully participate in the making of their own history" (Panich and Schneider 2014).

Among those working assiduously to subordinate the mission Indians was Fermín Francisco de Lasuén, the father president of Mission La Purísima, which was founded in 1787 on a site some 64 km (40 miles) north of Santa Barbara. More than a decade after its founding, however, Fray Lasuén continued to rue that the "uncultivated soil supports [the Chumash's] manner of life." As he wrote his superiors in 1801: "the greatest problem for the missionary…[is] how to transform a savage race such as these into a society that is human, Christian, civil, and industrious. This can only be accomplished by denaturalizing them" (Panich and Schneider 2014). To the extent that the Chumash refused to be "denaturalized," suggests how chaparral—and its bounty—had trumped the Church.

Even when Lasuén and his successors used indiscriminate force to compel the Chumash of the central coast to live in mission compounds, regulated their lives around the Catholic liturgical calendar, and demanded that they perform the back-breaking labor needed to plant, tend, and harvest European crops—a denaturalizing strategy pursued across the Spanish Borderlands (Resendez 2016)—the Chumash and their contemporaries negotiated some of the terms of their efforts. Such negotiations were most apparent when Euro-American agriculture and ranching failed to provide enough food; such failures were common across southern California. In response, the Chumash found sustenance in "the wild grain in the hills," and their continued ability to harvest it led a relieved Fray Fernandez of Mission La Purísima to praise "the goodness of God" for furnishing this food "on their native soil" (Panich and Schneider 2014). At other times, aggrieved natives resisted outright, as evident in the revolts that broke out at San Gabriel Mission (1771, 1785), San Diego Mission (1775), San Luis Obispo (1776), and at Santa Barbara, Santa Ynez, and La Purisima (1824), among other locations.

Brutal treatment could also impel the Chumash to flee to refuges in their home ground. Zenas Leonard, a clerk on an 1839 trapping expedition to the southern Sierra, came upon one group in what now is called Walker Pass. These Spanish speaking Chumash numbered about 100 and they farmed corn, hunted meat, and harvested other resources from chaparral (Leonard 1839). Scholars have also examined three refuge sites in the "colonial hinterlands" located deep in the interior ranges to the east of Santa Barbara (Bernard et al. 2014). Preliminary findings indicate that these remote locations were sited in a manner that seems consistent with the need to monitor Spanish troops sent to recapture them. Because these refugees left behind fewer botanical and zooarchaeological debitage than previous generations, scholars speculate that they moved frequently to minimize "their impact on existing populations that were essentially playing host to refugee communities" (Bernard et al. 2014). An added benefit was that these hideaways provided "opportunities for the construction and negotiation of indigenous identities in the wake of colonialism, thus serving as a key path to cultural persistence as well as transformation" (Bernard et al. 2014).

The possibility of this kind of cultural persistence was severely challenged following the US conquest of California in 1846–1847 and the unleashing of a genocidal assault on the native peoples of the new state (Lindsay 2012). State sponsored and vigilante violence, expropriation of treaty-granted land and water rights, and the spread of disease slashed indigenous populations by 80% in less than three decades, falling from 150,000 in 1846 to roughly 30,000 by 1873 (Madley 2016). Among the hardest hit were the native peoples who inhabited southern California's coastal and inland chaparral ecosystems. American ranchers and homesteaders highly prized these particular landscapes and used legal and illegal means to expropriate them; once secured, they intensified the production of logging, grazing, mining, and farming.

Two special agents to the US Commissioner of Indian Affairs, novelist Helen Hunt Jackson and conservationist Abbott Kinney, detailed these escalating depredations. Their 1883 report tracked the disruptive impact of the "robber whites who drove [the Indians] out" and accused the state and federal governments of ignoring this dislocation (Jackson and Kinney 1883). Yet Jackson's and Kinney's on-the-ground investigation also shows how native people managed to minimize some of the onslaught's impact. Choosing "remote and inaccessible spots," living in "small and isolated villages," some of which "being literally in the last tillable spot on the desert's edge or in mountain fastness," the remaining native peoples sought shelter on the distant margins of white society (Jackson and Kinney 1883). In returning to "their old wilderness homes," like their ancestors had done to escape Spanish oppression, they made similar use of chaparral's ecosystem services—its foodstuffs supplemented (or substituted for) whatever produce and grain they were able to harvest from the less than fertile soils available to them. Theirs was "a very poor living, it is true, but they are independent and self-respecting in it" (Jackson and Kinney 1883).

Recounting examples of native agency in the face of overwhelming force will not change the impact that historic enslavement, hard labor, brutal treatment, rape, and pillage had on the Chumash, Tongvan and Serrano, the Cahuilla and Kumeyaay, or the Miwok. Nor will a more complete understanding of what happened in the Golden State bring back the indigenous landscapes that sustained and harbored the First People of California. Yet by establishing the native peoples' ability to construct safe havens amid chaparral ecosystems that helped cushion the repeated blows from Spanish, Mexican, and US occupations, and by detailing their lived experiences within indigenous landscapes, might provide "critical information to contemporary tribes who are working to document their continued existence in tribal territories" (Panich and Schneider 2014), an important step in seeking federal recognition. If so, then this ethnobotanical, environmental, and cultural history might enable native people to secure what Jackson and Kinney chastised the federal government for failing to grant them: a "full measure of justice" (Jackson and Kinney 1883).

5.4 Late Nineteenth Century Conservation and Chaparral

Even as Jackson and Kinney argued for federal intervention on behalf of the uprooted and decimated native people of southern California, and placed their hopes on the White House and Congress establishing permanent reservations for Indian communities, their aspirations would be undercut by the emergence of the conservation movement. This movement asserted an exclusionary claim to the region's mountains, foothills, and canyons, and the resulting ecosystem services they generated. Indeed, Kinney would prove one of this new movement's most forceful promoters. In the early 1880s, he and like-minded peers embraced the idea that the elevated landscapes rising above San Diego, San Bernardino, Riverside, Los Angeles and Santa Barbara, must be protected for the natural resources they contained and the recreation they offered.

One resource in particular took priority—water. Its priority was predicated on the region's rapid transformation in the post-Civil War era. Spikes in the Euro-American population, and the arrival of the railroad which fed the boom in new residents, led to extensive urban development. At the same time that urbanites required larger amounts of clean water, so did the expanding citrus industry that was sprawling across inland valleys. The local sources of water once had been as clear as the snow capping the San Gabriel and San Bernardino Mountains in the winter months, but its clarity was increasingly muddied by three different economic uses that Kinney and other conservationists sought to regulate.

The first was logging. Although most of the San Bernardino, San Gabriel, and San Jacinto mountains are covered in chaparral, the San Bernardino and San Jacinto ranges contained significant amounts of accessible timber at higher altitudes. The San Gabriels, Abbot Kinney observed, were "too steep and inaccessible yet to be exploited, but on San Bernardino and San Jacinto large saw mills are continually at work, supplying the colonies below in the valleys with lumber" (Kinney 1888). By the late 1890s, these stands had been cut out, leaving "much of the crest forest a desolate wasteland" (Robinson 1989).

Sheep and cattle also chewed up forest understory and chaparral habitat. In 1894, the *Redlands Citrograph* reported that the "San Bernardino Mountains are thicker with sheep than locusts of olden times and twice as destructive. There is not a green shrub to be seen. The young trees, especially the young oak, are eaten down to the earth. The water streams are all demoralized" (quoted in Robinson 1989).

John Muir, while working as a shepherd in the central Sierra, witnessed a similar level of destruction. As he helped drive a "poor dust-choked flock" through the Sierran foothills on their way to Alpine meadows, the hungry animals found little to satiate their "ravenous appetites." Earlier flocks had mowed down the chaparral:

"scare a leaf, green or dry, was left; therefore the starving flock had to be hurried over the bare hot hills to the nearest green pastures, about twenty or thirty miles from here" (Muir 1911). Fire added a related complication. To increase browse in the Sierra, as well as in the foothills and mountains of southern California, ranchers and shepherds routinely burned chaparral, blazes that "raged unchecked for days" (Robinson 1989). The downstream consequences were immediately apparent to domestic and agricultural consumers in the flatlands. Ponds, streams, and rivers were polluted, thus befouling potable water supplies, and farmers complained about using these polluted waters on their crops. Furthermore, the denuding of the high country seemed to intensify winter flooding in the Central Valley and in southern California valleys.

Mining added to this growing list of woes. Although California's major mining operations occurred in the Sierra, there were minor strikes in the San Gabriels and San Bernardinos. Like their peers to the north, southern California's miners often used mercury and cyanide to extract gold from low grade ore; less frequently, they employed hydraulic-mining techniques to blast rock material into slurry that could be sifted for gold. The result was despoiled local waterways, as occurred in the mid-1890s in San Antonio Canyon in the San Gabriels. No sooner had the Hocumac Company's employees turned high-pressure hoses on the upper canyon's walls, then a stream of reddish-brown water rushed into the creek, flowed into the local mutual water company's diversion channels in the lower canyon, and made potable supplies undrinkable for several weeks. The San Antonio Water Company (SAWC), whose waterworks had been clogged by the silt, filed suit and the county court found in its favor, prohibiting hydraulic miners from "polluting or discoloring the water of the San Antonio Creek in any way" (Dawson 1958). Shortly thereafter, SAWC bought out the Hocumac Company and its water rights to protect its operations. Significantly, SAWC erected a sign at the canyon's mouth warning people not to cut any brush on its property. SAWC knew that chaparral was integral to the protection of its water supply system (Hackenberger and Miller 2017).

Abbot Kinney was also convinced that it was essential to protect chaparral in the southern Sierra and southern California. Indeed, he regularly spoke to conservation groups across the state on these issues. His focus in his talks, however, was on public lands not private property. Hoping to gain some measure of control over the exploitation of the forest, grass, and water on the public domain, he used his position as chair of the California State Forestry Board to push for tighter federal management of the state's mountainous environs. Because these "brush lands almost all belong to the Government, and, being of little direct value, will probably long remain its property" (Kinney 1888), he laid out an ecological case for why the chaparral biomes were of considerable value.

> "Every year disastrous fires sweep off great areas of this mountain covering. The Government sets no watch or takes no heed of its property, and the fires run into and destroy the forest as well as the brush. Every year, as a consequence, water rights are decreased in value, through the springs diminishing in summer, and torrents run more violently and bring down more sand and stones to scatter on the farms. The floods each wet year are more destructive than before. On each of these watersheds extensive destruction of trees and brush has taken place…by deliberately set and deliberately repeated fires. (Kinney 1888)".

Kinney was particularly concerned with maintaining the chaparral, whose growth was so thick it was "almost impenetrable" (Kinney 1888). Its density and ubiquity enhanced its ecological services. "In the mountains and canyons it furnishes food for the bees, and most important of all, it acts as a reservoir, in allowing the rains of the wet season time to seep into the soil and rock veins, to appear again in the dry season as springs in the low country" (Kinney 1888). Chaparral was a key contributor to and indicator of the region's rich biodiversity, ecological health, and economic sustainability.

There was only one possible resolution. "The necessity of the hour is the intelligent supervision of the forests and brush lands of California, with a view to their preservation," Kinney argued. Logging in the forests and the clearing away of chaparral should be "so reasonably regulated" as to insure their "reproductive power, and, above all, to maintain the forest influence on climate and secure the farmer the perennial character of the springs and streams necessary in the dry season for irrigation". His insights were in line with those his contemporaries, conservationists George Perkins Marsh and George Bird Grinnell, advocated. So was his sense of urgency. Without significant increase in federal management, California would suffer, he asserted: "The destruction of the forests in the southern counties means the destruction of the streams, and that means the destruction of the country" (California State Board of Forestry 1885).

5.5 Nature Appreciation and Southern California's Great Hiking Era

John Muir identified another reason to preserve these imperiled landscapes. His 1869 trek through the stripped-bare chaparral of the central Sierra had been dispiriting, leading him to exult when he and the flock finally reached more green and forested elevations, "We are now in the mountains and they are in us, kindling enthusiasm, making every nerve quiver, filling every pore and cell of us" (Muir 1911). But his appreciation for chaparral reached the same heights after he encountered its intact form on a hike in the San Gabriels. "Not even in the Sierra have I ever made the acquaintance of mountains more rigidly inaccessible" (Muir 1918). Here, "Mother Nature is most…thornily savage," and that was a good thing for this wilderness lover. "Chaparral constitutes [the mountains'] chief defense," he observed. "It swoops into every hollow and swells over ever ridge, gracefully complying with the varied topography, in shaggy, ungovernable exuberance, fairly dwarfing the utmost efforts of human culture out of sight and mind" (Muir 1918).

The intrepid hiker discovered just how ungovernable chaparral could be as he struggled up the loose-soiled terrain above Eaton Canyon, thick with a "bristly mane of chaparral" (Muir 1918). It was so thick in one place that he was "compelled to creep more than a mile on hands and knees" (Muir 1918), although days later, while ascending the eastern rim of the canyon, he found it "denser and more stubbornly bayoneted than ever". The only way forward was to follow "bear trails, where in some places I found tuffs of hair that had been pulled out in squeezing a way through." His own, clothes-tearing exertion "far overpaid all my pains" (Muir 1918).

Like the Sierra, which Muir had anointed the Range of Light, southern California's mountains illuminated his faith that the Great Outdoors would increase human health—bodily, aesthetically, and spiritually. As he once put it: "Thousands of tired, nerve-shaken, over-civilized people, are beginning to find out that going to the mountains is going home; that wildness is a necessity" (Muir 1901). It was necessary for him, too. Muir reportedly enthused about Mount San Jacinto that the 3302 m (10,833 ft) peak offered "the most sublime spectacle to be found anywhere on this earth!" (Heald 1963). Spectacular, too, was Eaton Canyon, which Muir dubbed the Yosemite of the San Gabriels. In its gurgling streams, thunderous cascades, and "rich profusion of wild flowers," tourists would encounter a "charming poem of wildness" (Muir 1918).

Yet this inspiring vision of the wild depended on a construction of nature that excluded people from living in its midst—particularly native peoples. Muir's troubling dismissal of their presence on the land, which many of his fellow conservationists around the country shared (Cronon 1997; Spence 1999), was evident in his account of an encounter with a band of Mono Indians in the high Sierra. As "hairy as bears and as crooked as summit pines," their faces appeared so "ancient and so undisturbed it might almost possess a geological significance. The older faces were, moreover, strangely blurred and divided into sections by furrows that looked like the cleavage joints of rocks, suggesting exposure on the mountains in a castaway condition for ages. Somehow they seemed to have no right place in the landscape, and I was glad to see them fading out of sight down the pass" (Muir 1894).

That said, Muir did not hesitate to encourage masses of over-civilized Euro-Americans into these same mountains. Certainly his essays about the Sierra and San Gabriels played a significant role in luring urbanites in large numbers into high country. Muir led countless camping expeditions into the Sierra to cultivate urban support for the protection and preservation of these rural terrain (Worster 2008). Southern California's rugged mountains, he cheered, were "quickly available retreats from dust and heat and care". Within a "few hours lowlanders can get well up into the sky and find refuge in hospitable camps and club-houses, where, while breathing reviving ozone, they may absorb the beauty about them" (Muir 1901). Many followed his prescription. In the last decades of the nineteenth century, each year more than 100,000 Los Angeles residents, along with a growing number of visitors who came west to enjoy the region's salubrious weather, followed Muir's upcountry trail (Hoffman 1976). By the end of the so called Great Hiking Era in the late 1930s, an estimated two million people a year recreated in the San Gabriels. The spike in visitors was only partly a result of the city's population explosion, which in 1880 stood at 11,000, reached 100,000 by the 1900 census, and 30 years later topped 1.2 million. Transporting that many people to the mountains required an extensive system of mass transit, and the Pacific Electric Railway streetcars had stops at a number of canyon mouths. One of these connected passengers to the

Mount Lowe Incline that would then carry them up to restaurants and lodges crowning mountaintop and ridgeline (Robinson 1977).

Other trailheads offered the hardy (or foolhardy) the opportunity to ride a mule up vertiginous trails with tight switchbacks. Early twentieth century humorist Mina Deane Halsey memorialized her one trek up to Mount Wilson. Reaching its 1737 m (5700 ft) summit, she laughed, was "the nearest station to Heaven yours truly ever expects to get" (Halsey 1912). Not that she wanted to repeat the ride up the mountain's rough-and-tumble terrain. "The trip up Mt. Wilson makes me heave many sighs. In fact, I heaved so many sighs for weeks after that trip, that I had a hard time making anyone believe I had a good time. But I did" (Halsey 1912). The ride down was even scarier. "It takes four—five—six or seven hours to get up the trail, and it only took me somewhere around forty minutes to come down. Of course most people don't hurry so on the down trip, but some things are forced upon us in this world, and that jackass of mine certainly knew his business" (Halsey 1912). Despite the self-deprecating jokes she wrung from her travails, Halsey conceded that there "were some wonderful sights along the way," including a glorious sunset (Halsey 1912).

Charles Francis Saunders also caught the sights while retracing John Muir's strenuous hike in Eaton Canyon—"the lapse of nearly half a century since Muir's visit to the spot has wrought little change in it" (Saunders 1923). Saunders reveled in "the exercise of scrambling over granite boulders, and the excitement of picking a hazardous way up precipitous slopes treacherous with shaly rock" (Saunders 1923). He found comfort in the "solitude of pure wilderness" and marveled at chaparral's tough, if remarkable, beauty. Following the rainy season, manzanita, currant, and gooseberry bloomed while the "dun slopes of chaparral break brightly into color, acre upon acre of massed white and blue. This is the flowering of wild lilacs, the most spectacular of spring miracles in the California mountains" (Saunders 1923). Even as Saunders' mimicked many of Muir's insights, he also made the case that one of chaparral's key ecosystem services was its profound cultural cachet.

5.6 Chaparral and the Creation of the National Forests

Chaparral's manifold biological, hydrological, and recreational values underlay the arguments that late nineteenth century conservationists in Santa Barbara, San Diego, Los Angeles, the Bay Area, and Sacramento deployed to protect the state's mountain ranges. Beginning in 1880, Abbott Kinney criticized those who used fire and mechanical means to clear away brush and chaparral, scrub oak, greasewood, and sagebrush, leaving watersheds bereft of verdure. He wrote countless newspaper articles and letters to the editor in statewide and local newspapers asserting that chaparral was crucial to watershed management. He reiterated this point in an open letter to Nathaniel Egleston, head of the Division of Forestry in the Department of Agriculture. In the 1885 missive, he chastised the agency for omitting California from its reports about the link between forest protection and valley irrigation.

Kinney informed the Washington bureaucrat that the Golden State "has greater material interests dependent on irrigation than any other" (Kinney 1885). The state also offered one of the last remaining opportunities "still open to the Government to withdraw the forest land and see to its exploitation and preservation" (Kinney 1885).

Mutual water companies, local chambers of commerce, public officials, recreation groups, hoteliers and outfitters also rallied around the mountains' interests and their own. The city of Santa Barbara, for example, had already secured property in the upper reaches of its watershed to protect its potable supplies and urged the federal government to undertake tighter management of the Santa Ynez Mountains for fire and water flows (Brown 1945). In San Diego County, over-grazing of and fire damage to local watersheds generated calls for protection of the Elfin Forests—an affectionate term for chaparral—that overlay the watersheds of the Palomar, Laguna, and San Jacinto Mountains (Newland 2008). Public opinion throughout California rallied quickly after the 1891 passage of the Forest Reserve Act, the legislation granted the president the power to withdraw portions of the public domain to create what were then called forest reserves (Godfrey 2005). Although the law did not indicate how these reserves would be managed, a failing that the 1897 Forest Management Act would rectify, this lacuna did not stop Californians from buttonholing B.F. Allen, special agent of the Interior Department investigating the state's potential for new reserves. So unified was their lobbying campaign that Allen cabled his superiors in Washington immediately to withdraw a four million acre swath of the Sierras, from Yosemite south through Kern County. First called the "Tulare Reserve" and now known as the Sierra National Forest, it incorporated much the chaparral biome in the central and southern extent of the range and it was set aside not as a park, but, as one of its proponents observed, to save water supply irrigation below and to preserve timber" (Godfrey 2005). Special gent Allen also advised the Secretary of the Interior that the San Gabriels should be set aside, along with the San Bernardino Mountains and portions of the Santa Ana. In rapid order, President Benjamin Harrison signed proclamations denoting the San Gabriel Timberland Reserve (December 20, 1891) and the Sierra Forest Reserve (February 14, 1893). That same February, he proclaimed reserves covering the Trabuco Canyon (now part of the Cleveland National Forest) and the San Bernardino Mountains (Godfrey 2005).

President Harrison's actions set the stage for a radical new conception of the purposes of the public domain. Hitherto, Congress' ambition had been to sell or give away millions of acres to homesteaders, farmers, loggers, miners, and railroad corporations to encourage settlement and development. By the late nineteenth century, as Californians' agitation demonstrated, this policy had gained an array of detractors. Engaged citizens pressed for a more robust nation-state that would regulate the public lands and the resources they contained. The idea that managing nature upstream to sustain human interests downstream, and that Washington could and should resolve local disputes over resource allocation and consumption, signaled a broader desire for a more effective national government. The call for making public life more orderly, rational, and manageable was a hallmark of Progressive Era reform and activism. Emblematic of this era's ethos was the establishment of the initial forest reserves, as well as the 1905 formation of the US Forest Service to manage them (Miller 2016a). In California, chaparral was a formative part of this powerful political discourse.

5.7 Transplanted Ideas

Chaparral did not hold this privileged position for long. Two institutions that emerged in response to the demand for more rigorous management and regulation of the region's mountains, the Los Angeles County Fire Department (LACFD, established in 1911) and the US Forest Service (1905), frequently considered this habitat to be an impediment to their managerial schemes.

Los Angeles County Forester Stuart J. Flintham, for instance, doubted chaparral's utility. With two degrees in forestry, his first job was with the US Forest Service, serving as a forest inspector in California. In 1907, he was appointed supervisor of the Stanislaus National Forest but within the next year Chief Gifford Pinchot had fired Flintham for his reported inability to get along with local communities. This negative characterization would have surprised those who hired Flintham in 1911 to run LA County's fledgling forestry office. They found him an engaging, hardworking, and innovative manager. Flintham's greatest impact, however, may have been in his pioneering efforts to control southern California's wildland fires (Miller 2012). Drawing on some of the lessons that he and his professional colleagues learned as a result of the devastating Big Burn of 1910, which had scorched nearly 1.3 million ha (~3 million acres) in Washington, Idaho, and Montana, Flintham devised a systematic approach to identifying, monitoring, and, where possible, suppressing wildfire. This included cutting hundreds of miles of firebreaks, procuring horses, mules, and vehicles to mount patrols during fire season, and constructing lookout towers to increase surveillance capabilities. The county forester's office first utilized the telephone to speed up communication between firefighters on the ground, and by the early 1920s had devised a mobile radio unit to coordinate its efforts. Even the local US Army air squadron was enlisted: its bi-planes provided aerial mapping services and conducted fire overflights. In 1924, Flintham's deft administration earned the praise of one of his Yale classmates, William B. Greeley, Chief of the US Forest Service. After inspecting the LA County forestry program, Greeley applauded the organization's efficiency and effectiveness (Miller 2012).

Greeley's applause is not that surprising, for the federal agency had been busy adopting southern California firefighting agencies' strategy of laying down firebreaks along ridgelines in the San Gabriel and San Bernardino mountains in hopes of increasing firefighters' chances of suppressing fires. In 1914, the Sierra and Sequoia national forests managers began to construct similar firebreaks in hopes of slowing and controlling chaparral fires in the Central Sierra. Fifteen years later, S.B. Show proposed a massive expansion of this earlier infrastructure. He dubbed it the Ponderosa Way, and it was to run 1106 km (687 miles) along the western front of the Sierra and Cascades, from Kern River in the south to the Pit River in the north. In 1933, capitalizing on New Deal dollars and Civilian Conservation Corps labor, Show's project got underway and within a year 708 km (440 miles) of the Ponderosa Way had been cleared. US Forest Service officials would discover, as had southern California firefighting agencies, that firebreaks were expensive to maintain and did not stop fires on their own. What they came to realize after the tough fire seasons in the 1920s and 1930s is that these were more effective as cleared space that firefighters could utilize "as a line of defense where high values were at stake" (Cermak 2005).

Not everyone was convinced that federal and local firefighting strategies were effective, regardless of technology or technique. In the charred aftermath of southern California's 1924 fire season (which was so bad that Greeley raced west to consult with his US Forest Service subordinates and Stuart Flintham [Los Angeles Times 1924]), the Mission Indian Federation of San Jacinto, California sent a letter to President Calvin Coolidge asserting that the native people could do a better job managing the local national forests, a claim that rested on their millennia-long experience with chaparral ecosystems. The Chief Executive countered by privileging scientific management and technical expertise: "the administration of the national forests presented a great many problems, and...called for the services of men of wide experience and training" (Godfrey 2005).

Further signifying the privilege granted scientific and technical expertise was LACFD's and the US Forest Service's conviction that "California's brush types could be readily converted to forest" (Radtke 1978). During the first three decades of the twentieth century, for example, LACFD planted tens of thousands of seedlings in the expectation that replacing the Mediterranean-type climate region shrubs with pine and fir would result in a decrease in the number of major fires and slow hillside erosion. Despite the high mortality of these non-native trees, enough "somehow survived to nourish the dream of converting the chaparral to forest as a by-product of watershed erosion control plantings" (Radtke 1978). The dream faded when Flintham's successor, Chief Spencer Turner, came to recognize that "resprouting chaparral strongly competed for moisture, light, and nutrients, and therefore caused great tree mortality" and that this endemic habitat was "a precious watershed cover that perhaps is fire dependent and best adapted to the site" (Radtke 1978). In 1930, Turner issued a restraining order: his foresters henceforth would "plant less, plant better" (Radtke 1978).

The US Forest Service did not follow LACFD's lead in this case. Despite knowing that non-native species had demonstrable difficulty adapting to the low fertility of local soils and the region's alternating weather patterns of drought and deluge, to say nothing of the speed of chaparral's postfire regeneration, the federal agency frequently endeavored to re-engineer the San Gabriels' chaparral based ground cover. Early in the twentieth century, for instance, it collaborated with the county's afforestation experiments, with predictable results: "In the 1920s, a million trees including exotic Canary Island pines were planted in the San Gabriel Mountains in a misguided effort to fix something that was not a problem—a predominance of native chaparral," the Chaparral Institute's Richard Halsey told the Los Angeles Times. "Most of those trees died because of drought" (Sahagun 2011).

Relatedly, consider how the federal agency responded in the aftermath of a 1960 blaze that roared through upper San Dimas Canyon (Miller 2016b). Once more, US Forest Service researchers experimented with converting chaparral to grassland, this time in the 6880 ha (17,000 acre) San Dimas Experimental Forest, which had been founded in 1933 as a "center for hydrologic research in mountainous watersheds" in the San Gabriel Mountains (USDA Forest Service PSW n.d.-a; Miller 2016b). The post-1960 fire research was concerned with determining "first-aid treatments aimed at reducing the damage" the fire had generated. In particular, it tested vegetative responses, including comparative "establishment and growth of artificially seeded species, the regrowth of native vegetation, and the effects of plant growth on watershed rehabilitation" (Corbett and Green 1965). Researchers randomly selected eight watersheds to seed with annual grasses and eight with perennial grasses, and four were left untouched as a control group (Corbett and Green 1965). What they (re)discovered was that native plant material is a fierce competitor. To restrain its competitive energy, the research team sprayed the grass-planted watersheds with herbicides, a strategy that comes with two ironies. The first is cultural: Rachel Carson's *Silent Spring* (1962) had just been published, a book that decried the very poisoning of America's waterways and life systems in which the agency was now complicit. The second is ecological: whatever the downstream consequences of the herbicidal spray, upslope the grasses' growth rate slowed over the project's life-time, and, despite the chemical assault the indigenous plants endured, they took root and won out. In the control-group watersheds, chaparral, Manzanita, buckwheat, ceanothus, deervetch, and morning glory—which began resprouting within 10 days of the fire—performed even better. After 4 years, these and other opportunistic plants had revegetated approximately 50% of the burned watersheds, a speed of recovery and density of cover that modern technology could not replicate. The researchers did not fully accept the evidence their studies revealed, instead, they insisted that under ideal conditions—ample rain, significant labor, and chemical applications—the "land managers would have been justified in trying to establish a grass crop." Still, they acknowledged, "this treatment should not be taken as a cure-all" (Corbett and Green 1965).

Forty-five years later, that data-driven insight about chaparral's competitive advantage resurfaced in the US Forest Service's initial response to the Station Fire, which charred upwards of 65,155 ha (161,000 acres) of the San Gabriel Mountains between August and October 2009. In theory, the official Station Fire Recovery Strategy (Angeles National Forest 2011) was to "listen to the land," to develop "ecosystem recovery actions that facilitate the natural recovery process to the greatest extent possible," and to use "field-based science and adaptive management to understand what the natural recovery cycle is, and then to use that information in designing, scheduling and monitoring recovery projects that work with nature." In practice, the US Forest Service initiated a massive reforestation project centered on the planting of Coulter pine. The project drew sharp criticism. "If they are planting big-cone Douglas fir (sites) with Coulter pine," argued USGS ecologist Jon Keeley, "that is not an appropriate form of ecosystem management" (Sahagun 2011). Chaparral Institute president Richard Halsey predicted failure: "The reality we live in is a Mediterranean climate, and there is just not enough water to create what they have in mind" (Sahagun 2012). One year later, 75% of the 900,000 seedlings planted in the burn areas had died (Sahagun 2012).

5.8 Resilient Chaparral: A Conclusion

Each attempt in the modern era to convert chaparral to forest or grassland has revealed instead the native habitat's endurance. Its success has reinforced the need for public land managers to work with chaparral not against it, a reconceptualization of its status that has received a boost from a series of interlocking policy decisions. The passage of the National Environmental Policy Act (1970) and the Endangered Species Act (1973), and other regulatory initiatives, compelled federal land management agencies to broaden the scope of their managerial missions to include the protection of wildlife and the promotion of biodiversity. To do so meant they also had to hire larger numbers of biologists and related disciplines to fulfill these new, mandated obligations (Miller 2012). In California, these changes led to a greater focus on chaparral as a critical habitat for a number of threatened and endangered species—including the iconic California Condor, whose "nesting occurs primarily in the rugged, chaparral covered mountains" of the state's central coast (US Fish and Wildlife Service 2016). Relatedly, in 1976 the San Dimas Experimental Forest, site of some of the most important baseline research into chaparral, became a UNESCO Biosphere Reserve. This designation reinforced the shift in attitude toward this plant community, for the primary objectives of this international system of reserves are "to identify and protect representative and unique segments of the world's biotic provinces as major centers for biotic and genetic preservation, ecological and environmental research, education, and demonstration" (Franklin 1977). There have been other markers of a renewed appreciation for chaparral's centrality as reflected in a series of chaparral restoration projects across California (see Chap. 15), and especially those mounted on the Los Padres National Forest (USDA Forest Service n.d.), the Angeles National Forest (USDA Forest Service 2014), and the San Bernardino National Forest (USDA Forest Service PSW, n.d.-b). These early twenty-first century initiatives have received White House sanction in the acts that established the Sand to Snow National Monument (Presidential Proclamation 2016) and the Berryessa Snow Mountain National Monument (Presidential Proclamation 2015), each of which cited chaparral as a key habitat that required the added protections that national monument designation offered. President Barack Obama made a similar claim in the official proclamation creating the San Gabriel Mountains National Monument:

> "The mountains harbor several of California's signature natural vegetation communities, including the drought tolerant and fire-adapted chaparral shrubland, which is the dominant community and includes scrub oaks, chamise, manzanita, wild lilac, and western mountain-mahogany. Mixed conifer forest is an associated vegetation community comprising Jeffrey pine, sugar pine, white fir, and riparian woodlands including white alder, sycamore, and

willow. These communities provide habitat for numerous native wildlife and insect species, including agriculturally important pollinators, the San Gabriel Mountains slender salamander, San Bernardino Mountain kingsnake, song sparrow, Peregrine falcon, mule deer, and Pallid bat. (Presidential Proclamation 2014; San Gabriel Mountains National Monument 2016)".

Chaparral's newfound status also has been shoveled into the landscape surrounding the headquarters of the Angeles National Forest in Arcadia, California. Completed in 2012, and sited so that it faces the southern slope of the San Gabriel Mountains, the LEED Gold-certified structure is set off by such plants as deer grass, manzanita, sage, ceanothus, and chamise, and is shaded by oak and sycamore (Los Angeles Daily News 2012). This choice of vegetation was designed to educate residents of the San Gabriel Valley (and beyond) about the possibility of replacing water-dependent, monocultural lawns with a diverse array of chaparral vegetation. Drought tolerant and resilient, this aesthetic should enable southern California property owners to weather what the US EPA predicts will be a century-long drying out of the southwestern region (US EPA 2014)—yet one more example of chaparral's symbolic importance, educative impact, and essential nature.

References

Anderson, M. K. 2005. Tending the wild: Native American knowledge and the management of California's natural resources. University of California Press, Berkeley, California, USA.

Angeles National Forest. 2011. Station fire restoration strategy. Revised. http://www.fs.usda.gov/Internet/FSE_DOCUMENTS/stelprdb5312770.pdf

Bernard, J., D. Robinson, and F. Sturt. 2014. Points of refuge in the south central California colonial hinterlands. Pages 154-171 *in* L. M. Panich and T. D. Schneider, editors. Indigenous landscapes and Spanish missions: new perspectives from archaeology and ethnohistory. University of Arizona Press, Tucson, Arizona, USA.

Brown, W. S. 1945. History of the Los Padres National Forest. USDA Forest Service, Regional Forest Office, San Francisco, California, USA.

California State Board of Forestry. 1885. Annual report (1885-1886). State Printing Office, Sacramento, California, USA.

Carson, R. 1962. Silent spring. Houghton Mifflin, Boston, Massachusetts, USA.

Cermak, R. W. 2005. Fire in the Forest: A History of Forest Fire Control on the National Forests in California, 1898-1956. USDA Forest Service: California. July. R5-FR-003. USA.

Corbett, E. S., and L. R. Green.1965. Emergency revegetation to rehabilitate burned watersheds in southern California. Research Paper PSW-RP-22. USDA Forest Service, Berkeley, California, USA.

Cronon, W. 1997. The trouble with wilderness. Or, getting back to the wrong nature. Pages 28-50 *in* Miller, C. and H. K. Rothman, editors. Out of the woods: essays in environmental history. University of Pittsburgh Press, Pittsburgh, Pennsylvania, USA.

Dawson, M. 1958. Mining in Upper San Antonio Canyon. Castle Press, Pasadena, California, USA.

Deitler, J., H. Gibson, and J. M. Potter, editors. 2016. Abundant harvests: the archaeology of industry and agriculture at San Gabriel Mission. University of Arizona Press, Tucson, Arizona, USA.

Franklin, J. F. 1977. The Biosphere Program in the United States. Science 195:262-67.

Fried, J. S., C. L. Bosinger, and D. Beardsley. 2004. Chaparral in southern and central coastal California in the Mid-1990's: area, ownership, condition, and change. Resource Bulletin PNW-RB-240. USDA Forest Service, Berkeley, California, USA.

Gamble, L. H. 2011. The Chumash world at European contact. University of California Press, Berkeley, California, USA.

Godfrey, A. 2005. The ever-changing view: a history of the national forests in California. Publication FR5-FR-004. USDA Forest Service. Pacific Southwest Region, California, USA.

Hackenberger, B., and C. Miller. 2017. Watershed politics: groundwater management and resource conservation in southern California's Pomona Valley. Journal of Urban History. https://doi.org/10.1177/0096144217692986

Halsey, M. D. 1912. A tenderfoot in southern California. J. J. Little & Ives Co., New York, New York, USA.

Heald, W. F. 1963. The lordly San Jacintos. Westways 55:16-17.

Hoffman, A. 1976. Mountain resorts and trail camps in southern California's great hiking era, 1884-1938. Southern California Quarterly 53:381-406.

Jackson, H., and A. Kinney. 1883. Report on the condition and needs of mission Indians of California. Government Printing Office, Washington, DC, USA .

Kinney, A. 1885. The forestry question in California. Pacific Rural Press 33:459.

Kinney, A. 1888. The forests of Los Angeles, San Diego, and San Bernardino counties, California. Report on the forest conditions in the Rocky Mountains and other papers. Government Printing Office, Washington, D.C., USA.

Leonard, Z. 1839. Narrative of the adventures of Zenas Leonard. D.W. Moore. Clearfield, Pennsylvania, USA.

Lindsay, B. C. 2012. Murder State: California's Native American genocide, 1846-1873. University of Nebraska Press, Lincoln, Nebraska, USA.

Los Angeles Daily News. 2012. December 11. New $12 million Angeles National Forest headquarters is dedicated. http://www.dailynews.com/article/ZZ/20121211/NEWS/121218463

Los Angeles Times. 1924. August 8. Forest head speeds west.

Madley, B. 2016. An American genocide: the United States and the California Indian catastrophe, 1846-1873. Yale University Press, New Haven, Connecticut, USA.

Miller, C. 2012. Public lands, public debates: a century of controversy. Oregon State University Press, Corvallis, Oregon, USA.

Miller, C. 2016a. America's great national forests, wildernesses, and grasslands. Rizzoli, New York, New York, USA.

Miller, C. 2016b. Not so golden state: sustainability vs. the California dream. Trinity University Press, San Antonio, Texas, USA.

Muir, J. 1894. The mountains of California. The Century Co., New York, New York, USA.

Muir, J. 1901. Our national parks. Houghton Mifflin, Boston, Massachusetts, USA.

Muir, J. 1911. My first summer in the sierra. Houghton Mifflin, Boston, Massachusetts, USA.

Muir, J. 1918. Steep trails. Houghton Mifflin, Boston, Massachusetts, USA.

Newland, J. D. 2008. Cleveland National Forest. Arcadia Publishing, Chicago, Illinois, USA.

Panich, L. M., and T. D. Schneider. 2014. Indigenous landscapes and Spanish missions: new perspectives from archaeology and ethnohistory. University of Arizona Press, Tucson, Arizona, USA.

Patterson, T. 2014. From acorns to warehouses: historical political economy of southern California's Inland Empire. Routledge, New York, New York, USA.

Presidential Proclamation. 2014. San Gabriel Mountains National Monument. https://www.whitehouse.gov/the-press-office/2014/10/10/presidential-proclamation-san-gabriel-mountains-national-monument

Presidential Proclamation. 2015. Berryessa Snow Mountain National Monument. https://www.whitehouse.gov/the-press-office/2015/07/10/presidential-proclamation-establishment-berryessa-snow-mountain-national

Presidential Proclamation. 2016. Snow to Sand National Monument. https://www.whitehouse.gov/the-press-office/2016/02/12/presidential-proclamation-establishment-sand-snow-national-monument

Radtke, K. 1978. Wildland planting & urban forestry: native & exotic, 1911-1977, Department of Forester and Fire Warden, County of Los Angeles, Los Angeles California, USA.

Resendez, A. 2016. The other slavery: the uncovered story of Indian enslavement in America. Houghton Mifflin, Boston, Massachusetts, USA.

Robinson, J. W. 1977. The San Gabriels: southern California mountain country. Golden West Books, San Marino, California, USA.

Robinson, J. W. 1989. The San Bernardinos: the mountain country from Cajon Pass to Oak Glen; two centuries of changing use, Big Santa Anita Historical Society, Arcadia, California, USA.

Sahagun, L. 2011. April 15. 3 million trees to be planted in the Station fire burn area. Los Angeles Times, Los Angeles, California, USA. http://articles.latimes.com/2011/apr/15/local/la-me-station-fire-20110415

Sahagun, L. 2012. April 7. Reforestation not taking hold in land burned by Station fire. Los Angeles Times, Los Angeles, California, USA. http://articles.latimes.com/2012/apr/07/local/la-me-dead-trees-20120408

San Gabriel Mountains National Monument. 2016. San Gabriel Mountains National Monument management plan environmental assessment. USDA Forest Service, Angeles National Forest, Arcadia, California, USA. http://a123.g.akamai.net/7/123/11558/abc123/forestservic.download.akamai.com/11558/www/nepa/101660_FSPLT3_3046119.pdf

Saunders, C. F. 1923. The southern Sierras of California. Houghton Mifflin, Boston, Massachusetts, USA.

Spence, M. D. 1999. Dispossessing the wilderness: Indian removal and the making of the National Parks. Oxford University Press, New York, New York, USA.

US EPA [US Environmental Protection Agency]. 2014. Climate impacts in the southwest. https://www3.epa.gov/climatechange/impacts/southwest.html

US Fish and Wildlife Service. 2016. California Condor Recovery Program. https://www.fws.gov/cno/es/calcondor/Condor.cfm

USDA Forest Service. 2014. News Release, ANF-06-FY15. US Forest Service, Angeles National Forest and San Gabriel Mountains National Monument, California, USA. http://www.fs.usda.gov/Internet/FSE_DOCUMENTS/stelprd3823865.pdf

USDA Forest Service. n.d.. Zaca and Piru fires restoration conservation strategy. http://www.nfwf.org/lpnf/Documents/lospadres_conservation-strategy.pdf

USDA Forest Service PSW [Pacific Southwest Region]. n.d.-a. San Dimas Experimental Forest: a vision. Miscellaneous PSW-MISC-78. USDA Forest Service, Pacific Southwest Research Station, Albany, California, USA. http://www.fs.fed.us/psw/publications/documents/psw_misc078/psw_misc078.pdf

USDA Forest Service PSW [Pacific Southwest Region]. n.d.-b. San Bernardino National Forest ecological restoration implementation plan. http://www.fs.usda.gov/Internet/FSE_DOCUMENTS/stelprdb5411441.pdf

Worster, D. 2008. A passion for nature: the life of John Muir. Oxford University Press, New York, New York, USA.

Chapter 6
Biogeochemical Cycling of Carbon and Nitrogen in Chaparral Dominated Ecosystems

G. Darrel Jenerette, Isaac W. Park, Holly M. Andrews, and Jennifer R. Eberwein

Abstract The controls and trajectories of biogeochemical dynamics in chaparral dominated ecosystems are highly diverse. This breadth of diversity in chaparral biogeochemical dynamics is caused by a combination of diverse biogeophysical drivers, self-organization, and sensitivity to historical legacies. The high topographic heterogeneity and decoupling of moisture inputs and energy inputs in the winter precipitation dominated climates imposes unique ecohydrologic adaptations that influence plant and microbial metabolic activities and distributions throughout chaparral landscapes. Chaparral biogeochemical dynamics, especially interactions among carbon, nitrogen, and hydrologic cycling, may have important implications for the ecosystem services and disservices provided by these ecosystems, including greenhouse gas emissions and pollutants to air and water, and influence the vulnerability of chaparral biological communities to global change drivers. These ecosystems may have a large capacity for carbon storage and sustained carbon sequestration at rates comparable with those of old-growth forests. In other places, chaparral ecosystems may have low pools of carbon and low rates of sequestration. While commonly considered nitrogen limited, chaparral dominated ecosystems may also have large nitrogen emissions through atmospheric and hydrologic pathways. Future trajectories of chaparral biogeochemical dynamics are also variable. Unprecedented interactions among global change drivers, including climate, nitrogen deposition, fire frequency, and invasion risks, are pushing many of these systems to tipping points of reorganization and type-conversion to grass dominated states with uncertain biogeochemical consequences. Management for maintaining ecosystem services associated with biogeochemical dynamics can be improved with growing opportunities provided by a suite of environmental sensors and next generation modeling approaches.

Keywords Biodiversity · Biogeochemistry · Carbon · Ecosystem · Global change · Invasion · Nitrogen · Tipping point

---

G. D. Jenerette (✉) · I. W. Park · H. M. Andrews · J. R. Eberwein
University of California, Riverside, CA, USA
e-mail: darrel.jenerette@ucr.edu

E. C. Underwood et al. (eds.), *Valuing Chaparral*, Springer Series on Environmental Management, https://doi.org/10.1007/978-3-319-68303-4_6

6.1 Introduction to Biogeochemical Cycling in Chaparral

Biogeochemical dynamics in chaparral dominated ecosystems are unusually complex and provide a valuable testbed for evaluating general ecosystem theories. What is known suggests important differences between chaparral and most other terrestrial ecosystems. Notably, the rates of C and N accumulation, transformations, and losses in chaparral dominated ecosystems vary dramatically in time and space. As such, chaparral ecosystems have a potential for long-term carbon (C) accumulation (Luo et al. 2007) and high rates of nitrogen (N) losses (Homyak et al. 2014). However, the actual rates of C and N accumulation, transformations, and losses in chaparral dominated ecosystems vary considerably in time and space. The sources of chaparral biogeochemical variability, in part, depend on underlying biogeophysical drivers, historical legacies, and self-organizing processes connecting hourly to millennial time-scales and local to landscape spatial scales (Fig. 6.1). These interacting drivers dictate, to a large degree, the availability of moisture, which has an overriding influence on chaparral biogeochemical functioning.

The chaparral landscapes of California show extensive variation in their biogeophysical template; the combination of geology, climate, and available species that sets the stage for biogeochemical cycling. Important factors include the distribution of bedrock origins, climate gradients spanning coastal to inland and latitudinal gradients, and rugged topography. Distributions of biological traits of plants and microbes are another source of biogeochemical diversity that can moderate or accentuate variation from the geophysical template. Historical legacies, past conditions that influence current functioning, further influence biogeochemical variation and include dramatic events such as wildfire or variation in precipitation across event to decadal time-scales. Historical legacies can influence biological responses to the geophysical template and biological capacity. A distinguishing feature of biogeochemical cycling in chaparral, in part because of the diversity of environmental conditions and ecological responses, is a high diversity in the size of specific material pools (e.g., above-ground biomass, roots, litter, soil) and similarly high diversity of the controls to fluxes among these pools. Furthermore, other components of

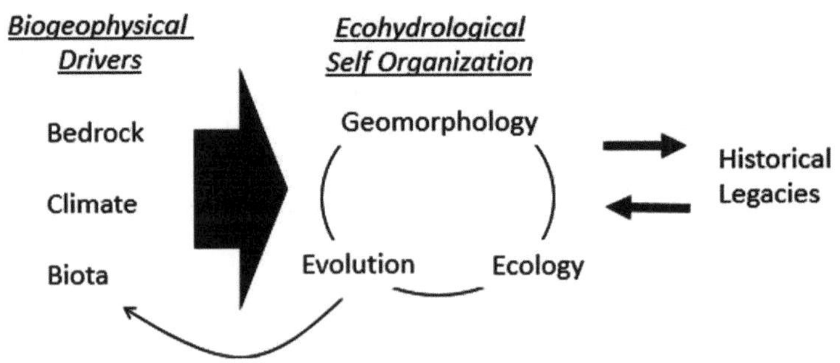

Fig. 6.1 Controls on chaparral biogeochemical cycling through external drivers, ecohydrological self-organization, and historical legacies

chaparral ecosystems can cause slowing or accelerating of the biogeochemical cycle. For example, ecosystem characteristics such as water availability during cooler periods and species with low rates of photosynthesis tend toward the slowing of biogeochemical cycles, while the rapid response of chaparral ecosystems to wetting and rapid soil movement tend toward accelerated biogeochemical cycling.

Future chaparral biogeochemical cycling will respond to inter-active global change drivers through several potentially interacting pathways. These drivers influence the biogeophysical template through changes in climate and nutrient inputs, biotic diversity, expansion of non-native species, and land use that removes and fragments chaparral communities. Furthermore, the combination of global change drivers may increase opportunities for community type change through conversion of native shrub to non-native grasses with unknown implications for the biogeochemical cycle. Understanding the spatial distribution and controls of biogeochemical dynamics in chaparral provides an essential basis for assessing future trajectories of these ecosystems and maximizing their services while providing valuable directions for improving the next generation of ecosystem and land surface models that include chaparral dynamics.

6.2 Controls of Biogeochemical Variation in Chaparral

6.2.1 Boundary Delineation of Chaparral Biogeochemistry

Ecosystem science operates from two often contrasting viewpoints: the Newtonian view where understanding broadly applicable general mechanisms is desired and the Darwinian view where understanding is derived from specific details of particular systems that are used to mechanistically build up from first principles (Harte 2002; Jenerette et al. 2012). In reconciling these alternate views, an ecosystems framework that relies on mass balanced fluxes of elements between different pools has been a fundamental perspective for examining biogeochemical dynamics that allows both for broad generality and detailed specificity (Odum 1983). A systems approach is used as a basis for biogeochemical and climate modeling and is leading to new developments in ecosystem theory (Wu and David 2002). Using a systems approach allows quantification of mass and energy balance and exchanges among distinct pools within the ecosystem and between the ecosystem and the environment (Odum 1983; Chapin et al. 2012). This framework allows an evaluation of how controls and general mechanisms vary among systems, while also facilitating detailed evaluation of individual systems.

An essential activity for any analysis within the ecosystem framework is boundary delineation. A fundamental challenge is that ecosystems are inherently open systems. Boundaries are necessarily imposed and can differ depending on the purpose of analysis. For chaparral biogeochemical cycling, boundaries can vary spatially, in both the vertical direction extending through the soil profile and vegetation and horizontally across the land surface, encompassing a diversity of landscape components. Boundaries also change temporally, spanning instantaneous snapshot to millennia. Vertically, boundaries typically include the upper few centimeters of soil profile to the top of the plant canopy that is readily accessible for measurements. However, extending the vertical profile deeper to the plant roots is desirable both to better characterize the active ecohydrological layer and its influence on plant dynamics, and the potential for storage of biogeochemical constituents. However, characterizing biogeochemical dynamics deeper in the soil poses logistical challenges and can be disruptive to the ecosystem. At the most complete scope, the critical zone—spanning the bottom of groundwater to the top of the trees, provides an increasingly used framework for bounding ecosystem processes (Amundson et al. 2007; Chorover et al. 2011). The full critical zone provides a comprehensive perspective for examining long-term biogeochemical dynamics resulting from coupled geomorphic and ecological dynamics. In the horizontal direction, several alternative boundary delineations are frequently used including watersheds, management units, and plant community extents (Fortin et al. 2000; Bailey 2004). Practically, much biogeochemical characterization occurs within individual sampling plots ranging from m^2 to ha. Temporally, most analyses are bounded by only a few years of data, however, a limited number of studies of chaparral biogeochemistry span a few decades. The fine scale nature of ecological research compared to the expected time-scales of ecohydrological changes poses important limitations that need addressing. While boundaries are inherently arbitrary, imposing boundaries to chaparral biogeochemistry is an essential task for quantifying rates of biogeochemical cycling with important consequences for estimates of pool sizes and exchange rates.

6.2.2 Geophysical Template for Chaparral Biogeochemistry

While the fundamental components of biogeochemical cycling in chaparral are consistent with general ecosystem theory, the unique characteristics of chaparral ecosystems limit our understanding of how biogeochemical cycle components, their magnitudes, and their drivers vary. Chaparral dominated ecosystems occur in Mediterranean-type climate (MTC) regions that receive primarily winter precipitation. This imbalance between energy inputs in the summer months and moisture in the winter months imposes selective pressure for unique ecohydrologic adaptations that influence plant and microbial metabolic activities and distributions throughout the landscape (Baldocchi and Xu 2007; Pumo et al. 2008). Interacting with this general climate influence, the distribution of chaparral across latitudinal, coastal to inland, and elevation gradients has implications for the distributions of precipitation, temperature, and potential evaporation. Geologic variation also leads to differences in soil conditions, erosion, and deposition that can influence soil moisture distributions and nutrient availability. Through erosion and deposition processes chaparral ecosystems are characterized by high topographic heterogeneity, which further impacts water availability. The rugged topography leads to complex patterns in water availability, through large variation in insolation and upslope accumulation, influencing water loss and inputs respectively. Large rates of soil movement associated with the characteristically steep slopes in chaparral dominated regions also lead to large variation in the capacity of soil to hold water.

Precipitation in MTC regions imposes a key constraint on biological processes—these systems are by definition water-limited and changes in water availability have direct and indirect consequences to biogeochemical cycling (e.g., Biederman et al. 2016). The effects of precipitation can occur within minutes, however, it can leave legacies persisting across seasons, years, and potentially even decades (Jenerette et al. 2008; Scott et al. 2009; Shen et al. 2016). Total annual precipitation is commonly the principal variable used to quantify ecosystem responses to moisture. However, the partitioning of precipitation between hydrologically distinct flow paths such as transpiration, soil evaporation, or runoff can also have a large influence on how the ecosystem and rates of biogeochemical cycling respond to precipitation distributions. Independent of the total annual precipitation, the variation in timing, intensity, and distribution of precipitation can have large effects on moisture partitioning and lead to contrasting influences on biogeochemical cycling (Huxman et al. 2004; Shen et al. 2008).

With seasonal precipitation in chaparral ecosystems dominated by winter rainfall, plant growth, and many associated ecosystem dynamics are generally more closely connected to winter rainfall patterns. In contrast, moisture from summer rainfall in June–September (a total average of <10 mm in southern California and <15 mm in northern California chaparral) is less likely to contribute to plant growth and generally is rapidly lost through soil evaporation. Within the winter season, variation in the distribution of rainfall can have large influences on plant driven biogeochemical cycling. For example, high intensity rainfall events can lead to increased runoff as soil infiltration capacity is exceeded (see Chap. 7). Alternatively, clusters of rainfall events can lead to increased runoff losses as soil water holding capacity is filled through successive wetting. Both situations reduce precipitation-related moisture available in the ecosystem. In contrast, small low-intensity rainfall events may effectively only wet the leaf or soil surface and be primarily lost through evaporation. Large gaps between rainfall events may be associated with increased proportions of moisture lost through soil evaporation. The effects of contrasting precipitation distributions can influence plant uptake of C and N and subsequent litter contributions of C and N to the soil. Pulse responses to cycles of drying and re-wetting can have a large influence on biogeochemical cycling in chaparral dominated ecosystems (Miller et al. 2005; Jenerette and Chatterjee 2012; Jenerette et al. 2012). All of these components of precipitation can influence the effective moisture in the soil and the capacity of biogeochemical processes.

Across the range of chaparral vegetation, multiple geographic gradients further influence local energy balance and moisture availability. The large latitudinal gradient of chaparral, spanning most of California (Fig. 1.1), is further associated with a large variation in ecohydrologic dynamics, with more moisture and cooler temperatures in northern chaparral communities compared to southern communities. Landscape position along coastal to inland gradients are similarly associated with climate differences, with coastal chaparral communities generally wetter and cooler than more inland communities. Variation in elevation can also influence local climate, with drier and hotter conditions at lower elevation and cooler and wetter conditions at higher elevations. These geographic gradients lead to a complex mosaic of climate distributions influencing chaparral biogeochemical cycling.

Somewhat independent of the climate gradients is variation in underlying geology including bedrock and mountain formation. Bedrock mineralogy can influence many biogeochemical processes, for example, soils from parent materials dominated with non-crystalline materials, e.g., andesite, may have higher rates of soil C mineralization than soils from parent materials dominated by crystalline materials, e.g., granite (Rasmussen et al. 2006). An extreme of geological influence can been seen in chaparral found on serpentine soils. These ecosystems show characteristic responses to low nutrient and high metal content through communities with slow growth, dwarfed stature, and reduced cover (Kruckeberg 1984; Gough et al. 1989) (see Chap. 1).

6.2.3 Biotic Diversity as a Driver of Chaparral Biogeochemistry

While individual communities are generally dominated by only a few locally prevalent species, total plant species diversity in chaparral vegetation throughout California is high. Recent estimates suggest chaparral harbors 1177 vascular plant species, a substantial proportion of the plant diversity in California (Keeley 2005; Halsey and Keeley 2016). The biogeochemical diversity includes both the potential for high variation within individual sites and large differences among sites throughout the entire range of chaparral ecosystems (see Chap. 2). Species variation across broad geophysical gradients further influence biogeochemical diversity within chaparral.

In addition to taxonomic biodiversity, functional biodiversity, the breadth of traits represented in the community, also has a critical role on the influence of chaparral species and consequently biogeochemical cycling. The species comprising chaparral communities have a large breadth of functional diversity (Ackerly 2004; Pivovaroff et al. 2016). Typically, functional diversity has been considered along suites of potentially distinct axes such as leaf photosynthetic rate, size, parental investment, and hydraulic capacity (Westoby 1998; Li et al. 2015; Diaz et al. 2016). Leaf axes of functional variation are primarily arrayed along a leaf economic spectrum of high maximum rates of photosynthesis and short leaf lifespan versus lower maximum rates of photosynthesis and longer lifespans (Wright et al. 2004). Leaf traits further influence soil biogeochemical cycling as the shorter-lived leaves are also associated with more rapid decomposition (Cornwell et al. 2008). Species typically dominant in chaparral communities are generally associated with characteristics of the slow-end of the leaf economic spectrum with low N concentration, low maximum rates of photosynthesis, and slow to decay litter. Another key plant characteristic that varies among chaparral species is rooting depth, which influences the zone of biologically available water with concomitant changes to hydrology and plant functioning (Scott et al. 2014).

In MTC regions, diverse trait combinations have evolved to address fundamental constraints of water limitation but also take advantage of the large ecohydrological diversity within chaparral dominated ecosystems. Drought adaptation strategies among chaparral species are diverse and include strategies that look to maximize one or combinations of vulnerabilities to cavitation, sapwood capacitance, deciduousness, photosynthetic stems, photosynthetic responses to leaf water potential, and hydraulic architecture (Ackerly 2004; Pivovaroff et al. 2016). Within chaparral ecosystems, many distinct plant community associations can be identified in addition to the dominant shrubs including grass, conifer forest, and riparian communities. These individual communities also vary in height, leaf, and rooting depth of the dominant organism. In addition, subdominant plant species also contribute to functional diversity of plant traits. For example, shrub communities have a breadth of plant heights spanning low growing herbs and taller trees and are characterized by long-lived leaves and low rates of photosynthesis. In contrast, many other interstitial members of the plant community, including annual grasses, have much shorter-lived leaves and higher rates of photosynthesis. Chaparral species also express extensive variation in rooting depth (Ichii et al. 2009), ability to form symbioses allowing N fixation (Delwiche et al. 1965), and resprouting capacity after a fire (Bell 2001). Together the combinations of different traits contribute to diverse biogeochemical dynamics in chaparral ecosystems.

6.2.4 Ecohydrological Self-Organization and Chaparral Biogeochemical Cycling

In the context of the biogeophysical drivers, a system of ecohydrological processes and non-linear feedbacks can mute or amplify environmental sensitivities through processes whose outcomes cannot be understood through independent evaluation of the parts. This "self-organizing" (for a general introduction see Levin 2000) process, for example specific trajectory of hydrologic flow-paths, species distributions, and fire dynamics, lead to complex biogeochemical dynamics spanning scales of individual site to entire chaparral landscapes. Variation in chaparral biogeochemical cycling and potential future trajectories are defined by both dynamics of individual component processes and system level feedbacks and interactions. Systems are understood through a web of inter-connected relationships between many distinct components, where the connections between components can be as important as dynamics of individual components. Feedbacks, resulting from the web of interconnected relationships between many distinct system components (Ashby 1956; Forrester 1961), are important aspects of system dynamics that can both magnify changes through positive feedbacks or moderate change through negative feedbacks. Feedbacks can lead to emergent properties through self-organization of a system in concert with its environment (Wu and Loucks 1995; Levin 1998; Holling and Gunderson 2002). Through self-organization biogeochemical dynamics are not predictable from individual component processes in isolation. Recognition of the importance of emergent properties has increased with growing evidence of alternative stable states and tipping points between regime changes occurring within ecosystems (Scheffer et al. 2001; Scheffer and Carpenter 2003), where a system can rapidly change between contrasting configurations. These "state changes" occur with only limited changes in the environment at a critical threshold, while otherwise the system is largely insensitive to environmental variation away from such thresholds. Thus, for the same underlying biogeophysical drivers contrasting biogeochemical patterns can develop in response to ecohydrological dynamics and historical legacies. Evaluating biogeochemical variation in the context of self-organization systems theory provides both a framework for identifying the relationships between discrete components and examining dynamics spanning from of steady-state stabilizing mechanisms to self-organized and adaptive responses to the environment that can lead to large and potentially rapid changes in biogeochemical dynamics.

Self-organizing ecohydrological processes that influence chaparral biogeochemical variation are primarily associated with changing geomorphological processes, ecological processes, and in the context of high rates of endemism within chaparral evolutionary processes (Jenerette et al. 2012). Geomorphological processes occur in the context of changes in underlying geophysical and climate drivers and the characteristics of precipitation and meteorological dynamics. Geomorphological processes contribute to the rugged topography of chaparral ecosystems and influence water storage and movement as well as soil formation and transport. The influence of aspect in response to geomorphology has a strong influence on biogeochemical cycling, with wetter north-facing slopes typically harboring more biomass than corresponding drier south-facing slopes (Parker et al. 2016). Hydrologic routing can also have strong influences on soil biogeochemical dynamics, where locations with large amounts of upslope contributing area, such as riparian areas, are often wetter than locations with low upslope contribution areas, such as ridges (Riveros-Iregui et al. 2012). Geomorphological dynamics are further coupled with plant and microbial communities that influence chaparral biogeochemical dynamics through ecological and evolutionary changes. Ecological dynamics lead to assembly of communities and distribution of functional traits. The resulting ecological interactions between biotic communities and abiotic environment, structure nutrient cycles, disturbance, and physical organization can have large influences on biogeochemical dynamics. Interactions among traits and ecosystem feedbacks can further influence biogeochemical cycling, imposing constraints to processes varying in space and time and directly influencing rates of nutrient transformation.

Disturbance dynamics, especially in context of fire, can rapidly change community composition and capacity. Fire distributions are closely related to self-organizing processes associated with fuel load accumulation and decay. Fire frequency distributions often follow power-law distributions distinct from the distributions of the drivers of fire occurrence, a key characteristic of self-organization processes (Malamud et al. 1998). Fire has long been associated with chaparral as a major disturbance that resets the ecosystem and maintains shrub dominated communities. By burning biomass, litter, and topsoil, fires emit a large amount of particulate material that can have relatively high carbon content of variable chemical composition. Intense burning of biomass coupled with high burn temperatures results in greater transportation and transformation of nutrients. While all fires alter nutrient cycling in chaparral, the extent to which nutrient transformations occur depends on the fire severity, extent, and return interval. Notably, if seedbanks are not destroyed, high severity fires can induce germination of many chaparral seedlings and promote quick recovery times of native herb and shrub species, allowing nutrient storage to begin again (Thanos and Rundel 1995).

6.2.5 *Historical Legacies and Their Influence on Chaparral Biogeochemistry*

Arising from dynamics of biogeophysical drivers and ecohydrological self-organization are legacies of past processes that can further influence chaparral biogeochemistry. Historical ecosystem legacies are processes that occurred in the past and have a discernable influence on current ecosystem functioning, including biogeochemistry. Disturbances, such as fire, are prominent events whose legacy on ecosystem functioning has immediate consequences (Moreno et al. 2013) and can propagate for decades. Much of the biogeochemical sensitivity to precipitation variation such as timing and distribution, reflects an influence of precipitation legacies from individual events to potentially decadal temporal scales on ecosystem functioning (Scott et al. 2009; Shen et al. 2016). For example, the length of time between precipitation events is a direct quantification of legacy effects. Such effects can propagate across seasons and between years, with increasing moisture in wet years potentially priming production or reducing nutrient availability (Jenerette et al. 2010; Shen et al. 2016). At inter-decadal scales, variation in precipitation may influence ecosystem C storage and elevate rates of C loss in subsequent dry periods (Scott et al. 2009; Shen et al. 2016). However, our understanding of how historical legacy effects influence chaparral biogeochemical cycling is currently limited and more research is needed.

6.2.6 *Synthesis of Variation in Chaparral Biogeochemical Cycling*

The magnitudes of material storage, transformation, and loss associated with biogeochemical cycling in chaparral vary dramatically in response to both differences in biogeophysical drivers, ecohydrological self-organization, and the influence of historical legacies (Fig. 6.1). Biogeochemical variation in chaparral ecosystems is high at both local and regional scales (Chatterjee and Jenerette 2011; Dahlin et al. 2013). The suite of biogeophysical drivers associated with below-ground geology, above-ground climate, and a diverse suite of biological species all provide constraints and capacities for biogeochemical dynamics. These drivers intersect with self-organizing ecohydrological processes that influence geomorphology, ecology, and evolutionary dynamics that have a strong direct influence on chaparral biogeochemical cycling. Historical contingencies arising from both dynamics of biogeophysical drivers and ecohydrological self-organization further impose constraints to biogeochemical cycling. A recurring thread through these processes is the resulting distribution and sensitivities to soil moisture conditions. In general, greater moisture availability, either through greater inputs from winter rains or reduced evaporative demands, will be associated with greater biogeochemical pools. Microenvironments created by geomorphology and organismal effects can contribute to variations in total pools of C, N, and other elements and fluxes. Interacting drivers on biogeochemical variation can be complex with non-linear effects that have large uncertainties. Recently, and increasingly in the future, anthropogenic influences are altering the dynamics of biogeophysical drivers, ecohydrological self-organization, and leaving legacies that may remain for millennia.

6.3 Key Biogeochemical Cycles in Chaparral: Carbon and Nitrogen

6.3.1 Overview of Biogeochemical Dynamics

Biogeochemical dynamics of chaparral ecosystems are characterized by continuous and discontinuous responses to environmental conditions, known as pulse dynamics, which are both associated with soil moisture variation. A biogeochemical pulse occurs when wetting of previously dry soil acts as a trigger for biogeochemical processes and then subsequent dynamics are decoupled from soil moisture availability but depend on subsequently limiting resources and biological capacity. Rates of biogeochemical cycling decline with lower water availability and through direct desiccation and reduced rates of resource supply to microbes. In contrast, higher water availabilities can also reduce biogeochemical cycling through reduced oxygen availability. Biogeochemical cycling may respond to isolated precipitation events that act as a trigger, but then subsequent dynamics may be then regulated by substrate availability and lead to a decoupling of biogeochemical dynamics from actual moisture availability. Pulse dynamics are frequently observed in dryland ecosystems and are particularly relevant to chaparral dominated landscapes (Baldocchi et al. 2006; Jarvis et al. 2007; Jenerette and Chatterjee 2012). Because of pulse dynamics and the importance of wetting dry soils, small precipitation events can have disproportionately large influences on soil biogeochemical cycling and thus the timing and distribution of wetting is also important in addition to the total amount of precipitation (Huxman et al. 2004; Jenerette et al. 2008).

Another overarching influence on chaparral biogeochemical dynamics is the distinct characteristic time-scales between microbial and plant dynamics, with microbial dynamics changing much faster than plant dynamics. These contrasts are most evident in pulse dynamics, where microbial processes may respond within hours and much more dramatically than plants, which typically respond at the scale of days. As a consequence, rain in summer seasons or the initiation of winter season precipitation may have a much larger influence on chaparral soil-driven biogeochemical cycling than wetting patterns favoring plant growth (Miller et al. 2005; Jenerette and Chatterjee 2012). The contrasting influence of microbes and plants on nutrients can lead to complex dynamics that appear to favor nutrient losses and recycling simultaneously, in contrast to general ecosystem expectations.

In the context of the drivers and processes influencing chaparral biogeochemical dynamics, C and N cycles are closely connected to many ecosystem processes and are directly related to multiple ecosystems services, including climate regulation and provisioning of clean water. These two cycles include components of biomass accumulation, photosynthesis, and what is commonly considered its main limiting nutrient following water, N. Likewise, C availability can influence some N cycling pathways and thus leads to two-way coupling between biogeochemical cycles and opportunities for feedbacks.

6.3.2 Carbon Cycling in Chaparral Ecosystems

Within an ecosystem many transformations of C occur between its initial fixation and eventual return to the atmosphere. In chaparral dominated ecosystems, C cycling features uptake through photosynthesis, transformations within the ecosystem, and losses through respiration with opportunities for deposition or erosion to bring new C into, or remove it from, the ecosystem (Fig. 6.2, see Box 6.1). Dominant pools of C include soil and plant components, with additional C contained in litter. The identification of long-term C sequestration, or positive Net Ecosystem Production (NEP), has become increasingly of interest, as a result effort has been directed to identifying mechanisms that promote the stabilization of C and prevent the return of C to the atmosphere as a greenhouse gas (GHG).

C uptake at the leaf is the point of entry of C into chaparral ecosystems. The total amount of photosynthesis, or gross primary production, is regulated by physical and biochemical characteristics of individual leaves and their extent and distribution within an ecosystem. Plants also respire C and the balance between photosynthesis and respiration is Net Primary Production (NPP) or the amount of plant growth. In response to regular water limitation, root allocation is a large component of plant C, which leads to deeper C inputs into the soil. Recent interest has been directed to C in plants that is readily mobilized and not used in structures, such as plant Non-Structural Carbohydrates (NSC) (Dietze et al. 2014). Increasing evidence from many woody plants suggests increasing NSC content is linked to increased growth, seed production, disease resistance, and drought resistance. Within some chaparral shrubs plant age may have a role in the distribution and allocation rates of the NSC pool (Sparks et al. 1993), however the dynamics and distribution of chaparral NSC is an emerging research need.

The transfer of plant C to soil heterotrophs and potential long-term sequestration occur through several pathways associated with both structural and non-structural C. Leaf and stem litter drop to the soil surface. Root death and root exudates provide plant C within the soil. The litter itself can be directly consumed or incorporated into the soil. In MTC ecosystems, as in most dryland ecosystems, processes of litter decomposition may be greatly accelerated by UV induced photodegradation (Austin

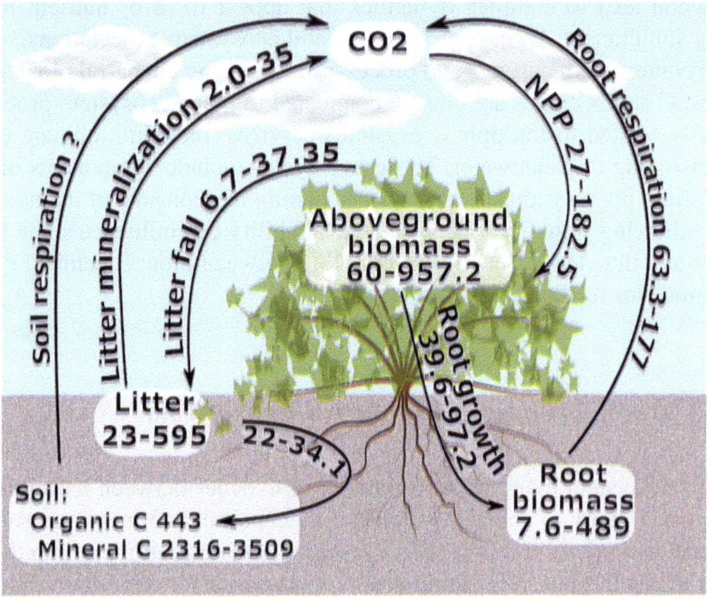

Fig. 6.2 Synthesis of existing literature reporting pools and fluxes of C in chaparral dominated ecosystems. The values are not exhaustive but represent an initial review of the range of variation in key C cycle pools and fluxes. All pools (above-ground biomass, root biomass, soil, litter are in units g C m^{-2}. All fluxes (litter mineralization, litter fall, NPP, root respiration root growth litter incorporation into soil) are in units g C m^{-1} $year^{-1}$. Units are re-scaled for comparison amongst different studies when units differed. Data were derived from published results (Harrison et al. 1971; Rundel and Parsons 1979; Mooney and Rundel 1979; Gray and Schlesinger 1981; Dunn et al. 1988; Ulery et al. 1995; Quideau et al. 1998; Miller et al. 2005; Li et al. 2006; Smith et al. 2007; Grünzweig et al. 2007; Pasquini and Vourlitis 2010; Drenovsky et al. 2013; Dickens and Allen 2014; Homyak et al. 2014; Hanan et al. 2016a; Vourlitis and Hentz 2016)

and Vivanco 2006; Gliksman et al. 2017). However, with a frequently closed canopy, the availability of UV at the litter layer may be limited. Living plant C and litter C can have turnover times of seasons to decades. In this sense, increasing C in living biomass or rapidly cycling litter may represent a short-term C sink—this C is primarily respired and does not contribute to long-term C sequestration. Stabilization, the conversion of organic C to a form highly resistant to decomposition, may have low rates but contribute to long-term C sequestration (Six et al. 2002). C stabilization can occur through physical and chemical processes, such as the formation of soil aggregates or humification. Through aggregate formation, soil C is physically protected within the soil matrix such that it is inaccessible to microbial degradation (Schmidt et al. 2011). Humification, conversion of soils along a continuum from chemically simple to large and complex organic molecules, is an alternate pathway of ecosystem C stabilization (Lehmann and Kleber 2015). Sequestration rates may be influenced by vegetation, as chaparral communities dominated by *Ceanothus* spp. sequestered more C than *Adanostoma* spp. dominated chaparral (Ulery et al. 1995; Quideau et al. 1998).

Fires have immediate and legacy effects on chaparral C cycling. Fire itself results in an immediate loss of C from the ecosystem, primarily as a fraction of above-ground plant biomass, litter, and organic C in the upper horizons of the soil. C losses from fire include particulate material and a variety of C trace gasses including CO_2, CO, and CH_4. Recently fire mediated charcoal production has been noted as a potential sequestration pathway, which converts comparatively labile plant C forms with residence times of decades to centuries to a form with residence times of millennia or longer (Heckman et al. 2013). Incompletely burned plant C can be returned to the litter layer and soil as black carbon, which can be highly stabilized and serve as an important pathway of long-term sequestered C. May et al. (2014) propose that chaparral fires emit the highest amounts of refractory black carbon compared to other ecosystems, further supporting the potential for this pathway for chaparral ecosystem to sequester C. Further, periodic fires may also increase the rate of soil C inputs within chaparral communities by producing large quantities of wood ash and charred woody material from which carbon in the form of calcite readily leaches into the soil (Ulery et al. 1995). Fire can lead to short-term reductions in soil C, although these can return to prefire levels within a year (Graham et al. 2016). While fire does lead to immediate losses of C through combustion, the long-term consequences of fire may primarily serve as mechanism for C sequestration. Thus, while

Box 6.1 Case Studies of Chaparral Carbon Dynamics

At present, the majority of information about chaparral has been obtained from widely distributed individual studies and concerted evaluation of specific sites and landscapes. The Deep Canyon transect in southern California, which spans a 2500 m (8202 ft) elevation gradient has been used for more than 4 decades of research in soil biogeochemistry across a transect of chaparral vegetation bounded at the lower end by a creosote dominated desert ecosystem and at the higher end by coniferous dominated subalpine forests (Hanawalt and Whittaker 1976; Jenerette and Chatterjee 2012). Exploring the local and regional variation in chaparral soil C emissions was the focus of a series of studies conducted along this transect. With increasing elevation, the climate is characterized by decreasing temperatures and increasing precipitation. Soil organic matter and nutrient content in the chaparral components of the transect generally increase with elevation (Hanawalt and Whittaker 1976; Chatterjee and Jenerette 2015). The spatial structure of soil organic matter and soil CO_2 emissions also changes throughout the transect. The microenvironment effects of local shrub canopies on distributions of organic matter, C, and CO_2 emission potentials decrease in magnitude, but increase with higher elevations (Chatterjee and Jenerette 2011). Throughout the transect, maximum rates of soil CO_2emissions from laboratory incubations were consistently above 35 °C and in chaparral ecosystem were above 45 °C. While soil C emissions generally increased with elevation, the pulse sensitivity of CO_2 emissions to wetting generally decreased with elevation, which led to a large scale negative relationship between soil C and pulse emissions. However, within sites at the same elevation, soil C emission pulses were positively related with soil C and pulse emissions. Thus, across an elevation gradient in chaparral dominated ecosystems a contrasting scale-effect of C and pulse dynamics were observed (Jenerette and Chatterjee 2012). This case study, demonstrates the large variation in chaparral biogeochemistry and the varying potential of pulse driven biogeochemical fluxes.

A second case study of chaparral C cycling examines whole ecosystem fluxes and net ecosystem C exchange within a single site in southern California. The Sky Oaks field station, featuring both old-growth chaparral and some recently burned areas, has served as a model system of chaparral with several connected long-term studies into C cycling. An evaluation of this well-studied site presents a valuable lesson about the complexity of chaparral C cycling. Data from the Sky Oaks site provides the most comprehensive evaluation of chaparral ecosystem C cycling variation and has led to improved understanding of the distinct controls on chaparral biogeochemical dynamics and the large variability of C in chaparral ecosystems. In an old-growth area at Sky Oaks, whole ecosystem CO_2 exchanges known as Net Ecosystem Exchange (NEE), have been measured through the eddy covariance approach and showed high C uptake rates, up to 155 g C m^{-2} $year^{-1}$ (Luo et al. 2007). These rates are comparable with uptake rates of other old-growth forests distributed globally. However, in dry years NEE led to no uptake or even net emissions back to the atmosphere at the annual scale. Within years, whole ecosystem C exchange showed large daily variation, spanning effectively 0 g C m^{-2} to above 4.5 g C m^{-2} $year^{-1}$ in two distinct chaparral communities (Stylinski et al. 2002; Sims et al. 2006). Spatially mapping CO_2 fluxes suggests equally large spatial variation within local landscapes of chaparral dominated ecosystems (Fuentes et al. 2006). The complex temporal and spatial dynamics of chaparral C at this single site highlights both the large potential for C uptake, especially from old-growth stands, to net emissions depending on moisture availability. While likely other chaparral communities show similar magnitudes of variation in C exchanges, the Sky Oaks site provides a novel window into C dynamics through the long-term records that are not available elsewhere.

typically fire is associated with large emissions of C from an ecosystem, perhaps surprisingly fire may also be associated with long-term sequestration of C into highly recalcitrant C or black C. The mechanisms of organic C transformation into long-term stable forms are highly uncertain and are a current area of active research (Stewart et al. 2008; Schmidt et al. 2011).

6.3.3 Nitrogen Cycling in Chaparral Ecosystems

While N is the most abundant element in the atmosphere it occurs in the highly stable form of dinitrogen (N_2, Schlesinger and Bernhardt 2013). Reactive N that can be used by organisms is comparatively scarce. As with many dryland ecosystems, chaparral productivity and many biogeochemical processes have been historically considered secondarily N limited after water limitation. N limited systems are characterized by several factors promoting nutrient conservation, including long leaf life-times, nutrient-poor sclerophyllic tissues, and nutrient resorption during senescence. The resulting low-quality litter increases N immobilization during decomposition and lower rates of mineralization. Slow soil cycling of N and increased losses contribute to poor soil fertility and tighter N recycling within plants. For example, many evergreen chaparral shrubs uptake N during the fall or winter and store in old tissues for new growth in the spring (Mooney and Rundel 1979). Internal cycling and outputs reflect processes associated with N limitation, although they also suggest processes, including potential for high rates of N losses that contrast with predictions of N limitations (Fig. 6.3). Resolving this apparent dichotomy has been an important source of recent progress in chaparral biogeochemistry.

Inputs of reactive N, including mineralized and organic forms, are a key step in chaparral N cycling. Reactive N constitutes the pools of N critical to ecosystem dynamics and is distributed through many pools in chaparral dominated ecosystems analogous to C pools (Fig. 6.3). An important source of chaparral N is the direct conversion of N_2 into the bioavailable form of ammonia (NH_3^+) through the process of nitrogen fixation, mainly through the action of symbiotic bacteria and to a lesser extent by free-living bacteria or lightning. Several chaparral species, including *Ceanothus* spp., western redbud (*Cercis occidentalis*), desert false indigo (*Amorpha fruticose*), and chaparral pea (*Pickeringia montana*) form N fixing symbioses and contribute to N inputs to chaparral ecosystems (Delwiche et al. 1965; Ulery et al. 1995). N fixation in addition to atmospheric inputs, which have historically been low but are now elevated for many chaparral ecosystems, set the stage for chaparral N biogeochemical cycling. Rates of N deposition to chaparral ecosystems can vary widely and may depend on both atmospheric concentrations of reactive N as well as endogenous factors of species traits and biomass. For the same atmospheric reactive N concentrations, deposition increases with both increases in leaf-specific deposition velocity and increased leaf area (Bytnerowicz et al. 2015). Thus, while atmospheric reactive N concentrations contribute to N inputs, ecosystem processes including fixation and deposition also have roles in determining N inputs.

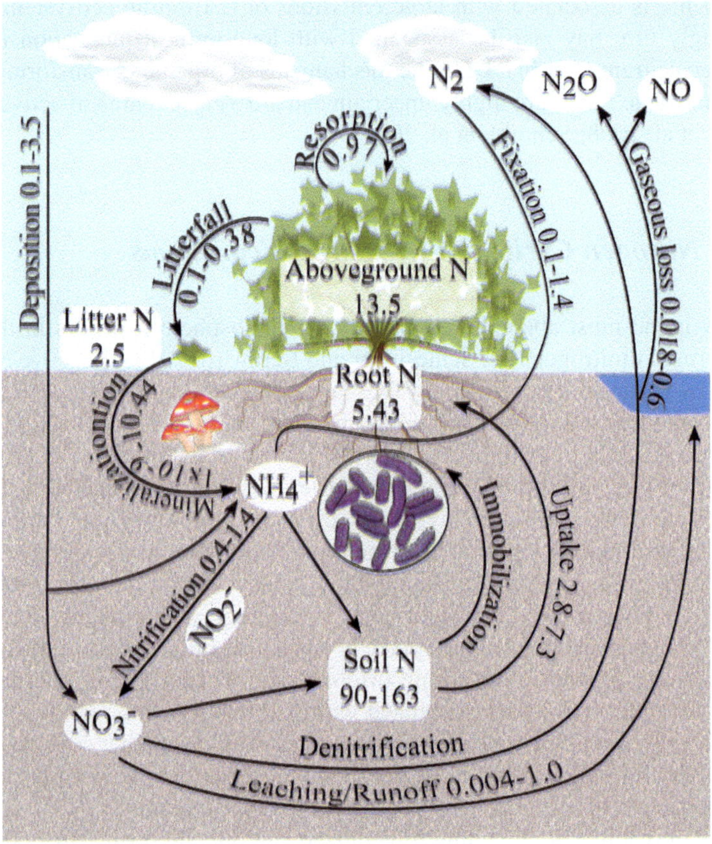

Fig. 6.3 Synthesis of existing literature reporting pools and fluxes of N in chaparral dominated ecosystems. The values are not exhaustive but represent an initial review of the range of variation in key C cycle pools and fluxes. All pools (above-ground N, root N, soil, litter N, are in units g C m^{-2}. All fluxes (litter ammonification, nitrification, immobilization, denitrification, leaching are in units g C m^{-1} $year^{-1}$. Units are re-scaled for comparison amongst different studies when units differed. Data were derived from published results (Kummerow et al. 1978; Mooney and Rundel 1979; Gray and Schlesinger 1981; DeBano et al. 1984; Riggan et al. 1985; Bytnerowicz et al. 1987; Ulery et al. 1995; Fenn et al. 1996; Quideau et al. 1998; Fenn et al. 2003; Miller et al. 2005; Li et al. 2006; Meixner et al. 2006; Grünzweig et al. 2007; Pasquini and Vourlitis 2010; Vourlitis and Fernandez 2012; Drenovsky et al. 2013; Dickens and Allen 2014; Homyak et al. 2014; Hanan et al. 2016a; Vourlitis and Hentz 2016)

Following fixation or atmospheric inputs, if not directly incorporated into plant biomass, NH_3^+ is then converted to nitrate (NO_3^-) via nitrite (NO_2^-) in the two-step process of nitrification carried out by aerobic chemoautotrophs. Both $NH_3^+$and NO_3^- are available for assimilation by plants and soil microorganisms into organic matter. The process of mineralization returns organic N to plant and microbial available N pools by converting soil organic matter back to ammonia. Simultaneously, soil conditions in chaparral tend to be coarse-textured, quickly draining, and aerobic, with high base cation saturation and low organic matter, which generally lead to high rates of nitrification in relation to mineralization (Fenn et al. 2003). Nevertheless, in some sites, such as a coastal old-growth sites, low rates of net nitrification have been observed (Hanan et al. 2016a) further highlighting the diversity of chaparral biogeochemistry. Species traits, such as higher leaf N content and low concentrations of secondary compounds, can increase N inputs to the soil and rates of biogeochemical cycling through litter interactions with soil microbes. Some shrubs (e.g., chamise [*Adenostema fasciculatum*]) promote higher concentrations of nitrogenous compounds in soil than others (e.g., desert ceanothus [*Ceanothus greggii*]), presumably because some shrubs cultivate nitrifying bacteria more than others (Fenn et al. 1993).

Mineral N in the soil that is not immobilized in microbial or plant biomass may be lost from the ecosystem primarily through leaching, volatilization, and trace-gas emission. Nitrate is highly mobile in soil, and coupled with low rates of biotic activity during the winter rainy season, considerable leaching loss can occur. Nitrate leaching becomes a significant form of N export from chaparral dominated ecosystems at deposition rates above 25 kg N ha^{-1} $year^{-1}$ (Fenn et al. 1996). While leaching is relatively low in undisturbed and unpolluted chaparral, losses are generally highest following the first winter rains. N can also be lost to the atmosphere through the process of denitrification where oxidized nitrogen is used as an alternative electron donor in heterotrophic anaerobic respiration resulting in the sequential production of nitric oxide (NO), nitrous oxide (N_2O) and N_2. The emissions of NO and N_2O are conceptualized by the hole-in-the-pipe model developed by Firestone and Davidson (1989), where the size of the pipe describes the total amount of nitrogen moving through the system and the "holes" in the pipe represent the relative amount of NO or N_2O lost through gaseous production. The hole size is controlled mainly by soil water content, influencing both oxygen availability and diffusion of gases from the soil, facilitating production or consumption. NO and presumably N_2O emissions from chaparral soil are generally are low but can be large upon wetting of dry soils (Homyak and Sickman 2014).

Fire can lead to immediate N losses to the atmosphere and have longer term influences on chaparral N biogeochemistry. Following fires, stream export of N can be elevated up to 40 times that of unburned areas (Riggan et al. 1994), suggesting both increased mobility of N and reduced plant uptake. In the first growing season after fire, new leaf and stem growth contains higher N content than growth before fire or in subsequent growing seasons owing to higher mineral N availability in the soil (Franco-Vizcaino and Sosa-Ramirez 1997). Spikes in nitrogen can also induce germination of certain chaparral seeds postfire (Thanos and Rundel 1995; Keeley and Fotheringham 1997). However, elevated N cycling has been shown to dissipate within 7 months of the fire (de Koff et al. 2006). High severity fires can also result in higher emissions of volatilized N, NO_x and N_2O into the atmosphere (DeBano and Conrad 1978). Although soil N is reduced postfire, burned biomass can lead to a spike of soil NO_3^- (Christensen 1973). This spike subsequently increases nitrification rates and results in a net loss of N from chaparral systems as NO_x and N_2O emissions into the atmosphere and NO_3^- into streamflow. Char legacies from previous fires may increase rates of N immobilization into microbial biomass (Hanan et al. 2016b). As with soil C, fire can lead to short-term reductions in soil N, which also can return to prefire levels within a year (Graham et al. 2016).

Because chaparral exists under the unique constraints of a Mediterranean-type climate, soils are subject to long periods of hot and dry conditions. As a consequence of this regular drying, pulse-driven dynamics can also contribute to nitrogen cycling and associated losses via leaching and trace-gas emissions (Fierer and Schimel 2002; Miller et al. 2005; Oikawa et al. 2015; Liang et al. 2016). Soil C and N accumulate during the dry summer months when microbial and plant activity is limited by soil moisture. Precipitation results in a large flush of C and N resources that can produce a substantial pulse in soil gaseous emissions. These wetting-induced pulses can contribute to the annual budget of C and N chaparral in arid systems. For example, NO emissions from chaparral ecosystem can be large, exceeding 350 ng m^{-2} s^{-1} compared to unwetted values in the summer (near zero) (Homyak and Sickman 2014). Thus, in contrast to an expectation of N limitation, the potential for chaparral ecosystems to also exhibit high N leakiness, typically associated with N saturation, is also evident (Fenn et al. 1996; Homyak et al. 2014).

The potential for high N losses have led to revisions of standard ecosystem concepts of nutrient limitation and nutrient saturation in chaparral dominated ecosystems. Microbial activity can more rapidly immobilize N at the beginning of the rain season, but plant uptake can outcompete microbial nitrogen cycling towards the end of the rain season, resulting in a transition from nitrogen loss to nitrogen retention (Homyak et al. 2014, 2016). The often low N input rates and presence of N conserving strategies by chaparral communities, including plant nutrient reabsorption, litter N immobilization, deep rooting, and supporting N fixation, which conserve N within chaparral ecosystems, suggests N limitation. In contrast, the regular large emissions suggest N saturation. Working towards a more comprehensive theory of ecosystem N cycling that allows for simultaneous indicators of both N limitation and saturation has been a source of recent interest. Recent findings suggest N cycling rates and the importance of immobilization and loss pathways may dominate following initial rain at the beginning of the winter season but then plant activity immobilizes N later in the season with the initiation of plant growth (Homyak et al. 2016).

In addition to the nutrient consequences for chaparral ecosystems, N losses to gaseous production of NO and N_2O can have multiple environmental consequences. NO contributes to the production of ozone, representing a significant health concern for people and plants and contributes to acid rain and N redeposition to other ecosystems. N_2O is a potent greenhouse gas, which is ~300 times more effective than CO_2 at depleting ozone in the upper atmosphere. NO_3^- losses to streamflow can lead to deterioration of downstream water quality (Fenn et al. 1998). In addition, many of the chaparral ecosystems in California, primarily inland areas, are subjected to elevated rates of anthropogenic nitrogen deposition, leading to stream eutrophication, and the hydrologic export of nitrate, all of which are indicative of nitrogen saturation in these systems.

6.4 Anthropogenic Influences and Future Trajectory of Chaparral Biogeochemistry

6.4.1 Nitrogen Deposition and Chaparral Biogeochemistry

Anthropogenic activities have resulted in inputs of N into natural ecosystems that have surpassed biological fixation (Galloway et al. 2008). In the western United States and the regions of chaparral dominated ecosystems in North America, the largest sources of N deposition are transportation, agriculture, and industry, typically resulting in deposition rates of 1–4 kg N ha^{-1} $year^{-1}$ (Fenn et al. 2003). This number can be as high as 30–90 kg N ha^{-1} $year^{-1}$ in chaparral dominated communities near urban and agricultural areas (Fenn et al. 2003; Bytnerowicz et al. 2015). This dramatic increase in N deposition can result in extensive consequences to chaparral ecosystems, such as species composition shifts, soil acidification, and eutrophication of aquatic systems (see Box 6.2). In southern California, which experiences some of the highest rates of N deposition in North America (Bytnerowicz et al. 2015), research has linked N deposition with an increase in the establishment of non-native annual plant species and consequently, greater risk of fire (Fenn et al. 2003; Rao et al. 2010). While N deposition can lead to increases in chaparral ecosystem C and N pools (Vourlitis et al. 2007b), the effects are complex. Critical loads

Box 6.2: Case Study of Nitrogen Fertilization Experiment

One of the most comprehensive and long-term evaluations of the effects of experimental N fertilization to shrublands of southern California, including the chaparral dominated Sky Oaks field site, has been conducted by Vourlitis and colleagues (Vourlitis et al. 2007a; Vourlitis and Hentz 2016) (Fig. 6.4). This experiment has shown complex biogeochemical responses to high rates of N amendments that continue to change over more than a decade of N additions (Vourlitis and Hentz 2016). Immediately following N amendments, large increases in mineral N was observed (Vourlitis et al. 2007a), however the mineral N increases did not translate into immediate changes to total ecosystem N or C content. Several complex biogeochemical pathways lead to multiple indirect responses to N addition in chaparral dominated ecosystems (Vourlitis and Fernandez 2015). For example, soils from the experimental plots showed patterns of reduced rates of litter decomposition and N demand, which may affect C storage and N availability following increased N deposition (Biudes and Vourlitis 2012). Alternatively, increases in N trace gas emissions following N deposition may reflect an elevated N loss pathway counteracting the effects of N deposition to the ecosystem (Vourlitis et al. 2015). The high N leakage potentials determined from laboratory incubations are consistent with findings after 4 years following of N additions that no changes in either ecosystem C inputs or N storage was observed (Vourlitis et al. 2009).

Fig. 6.4 Sky Oaks Field Station, location of nitrogen addition experiment plots taken from drone at 15 m high (49 ft) in Fall 2016. Photo by George Vourlitis

Overlaying the experimental manipulation, the effects of N addition on biogeochemical cycling was strongly coupled to precipitation patterns, with greater effects observed in wetter years further complicated by historical legacy effects (Vourlitis 2012). After 4 years of fertilization, increases in above-ground C and N were observed and the ecosystem continues responding to N additions up to a decade following N amendments (Vourlitis and Hentz 2016). Throughout this period, N loss pathways were also elevated, again highlighting the complex N dynamics in chaparral ecosystems that show characteristics both of N limitation and N saturation. Further examination of plant leaf and hydraulic responses have shown large differences among species in their responses, with nitrogen addition leading to increased leaf N content and increasing hydraulic conductivity, stomatal conductance, and leaf carbon isotopic composition while decreasing wood density in several species (Pivovaroff et al. 2016). These plant responses to nitrogen generally increase photosynthetic rates and water movement through the organism and have clear implications for ecosystem drought sensitivity and may influence future community assembly. This prediction has been supported through recent findings of species shifts in response to fertilization and increasing the potential of non-native grass invasion (Pasquini and Vourlitis 2010; Vourlitis 2017). This long-term experiment shows that even after a decade of elevated N, chaparral dominated ecosystems can continue exhibiting dynamics of both N limitation, through positive effects on C accumulation, and N saturation, through high potential rates of N leakage. Future biogeochemical trajectories may depend on community reorganization dependent on nutrient availability and drought tolerances.

of N deposition for chaparral have been assessed at 10 kg N ha^{-1} $year^{-1}$ for small catchments (4–10 ha or 9.9–24.7 acres) and 14 kg N ha^{-1} $year^{-1}$ for intermediate and large catchments (>10 ha or 24.7 acres) (Fenn et al. 2010). Currently, ~15% of chaparral land area in California is in exceedance of 10 kg N and 3.3% of chaparral land for the 14 kg N thresholds. In contrast, if a critical load of 5.5 kg N ha^{-1} $year^{-1}$ is used, the level identified leading to a shift in chaparral lichen communities to eutrophic communities, then 53% of chaparral areas exceed this threshold (Fenn et al. 2010). The impact of changes in N can be seen in the modification of lichen communities which experience an increasing dominance by nitrophytes. Increases in N deposition have been associated with more than three orders of magnitude greater NO_3^- export from chaparral ecosystems (Riggan et al. 1985), with stream [NO_3^-] reaching 1120 μmol_c L^{-1} from catchments dominated by chaparral in a high deposition region after a recent fire (Riggan et al. 1994) compared to very low detection levels in pristine streams. While chaparral communities are historically considered resistant to the effects of atmospheric pollution (Allen et al. 2005a; Vourlitis and Pasquini 2009), increased atmospheric N deposition may significantly alter the soil chemistry, resulting in a modification of the competitive balance between chaparral shrubs and non-native herbs (Rao et al. 2009; Fenn et al. 2010; Vourlitis 2017). The effects of increased deposition can influence the biogeophysical drivers of chaparral and further influence self-organizational processes within the ecosystem.

6.4.2 *Grazing and Chaparral Biogeochemistry*

The influence of grazing on C and N has long been recognized. Studies of grazing in chaparral indicate that soil C and N contents can be significantly reduced by grazing (Brejda 1997; Milchunas 2006). Grazing induced C and N reductions are consistent with other disturbances such as fire. Nevertheless, grazing has also, in some cases, been found to increase soil C and N (Brejda 1997), contributing to the biogeochemical diversity of chaparral dominated ecosystems. Grazing-induced changes in soil N and C primarily occur in litter and surface soil layers due to interception of foliage by grazers before leaf senescence (Severson and DeBano 1991). However, studies of grazing associated reductions in soil N undertaken in chaparral shrublands in Arizona found reductions were greatest under the canopy of nitrogen-fixing *Ceanothus* shrubs than under other shrubs, likely due to preferential browsing of that taxa by goats and other grazers (Sidahmed et al. 1981; Knipe 1983; Severson and DeBano 1991). While goats typically exhibit grazing preferences similar to native grazers and cattle (Milchunas 2006), the preferences of different grazer species can differ substantially, with some grazing heavily on *Ceanothus* (Knipe 1983). Thus, the effects of grazing on N fixing species, and associated impacts on N cycling and storage, may depend on the preferences of local ungulates.

In addition to the direct impacts on soil chemistry through browsing and litter reduction, heavy grazing has been observed to reduce shrub cover while also preventing colonization of the subsequent bare soil by perennial forbs and grasses. In the absence of this re-colonization, previous shrub cover is converted to an annual dominated cover that is more capable of withstanding heavy grazing pressures (Severson and DeBano 1991). Nevertheless, examination of different levels of grazing in a watershed in Arizona detected no difference in runoff or erosion (Rich and Reynolds 1963). Thus grazing can have direct influence on above-ground C and N pools through consumption, and may have multiple indirect effects through changes in soil structure and plant species composition. Yet, many aspects of chaparral biogeochemistry may be resilient to grazing. The diversity of chaparral responses to grazing may depend on moisture availability, topography, local species characteristics, and historical legacies.

6.4.3 Fire and Chaparral Biogeochemistry

Fires and fire regimes are an interaction between biophysical drivers and ecohydrological self-organization that creates potentially long-lived legacies. In recent times, chaparral fires are influenced by human activities both through increased fire suppression and ignition sources with a net effect of reducing the fire interval (Syphard et al. 2007). More frequent fires can influence plant community composition by first reducing the competitive abilities of obligate seeders compared to resprouters (Franklin et al. 2005) which, in turn, may have large indirect effects on biogeochemical cycling. In addition to the direct loss of biomass and litter pools through burning, differences in fire history and fire characteristics can have large consequences to chaparral biogeochemistry. Connected to fire, fuel management practices such as the creation of fuelbreaks and fire breaks are formed by crushing or otherwise denuding strips of land of shrub cover (Green 1977), to slow fires and facilitate access by firefighters and equipment into otherwise inaccessible areas. Such disturbances can alter soil chemistry substantially. Firebreaks are also colonized predominantly by non-native species and may then function as sources for further invasion into adjacent areas, with a variety of impacts on local biogeochemistry (Zink et al. 1995; Keeley et al. 2005; Mayberry 2011; Syphard et al. 2014).

6.4.4 Land Use and Development

In recent decades, urban development has increasingly encroached on chaparral landscapes (Syphard et al. 2007). The conversion and fragmentation of chaparral has multiple implications, both due to the direct loss of chaparral habitat and due to the downstream effects of developed landscapes on adjacent chaparral. As chaparral habitat is developed, the removal of deep-rooted shrubs and replacement by largely impermeable surfaces typically results in significant increases in runoff and stormwater discharge (White and Greer 2006; Warrick and Rubin 2007, see Chap. 7).

Ironically, urbanization can lead to some improvements in water quality, as sediment discharge often remains similar to pre-development levels, while water discharge increases dramatically, thereby diluting suspended sediment loads (Warrick and Rubin 2007). In addition, as housing development encroaches into chaparral vegetation, the natural fire cycle has increasingly become a major hazard for human safety and property loss (Rodrigue 1993), leading to a variety of management strategies intended to mitigate risks to life and property (Gardner et al. 1987; Gill and Stephens 2009). Many of these strategies may exacerbate the effects of such development on native chaparral vegetation, and also impact the biogeochemistry of the surrounding landscape through denudation and conversion of native vegetation, leaching of heavy metals, and other processes.

Associated with encroaching urbanization into chaparral systems, is the creation of roads for access and alternative evacuation routes in case of fire (Cova 2005). Roads have been found to act as corridors for non-native plant species and increases the susceptibility of adjacent areas to invasion (Lambrinos 2006; Davies et al. 2013), which may in turn alter soil sequestration of both C and N. In addition, roadside soils have been observed to exhibit increased deposition of reactive N owing to automobile exhaust (Fenn et al. 1996; Rossi et al. 2015) as well as increased heavy metal content (Rossi et al. 2015).

6.4.5 Atmospheric CO_2 Concentrations and Chaparral Biogeochemistry

Another important influence on the biogeochemistry of chaparral systems is the direct effect of rising CO_2 concentrations which not only influence climate but also have a direct effect on many biogeochemical processes. To experimentally evaluate C cycle sensitivities to CO_2 concentrations in chaparral dominated landscapes a Free Air CO_2 Enrichment (FACE) experiment was established in 1995 at the Sky Oaks field site (Roberts et al. 1998). Within 3.5 years, the chaparral ecosystems had increased rates of CO_2 uptake and increased storage of soil C (Treseder et al. 2003). The effect of increased CO_2 in particular led to increases in the C content of water stable aggregates, which was hypothesized to have long turnover periods and form a sequestration pathway. In contrast microbial turnover may have been faster and potentially limited by N availability (Allen et al. 2005b). Nevertheless, after 6 years of elevated CO_2, a surprising result showed that increasing CO_2 concentrations led to reductions in soil C content (Del Galdo et al. 2006). This response was associated with decreases in the fraction of large particles within the soil, bound in part by organic materials (macroaggregates) at the highest CO_2 levels, which implied reduced physical protection of soil organic matter. Based on these findings, future CO_2 atmospheric concentrations may impose a positive feedback from chaparral soils as they will concomitantly increase emissions (Del Galdo et al. 2006; Trueman et al. 2009).

6.5 Potential Chaparral Type-Conversion and Biogeochemical Cycles

Historically, much effort has been directed to type-converting chaparral either to grass dominated communities for grazing or coniferous forest for timber harvesting and more recently, a variety of factors are causing type-conversion of chaparral to non-native grasslands. Evidence suggests these regime shifts of chaparral communities from shrub dominated to grass dominated may have a large influence on biogeochemical cycling (Park et al. in press).

Coniferous forests, while unlikely to encroach into chaparral under normal circumstances (Burns and Sauer 1992), have invaded into chaparral following extensive fire suppression (Nagle and Taylor 2005; Lauvaux et al. 2016). Further, conifers have been planted in areas dominated by chaparral after fires, primarily to increase the area of conifer forest for future timber harvest on federal lands (Burns and Sauer 1992). Comparisons of soil C sequestration between chaparral and coniferous forests have shown mixed conclusions regarding the effects of such vegetation conversion on C and N storage. Some studies indicate chaparral sequesters more C and N within the soil A-horizon than coniferous forests (Ulery et al. 1995). Similarly, comparisons of chaparral to oak or conifer dominated forests also found that chaparral sequestered greater amounts of soil C and N than pine forests (Ulery et al. 1995; Quideau et al. 1998). In contrast to these findings, other studies on shrublands in Israel (Grünzweig et al. 2007) found far less above- and below-ground carbon sequestration than pine and juniper forests. Evidence suggests a combination of increased N use efficiency, enhanced Soil Organic Carbon (SOC) protection, and reduced decomposition rates following an increase in forest species. This latter finding implies the potential for increases in biogeochemical pools when chaparral shrublands are converted to coniferous forests.

In contrast, the potential for non-native grass invasions is posing a much greater threat to chaparral ecosystems. Although chaparral has historically been considered resistant to plant invasions (Minnich and Bahr 1995; Allen et al. 2005b), increasing drought intensity and anthropogenic disturbances such as altered fire regimes, encroaching urbanization, and increased dissection of chaparral landscapes may push many of these systems to tipping points of shrub mortality and possible type-conversion to invaded grasslands (Hamilton 1997; Keeley and Brennan 2012; Syphard et al. 2013; Meng et al. 2014). Roadsides, firebreaks, and other anthropogenic features within chaparral are also typically dominated by non-native herbs (Zink et al. 1995; Giessow and Zedler 1996; Lambrinos 2006; Davies et al. 2013). In the past, managers have attempted to forcibly convert chaparral to grasslands in efforts to increase water yield, improve fire control, or increase grazing habitat (Bentley 1967; Cable 1975; Anderson et al. 1976). These conversion efforts, completed through the extensive use of physical removal, active seeding, and chemical application, have been somewhat successful and have now contributed to the invasion of non-native annual grasses. Grazing has also been used as a mechanism to impose type-conversion of chaparral into grasslands (Knipe 1982, 1983; Severson and DeBano 1991).

Conversion of coastal sage scrub (CSS) communities into non-native grasslands has been documented to increase sequestration of both N and C (Wolkovich et al. 2010), in contrast, chaparral dominated communities have experienced substantial reductions in soil C sequestration after type-conversion to non-native grasslands (Williamson et al. 2004a). The reductions are in part associated with greater plant cover and depth of rooting in chaparral species than grasses, along with different phenology, and litter chemistry. Soil C reductions occur not only at soil depths below the root horizon of the shallower-rooted grasses, but also in the shallower soil layers in which grasses may be expected to deposit the majority of soil C. Such increases in surface C may increase rates of decomposition through greater exposure to photodegradation and sensitivity to small rainfall events.

In addition to reductions in C and N sequestration with the invasion of non-native grasses, these grasses also intercept far less moisture than chaparral shrubs (Rowe 1963; Corbett and Crouse 1968; Williamson et al. 2004a) and do not capture water at depths below their roots (Williamson et al. 2004b), which are shallow in comparison to many chaparral shrub species. As a consequence, chaparral sites have also been observed to exhibit large increases in throughflow and downstream soil moisture (Hibbert 1971; Davis 1984) as well as possible groundwater recharge (Williamson et al. 2004b) after conversion into invaded grasslands. In aggregate, the hydrologic effects of conversion can even convert streams from intermittent to perennial (Orme and Bailey 1970). The increase in throughflow leads to reduced water available for plants to use in conjunction with photosynthesis and can directly reduce total ecosystem C inputs. Further, when coupled with a reduced ability of grasses to anchor soils, invaded grassland invasion often leads to substantial increases in erosion and soil slippage compared to intact chaparral (Rice et al. 1969). Increased erosion associated with the conversion of chaparral into grasslands may further exacerbate the reduced ability of grass cover to sequester C into the soil by exposing lower soil layers to weathering, and increasing the rate of C cycling from the deeper soil layers back into the atmosphere.

Conversion of chaparral to grasslands has also been observed to increase the nitrate content of stream runoff (Longstreth and Patten 1975). Runoff nitrate content has been observed to increase by up to 100-fold in the years immediately following conversion (Davis and DeBano 1986; Davis 1987), likely due to the decomposition of above- and below-ground organic matter from the remnants of displaced shrubs. Such effects are also long-lasting with studies estimating nitrate runoff to remain 14-fold or greater above pre-conversion levels a decade after conversion (Davis 1984).

Increases in downslope moisture availability after conversion of chaparral watersheds to grasses has also been observed to produce substantial effects on downstream vegetation. Examinations of riparian areas downstream of formerly chaparral sites that had been recently converted were found to have tripled the density of their riparian vegetation due to increased streamflow and flow duration (DeBano et al. 1984). Moisture availability is also a critical component in determining the success of postfire chaparral re-colonization, with low moisture benefitting non-native herbs and grasses over native shrubs (Frazer and Davis 1988; Keeley et al. 2005). Thus, conversion of upstream watersheds from chaparral into grasslands possibly will increase resistance and resilience of downstream chaparral to invasion.

6.6 Toward Improved Monitoring and Modeling of Chaparral Biogeochemical Cycles

6.6.1 Environmental Sensing Approaches for Assessing Chaparral dominated Biogeochemical Cycles

The ability to measure and model the dynamics and trajectories of chaparral biogeochemical cycles is rapidly increasing in sophistication. Many new in-situ and remote sensing tools are becoming available that allow multiple temporal scale resolutions—from minutes to decades—of many important biogeochemical dynamics. These new monitoring technologies are providing extensive information to improve models and their prediction capabilities. Historically, the primary approach for quantifying ecosystem biogeochemical dynamics were through measurements of changes in the dominant pools, typically at annual scales for production and longer scales for soils. This research in chaparral dominated ecosystems requires extensive fieldwork in environments that are challenging to sample due to the continuous low level woody coverage, rugged topography, and deep rocky soils. Advanced environmental sensing platforms, both remote imaging and in-situ sensors, are providing new methods for quantifying variation in many factors closely connected to biogeochemical cycling in chaparral, although deep soil characterization remains a challenge.

Remote imaging platforms allow new methods for mapping chaparral biogeochemical variation for C, N, and potentially other elements using hyperspectral approaches from airborne and satellite sensors which record information across hundreds of spectral bands (Serrano et al. 2002; Dahlin et al. 2013). Resulting hyperspectral imagery have been used to develop estimates of carbon exchange in chaparral dominated ecosystems (Fuentes et al. 2006). Detailed information on canopy cover and vertical biomass distributions can also be collected through active remote sensing approaches using Light Detection and Ranging Laser (LIDAR) imagery (Garcia et al. 2015). Finally, thermal imaging of land surfaces is providing new opportunities to evaluate metabolic activity and water dynamics, especially evapotranspiration (Sims et al. 2008; Kustas and Anderson 2009), which is currently the focus of studies in chaparral, forest, and grassland ecosystems. When compared with field measurements, thermal imaging provides Other satellite-based sensors, such as the Ozone Monitoring Instrument (OMI) allow the characterization of atmospheric pollution including ozone and NO_2 (Levelt et al. 2006) that may affect chaparral biogeochemistry. In the future, opportunities for the remote characterization of chaparral will likely expand through efforts such as the combined imaging spectrometer and multi-band thermal sensors in the HyspIRI mission (Roberts et al. 2015) and the combination of different instruments such as Sentinal (European Space Agency) and Landsat (US National Aeronautical and Space Administration) to create a virtual satellite constellation providing multiple viewing angles and imaging times.

The advances in remote sensing are concomitant with advancements in developing automated in-situ sensor measurements. In-situ environmental sensors can now provide minute and hourly scale resolution information on many components of biogeochemical cycling (Rundel et al. 2009). Repeating ground-based cameras are providing extensive information on the dynamics of fine scale heterogeneity occurring both above-ground (Sonnentag et al. 2012) and in the soil (Allen and Kitajima 2013). Embedded sensors are also allowing the evaluation of many biogeochemical transformations, including whole ecosystem carbon and energy exchanges, soil metabolism, and whole plant water fluxes. Expanded applications of environmental sensor approaches in chaparral dominated ecosystems is a clear research priority for improved understanding of biogeochemical dynamics in this unusual ecosystem.

6.6.2 Looking Toward the Next Generation of Chaparral Biogeochemical Models

To synthesize understanding of chaparral dominated biogeochemical interactions and improve predictive skill of biogeochemical trajectories new modeling approaches are necessary. The performance of current ecosystem, land surface, and biogeochemistry models are generally applied to forested ecosystems and minimally evaluated for chaparral dominated ecosystems. Broadly used ecosystem models applicable to global ecosystem distributions generally perform poorly in regions of extensive water limitation (Niu et al. 2014) and likely even more so in MTC's where energy and moisture are disconnected. Even when specifically parameterized for chaparral dominated ecosystems model performance is limited (Li et al. 2006) or not evaluated (Tague et al. 2009). Ecosystem and land surface models generally miss important pulse driven dynamics and have poor representation of precipitation-driven phenology. Current models also have limited abilities to explicitly incorporate the large biological and geomorphological diversity characteristic of chaparral dominated ecosystems that can lead to spatial hotspots and temporal pulses of biogeochemical cycling.

To improve biogeochemical models, lessons can be learned from other dryland specific models. Ecosystem models developed specifically for drylands can obtain reasonable matches with field data (Shen et al. 2008, 2016) although even dryland specific models have not been tested in chaparral dominated ecosystems. Pulse driven biogeochemical cycling has characteristics that are distinct from non-pulse dynamics and generally excluded from ecosystem or land surface models. Simplified models of pulse driven biogeochemical dynamics can be effective using either theoretical (Jenerette et al. 2008) or empirical (Li et al. 2010) justifications, which can then provide pulse functionality within more general models. Recently, more complex pulse models have expanded in their representation of underlying mechanisms, including the formation and consumption of a labile C pool, and predictive abilities (Oikawa et al. 2014) and fit within broader modeling goals of a better microbial representation within ecosystem models. Addressing two research priorities will help advance chaparral modelling efforts. One is the accurate inclusion of precipitation and its relationship with phenology (Jenerette et al. 2010). The underlying controls on phenology are not well understood and the variation among species in phenological timing and responsiveness to meteorological dynamics can be large. Second, an improved representation of topographic heterogeneity is needed to better account for water dynamics and fine scale hotspots of biogeochemical cycling within chaparral dominated regions (Tague et al. 2009).

New biogeochemical models that build from theoretical underpinnings and combine data from new remote sensing techniques with extensive empirical evidence will provide a roadmap for improved understanding of chaparral dominated biogeochemical dynamics. The large uncertainties in models currently applied to chaparral ecosystems suggest a need for improved models that can better predict interactive element cycles in response to multiple global change drivers. At the same time the uniqueness of chaparral ecosystems provides an important testbed for a more comprehensive theory of ecosystem dynamics.

6.7 Conclusion

The breadth of chaparral biogeochemical diversity in part contributes to its high potential for continued biogeochemical sequestration storage over long time periods as well as rapid fluxes over short periods. Interactions among biophysical drivers, self-organization, and historical legacies generate large diversity in chaparral ecosystem dynamics. The consequences of ecosystem diversity are broad distributions in the rates of biogeochemical cycling. Current and future distributions of chaparral biogeochemical cycling will likely respond to multiple interactive anthropogenic drivers including increased nitrogen inputs, grazing, increased fire frequencies, more fragmented landscapes, and increasing CO_2 concentrations. These changes can influence the biogeophysical template, self-organization, and the influence of historical legacies through direct and indirect pathways. To better quantify the ecosystem services provided by chaparral and their potential future trajectories, new measurement and modeling opportunities are becoming available. Monitoring through in-situ and remote sensing techniques substantially increases the amount of data in both spatial and temporal resolution and extent than can be obtained from time-intensive field sampling. New modeling approaches are reducing uncertainties in extending findings spatially and improving projections of future biogeochemical cycling. Validating sensor measurements and model projections will still require direct field measurements, but this work will allow a larger source of information than fieldwork alone. A greater emphasis on measuring biogeochemical cycles in chaparral in combination with traditional fieldwork, new sensor tools, and advanced modeling can provide information to resolve uncertainties in ecosystem theory and to assess current and future provisioning of chaparral ecosystem services, such as climate change moderation through C sequestration or clean water provisioning through limited NO_3 losses to streams.

References

Ackerly, D. 2004. Functional strategies of chaparral shrubs in relation to seasonal water deficit and disturbance. Ecological Monographs 74:25-44.

Allen, M. F., and K. Kitajima. 2013. *In situ* high-frequency observations of mycorrhizas. New Phytologist 200:222-228.

Allen, M. F., J. N. Klironomos, K. K. Treseder, and W. C. Oechel. 2005a. Responses of soil biota to elevated CO_2 in a chaparral ecosystem. Ecological Applications 15:1701-1711.

Allen, E. B., A. G. Sirulnik, L. Egerton-Warburton, S. N. Kee, A. Bytnerowicz, P. E. Padgett, P. J. Temple, M. E. Fenn, M. A. Poth, and T. Meixner. 2005b. Air pollution and vegetation change in southern California coastal sage scrub: a comparison with chaparral and coniferous forest. Pages 79-95 *in* B. E. Kus and J. L. Beyers, editors. Planning for biodiversity: bringing research and management together. General Technical Report PSW-GTR-195. Pacific Southwest Research Station, USDA Forest Service, Albany, California, USA.

Amundson, R., D. D. Richter, G. S. Humphreys, E. G. Jobbagy, and J. Gaillardet. 2007. Coupling between biota and earth materials in the Critical Zone. Elements 3:327-332.

Anderson, H. W., M. D. Hoover, and K. G. Reinhart. 1976. Forests and water: effects of forest management on floods, sedimentation, and water supply. General Technical Report GTR-PSW-018. USDA Forest Service, Pacific Southwest Forest and Range Experiment Station, Berkeley, California, USA.

Ashby, W. R. 1956. An introduction to cybernetics. Chapman and Hall Ltd., New York, New York, USA.

Austin, A. T., and L. Vivanco. 2006. Plant litter decomposition in a semi-arid ecosystem controlled by photodegradation. Nature 442:555-558.

Bailey, R. G. 2004. Identifying ecoregion boundaries. Environmental Management 34:S14-S26.

Baldocchi, D., J. W. Tang, and L. K. Xu. 2006. How switches and lags in biophysical regulators affect spatial-temporal variation of soil respiration in an oak-grass savanna. Journal of Geophysical Research: Biogeosciences 111:G02008.

Baldocchi, D. D., and L. K. Xu. 2007. What limits evaporation from Mediterranean oak woodlands—the supply of moisture in the soil, physiological control by plants, or the demand by the atmosphere? Advances in Water Resources 30:2113-2122.

Bell, D. T. 2001. Ecological response syndromes in the flora of southwestern Western Australia: fire resprouters versus reseeders. Botanical Review 67:417-440.

Bentley, J. R. 1967. Conversion of chaparral areas to grassland: techniques used in California. USDA Agriculture Handbook number 328. USDA, Washington D.C., USA.

Biederman, J. A., R. L. Scott, M. L. Goulden, R. Vargas, M. E. Litvak, T. E. Kolb, E. A. Yepez, W. C. Oechel, P. D. Blanken, T. W. Bell, J. Garatuza-Payan, G. E. Maurer, S. Dore, and S. P. Burns. 2016. Terrestrial carbon balance in a drier world: the effects of water availability in southwestern North America. Global Change Biology 22:1867-1879.

Biudes, M. S. and G. L. Vourlitis. 2012. Carbon and nitrogen mineralization of a semiarid shrubland exposed to experimental nitrogen deposition. Soil Science Society of America Journal 76:2068-2073.

Brejda, J. T. 1997. Soil changes following 18 years of protection from grazing in Arizona chaparral. Southwestern Association of Naturalists 42:478-487.

Burns, C., and J. Sauer. 1992. Resistance by natural vegetation in the San Gabriel Mountains of California to invasion by introduced conifers. Global Ecology and Biogeography Letters 2:46-51.

Bytnerowicz, A., R. F. Johnson, L. Zhang, G. D. Jenerette, M. E. Fenn, S. L. Schilling, and I. Gonzalez-Fernandez. 2015. An empirical inferential method of estimating nitrogen deposition to Mediterranean-type ecosystems: the San Bernardino Mountains case study. Environmental Pollution 203:69-88.

Bytnerowicz, A., P. R. Miller, D. M. Olszyk, P. J. Dawson, and C. A. Fox. 1987. Gaseous and particulate air-pollution in the San Gabriel Mountains of southern California. Atmospheric Environment 21:1805-1814.

Cable, D. R. 1975. Range management in the chaparral type and its ecological basis: the status of our knowledge. Research Paper RM-155. USDA Forest Service, Rocky Mountain Forest and Range Experiment Station, Fort Collins, Colorado, USA.

Chapin, F. S., P. A. Matson, and H. A. Mooney. 2012. Principles of terrestrial ecosystem ecology. Springer, New York, New York, USA.

Chatterjee, A., and G. D. Jenerette. 2011. Spatial variability of soil metabolic rate along a dryland elevation gradient. Landscape Ecology 26:1111-1123.

Chatterjee, A., and G. D. Jenerette. 2015. Variation in soil organic matter accumulation and metabolic activity along an elevation gradient in the Santa Rosa Mountains of Southern California, USA. Journal of Arid Land 7:814-819.

Chorover, J., P. A. Troch, C. Rasmussen, P. D. Brooks, J. D. Pelletier, D. D. Breshears, T. E. Huxman, S. A. Kurc, K. A. Lohse, J. C. McIntosh, T. Meixner, M. G. Schaap, M. E. Litvak, J. Perdrial, A. Harpold, and M. Durcik. 2011. How water, carbon, and energy drive Critical Zone evolution: the Jemez-Santa Catalina Critical Zone Observatory. Vadose Zone Journal 10:884-899.

Christensen, N. L. 1973. Fire and the nitrogen cycle in California chaparral. Science 181:66-67.

Corbett, E. S., and R. P. Crouse. 1968. Rainfall intrception by annual grass and chaparral: losses compared. PSW-RP-48. USDA Forest Service, Pacific Southwest Forest and Range Experiment Station, Berkeley, California, USA.

Cornwell, W. K., J. H. C. Cornelissen, K. Amatangelo, E. Dorrepaal, V. T. Eviner, O. Godoy, S. E. Hobbie, B. Hoorens, H. Kurokawa, N. Perez-Harguindeguy, H. M. Quested, L. S. Santiago, D. A. Wardle, I. J. Wright, R. Aerts, S. D. Allison, P. van Bodegom, V. Brovkin, A. Chatain, T. V. Callaghan, S. Diaz, E. Garnier, D. E. Gurvich, E. Kazakou, J. A. Klein, J. Read, P. B. Reich, N. A. Soudzilovskaia, M. V. Vaieretti, and M. Westoby. 2008. Plant species traits are the predominant control on litter decomposition rates within biomes worldwide. Ecology Letters 11:1065-1071.

Cova, T. J. 2005. Public safety in the urban-wildland interface: should fire-prone communities have a maximum occupancy? Natural Hazards Review 6:99-108.

Dahlin, K. M., G. P. Asner, and C. B. Field. 2013. Environmental and community controls on plant canopy chemistry in a Mediterranean-type ecosystem. Proceedings of the National Academy of Sciences of the United States of America 110:6895-6900.

Davies, K. W., A. M. Nafus, and M. D. Madsen. 2013. Medusahead invasion along unimproved roads, animal trails, and random transects. Western North American Naturalist 73:54-59.

Davis, E. A. 1984. Conversion of Arizona chaparral to grass increases water yield and nitrate loss. Water Resources Research 20:1643-1649.

Davis, E. A. 1987. Chaparral conversion and streamflow: nitrate increase is balanced mainly by a decrease in bicarbonate. Water Resources Research 23:215-224.

Davis, E. A., and L. F. DeBano. 1986. Nitrate increases in soil water following conversion of chaparral to grass. Biogeochemistry 2:53-65.

DeBano, L. F., J. J. Brejda, and J. H. Brock. 1984. Enhancement of riparian vegetation following shrub control in Arizona chaparral. Journal of Soil and Water Conservation 39:317-320.

DeBano, L. F., and C. E. Conrad. 1978. The effect of fire on nutrients in a chaparral ecosystem. Ecology 59:489–497.

de Koff, J. P., R. C. Graham, K. R. Hubbert, and P. M. Wohlgemuth. 2006. Prefire and postfire erosion of soil nutrients within a chaparral watershed. Soil Science 171.12:915-928.

Del Galdo, I., W. C. Oechel, and M. F. Cotrufo. 2006. Effects of past, present and future atmospheric CO_2 concentrations on soil organic matter dynamics in a chaparral ecosystem. Soil Biology and Biochemistry 38:3235-3244.

Delwiche, C. C., P. J. Zinke, and C. M. Johnson. 1965. Nitrogen fixation by ceanothus. Plant Physiology 40:1045-1047.

Diaz, S., J. Kattge, J. H. C. Cornelissen, I. J. Wright, S. Lavorel, S. Dray, B. Reu, M. Kleyer, C. Wirth, I. C. Prentice, E. Garnier, G. Bonisch, M. Westoby, H. Poorter, P. B. Reich, A. T. Moles, J. Dickie, A. N. Gillison, A. E. Zanne, J. Chave, S. J. Wright, S. N. Sheremet'ev, H. Jactel, C. Baraloto, B. Cerabolini, S. Pierce, B. Shipley, D. Kirkup, F. Casanoves, J. S. Joswig, A. Gunther, V. Falczuk, N. Ruger, M. D. Mahecha, and L. D. Gorne. 2016. The global spectrum of plant form and function. Nature 529:167-173.

Dickens, S. J. M., and E. B. Allen. 2014. Exotic plant invasion alters chaparral ecosystem resistance and resilience pre- and post-wildfire. Biological Invasions 16:1119-1130.

Dietze, M. C., A. Sala, M. S. Carbone, C. I. Czimczik, J. A. Mantooth, A. D. Richardson, and R. Vargas. 2014. Nonstructural carbon in woody plants. Annual Review of Plant Biology 65:667-687.

Drenovsky, R. E., C. E. Koehler, K. Skelly, and J. H. Richards. 2013. Potential and realized nutrient resorption in serpentine and non-serpentine chaparral shrubs and trees. Oecologia 171:39-50.

Dunn, P. H., S. C. Barro, W. G. Wells II, M. A. Poth, P.M. Wohlgemuth, and C. G. Colver. 1988. The San Dimas experimental forest: 50 years of research. General Technical Report PSW-104. Pacific Southwest Region, USDA Forest Service, Berkeley, California, USA.

Fenn, M. E., E. B. Allen, S. B. Weiss, S. Jovan, L. H. Geiser, G. S. Tonnesen, R. F. Johnson, L. E. Rao, B. S. Gimeno, F. Yuan, T. Meixner, and A. Bytnerowicz. 2010. Nitrogen critical loads and management alternatives for N-impacted ecosystems in California. Journal of Environmental Management 91:2404-2423.

Fenn, M. E., J. S. Baron, E. B. Allen, H. M. Rueth, K. R. Nydick, L. Geiser, W. D. Bowman, J. O. Sickman, T. Meixner, D. W. Johnson, and P. Neitlich. 2003. Ecological effects of nitrogen deposition in the western United States. Bioscience 53:404-420.

Fenn, M. E., M. A. Poth, J. D. Aber, J. S. Baron, B. T. Bormann, D. W. Johnson, A. D. Lemly, S. G. McNulty, D. E. Ryan, and R. Stottlemyer. 1998. Nitrogen excess in North American ecosystems: Predisposing factors, ecosystem responses, and management strategies. Ecological Applications 8:706–733.

Fenn, M. E., M. A. Poth, P. H. Dunn, and S. C. Barro. 1993. Microbial N and biomass, respiration, and N-mineralization in soils beneath 2 chaparral species along a fire-induced age gradient. Soil Biology and Biochemistry 25:457-466.

Fenn, M. E., M. A. Poth, and D. W. Johnson. 1996. Evidence for nitrogen saturation in the San Bernardino Mountains in southern California. Forest Ecology and Management 82:211-230.

Fierer, N., and J. P. Schimel. 2002. Effects of drying-rewetting frequency on soil carbon and nitrogen transformations. Soil Biology and Biochemistry 34:777-787.

Firestone, M. K., and E. A. Davidson. 1989. Microbiological basis of NO and N20 production and consumption in soil. Pages 7-21 *in* M. O. Andreae and D. S. Schimel. Exchange of trace gases between terrestrial ecosystems and the atmosphere. John Wiley and Sons, New York, New York, USA.

Forrester, J. W. 1961. Industrial dynamics. MIT Press, Cambridge, Massachusetts, USA.

Fortin, M. J., R. J. Olson, S. Ferson, L. Iverson, C. Hunsaker, G. Edwards, D. Levine, K. Butera, and V. Klemas. 2000. Issues related to the detection of boundaries. Landscape Ecology 15:453-466.

Franco-Vizcaino, E., and J. Sosa-Ramirez. 1997. Soil properties and nutrient relations in burned and unburned Mediterranean-climate shrublands of Baja California, Mexico. Acta Oecologica-International Journal of Ecology 18:503-517.

Franklin, J., A. D. Syphard, H. S. He, and D. J. Mladenoff. 2005. Altered fire regimes affect landscape patterns of plant succession in the foothills and mountains of southern California. Ecosystems 8:885-898.

Frazer, J. M., and S. D. Davis. 1988. Differential survival of chaparral seedlings during the first summer drought after wildfire. Oecologia 76:215-221.

Fuentes, D. A., J. A. Gamon, Y. F. Cheng, H. C. Claudio, H. L. Qiu, Z. Y. Mao, D. A. Sims, A. F. Rahman, W. Oechel, and H. Y. Luo. 2006. Mapping carbon and water vapor fluxes in a chaparral ecosystem using vegetation indices derived from AVIRIS. Remote Sensing of Environment 103:312-323.

Galloway, J. N., A. R. Townsend, J. W. Erisman, M. Bekunda, Z. C. Cai, J. R. Freney, L. A. Martinelli, S. P. Seitzinger, and M. A. Sutton. 2008. Transformation of the nitrogen cycle: recent trends, questions, and potential solutions. Science 320:889-892.

Garcia, M., J. Gajardo, D. Riano, K. G. Zhao, P. Martin, and S. Ustin. 2015. Canopy clumping appraisal using terrestrial and airborne laser scanning. Remote Sensing of Environment 161:78-88.

Gardner, P. D., H. J. Cortner, and K. Widaman. 1987. The risk perceptions and policy response toward wildland fire hazards by urban home-owners. Landscape and Urban Planning 14:163-172.

Giessow, J., and P. H. Zedler. 1996. The effects of fire frequency and firebreaks on the abundance and species richness of exotic species in coastal sage scrub. *In* J. E. Lovich, J. Randall, and M. D. Kelly, editors. Proceedings of the 1996 California Exotic Pest Council 1996 Symposium, Vol 2, San Diego, California 1996. Available from: http://www.cal-ipc.org/symposia/archive

Gill, A. M., and S. L. Stephens. 2009. Scientific and social challenges for the management of fire-prone wildland-urban interfaces. Environmental Research Letters 4:034014.

Gliksman, D., A. Rey, R. Seligmann, R. Dumbur, O. Sperling, Y. Navon, S. Haenel, P. De Angelis, J. A. Arnone, and J. M. Grunzweig. 2017. Biotic degradation at night, abiotic degradation at day: positive feedbacks on litter decomposition in drylands. Global Change Biology 23:1564-1574.

Gough, L. P., G. R Meadows, L. L. Jackson, and S. Dudka. 1989. Biogeochemistry of a highly serpentinized, chromite-rich ultramafic area, Tehama County, California. US Geological Survey Bulletin 1901. The US Government Printing Office, Washington D.C., USA.

Graham, R. C., L. M. Egerton-Warburton, P. F. Hendrix, P. J. Shouse, J. L. Johnson-Maynard, S. A. Quideau, P. D. Sternberg, J. A. Jobes, and J. M. Breiner. 2016. Wildfire effects on soils of a 55-year-old chaparral and pine biosequence. Soil Science Society of America Journal 80:376-394.

Gray, J. T., and W. H. Schlesinger. 1981. Nutrient cycling in Mediterranean-type ecosystems. Pages 259-285 in P. C. Miller, editor. Resource use by chaparral and matorral: a comparison of vegetation function in two Mediterranean-type ecosystems. Springer, Berlin, Germany.

Green, L. R. 1977. Fuel reduction without fire—current technology and ecosystem impact. Pages 163-171 *in* Proceedings of the symposium of environmental consequences of fire and fuel management in Mediterranean ecosystems. General Technical Report WO-3. USDA Forest Service, Washington D.C., USA.

Grünzweig, J. M., I. Gelfland, and D. Yakir. 2007. Biogeochemical factors contributing to enhanced carbon storage following afforestation of a semi-arid shrubland. Biogeosciences Discussions, European Geosciences Union 2111-2145.

Halsey, R. W., and J. E. Keeley. 2016. Conservation issues: California chaparral. *In* S. A. Elias, D. Alterton, J. K. Cochran, D. A. Dellasala, F. Funiciello, M. I. Goldstein, K. Lajtha, S. Marshall, and P. Mineau. Reference module in earth systems and environmental sciences. Elsevier Publications Inc., Amsterdam, Netherlands. https://doi.org/10.1016/B978-0-12-409548-9.09584-1.

Hamilton, J. G. 1997. Changing perceptions of pre-European grasslands in California. Madroño 44:311-333.

Hanan, E. J., C. M. D'Antonio, D. A. Roberts, and J. P. Schimel. 2016a. Factors regulating nitrogen retention during the early stages of recovery from fire in coastal chaparral ecosystems. Ecosystems 19:910-926.

Hanan, E. J., J. P. Schimel, K. Dowdy, and C. M. D'Antonio. 2016b. Effects of substrate supply, pH, and char on net nitrogen mineralization and nitrification along a wildfire-structured age gradient in chaparral. Soil Biology and Biochemistry 95:87-99.

Hanawalt, R. B., and R. H. Whittaker. 1976. Altitudnally coordinated patterns of soils and vegetation in San Jacinto Mountains, California. Soil Science 121:114-124.

Harrison, A. A. T., E. Small, and H. A. Mooney. 1971. Drought relationships and distribution of two Mediterranean-climate California plant communities. Ecology 52:869-875.

Harte, J. 2002. Toward a synthesis of the Newtonian and Darwinian worldviews. Physics Today 55:29-34.

Heckman, K., J. Campbell, H. Powers, B. Law, and C. Swanston. 2013. The influence of fire on the radiocarbon signature and character of soil organic matter in the Siskiyou National Forest, Oregon, USA. Fire Ecology 9:40-56.

Hibbert, A. R. 1971. Increases in streamflow after converting chaparral to grass. Water Resources Research 7:71-80.

Holling, C. S., and L. Gunderson. 2002. Resilience and adaptive cycles. Pages 25-62 *in* L. Gunderson and C. S. Holling, editors. Panarchy: understanding transformations in human and natural systems. Island Press, Washington D.C., USA.

Homyak, P. M., J. C. Blankinship, K. Marchus, D. M. Lucero, J. O. Sickman, and J. P. Schimel. 2016. Aridity and plant uptake interact to make dryland soils hotspots for nitric oxide (NO) emissions. Proceedings of the National Academy of Sciences of the United States of America 113:E2608-E2616.

Homyak, P. M., and J. O. Sickman. 2014. Influence of soil moisture on the seasonality of nitric oxide emissions from chaparral soils, Sierra Nevada, California, USA. Journal of Arid Environments 103:46-52.

Homyak, P. M., J. O. Sickman, A. E. Miller, J. M. Melack, T. Meixner, and J. P. Schimel. 2014. Assessing nitrogen-saturation in a seasonally dry chaparral watershed: limitations of traditional indicators of N-saturation. Ecosystems 17:1286-1305.

Huxman, T. E., K. A. Snyder, D. Tissue, A. J. Leffler, K. Ogle, W. T. Pockman, D. R. Sandquist, D. L. Potts, and S. Schwinning. 2004. Precipitation pulses and carbon fluxes in semiarid and arid ecosystems. Oecologia 141:254-268.

Ichii, K., W. L. Wang, H. Hashimoto, F. H. Yang, P. Votava, A. R. Michaelis, and R. R. Nemani. 2009. Refinement of rooting depths using satellite-based evapotranspiration seasonality for ecosystem modeling in California. Agricultural and Forest Meteorology 149:1907-1918.

Jarvis, P., A. Rey, C. Petsikos, L. Wingate, M. Rayment, J. Pereira, J. Banza, J. David, F. Miglietta, M. Borghetti, G. Manca, and R. Valentini. 2007. Drying and wetting of Mediterranean soils stimulates decomposition and carbon dioxide emission: the "Birch effect". Tree Physiology 27:929-940.

Jenerette, G. D., G. A. Barron-Gafford, A. J. Guswa, J. J. McDonnell, and J. C. Villegas. 2012. Organization of complexity in water limited ecohydrology. Ecohydrology 5:184-199.

Jenerette, G. D., and A. Chatterjee. 2012. Soil metabolic pulses: water, substrate, and biological regulation. Ecology 93:959-966.

Jenerette, G. D., R. L. Scott, and T. E. Huxman. 2008. Whole ecosystem metabolic pulses following precipitation events. Functional Ecology 22:924-930.

Jenerette, G. D., R. L. Scott, and A. R. Huete. 2010. Functional differences between summer and winter season rain assessed with MODIS-derived phenology in a semiarid region. Journal of Vegetation Science 21:16-30.

Keeley, J. E. 2005. Fire as a threat to biodiversity in fire-type shrublands. Pages 97-106 *in* B. E. Kus and J. L. Beyers, technical coordinators. Planning for biodiversity—bringing research and management together. General Technical Report PSW-GTR-195. US Forest Service, Pacific Southwest Research Station, Albany, California, USA.

Keeley, J. E., M. Baer-Keeley, and C. J. Fotheringham. 2005. Alien plant dynamics following fire in mediterranean-climate california shrublands. Ecological Applications 15:2109-2125.

Keeley, J. E., and T. J. Brennan. 2012. Fire-driven alien invasion in a fire-adapted ecosystem. Oecologia 169:1043-1052.

Keeley, J. E., and C. J. Fotheringham. 1997. Trace gas emissions and smoke-induced seed germination. Science 276:1248-1250.

Knipe, O. D. 1982. Angora goats for conversion of Arizona chaparral: early results. General Technical Report PSW-GTR-59. USDA Forest Service, Pacific Southwest Forest and Range Experiment Station, Berkeley, California, USA.

Knipe, O. D. 1983. Effects of Angora goat browsing on burned-over Arizona chaparral. Rangelands 5:252-255.

Kruckeberg, A. R. 1984. California serpentines, flora, vegetation, geology, soils, and management problems. University of California Press, Berkeley, California, USA.

Kummerow, J., J. V. Alexander, J. W. Neel, and K. Fishbeck. 1978. Symbiotic nitrogen-fixation in Ceanothus roots. American Journal of Botany 65:63-69.

Kustas, W. and M. Anderson. 2009. Advances in thermal infrared remote sensing for land surface modeling. Agricultural and Forest Meteorology 149:2071-2081.

Lambrinos, J. G. 2006. Spatially variable propagule pressure and herbivory influence invasion of chaparral shrubland by an exotic grass. Oecologia 147:327-334.

Lauvaux, C. A., C. N. Skinner, and A. H. Taylor. 2016. High severity fire and mixed conifer forest-chaparral dynamics in the southern Cascade Range, USA. Forest Ecology and Management 363:74–85

Lehmann, J., and Kleber, M. 2015. The contentious nature of soil organic matter. Nature 528:60-68.

Levelt, P. F., G. H. J. Van den Oord, M. R. Dobber, A. Malkki, H. Visser, J. de Vries, P. Stammes, J. O. V. Lundell, and H. Saari. 2006. The ozone monitoring instrument. Ieee Transactions on Geoscience and Remote Sensing 44:1093-1101.

Levin, S. A. 1998. Ecosystems and the biosphere as complex adaptive systems. Ecosystems 1:431-436.

Levin, S. A. 2000. Fragile dominion: complexity and the commons. Perseus Books, New York, New Yok, USA.

Li, L., M. L. McCormack, C. G. Ma, D. L. Kong, Q. Zhang, X. Y. Chen, H. Zeng, U. Niinemets, and D. L. Guo. 2015. Leaf economics and hydraulic traits are decoupled in five species-rich tropical-subtropical forests. Ecology Letters 18:899-906.

Li, X. Y., T. Meixner, J. O. Sickman, A. E. Miller, J. P. Schimel, and J. M. Melack. 2006. Decadal-scale dynamics of water, carbon and nitrogen in a California chaparral ecosystem: DAYCENT modeling results. Biogeochemistry 77:217-245.

Li, X. Y., A. E. Miller, T. Meixner, J. P. Schimel, J. M. Melack, and J. O. Sickman. 2010. Adding an empirical factor to better represent the rewetting pulse mechanism in a soil biogeochemical model. Geoderma 159:440-451.

Liang, L. L., D. A. Grantz, and G. D. Jenerette. 2016. Multivariate regulation of soil CO_2 and N_2O pulse emissions from agricultural soils. Global Change Biology 22:1286-1298.

Longstreth, D. T., and D. T. Patten. 1975. Conversion of chaparral to grass in Central Arizona: Effects on selected ions in watershed runoff. American Midland Naturalist 93:25-34.

Luo, H. Y., W. C. Oechel, S. J. Hastings, R. Zulueta, Y. H. Qian, and H. Kwon. 2007. Mature semiarid chaparral ecosystems can be a significant sink for atmospheric carbon dioxide. Global Change Biology 13:386-396.

Malamud, B. D., G. Morein, and D. L. Turcotte. 1998. Forest fires: an example of self-organized critical behavior. Science 281:1840-1842.

May, A. A., G. R. McMeeking, T. Lee, J. W. Taylor, J. S. Craven, I. Burling, A. P. Sullivan, S. Akagi, J. L. Collett, Jr., M. Flynn, H. Coe, S. P. Urbanski, J. H. Seinfeld, R. J. Yokelson, and S. M. Kreidenweis. 2014. Aerosol emissions from prescribed fires in the United States: a synthesis of laboratory and aircraft measurements. Journal of Geophysical Research: Atmospheres 119:11826-11849.

Mayberry, J. A. 2011. Community wildfire preparedness: balancing community safety and ecosystem sustainability in southern California chaparral. Humboldt State University.

Meixner, T., M. E. Fenn, P. Wohlgemuth, M. Oxford, and P. Riggan. 2006. N saturation symptoms in chaparral catchments are not reversed by prescribed fire. Environmental Science and Technology 40:2887-2894.

Meng, R., P. E. Dennison, C. M. D'Antonio, and M. A. Moritz. 2014. Remote sensing analysis of vegetation recovery following short-interval fires in Southern California shrublands. PLoS ONE 9:e110637.

Milchunas, D. G. 2006. Responses of plant communities to grazing in the southwestern United States. General Technical Report RMRS-GTR-169. USDA Forest Service, Rocky Mountain Research Station, Fort Collins, Colorado, USA.

Miller, A. E., J. P. Schimel, T. Meixner, J. O. Sickman, and J. M. Melack. 2005. Episodic rewetting enhances carbon and nitrogen release from chaparral soils. Soil Biology and Biochemistry 37:2195-2204.

Minnich, R. A., and C. J. Bahr. 1995. Wildland fire and chaparral succession along the California Baja-California boundary. International Journal of WIldland Fire 5:13-24.

Mooney, H. A., and P. W. Rundel. 1979. Nutrient relations of the evergreen shrub, *Adenostoma fasciculatum*, in the California chaparral. Botanical Gazette 140:109-113.

Moreno, J. M., I. Torres, B. Luna, W. C. Oechel, and J. E. Keeley. 2013. Changes in fire intensity have carry-over effects on plant responses after the next fire in southern California chaparral. Journal of Vegetation Science 24:395-404.

Nagle, T. A., and A. H. Taylor. 2005. Fire and persistence of montane chaparral in mixed conifer forest landscapes in the northern Sierra Nevada, Lake Tahoe Basin, California, USA. Journal of the Torrey Botanical Society 132:442-457.

Niu, G.-Y., C. Paniconi, P. A. Troch, R. L. Scott, M. Durcik, X. Zeng, T. Huxman, and D. C. Goodrich. 2014. An integrated modelling framework of catchment-scale ecohydrological processes: 1. Model description and tests over an energy-limited watershed. Ecohydrology 7:427-439.

Odum, H. T. 1983. Systems ecology: an introduction. John Wiley & Sons, New York, New York, USA.

Oikawa, P. Y., C. Ge, J. Wang, J. R. Eberwein, L. L. Liang, L. A. Allsman, D. A. Grantz, and G. D. Jenerette. 2015. Unusually high soil nitrogen oxide emissions influence air quality in a high-temperature agricultural region. Nature Communications 6:1-10.

Oikawa, P. Y., D. A. Grantz, A. Chatterjee, J. E. Eberwein, L. A. Allsman, and G. D. Jenerette. 2014. Unifying soil respiration pulses, inhibition, and temperature hysteresis through dynamics of labile soil carbon and O_2. Journal of Geophysical Research: Biogeosciences 119:521-536.

Orme, A. R., and R. G. Bailey. 1970. The effect of vegetation conversion and flood discharge on stream channel geometry: a case of southern California watersheds. Proceedings of the American Association of Geographers 2:101-106.

Park, I. W., J. Hooper, J. M. Flegal, and G. D. Jenerette. 2018. Drivers of herbaceous cover throughout chaparral shrublands: Insights from a novel remote sensing method. Diversity and Distributions, in press.

Parker, V.T., R. B. Pratt, and J. E. Keeley. 2016. Chaparral. Pages 479-507 *in* H. Mooney and E. Zavaleta, editors. Ecosystems of California. University of California Press, Berkeley, California, USA.

Pasquini, S. C., and G. L. Vourlitis. 2010. Post-fire primary production and plant community dynamics in chaparral stands exposed to varying levels of nitrogen deposition. Journal of Arid Environments 74:310-314.

Pivovaroff, A. L., L. S. Santiago, G. L. Vourlitis, D. A. Grantz, and M. F. Allen. 2016. Plant hydraulic responses to long-term dry season nitrogen deposition alter drought tolerance in a Mediterranean-type ecosystem. Oecologia 181:721-731.

Pumo, D., F. Viola, and L. V. Noto. 2008. Ecohydrology in Mediterranean areas: a numerical model to describe growing seasons out of phase with precipitations. Hydrology and Earth System Sciences 12:303-316.

Quideau, S. A., R. C. Graham, O. A. Chadwick, and H. B. Wood. 1998. Organic carbon sequestration under chaparral and pine after four decades of soil development. Geoderma 83:227–242.

Rao, L. E., E. B. Allen, and T. Meixner. 2010. Risk-based determination of critical nitrogen deposition loads for fire spread in southern California deserts. Ecological Applications 20:1320-1335.

Rao, L. E., D. R. Parker, A. Bytnerowicz, and E. B. Allen. 2009. Nitrogen mineralization across an atmospheric nitrogen deposition gradient in Southern California deserts. Journal of Arid Environments 73:920-930.

Rasmussen, C., R. J. Southard, and W. R. Horwath. 2006. Mineral control of organic carbon mineralization in a range of temperate conifer forest soils. Global Change Biology 12:834-847.

Rice, R. M., E. S. Corbett, and R. G. Bailey. 1969. Soil Slips related to vegetation, topography, and soil in southern California. Water Resources Research 5:647-659.

Rich, L. R., and H. G. Reynolds. 1963. Grazing in relation to runoff and erosion on some chaparral watersheds of central Arizona. Society for Range Management 16:322-326.

Riggan, P. J., R. N. Lockwood, and E. N. Lopez. 1985. Deposition and processing of airborne nitrogen pollutants in Mediterranean-type ecosystems of southern California. Environmental Science and Technology 19:781-789.

Riggan, P. J., R. N. Lockwood, P. M. Jacks, C. G. Colver, F. Weirich, L. F. Debano, and J. A. Brass. 1994. Effects of fire severity on nitrate mobilization in watersheds subject to chronic atmospheric deposition. Environmental Science and Technology 28:369-375.

Riveros-Iregui, D. A., B. L. McGlynn, R. E. Emanuel, and H. E. Epstein. 2012. Complex terrain leads to bidirectional responses of soil respiration to inter-annual water availability. Global Change Biology 18:749-756.

Roberts, D. A., P. E. Dennison, K. L. Roth, K. Dudley, and G. Hulley. 2015. Relationships between dominant plant species, fractional cover and land surface temperature in a Mediterranean ecosystem. Remote Sensing of Environment 167:152-167.

Roberts, S. W., W. C. Oechel, P. J. Bryant, S. J. Hastings, J. Major, and V. Nosov. 1998. A field fumigation system for elevated carbon dioxide exposure in chaparral shrubs. Functional Ecology 12:708-719.

Rodrigue, C. M. 1993. Home with a view: chaparral fire hazard and the social geographies of risk and vulnerability. California Geographical Society 33:29-42.

Rossi, R. J., D. J. Bain, G. D. Jenerette, L. W. Clarke, and K. Wilson. 2015. Responses of roadside soil cation pools to vehicular emission deposition in southern California. Biogeochemistry 124:131-144.

Rowe, P. B. 1963. Streamflow increases after removing woodland-riparian vegetation from a southern california watershed. Journal of Forestry 61:365-370.

Rundel, P. W., E. A. Graham, M. F. Allen, J. C. Fisher, and T. C. Harmon. 2009. Environmental sensor networks in ecological research. New Phytologist 182:589-607.

Rundel, P. W., and D. J. Parsons. 1979. Structural changes in chamise (*Adenostoma fasciculatum*) along a fire-induced age gradient. Journal of Range Management 32:462-466.

Scheffer, M., S. Carpenter, J. A. Foley, C. Folke, and B. Walker. 2001. Catastrophic shifts in ecosystems. Nature 413:591-596.

Scheffer, M., and S. R. Carpenter. 2003. Catastrophic regime shifts in ecosystems: linking theory to observation. Trends in Ecology and Evolution 18:648-656.

Schlesinger, W. H., and E. S. Bernhardt. 2013. Biogeochemistry: an analysis of global change. Academic Press, Waltham, Massachusetts, USA.

Schmidt, M. W. I., M. S. Torn, S. Abiven, T. Dittmar, G. Guggenberger, I. A. Janssens, M. Kleber, I. Kogel-Knabner, J. Lehmann, D. A. C. Manning, P. Nannipieri, D. P. Rasse, S. Weiner, and

S. E. Trumbore. 2011. Persistence of soil organic matter as an ecosystem property. Nature 478:49-56.

Scott, R. L., T. E. Huxman, G. A. Barron-Gafford, G. D. Jenerette, J. M. Young, and E. P. Hamerlynck. 2014. When vegetation change alters ecosystem water availability. Global Change Biology 20:2198-2210.

Scott, R. L., G. D. Jenerette, D. L. Potts, and T. E. Huxman. 2009. Effects of seasonal drought on net carbon dioxide exchange from a woody-plant-encroached semiarid grassland. Journal of Geophysical Research: Biogeosciences 114:G04004.

Serrano, L., J. Penuelas, and S. L. Ustin. 2002. Remote sensing of nitrogen and lignin in Mediterranean vegetation from AVIRIS data: decomposing biochemical from structural signals. Remote Sensing of Environment 81:355-364.

Severson, K. E., and L. F. Debano. 1991. Influence of Spanish goats on vegetation and soils in Arizona chaparral. Journal of Range Management 44:111-117.

Shen, W., G. D. Jenerette, D. Hui, R. P. Phillips, and H. Ren. 2008. Effects of changing precipitation regimes on dryland soil respiration and C pool dynamics at rainfall event, seasonal, and interannual scales. Journal of Geophysical Research: Biogeosciences 113:G03024.

Shen, W., G. D. Jenerette, D. Hui, and R. L. Scott. 2016. Precipitation legacy effects on dryland ecosystem carbon fluxes: direction, magnitude and biogeochemical carryovers. Biogeosciences 13:425-439.

Sidahmed, A. E., J. G. Morris, and S. R. Radosevich. 1981. Summer diet of Spanish goat grazing chaparral. Journal of Range Management 34:1981.

Sims, D. A., H. Y. Luo, S. Hastings, W. C. Oechel, A. F. Rahman, and J. A. Gamon. 2006. Parallel adjustments in vegetation greenness and ecosystem CO_2 exchange in response to drought in a Southern California chaparral ecosystem. Remote Sensing of Environment 103:289-303.

Sims, D. A., A. F. Rahman, V. D. Cordova, B. Z. El-Masri, D. D. Baldocchi, P. V. Bolstad, L. B. Flanagan, A. H. Goldstein, D. Y. Hollinger, L. Misson, R. K. Monson, W. C. Oechel, H. P. Schmid, S. C. Wofsy, and L. Xu. 2008. A new model of gross primary productivity for North American ecosystems based solely on the enhanced vegetation index and land surface temperature from MODIS. Remote Sensing of Environment 112:1633-1646.

Six, J., R. T. Conant, E. A. Paul, and K. Paustian. 2002. Stabilization mechanisms of soil organic matter: implications for C-saturation of soils. Plant and Soil 241:155-176.

Smith, S. V, S. H. Bullock, A. Hinojosa-Corona, E. Franco-Vizcaino, M. Escoto-Rodriguez, T. G. Kretzschmar, L. M. Farfan, and J. M. Salazar-Cesena. 2007. Soil erosion and significance for carbon fluxes in a mountainous Mediterranean-climate watershed. Ecological Applications 17:1379-1387.

Sonnentag, O., K. Hufkens, C. Teshera-Sterne, A. M. Young, M. Friedl, B. H. Braswell, T. Milliman, J. O'Keefe, and A. D. Richardson. 2012. Digital repeat photography for phenological research in forest ecosystems. Agricultural and Forest Meteorology 152:159-177.

Sparks, S. R., W. C. Oechel, and Y. Mauffette. 1993. Photosynthate allocation patterns along a fire-induced age sequence in 2 shrub species from the California chaparral. International Journal of Wildland Fire 3:21-30.

Stewart, C. E., A. F. Plante, K. Paustian, R. T. Conant, and J. Six. 2008. Soil carbon saturation: linking concept and measurable carbon pools. Soil Science Society of America Journal 72:379-392.

Stylinski, C. D., J. A. Gamon, and W. C. Oechel. 2002. Seasonal patterns of reflectance indices, carotenoid pigments and photosynthesis of evergreen chaparral species. Oecologia 131:366-374.

Syphard, A. D., T. J. Brennan, and J. E. Keeley. 2014. The role of defensible space for residential structure protection during wildfires. International Journal of WIldland Fire 23:1165-1175.

Syphard, A. D., K. C. Clarke, and J. Franklin. 2007. Simulating fire frequency and urban growth in southern California coastal shrublands, USA. Landscape Ecology 22:431-445.

Syphard, A. D., H. M. Regan, J. Franklin, R. M. Swab, and T. C. Bonebrake. 2013. Does functional type vulnerability to multiple threats depend on spatial context in Mediterranean-climate ecosystems? Diversity and Distributions 19:1263-1274.

Tague, C., L. Seaby, and A. Hope. 2009. Modeling the eco-hydrologic response of a Mediterranean type ecosystem to the combined impacts of projected climate change and altered fire frequencies. Climatic Change 93:137-155.

Thanos, C. A., and P. W. Rundel. 1995. Fire-followers in chaparral: nitrogenous compounds trigger seed-germination. Journal of Ecology 83:207-216.

Treseder, K. K., L. M. Egerton-Warburton, M. F. Allen, Y. F. Cheng, and W. C. Oechel. 2003. Alteration of soil carbon pools and communities of mycorrhizal fungi in chaparral exposed to elevated carbon dioxide. Ecosystems 6:786-796.

Trueman, R. J., L. Taneva, M. A. Gonzalez-Meler, W. C. Oechel, and H. BassiriRad. 2009. Carbon losses in soils previously exposed to elevated atmospheric CO_2 in a chaparral ecosystem: potential implications for a sustained biospheric C sink. Journal of Geochemical Exploration 102:142-148.

Ulery, A. L., R. C. Graham, O. A. Chadwick, and H. B. Wood. 1995. Decadal-scale changes in of soil carbon, nitrogen, and exchangable cations under chaparral and pine. Geoderma 65:121-134.

Vourlitis, G. L. 2012. Aboveground net primary production response of semi-arid shrublands to chronic experimental dry-season N input. Ecosphere 3(3):22.

Vourlitis, G. L. 2017. Chronic N enrichment and drought alter plant cover and community composition in a Mediterranean-type semi-arid shrubland. Oecologia 184:267-277.

Vourlitis, G. L., C. DeFotis, and W. Kristan. 2015. Effects of soil water content, temperature and experimental nitrogen deposition on nitric oxide (NO) efflux from semiarid shrubland soil. Journal of Arid Environments 117:67-74.

Vourlitis, G. L., and J. S. Fernandez. 2012. Changes in the soil, litter, and vegetation nitrogen and carbon concentrations of semiarid shrublands in response to chronic dry season nitrogen input. Journal of Arid Environments 82:115-122.

Vourlitis, G. L., and J. S. Fernandez. 2015. Carbon and nitrogen mineralization of semi-arid shrubland soils exposed to chronic nitrogen inputs and pulses of labile carbon and nitrogen. Journal of Arid Environments 122:37-45.

Vourlitis, G. L., and C. S. Hentz. 2016. Impacts of chronic N input on the carbon and nitrogen storage of a postfire Mediterranean-type shrubland. Journal of Geophysical Research: Biogeosciences 121:385-398.

Vourlitis, G. L., S. Pasquini, and G. Zorba. 2007a. Plant and soil N response of southern Californian semi-arid shrublands after 1 year of experimental N deposition. Ecosystems 10:263-279.

Vourlitis, G. L., and S. C. Pasquini. 2009. Experimental dry-season N deposition alters species composition in southern Californian mediterranean-type shrublands. Ecology 90:2183-2189.

Vourlitis, G. L., S. C. Pasquini, and R. Mustard. 2009. Effects of dry-season N input on the productivity and N storage of Mediterranean-type shrublands. Ecosystems 12:473-488.

Vourlitis, G. L., G. Zorba, S. C. Pasquini, and R. Mustard. 2007b. Chronic nitrogen deposition enhances nitrogen mineralization potential of semiarid shrubland soils. Soil Science Society of America Journal 71:836-842.

Warrick, J. A., and D. M. Rubin. 2007. Suspended-sediment rating curve response to urbanization and wildfire, Santa Ana River, California. Journal of Geophysical Research 112:F02018.

Westoby, M. 1998. A leaf-height-seed (LHS) plant ecology strategy scheme. Plant and Soil 199:213-227.

White, M. D., and K. A. Greer. 2006. The effects of watershed urbanization on the stream hydrology and riparian vegetation of Los Peñasquitos Creek, California. Landscape and Urban Planning 74:125-138.

Williamson, T. N., R. C. Graham, and P. J. Shouse. 2004a. Effects of a chaparral-to-grass conversion on soil physical and hydrologic properties after four decades. Geoderma 123:99-114.

Williamson, T. N., B. D. Newman, R. C. Graham, and P. J. Shouse. 2004b. Regolith water in zero-order chaparral and perennial grass watersheds four decades after vegetation conversion. Vadose Zone Journal 3:1007-1016.

Wolkovich, E. M., D. A. Lipson, R. A. Virginia, K. L. Cottingham, and D. T. Bolger. 2010. Grass invasion causes rapid increases in ecosystem carbon storage in a semiarid shrubland. Global Change Biology 16:1351-1365.

Wright, I. J., P. B. Reich, M. Westoby, D. D. Ackerly, Z. Baruch, F. Bongers, J. Cavender-Bares, T. Chapin, J. H. C. Cornelissen, M. Diemer, J. Flexas, E. Garnier, P. K. Groom, J. Gulias, K. Hikosaka, B. B. Lamont, T. Lee, W. Lee, C. Lusk, J. J. Midgley, M. L. Navas, U. Niinemets, J. Oleksyn, N. Osada, H. Poorter, P. Poot, L. Prior, V. I. Pyankov, C. Roumet, S. C. Thomas, M. G. Tjoelker, E. J. Veneklaas, and R. Villar. 2004. The worldwide leaf economics spectrum. Nature 428:821-827.

Wu, J. G., and J. L. David. 2002. A spatially explicit hierarchical approach to modeling complex ecological systems: theory and applications. Ecological Modelling 153:7-26.

Wu, J. G., and O. L. Loucks. 1995. From balance of nature to hierarchical patch dynamics: a paradigm shift in ecology. Quarterly Review of Biology 70:439-466.

Zink, T. A., M. F. Allen, B. Heinde-Tenhunen, and E. B. Allen. 1995. The effect of a disturbance corridor on an ecological reserve. Restoration Ecology 3:304-310.

Chapter 7
Sediment Delivery, Flood Control, and Physical Ecosystem Services in Southern California Chaparral Landscapes

Peter M. Wohlgemuth and Keith A. Lilley

Abstract Southern California chaparral environments, with steep mountain slopes, semi-arid climate, and non-cohesive soils, are very erosive landscapes. Wildfire is the dominant ecological disturbance event in chaparral and it greatly accelerates flooding and erosion, which are directly and/or indirectly related to the loss of the protective vegetation. Since the 1920s, dams and debris basins have been constructed by public works agencies to protect the growing population and infrastructure of southern California by intercepting and impounding flows of water and debris. Dams also capture stream runoff for supplying water to downstream agriculture and urban populations. Major sediment inflows into dams and debris basins following fire can reduce capacity and threaten the ability to provide flood control and water supply. Chaparral provides physical ecosystem services that aid in flood hazard reduction, sediment retention, and the supply of water as well as protecting habitat for endangered species and soil quality.

Keywords Chaparral · Erosional processes · Flood control · Physical ecosystem services · Sediment delivery · Stormwater conservation

7.1 Introduction

Chaparral vegetation covers the foothills and lower mountain hillslopes over an extensive area of southern California (see Fig. 1.1). The physical characteristics of geology, topography, and soils in this region produce very erosive landscapes, a

P. M. Wohlgemuth (✉)
USDA Forest Service, Pacific Southwest Research Station, Riverside, CA, USA
e-mail: pwohlgemuth@fs.fed.us

K. A. Lilley
County of Los Angeles, Department of Public Works, Alhambra, CA, USA

E. C. Underwood et al. (eds.), *Valuing Chaparral*, Springer Series on Environmental Management, https://doi.org/10.1007/978-3-319-68303-4_7

condition that is further exacerbated by repeated wildfire that is occurring increasingly more frequently. The potential hazards fostered by this wildfire disturbance, such as flooding and accelerated sediment fluxes, threaten downstream human communities, prompting public works agencies to build flood control, sediment retention, and stormwater conservation networks that are expensive to construct and maintain. Intact native chaparral landscapes offer watershed protection by naturally reducing floods and retaining sediment. Other physical ecosystem services in chaparral environments also include water supply, protection of threatened and endangered species and their habitats, and protection of soil quality.

7.1.1 Sediment Fluxes in Southern California Chaparral

Sediment flux is the movement of granular rock and organic material through a watershed from hillslopes, through stream and river systems, and ultimately to the ocean. Southern California has some of the highest erosion rates, and hence sediment fluxes, in the world—comparable to the Himalayas or the Andes (Jansson 1988). The current landscape reflects the removal of soil and sediment from the uplands and deposition of this material in the adjacent lowlands over the last several millennia. Local erosion depends on many factors including topography, geology, soils, vegetation cover, and rainfall patterns.

Southern California is a zone of active tectonics associated with crustal plate interactions along the San Andreas Fault system, resulting in mountain-building in the Coast Ranges, the Peninsular Ranges, and the Transverse Ranges. Studies have shown that local mountain chains are uplifting at rates as high as 7.6 mm per year, despite the high erosion rates (Scott and Williams 1978; Spotila et al. 2002). This ongoing tectonic uplift has produced steep topography and exposed fractured, highly erodible crystalline basement rocks at the earth's surface. In chaparral environments, this erosive rock material, combined with any overlying soft sedimentary strata, weather to thin, coarse-textured, azonal soils (Dunn et al. 1988). These soils are typically non-cohesive, meaning that they lack clays and organic glues that bind the individual soil granules together. Southern California is classified as a Mediterranean-type climate region, consisting of cool, wet winters followed by long summer droughts. This amount and pattern of precipitation combined with the shrubland vegetation cover in the region maximize erosion (Langbein and Schumm 1958). Chaparral vegetation is drought tolerant, due in part to its small, sclerophyllous leaves (see Chap. 1). However, the waxy, resinous substances that allow chaparral to endure the long, dry summers also render it very prone to wildfire (Rice 1974).

Wildfire has a profound effect on sediment fluxes in chaparral. Fire has been a part of the southern California landscape since before recorded history and is the disturbance event which drives many of the environmental responses in chaparral ecosystems (Sugihara and Barbour 2006). Fire is also responsible for much of the erosion experienced across the landscape. With the combustion of the vegetation canopy and the surface litter layer, along with fire-induced changes in soil properties, erosion on hillslopes greatly increases (Wohlgemuth et al. 2009). Hillslope hydrology is also affected, increasing water delivery to the stream channels that may increase streamflow magnitude and initiate a range of sediment flux responses, including debris flows or slurries of mud, which can threaten both local flora and fauna as well as downstream human communities. Natural fire recurrence in chaparral ecosystems is 50–100 years (Minnich 1989), but fire frequency has increased significantly over the last century related to the large numbers of anthropogenic ignitions (Safford 2007).

Postfire natural hazards to human endeavors have been recognized in southern California since the late 1800s (Munns 1919; Wohlgemuth et al. 2009). Flooding and accelerated erosion from burned areas can threaten life, property, and community infrastructure at the wildland-urban interface. Residents and campers have been killed by flash floods and debris torrents, in many cases at significant distances downstream from the location of the fire. Homes and businesses have been inundated with debris originating from adjacent or nearby burns. Roads, bridges, utility lines, pipelines, and communication lines have all been damaged by both flood and mud. To mitigate these foregoing hazards, flood control and protection from transported sediment has been the goal of public works agencies in southern California since at least the 1920s (Wohlgemuth et al. 2009).

7.1.2 Flood Control and Stormwater Conservation in Chaparral

Southern California has historically been subjected to large but infrequent flooding events that occasionally shifted the flow paths of the region's rivers and streams. Southern California has also been subjected to long drought periods in between the large flooding events. Not only does this lead to a higher probability of wildfire but also an increase in postfire flooding when storms return.

The flood control and stormwater conservation history in Los Angeles County is representative of southern California as a whole. For example, a devastating storm hit the Los Angeles region in February 1914, causing widespread flooding, erosion damage, and deposition of over three million cubic meters of silt in the ports of Los Angeles and Long Beach. More than 50 people died and there was over $10 million worth of damage (approximately $237 million in 2015 dollars). In the aftermath of the flood, the State Legislature, responding to requests for action from Los Angeles County leaders, created the Los Angeles County Flood Control District in June 1915, the first flood control district in the State of California. The District was given the dual mission:

"…to provide for the control and conservation of the flood, storm and other waste waters of said district, and to conserve such waters for the beneficial and useful purposes by spreading, storing, retaining or causing to percolate into the soil within said district, or to save or conserve in any manner, all or any of such waters, and to protect from damage from such flood or stormwaters, the harbors, waterways, public highways, and property in said district" (Los Angeles County Flood Control Act, Section 2 of 1915).

Table 7.1 Southern California Flood Control Districts and dates of establishment

Agency	Established
Los Angeles Flood Control District	1915
Orange County Flood Control District	1927
San Bernardino County Flood Control District	1939
Ventura County Flood Control District	1944
Riverside County Flood Control and Water Conservation District	1945
Santa Barbara County Flood Control and Water Conservation District	1955
San Diego County Flood Control District	1966

Eventually, other areas in southern California followed suit as their populations, agricultural development, and urban development expanded, and major storms affected their populations and assets (Table 7.1).

Meanwhile, as far back as the 1890s the federal government recognized that stands of intact vegetation have a stabilizing effect on streamflow. This tends to prevent erosion and siltation, thereby maintaining the navigability of streams and water supply for power, irrigation, and urban use, and reducing devastating floods. It was also recognized that loss of vegetation from fires, logging, and mining activities had significant adverse impacts on this stabilizing effect. Legislation was passed to establish forest reserves, with the specific direction to make:

"provisions for the protection [of forest reservations] against destruction by fire and depredations..." (Organic Administration Act of 1897).

Among the earliest forest reserves was the San Gabriel Timberland Reserve (subsequently the Angeles National Forest), which was set aside primarily for the protection of irrigation projects and municipal water supplies (see Chap. 15).

The newly formed Los Angeles County Flood Control District started work in the 1920s, building several dams in the mountains (that would eventually total 90 across the county), serving a dual flood control and stormwater capture function, and building flood control channels in the valleys and coastal plains below. In subsequent plans, the District laid out the basis for a future water conservation system, pioneering the establishment of large scale artificial groundwater recharge facilities (spreading grounds) at locations adjacent to the major flood control channels. The intention of these spreading grounds, consisting of shallow basins with sand and gravel bottoms, was to enable diverted stormwater to percolate into the underlying groundwater aquifers, thereby providing a reliable source of water that could be pumped for agricultural, residential, and industrial use. The District partnered with both the US Army Corps of Engineers on flood control construction and the US Department of Agriculture Forest Service for upstream mountain erosion control to reduce the movement of eroded material in and below the Angeles National Forest. This had the dual purpose of protecting water supplies and also reducing flood risk to downstream development.

With the formation of flood control districts in southern California came the philosophy for flood control and water conservation, again illustrated by the actions of the Los Angeles County Flood Control District. Officials in southern California recognized early on the need to conserve water flowing from the mountains to the floodplains. As such, major flood control dams were also built with water conservation in mind. Additional water conservation was to be accomplished by in-stream percolation through the earth-bottom channels where feasible. In subsequent years, more sophisticated water conservation methods were developed after infrastructure was completed to deal with the worst flood threats. For example, the Los Angeles County plan of flood control consisted of several elements:

- The construction of major dams in the mountainous upper reaches of the main drainage systems to contain peak stormwater discharge and to regulate downstream flows. By reducing peak flows at the headwaters of major drainage systems, the downstream channels could be designed for lower flow rates.
- The straightening and stabilization of major natural river channels throughout the entire coastal plain to facilitate rapid drainage. The philosophy was to reduce flood damage by directing stormflows away from urban areas and to the ocean as quickly as possible.
- The development of standards for the design of flood control infrastructure to ensure an adequate level of regional flood protection. Eventually the District based its design standards on the major storms that occurred in 1938 and 1943. In short, its major infrastructure calls for safely handling the runoff from a once in 50-year storm event (a probability based on rainfall intensity) falling on a previously burned watershed that has 4 years of postfire vegetation recovery, the soil of which has been saturated by 3 days of antecedent storms (Los Angeles County 2006).
- The decision was made to leave many of the channels with "soft" bottoms (natural un-armored channel beds) which allowed for the flood waters, and water released from the dams to percolate through into the underlying groundwater aquifers.
- The development of dedicated, off-channel spreading grounds to allow stormwater that would otherwise flow to the ocean to be diverted for increased percolation into groundwater aquifers. The aim was to increase water supply for a growing population.

As southern California agencies began developing and operating their flood control and water conservation infrastructure, it became abundantly clear that impacts from not only stormwater but also sediment—particularly that which was associated with "fire-flood sequences"—needed to be addressed. Originally, small check dams were constructed in mountain areas to control erosion and debris flows downstream of burned areas (Eaton 1932). However, debris basins—retention structures at canyon mouths dedicated to capturing debris flows and sediment before they reach foothill communities—were soon recognized also to be critical for dealing with the effects of postfire flooding in the region (Los Angeles County 1931).

It took a large scale fire in 1933 and subsequent debris flows on New Year's Day 1934 in the foothill communities of Altadena, La Canada Flintridge, La Crescenta, and Tujunga in Los Angeles County to provide the impetus for the construction of debris basins. In La Crescenta alone, the debris flow, which occurred in little over 1 h, destroyed previously installed check dams, deposited over 450,000 m^3 of debris, claimed 40 lives, and destroyed 400 structures. The estimated cost of the damages was $6 million (approximately $106 million in 2015 dollars). During the mid-1930s, the District constructed its first debris basins with the intention of providing a more permanent solution to protect downstream communities.

7.2 Sediment Delivery in Chaparral

Sediment delivery is the transfer of soil and rock material from one component of the landscape to another: from the hillslopes to the stream channels and along the stream networks to a watershed outlet (i.e., a higher order stream channel, a lake, or the sea). On chaparral covered hillslopes, sediment is delivered by both hydrologic and gravity-driven erosion processes. At a larger scale, in headwater channels and larger streams, both classic fluvial transport and debris flows are responsible for sediment transfer. Fire enhances these delivery processes and continues to affect sediment movement for years into the future.

7.2.1 Hillslope Delivery Processes

Hillslopes are the source of nearly all the sediment moving through any landscape. Sediment generated on the hillslopes by the weathering of rock material is transported downhill by the various agents of erosion (wind, water, and gravity) and accumulates in the riparian zone at the base of the hill. In southern California these mechanisms of transport are amplified in the steep topography caused by active tectonics, while the forces of resistance are minimized by the non-cohesive soils that are typically dry. Hillslope sediment transport can be classified into two types: (1) surface erosion, either by dry ravel or water-borne processes, and (2) mass erosion, dominated by landslides.

7.2.1.1 Surface Erosion

The first type of surface erosion, dry ravel, refers to the dry, unconsolidated movement of granular material down a hillslope solely under the influence of gravity (Rice 1974). Dry ravel always requires a trigger mechanism, but these may be subtle, such as animal movement across a hillslope, microearthquakes, wind whipping the low-lying branches of shrubs onto the ground surface, or even differential thermal expansion of soil granules on a hot summer day. Ravel is a ubiquitous process, especially on steeper slopes (Gabet 2003b), but is often overlooked or unappreciated. However, studies have shown that dry season surface erosion can equal or exceed hydrologically dominated wet season surface erosion in some chaparral landscapes (Anderson et al. 1959). Dry ravel can be a primary mechanism for delivering hillslope sediment to headwater channels (Florsheim et al. 2015).

The second type of surface erosion consists of water-borne processes on hillslopes, including rainsplash and overland flow. Raindrops can dislodge and transport loose soil grains and preferentially transport them downhill. As rain persists, water accumulates on the ground surface, where much is absorbed by the organic material, with the litter layer acting a huge sponge. The remainder will infiltrate into the soil mantle, eventually percolating to the groundwater table. However, if the rain is falling faster than the soil infiltration rate, or if the ground becomes fully saturated, the excess moves across the ground surface as overland flow. Initially, this flow is as a thin film of sheetwash, but the differential resistances on the hillslope organize this runoff into distinct microchannels or rills. However, overland flow is very rare on unburned chaparral hillslopes with a canopy and litter cover, consequently water only reaches the stream channels by relatively slow flow paths through the soil mantle (Rice 1982).

7.2.1.2 Mass Erosion

Mass movement events—such as landslides, the failure of whole hillsides—can produce tremendous devastation. However, large landslides are relatively rare events and are almost always associated with heavy rain lubricating zones of geological weakness, such as slip planes in sedimentary rock, secondary dikes and sills in crystalline rock, or at the boundary between very different rock formations (Selby 1993). More common are small landslides or shallow "soil slips" produced by high intensity or persistent rain, where the weight of the extra water overcomes the balance of forces including the root strength provided by the vegetation (Rice et al. 1969; Campbell 1975). Debris flows (often called "mud flows") are a mass movement process that may be initiated as water mixes with landslide debris during high intensity rainfall events. Debris flows may originate on hillslopes and extend into steep headwater channels that have slopes greater than about 20 degrees (Prancevic et al. 2014). While mass erosion from large landslides, multiple soil slips, or debris flows may be spectacular, surface erosion by dry ravel is more pervasive. Both contribute to the sediment delivery from chaparral hillslopes.

7.2.2 Watershed Delivery Processes

Once soil and sediment material is delivered from the hillslopes to the channel networks, stream transport processes rout it through the riparian zone and eventually to a watershed outlet. In perennial streams, this transport is accomplished more or less continuously by the traditional fluvial processes described below. In ephemeral or intermittent streams—those that flow only in certain years or at certain times of the year—the same fluvial processes apply but transport is episodic, allowing hillslope material to accumulate in the channels between runoff and erosion events. In the chaparral environments of southern California, periods of low to moderate streamflow are punctuated irregularly by large storms that flush loose sediment out of the riparian networks, and the cycle of channel filling by hillslope delivery begins anew.

7.2.2.1 Fluvial Transport and Storage

The movement of sediment by rivers and streams is governed by the size of the streambed material, the channel gradient, and the volume of streamflow or discharge. Particles are entrained when shear stresses overcome the resistances of mass and inertia, with larger particles requiring more force for entrainment and subsequent transport than smaller ones (Leopold et al. 1964). Once entrained, sediment is transported by the force of the moving water. Some of the substances in transport can be chemically dissolved by the water and are then carried as solute load or in solution. Very fine particles (silts and clays) are transported as suspended sediment within the water column. Heavier sands are bounced along the streambed and heavier gravels and cobbles slide or roll as bedload. If stream power decreases because of increased resistances, a reduction in slope, or a drop in discharge, particles will begin to settle back to the bed and banks. The material will remain in temporary storage until the forces increase, and entrainment and transportation commences again. Thus sediment delivery in rivers and streams is usually episodic, especially for the larger particles.

The primary source of sediment that is entrained and transported by rivers and streams is from the alluvium stored in the channel bed and banks, originally delivered from the hillslopes. The transport of this material can result in bed and bank scour (unless the stream channel is carved into bedrock) which occurs along preferential flow paths where velocity is maximized. The streambed can be vertically downcut or incised, while the stream banks can be horizontally or laterally eroded.

7.2.2.2 Debris-Laden Flows

In contrast to the routine sediment transport of sediment along a stream channel, hyper-concentrated flows and debris flows are special forms of fluvial transport that occur when there is an excessive supply of sediment delivered to the stream channels, such as after a landslide. Sediment transport processes in hyper-concentrated flows are similar to water flows described above, whereby the force of the fluid mobilizes and transports sediment grains until the flow recedes, leaving channels filled with sediment that is easily entrained by later stormflows (Florsheim et al. 1991).

Debris flows are a type of mass movement that are usually confined to steep headwater stream channels. Although they can be generated when the debris from a landslide event reaches a creek or stream, debris flows may also form by the process of bulking, where more and more sediment is added to the mix such that the behavioral properties of the flow are no longer fluid. The resulting mixture in a debris flow often includes organic debris and has the consistency of wet concrete. Debris flows, though relatively rare, have tremendous erosive power in steep channels and can scour substantial sections within the riparian zone, transporting particles the size of small boulders long distances down the stream channel (Cannon and Gartner 2005).

7.2.3 *Fire Effects on Sediment Delivery*

Fire radically changes the rates of sediment delivery in chaparral ecosystems. With combustion of the vegetation and the litter layer, raindrop interception greatly decreases and water cannot be absorbed into the now-absent organic surface material. Moreover, fire causes physical and chemical changes in the soil. Organic particles are consumed in the uppermost soil layer, leaving voids that can collapse to increase the soil bulk density. Partial combustion of chaparral duff and the re-deposition of the vaporized waxy substances from the leaves into the cooler soil below can create a hydrophobic or water-repellent layer that blocks or greatly reduces infiltration (DeBano 1981). These changes reduce the resistances to sediment transport on hillslopes thereby enhancing the agents of erosion that govern both dry season and wet season sediment delivery from hillslopes to channels in chaparral environments. In the stream channel itself, fire can consume large organic debris, removing these stable components that may affect habitat and geomorphology. However, most of the alterations in postfire stream channel and watershed response derive from the massive changes on the hillslopes that deliver extra water and sediment from the hillslopes to the riparian networks.

7.2.3.1 Processes

Dry ravel transport rates greatly increase in postfire chaparral. Ravel can actually begin while the fire is still burning as organic barriers on the ground surface are consumed, liberating the soil and sediment trapped behind vegetation that dammed this material on the hillslopes. Immediately after a fire, ravel is pervasive on steep slopes, as the hyper-sensitive ground surface will initiate a cascade of soil material with the slightest disturbance. Ravel may continue for days or weeks after a fire until loose sediment on the burned hillslope is stabilized. Wind or other disturbances can re-initiate ravel transport.

Water-borne erosion processes on hillslopes are also magnified following a fire. With the removal of the vegetation canopy and litter layer, the mineral soil is subjected to unimpeded raindrop impacts, vastly improving splash efficacy. In addition, the spongy litter is lost, thereby reducing the water-holding capacity of the soil mantle and producing water repellent soil material a few centimeters below the surface, effectively eliminating infiltration and percolation. Consequently, water moves across the ground surface as overland flow which, on unprotected soil, forms rills that scour sediment off the hillslopes. Moreover, once the rill networks are created, subsequent rainfall is quickly conveyed off the hillslopes along these rills to the stream channels below (Wells 1981).

Contrary to popular notions, landsliding and soil slips are not enhanced in chaparral environments by fire. Rather, with the loss of infiltration and the increase in overland flow, less water is available to percolate to zones of geologic weakness that would initiate mass movement events. However, saturation of the upper soil layer resulting from the production of a water repellent layer can generate thin debris flows on the hillslopes that can greatly increase sediment delivery (Gabet 2003a).

Fluvial transport, bed and bank scour, and sediment storage in streams are all affected by fire. Reduced infiltration on the hillslopes promotes more overland flow and consequently greater runoff in the stream channels (Loaiciga et al. 2001). The initial sediment response of most stream channels in postfire chaparral is to fill with sediment from adjacent slopes, first by dry ravel and then by overland flow. Once the easily mobilized sediment is removed from the hillslopes, subsequent storms produce mostly water that tends to scour the newly formed channel deposits (Florsheim et al. 1991). Postfire channel sedimentation will preferentially fill stream pools, smoothing both the longitudinal and transverse river profiles (Keller et al. 1997).

Compared to unburned watersheds, debris flows are far more common in the postfire chaparral environment. Formed by the extra water generated on the hillslopes mobilizing the stored sediment in the stream channels, postfire debris flows scour the headwaters (further bulking the flows with sediment) then deposit the rocks and mud in large lobes where the channel gradient sufficiently flattens, often creating temporary barriers or dams (Wells 1987). However, despite their impressive ability to deliver sediment, debris flows following fire are still relatively rare events. Postfire sediment transport and channel flushing are more likely to occur by normal fluvial or hyper-concentrated flow processes than by high-magnitude but infrequent debris flows (Keller et al. 1997).

7.2.3.2 Fire Recovery

Accelerated postfire sediment delivery on chaparral hillslopes will eventually attenuate to prefire background levels as the area recovers. Postburn vegetation re-growth in chaparral ecosystems is very rapid, and sprouting from root crowns can occur within weeks of the fire. The initial herbaceous canopy can shield the ground surface and help stabilize hillslope sediments (Barro and Conard 1991). However, the decline in erosion also results from the depletion of the easily mobilized surface sediment in the first year postfire and the subsequent exposure of less erodible material from deeper in the soil column (Wohlgemuth 2006). Sediment delivery from chaparral hillslopes can recover to preburn rates in as little as two years (Wohlgemuth et al. 1998; Wohlgemuth 2015), although there are examples of somewhat longer recovery periods (Krammes 1965).

As with hillslopes, sediment yields at the broader watershed scale are greatest immediately after a fire, then decline to preburn rates as the catchments recover. Time to recovery depends on specific site factors, the severity of the fire, and rainfall patterns. As noted above, sediment delivery from the hillslopes can return to normal in just a few years. However, given the episodic nature of channel transport associated with flow fluctuations and the capacity for internal temporary sediment storage, watershed sediment yields may remain elevated for longer time periods. Based on vegetation re-growth and channel morphology, Davis et al. (1989) determined that burned watersheds initially recovered to prefire conditions after about 5 years, while Rowe et al. (1954) estimated postfire watershed recovery to be 8–10 years.

7.3 Flood Control and Water Supply Infrastructure

Flood control and water conservation infrastructure has evolved over the years. This is a result of structural improvements and more accurate weather forecasting, combined with the ability for real-time monitoring of rainfall rates, reservoir conditions, and channel flow rates. Historically, flood control and reducing the risk of flood damage to downstream communities has always held precedence over water conservation operations in southern California. However, increased population, more frequent droughts, and reduced availability of imported water have increased the focus on enhancing infrastructure and operations to increase capture of local stormwater runoff. These activities help ensure a more sustainable local water supply. New technologies make it possible to optimize stormwater runoff to capture it for local and regional water storage. For example, by using telemetry systems and collecting real-time weather and channel flow data, dam outflows can be limited to maximize water capture while still reducing downstream flooding. Similarly, after the storm event, dams can release water at a volume tailored to the rate of percolation in the spreading grounds. Given the ability to control water releases, dams can be filled to capacity and drained to their minimum pools more than once during a wet storm season.

7.3.1 Structures

The flood protection and water conservation systems in southern California are made up of several components. These components each perform their own unique functions to protect lives, property and infrastructure from flooding, and to contribute to their region's water supplies.

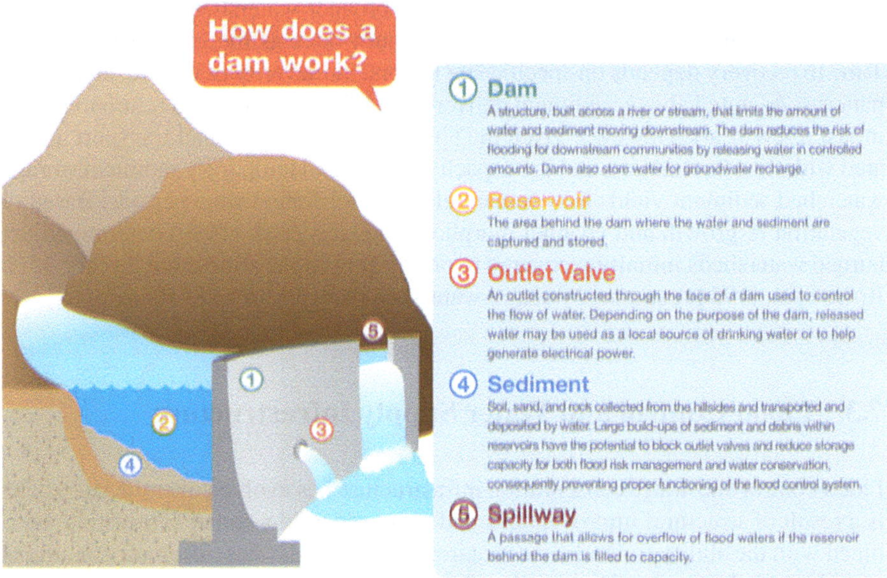

Fig. 7.1 Schematic diagram of a flood control dam describing different components (County of Los Angeles, Department of Public Works, www.lasedimentmanagement.com)

7.3.1.1 Dams

There are over 282 major dams in southern California. Dams, and the reservoirs behind them, can serve single or multiple functions including flood control, debris control, and stormwater capture for water supply. Many also provide recreational benefits within and downstream of the facilities. The majority of these dams provide flood control, but some were built specifically for water supply storage, with water pumped to them, and are not located in watercourses.

Dams can be made out of earth fill (rock and soil with a clay core) or concrete with a variety of different forms (Fig. 7.1). Flood control dams in southern California typically range from 7.5 m (24.6 ft) high to over 100 m (328.1 ft) high with reservoir capacities from approximately 62,000 m^3 (50.3 acre-ft) to as much as 65.8 million m^3 (53,345 acre-ft), although most are of the smaller size. Flood control dams are typically located in mountainous areas where narrow canyons provide suitable locations to block the watercourse of large upstream watersheds. The size of watersheds identified for dam placement can range from 500 to 45,000 ha (1236–111,197 acres).

In comparison to dams of similar heights on watersheds with gentler slopes (e.g., in northern California), southern California dams do not have huge storage capacities and a single major storm has the potential to completely fill a reservoir. As a consequence, dam operations need to constantly focus on evaluating storm conditions using advanced weather forecasting tools and real-time data. Based on this information, water is released through valves to decrease reservoir levels in antici-

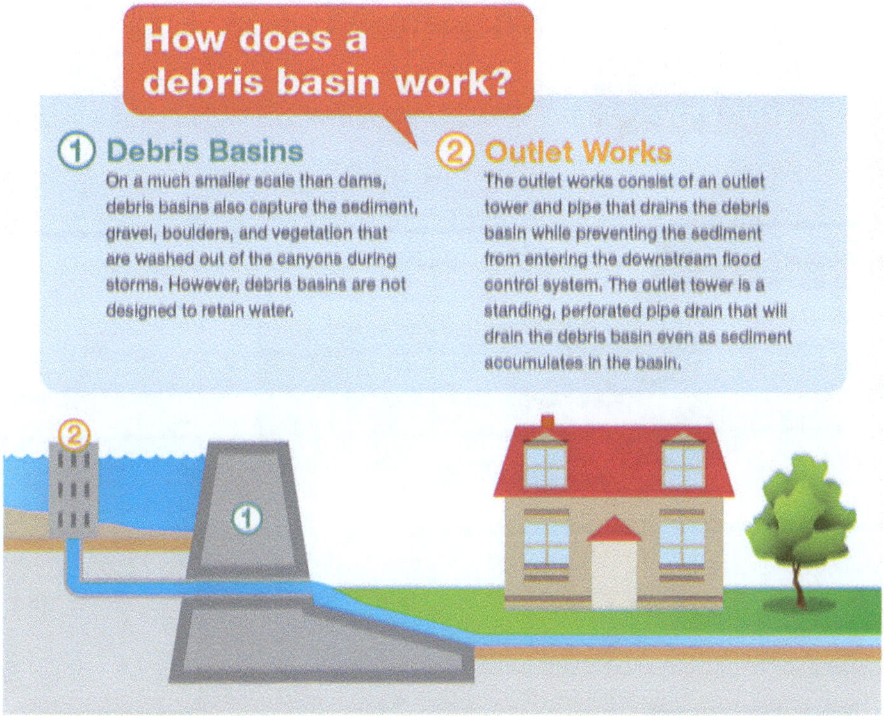

Fig. 7.2 Schematic diagram of a debris basin describing different components (County of Los Angeles, Department of Public Works, www.lasedimentmanagement.com)

pation of peak storm inflow rates. The goal is to remain in control of the dam and prevent the overflow of water through the spillway by managing the reservoir level through controlled releases. From the dams in Los Angeles County, spillway flows occur approximately only once every 7–10 years.

7.3.1.2 Debris Basins

Debris basins serve the single function of capturing sediment and debris from the upstream undeveloped watershed while allowing the stormflows to continue downstream. They are essentially small dams that typically have no valves to operate (Fig. 7.2). There are over 200 debris basins in southern California, usually 4.5–15 m (14.8–49.2 ft) high with less than 62,000 m^3 (50.3 acre-ft) of capacity but can be as large as 308,400 m^3 (250 acre-ft). Debris basins are typically formed by constructing an earthen or concrete dam near the mouth of a canyon and excavating a basin upstream to collect the debris-laden water. The heavier material such as boulders, rocks, and gravel are deposited on the bottom of the basin, while the settled, clear water flows out of the debris basin through the outlet tower and discharges to the downstream channel (Fig. 7.2).

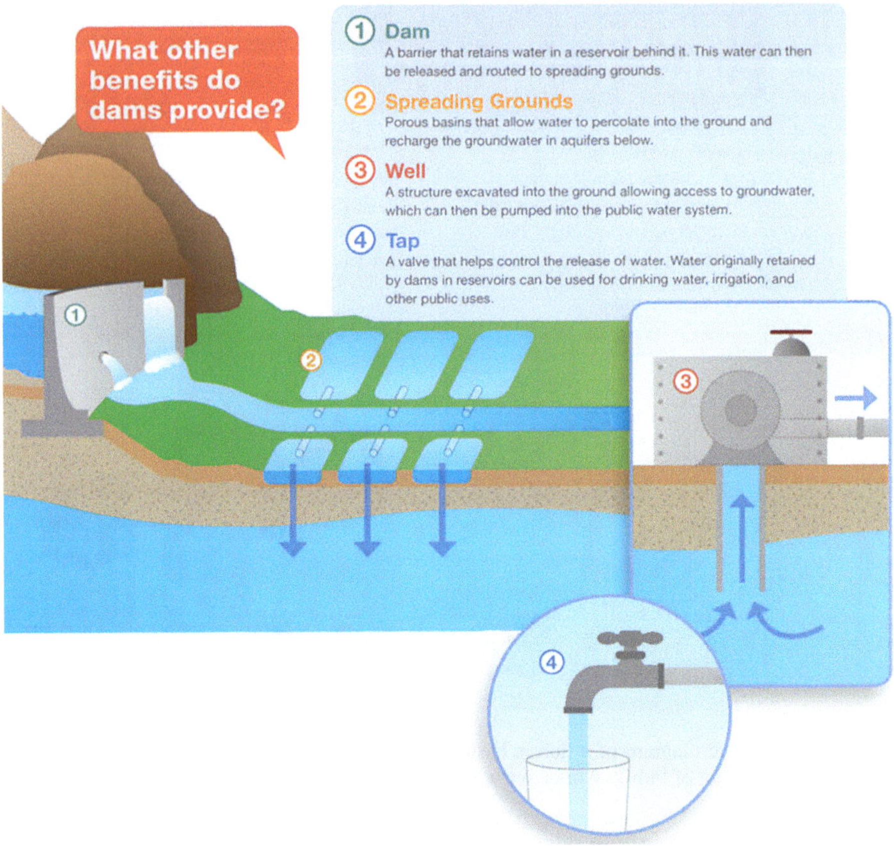

Fig. 7.3 Schematic diagram of a spreading ground describing different components (County of Los Angeles, Department of Public Works, www.lasedimentmanagement.com)

7.3.1.3 Spreading Grounds

Spreading grounds consist of one or more basins formed with earthen berms or levees where flow is directed by gates and weirs to the individual basins to percolate through layers of sands and gravels into the subsurface aquifer (Fig. 7.3). This recharges the groundwater basin which can later be pumped for water supply. In Los Angeles County, the largest spreading ground covers 93 ha (230 acres) of land and includes over 20 basins. It should be noted that many of these facilities not only recharge the groundwater basins with local storm runoff and stormwater captured by the dams, but also with imported water purchased by the local water entities for the purpose of groundwater replenishment. In addition, starting in the 1960s, they are also recharged with recycled water supplied by local wastewater treatment facilities. In Los Angeles County, the Flood Control District began constructing spreading grounds in the 1920s and now operates 27 such facilities which collectively recharge on average more than 370,000 m^3 (300 acre-ft) per year of water to the underlying aquifers.

7.3.2 Sediment Impacts on Flood Control Structures

Southern California experiences fire-flood cycles, where fires periodically burn mountainous areas and torrential rains result in debris flows pouring out of the mountains, which either inundate flatlands or are captured in flood control or water conservation infrastructure. As sediment accumulates behind dams and debris basins, the capacity of these facilities is diminished and outlet works are blocked. This compromises their ability to reduce flood risks and provide stormwater capture for water supply, as well as physically disrupting the functioning of the facility. As a consequence, stormflows cannot be attenuated and released to match downstream channel capacities and stormwater can fill the reservoir until uncontrolled spillway flows occur at much higher flow rates. In addition, blocked outlet works render the reservoir unable to be drained which is extremely important in the unlikely event of a dam safety emergency. Finally, it should be noted that sediment is heavier than water and in some cases would produce unacceptable forces on a dam or debris basin during a seismic event.

Blocked or partially blocked outlet works similarly preclude the ability to release captured stormwater for recharge to downstream spreading grounds, thereby reducing local water supply. Moreover, the increased turbidity of the released sediment-laden water may prevent distributing the water at spreading grounds due to concerns of plugging the sands and gravels with fine material and reducing percolation. Also, to slow the sediment accumulation near the outlet works, a "buffer pool" is created at the upper end of the reservoir to encourage sedimentation. This pool further reduces capacity for capturing stormwater for the regional water supply. Maintaining capacity through sediment management and sediment removal projects is therefore paramount to the sustained operations of these facilities.

7.3.3 Required Maintenance and Sediment Removal Projects

Debris removal from dams and debris basins is critical to ensure flood or debris protection for homes and infrastructure downstream of the facility and to maximize stormwater capture. The debris volume that accumulates in reservoirs in the mountains and along the foothills of southern California is tremendous. In 2012 it was estimated that it may be necessary to remove as much as 49 million m^3 (39,725 acre-ft) of sediment from these facilities over the next 20 years (for comparison, the Rose Bowl in Pasadena would hold about 300,000 m^3 (243 acre-ft) (LACDPW Sediment Management Plan, March 2013 [http://dpw.lacounty.gov/lacfcd/sediment/stplan.aspx]). The amount of material removed from a debris basin can range from 750 to 225,000 m^3 (0.6–182.4 acre-ft) and can take from 2 days to several months depending on their size and how quickly the sediment can dry so that it can be handled and transported.

7.3.3.1 Dams

Large dams in California are regulated by the California Department of Water Resources' Division of Safety of Dams (DSOD). DSOD requires regular monitoring and reporting on the structural and functional integrity of the dams and, when necessary, rehabilitation of dams to meet strict state and federal seismic, spillway capacity, and emergency dewatering standards. In addition, in areas with significant debris flow issues, like Los Angeles County, maintenance includes the costly periodic removal of accumulated debris from the reservoir.

Removal of sediment can be in the form of sluicing (i.e., washing the sediment out of the reservoir to the streams below), dry excavation, or dredging (i.e., excavation in standing water). Sluicing, which historically was performed outside of southern California's storm season, has the lowest cost, but has fallen out of favor due to concerns about impacts to downstream habitat and water quality. However, sluicing in the storm season is now being considered due to the desire to restore the natural balance of sediment in the habitats of major streams and slow down the rate of debris accumulation. In contrast, dry excavation requires substantial or complete draining of the reservoir, and transporting the material by conveyor belt, trucks, or combination thereof to a disposal site. Finally, dredging lessens the amount of reservoir draining, but also requires large dewatering areas for the wet sediment/water slurry to dry in addition to a final disposal site, which can be many kilometers away from the reservoir.

Needless to say, all these techniques for sediment removal from dams are very expensive. For example, in Los Angeles County, during the last 25 years, dry excavation has been the preferred clean-out method at the reservoirs. The excavation of 250,000 m^3 of sediment from Santa Anita Reservoir in 2011/2012 and transport via conveyor belt to a disposal site 3 km (1.9 miles) away cost $7.5 million (almost $30 per cubic meter [0.0008 acre-ft]). Similarly, the proposed removal of 1.8 million m^3 (1459.3 acre-ft) of sediment from Devils Gate Reservoir and associated truck transport to nearby landfills is estimated to cost $65 million (about $36 per cubic meter).

7.3.3.2 Debris Basins

Maintenance of these facilities consists of regular monitoring of the debris dam's condition, annual mowing (vegetation removal) of the basin/inlet bottom, and periodic clean-outs of the sediment. Several of the larger debris dams are under the regulatory jurisdiction of the State's DSOD and thus have more monitoring and reporting requirements than the smaller ones. Debris basin/inlet bottoms are mowed to ensure debris flows enter the basin and debris settles in a manner that maximizes the capacity of the basin. Mowing also facilitates clean-out of the structures and reduces the organic content of the debris being removed, thus making it easier and less expensive to find places for disposal. According to the Los Angeles County Sediment Management Strategic Plan, the cost of non-emergency debris basin clean-outs can be about $9.37 per cubic meter and as much as $80 per cubic meter during emergencies. Factors that influence these costs include the need to work 24-h shifts to restore capacity as soon as possible, handling material multiple times (if it is too wet and cannot be transported on roads, it can require stockpiling and drying and/or mixing with other material), and the distance to the placement site. Following the historic 2009 Station Fire, the 2009–2010 storm season delivered 933,280 m^3 (757 acre-ft) of sediment into 30 Los Angeles County debris basins. The cost to remove the sediment was $24,159,000 ($24.60 per cubic meter).

7.3.3.3 Spreading Grounds

Maintenance of the spreading grounds consists of keeping the intake clear, keeping the diversion works in the channel and the inter-basin gates in good working order, removing sediment and opportunistic vegetation from the settling basin and spreading basins, and scarifying the basin bottoms to re-open up the top layer of the soil on the basin bottoms.

Without removal of sediment from upstream dams and debris basins, more material (typically finer grain sizes) is washed downstream. This increases the likelihood of sediment moving into the percolation basins beyond the settling basins and further increasing the maintenance needs at the spreading grounds. In areas where burned watersheds do not drain into dams, the storm runoff delivered to downstream spreading grounds is more turbid. Even after heavier sediment drops out of the flows when they exit the canyons or debris basins, finer material can still make its way through the valleys below. Stormflows can be so silt-laden that the operators of spreading grounds find it necessary to divert flows so they do not enter the spreading grounds in order to prevent clogging of the spreading basin bottoms.

7.3.3.4 Sediment Placement Sites

The disposal of sediment removed from reservoirs and debris basins has always been challenging. The haul distances from facilities to landfills can be very long (over 30 km or 18.6 miles), and costs can also be very expensive (over $11 per cubic meter). As far back as the 1950s, the public works agencies explored the feasibility of having local gravel quarries utilize the sediment from the reservoirs and thus reduce the disposal costs. Although some of the sediment had been found to be of suitable commercial quality, the reservoirs' remote mountain locations were beyond the haul distances that quarry operators consider to be economically viable. As a result, the flood control districts acquired their own hillslope properties or obtained Special Use permits from the US Forest Service for properties in the mountains to serve as sediment placement sites. These sites serve as storage facilities for sediment from reservoirs and debris basins. However, the sediment placement sites are of finite capacity, and finding new sediment placement site locations is difficult. In 2013 Los Angeles County completed a Sediment Management Strategic Plan to explore alternatives which can reduce the environmental and social impacts of sediment management. The plan identified utilizing a mix of local gravel quarries, many of which are now nearly mined out, and landfills needing a daily cover of sediment to seal the waste. The plan would also use the existing sediment placement sites to accommodate occasions when sensitive resources are involved, the quarries are unable to economically process the sediment, or only small amounts are required for covering landfills.

7.4 Physical Ecosystem Services in Chaparral

Chaparral environments provide several physical ecosystem services, most of which are not appreciated until the vegetation is removed. Intact stands of chaparral reduce floods, retain sediment, provide clean water through the infiltration and replenishment of aquifers, protect threatened and endangered species and their habitats, and protect soil quality. These services are temporarily eliminated if the chaparral is removed, as after a large wildfire. However, these services could be permanently impaired if the chaparral is significantly degraded by persistent drought associated with climate change, by alterations in land use, or following invasion by non-native species associated with high frequency fires.

7.4.1 Flood Hazard Reduction Services

Chaparral ecosystems contribute to flood hazard reduction by reducing runoff both on the hillslopes and in the stream networks. When the vegetation canopy and litter layer are in place, overland flow on the hillslopes is rare and subsurface flow to the stream channels through the soil mantle is slow. With the removal of the canopy and the litter layer by fire, along with the attendant changes in soil properties, overland flow on the hillslopes is pervasive and delivery to the streams via the surface network of rills is rapid. Streams and rivers convey the floodwaters generated by large storms to the watershed outlet. In unburned chaparral catchments substantial flood events only occur about once a decade and only after the largest storms. Following a fire, large flood events are commonplace and can occur even after moderate rains (Rowe et al. 1954).

Increased floods in chaparral environments after fire are well documented. In watersheds with extensive prefire stream records in southern California, fire increased flow volumes by 3 to 5 times compared to an adjacent unburned catchment (Sinclair and Hamilton 1955). Using burned and unburned soil infiltrometer tests and a calibrated watershed model, Nasseri (1988) showed that moderate storms could produce floods of considerable magnitude in a burned chaparral watershed. Comparing similar storms both before and after a fire in chaparral, Krammes and Rice (1963) documented postfire peak flows that were 200–800 times greater than unburned levels.

The consequences of postfire flooding in chaparral depend on the downstream values at risk. In natural landscapes, postfire flooding can scour the channel bed and banks, re-arrange the position of boulders and large logs, and remove streamside vegetation. This can alter the geomorphology of the riparian zone and the habitat of the flora and fauna that occupy this niche. However, at the wildland-urban interface, the hazards of postfire flooding and the potential damage to human endeavors can be extensive.

One approach for estimating the value of chaparral for flood hazard reduction can be roughly calculated as the likelihood that assets would be impacted based on the difference in prefire and postfire threshold storm recurrence intervals. Assuming that flooding from a 10-year storm immediately following a fire would cause the same damage to property and infrastructure as a 50-year storm in unburned vegetation and assuming a five year recovery period after a fire described by an exponential decay curve (Rowe et al. 1954), the values at risk would be 7.5 times more likely to be affected by flooding after a fire than if the watershed had not burned. This increased likelihood of postfire flooding with the removal of chaparral vegetation is the impetus for the dams built and maintained by various flood control districts to protect life, property and public infrastructure in downstream human communities.

7.4.2 *Sediment Retention Services*

Chaparral environments retain sediment by reducing the surface erosion on the hillslopes and by reducing scour and transport in the stream channels. While some erosion is inevitable for granular material on an inclined surface, this transport is minimized under intact chaparral. If the vegetation is removed, the extra water and sediment generated on the bare hillslopes is quickly delivered to the streams which scours the bed and banks, increases transport capacity, and perhaps produces debris flows by mobilizing the loose sediment stored in the channels.

Increased postfire sediment yield in chaparral is also well documented, although rates can vary considerably depending on topography, soils, fire severity, and precipitation patterns. Rowe et al. (1954) determined that burned chaparral catchments produced an average of 35 times more sediment than those same watersheds prior to fire, similar to rates reported by Wohlgemuth (2015). Pase and Ingebo (1965) found that postfire sediment yields in chaparral can be 100 times greater than those from nearby unburned watersheds. In an extreme case, Kraebel (1934) documented burned catchments that experienced 1000 times more sediment yield than comparable unburned watersheds.

The consequences of postfire sediment yield in chaparral also depend on the downstream values at risk. In natural landscapes, postfire sediment transport in streams can incise the upstream sections of the channels and deposit this material further downstream. This incision or filling will modify the geomorphology and the habitat along the streams (Keller et al. 1997). However, at the wildland-urban interface, the risk of postfire sedimentation and the potential damage to the human built environment can be tremendous (Campbell 1975).

One method for estimating the value of chaparral for sediment retention can be roughly calculated as the likelihood that assets would be impacted based on the difference in prefire and postfire threshold sediment pulse recurrence intervals. Because sediment load is not just a function of storm size but also of sediment availability, a larger prefire storm size is necessary to compare to postfire sediment transport. Assuming that the sediment delivered from a 10-year storm immediately following a fire would cause the same damage as a 100-year storm in unburned vegetation and assuming a five year fire recovery period described by an exponential decay curve (Rowe et al. 1954), the values at risk would be 18 times more likely to be affected by sediment after a fire than if the watershed had not burned. This huge increase in vulnerability to postfire sedimentation with the loss of chaparral vegetation is the rationale for the construction and maintenance of dams and debris basins by the various public works agencies to protect downstream values at risk.

7.4.3 Water Supply Services

Chaparral ecosystems supply clean water from the outlets of larger watersheds for agricultural, domestic, and industrial uses. Rain falling on unburned chaparral gets absorbed into the litter layer, infiltrates into the soil mantle, and percolates through the fractured substrate to the groundwater table. During this slow flowing through subsurface pathways, the water is filtered of all but the tiniest soil grains and most of the contaminants derived from organic decay, unusual geology, or atmospheric deposition. Excess water not used by the hillslope shrubs or the riparian vegetation, or necessary for downstream resources, is then available for human use. As we have seen, fire increases the quantity of water delivered from the hillslopes, contributing to flooding in lower stream sections. However, fire also modifies the character of the water itself, affecting the turbidity and water chemistry, to the detriment of the anticipated water supply. After a fire, water from the hillslopes now reaches the streams primarily over the ground surface, bypassing the subsurface filtering. Thus, postfire streamflow is often laden with ash and fine sediments. Moreover, nutrients and soluble compounds from the ash are also flushed or leached from hillslope sources to become part of the water column. Therefore, although more water is generated from postfire environments, most of this water is unusable without expensive filtration. Settling in spreading grounds is not an option because those facilities will become clogged with the ash and fine sediments.

Lost stormwater supply from burned watersheds has been measured by local public works agencies. One reservoir in Los Angeles County received more than 1.7 million m^3 (1378.2 acre-ft) of sediment after a recent fire. This reduced the ability to capture stormwater annually by 617,000 m^3 (500.2 acre-ft). A similar reservoir experienced a series of fires in its contributing area and saw its capacity reduced annually by over 2.5 million m^3 (2026.8 acre-ft) The value of the lost water from these two reservoirs alone is $1.82 million per year based on the cost to replace this stormwater by purchasing imported water from the Metropolitan Water District (2016 rate of $708 per 1200 m^3 [0.97 acre-ft] for untreated water [www.mwdh20.com/whoweare/management/finincial-information]).

7.4.4 Threatened and Endangered Species Protection Services

Chaparral ecosystems are comprised of many species of flora and fauna, a number of which are considered to be threatened with extinction or extirpation (see Chaps. 2 and 3). Most of these species occupy the riparian zone but some reside on the hillslopes. All species in chaparral have adapted to a dynamic environment of extremes of heat and cold as well as flood and drought. Moreover, wildfire within the range of natural frequency is the disturbance event to which chaparral environments are adjusted. However, fire and the postfire hydrologic and sediment response pose an additional problem for threatened and endangered species at the wildland-urban interface. In the past, if postfire flooding and accelerated erosion eliminated a particular organism from a local creek, the area could be re-colonized from nearby unburned refugia. But if human development has fragmented the habitat and removed the corridors of connectivity, the threatened and endangered species could be permanently removed.

Intact chaparral reduces erosion, thereby protecting species both on the hillslopes and in the stream channels. When these services are interrupted, such as after a fire, hillslope erosion (either by dry ravel or overland flow) can remove small plants and strip away the seedbank. In the streams, aquatic organisms are subject to increased turbidity (affecting light transmission) and potentially detrimental changes in water chemistry. In addition, when the stream gradient flattens or the resistances increase, sediment load settles on the streambed or in pools, smothering any biota.

The value of chaparral to species protection is difficult to assess, but the costs of species recovery plans, monitoring, and breeding programs can be significant. Moreover, the consequences of reduced protection to threatened and endangered species when these services are not provided is the loss of individual species, altered food webs, and changes in overall species composition.

7.4.5 Soil Quality Protection Services

Chaparral cover reduces erosion, thus protecting the quality of the soil resource. Soils form as weathered rock material combines with decaying organic matter, often developing layers or horizons over time. However, soil material in mountainous southern California landscapes dominated by chaparral forms quickly but is transported rapidly by the agents of erosion. As a result, soil material lacks distinct horizons because of the low residence times on steep hillslopes. Chaparral soils are typically shallow with low fertility. Fire can liberate nutrients bound up in the organic litter cover, producing ash and char on the ground surface that can increase soil fertility. However, much of this material is removed by dry ravel and overland flow, and the residual soil is impoverished in nutrients compared to the unburned condition. With the stripping of the organic-rich topsoil, material of lower fertility is exposed at the soil surface.

The loss of chaparral by land use change also reduces soil protection. In the past, some areas of chaparral have been intentionally type-converted to grasslands to improve water yield or to expand grazing opportunities. However, with the loss of the root strength provided by chaparral shrubs, the grassy areas are susceptible to shallow soil slips (Rice et al. 1969; De Graff 1979), especially in high rainfall years. These small mass movements on formerly stable hillslopes contribute directly to immediate hillslope sediment delivery, while the scars on the hillslopes continue to be a source of soil erosion into the future.

In short, the consequences of diminished protection of the soil resource in post-fire chaparral environments is the loss of organic matter and lower soil productivity. This results in reduced plant growth and less forage for associated fauna. However, quantifying the value of chaparral for the protection of soil quality is difficult.

7.5 Summary

Southern California's chaparral environments are very erosive and these high levels of background erosion are further compounded by repeated wildfire. Erosion and sedimentation are responsible for creating the natural landscape over geologic time, but these processes are hazardous to people and the built environment. This has prompted public works agencies to construct dams, debris basins, and flood control channels to protect life, property, and infrastructure from the flooding and accelerated erosion that are invariably generated from burned watersheds as a consequence of the loss of chaparral vegetation. Chaparral environments provide many physical ecosystem services, most of which are under-appreciated by modern society (see Chap. 5). However, as burgeoning population centers continue to impinge on chaparral landscapes, the awareness of chaparral will continue to grow, as will a greater recognition of the value of chaparral ecosystems.

References

Anderson, H. W., G. B. Coleman, and P. J. Zinke. 1959. Summer slides and winter scour: dry-wet erosion in southern California mountains. Technical Paper 36. USDA Forest Service, Pacific Southwest Forest and Range Experiment Station, Berkeley, California, USA.

Barro, S. C., and S. G. Conard. 1991. Fire effects on California chaparral systems: an overview. Environment International 17:135–149.

Campbell, R. H. 1975. Soil slips, debris flows, and rainstorms in the Santa Monica Mountains and vicinity, southern California. Professional Paper 851. US Geological Survey, Washington, D.C., USA.

Cannon, S. H., and J. E. Gartner. 2005. Wildfire-related debris flow from a hazards perspective. Pages 363–385 *in* M. Jakob, and O. Hungr, editors. Debris-flow hazards and related phenomenon. Springer Praxis Books, Berlin, Germany.

Davis, F. W., E. A Keller, A. Parikh, J. L. Florsheim. 1989. Recovery of the chaparral riparian zone after wildfire. Pages 194–203 *in* D.L. Abell. Proceedings of the California Riparian Systems Conference: Protection, Management, and Restoration for the 1990s, Davis, California, September 22–24, 1988. General Technical Report PSW-GTR-110. USDA Forest Service, Pacific Southwest Forest and Range Experiment Station, Berkeley, California, USA.

DeBano, L. F. 1981. Water repellent soils: a state-of-the-art. General Technical Report PSW-GTR-46. USDA Forest Service, Pacific Southwest Forest and Range Experiment Station, Berkeley, California, USA.

De Graff, J. V. 1979. Initiation of shallow mass movement by vegetation-type conversion. Geology 7:426–429.

Dunn, P. H., S. C. Barro, W. G. Wells II, M. A. Poth, P. M. Wohlgemuth, and C. G. Colver. 1988. The San Dimas Experimental Forest: 50 years of research. General Technical Report PSW-GTR-104. USDA Forest Service, Pacific Southwest Forest and Range Experiment Station, Berkeley, California, USA.

Eaton, E. C. 1932. San Gabriel project check dams: report to the Los Angeles County Board of Supervisors from the Los Angeles County Flood Control District. Unpublished Report.

Florsheim, J. L., A. Chin, L. S. O'Hirok, and R. Storesund. 2015. Short-term post-wildfire dry-ravel processes in a chaparral fluvial environment. Geomorphology 252:32–39.

Florsheim, J. L., E. A. Keller, and D. W. Best. 1991. Fluvial sediment transport in response to moderate storm flows following chaparral wildfire, Ventura County, southern California. Geological Society of America Bulletin 103:504–511.

Gabet, E. J. 2003a. Post-fire thin debris flows: sediment transport and numerical modelling. Earth Surface Processes and Landforms 28:1341–1348.

Gabet, E. J. 2003b. Sediment transport by dry ravel. Journal of Geophysical Research 108:1978–2012.

Jansson, M. B. 1988. A global survey of sediment yield. Geografiska Annaler 70:81–98.

Keller, E. A, D. W. Valentine, and D. R. Gibbs. 1997. Hydrological response of small watersheds following the southern California Painted Cave Fire of June 1990. Hydrological Processes 11:401–414.

Kraebel, C. J. 1934. The La Crescenta flood. American Forests 40:251–254, 286–287.

Krammes, J. S. 1965. Seasonal debris movement from steep mountainside slopes in southern California. Pages 85–89 *in* Proceedings of the Federal Interagency Sedimentation Conference, Jackson, Mississippi, January 28–February 1, 1963. Miscellaneous Publication 970, US Department of Agriculture, Agricultural Research Service, Washington, D.C., USA.

Krammes, J. S., and R. M. Rice. 1963. Effect of fire on the San Dimas Experimental Forest. Pages 31–34 *in* Proceedings of Arizona's 7th Annual Watershed Symposium, Phoenix, Arizona, September 18, 1963. Arizona Water Comission, Phoenix, Arizona, USA.

Langbein, W. B., and S. A. Schumm. 1958. Yield of sediment in relation to mean annual precipitation. Transactions of the American Geophysical Union 39:1076–1084.

Leopold, L. B., M. G. Wolman, and J. P. Miller. 1964. Fluvial processes in geomorphology. W.H. Freeman, San Francisco, California, USA.

Loaiciga, H. H., D. Pedreros, and D. Roberts. 2001. Wildfire-streamflow interactions in a chaparral watershed. Advances in Environmental Research 5:295–305.

Los Angeles County. 1931. Comprehensive Plan for Flood Control and Water Conservation—Present Conditions and Immediate Needs. Los Angeles County Flood Control District, Los Angeles, California, USA.

Los Angeles County. 2006. Hydrology Manual. Department of Public Works, Water Resources Division, Los Angeles, California, USA.

Minnich, R. A. 1989. Chaparral fire history in San Diego County and adjacent northern Baja California: an evaluation of natural fire regimes and the effects of suppression management. Pages 37–47 *in* S. C. Keeley, editor. The chaparral: paradigms reexamined. Natural History Museum of Los Angeles County, Los Angeles, California, USA.

Munns, E. N. 1919. The control of flood water in southern California. Journal of Forestry 17:423–429.

Nasseri, I. 1988. Frequency of floods from a burned chaparral watershed. Pages 68–71 *in* N. Berg, editor. Proceedings of the symposium on fire and watershed management, Sacramento, California, October 26–28, 1988. General Technical Report PSW-GTR-109. USDA Forest Service, Pacific Southwest Forest and Range Experiment Station, Berkeley, California, USA.

Pase, C. P., and P. A. Ingebo. 1965. Burned chaparral to grass: early effects on water and sediment yields from two granitic soil watersheds in Arizona. Pages 8–11 *in* Proceedings of Arizona's 9th Annual Watershed Symposium, Tempe, Arizona, September 22, 1965. Arizona Water Commission, Phoenix, Arizona, USA.

Prancevic, J. P., M. P. Lamb, and B. M Fuller. 2014. Incipient sediment motion across the river to debris-flow transition. Geology 42:191–194.

Rice, R. M. 1974. The hydrology of chaparral watersheds. Pages 27–34 *in* M. Rosenthal, editor. Proceedings of a symposium on Living with the Chaparral, Riverside, California, March 30–31, 1973. Sierra Club, San Francisco, California, USA.

Rice, R. M. 1982. Sedimentation in the chaparral: how do you handle unusual events? Pages 39–49 *in* F. J. Swanson, R. J. Janda, T. Dunne, and D. N. Swanston, editors. Sediment budgets and routing in forested drainage basins. General Technical Report PNW-GTR-141. USDA Forest Service, Pacific Northwest Forest and Range Experiment Station, Portland, Oregon, USA.

Rice, R. M., E. S. Corbett, and R. G. Bailey. 1969. Soil slips related to vegetation, topography, and soil in southern California. Water Resources Research 5:647–659.

Rowe, P. B., C. M. Countryman, and H. C. Storey. 1954. Hydrologic analysis used to determine effects of fire on peak discharge and erosion rates in southern California watersheds. USDA Forest Service, California Forest and Range Experiment Station, Berkeley, California, USA.

Safford, H. D. 2007. Man and fire in southern California; doing the math. Fremontia 35:25–29.

Scott, K. M., and R. P. Williams. 1978. Erosion and sediment yields in the Transverse Ranges, southern California. Professional Paper 1030, US Geological Survey, Washington, D.C., USA.

Selby, M. J. 1993. Hillslope materials and processes. Oxford University Press, Oxford, UK.

Sinclair, J. D., and E. L. Hamilton. 1955. Streamflow reactions of a fire-damaged watershed. Proceedings of the American Society of Civil Engineers, Hydraulics Division 81:1–17.

Spotila, J. A., M. A. House, A. E. Blythe, N. A. Niemi, and G. C. Bank. 2002. Controls on the erosion and geomorphic evolution of the San Bernardino and San Gabriel Mountains, southern California. Geological Society of America, Special Papers 365:205–230.

Sugihara, N. G., and M. G. Barbour. 2006. Fire and California vegetation. Pages 1–9 *in* N. G. Sugihara, J. W. van Wagtendonk, K. E. Shaffer, J. Fites-Kaufman, and A. E. Thode, editors. Fire in California's ecosystems. University of California Press, Berkeley, California, USA.

Wells, W. G., II. 1981. Some effects of brushfires on erosion processes in coastal southern California. Pages 305–342 *in* T. R. H. Davies, and A. J. Pearce, editors. Proceedings of a Symposium on Erosion and Sediment Transport in Pacific Rim Steeplands, Christchurch, New

Zealand, January 26–31, 1981. International Association of Hydrological Sciences Publication No. 132. Washington, D.C., USA.

Wells, W. G., II. 1987. The effects of fire on the generation of debris flows in southern California. Geological Society of America, Reviews in Engineering Geology 7:105–114.

Wohlgemuth, P. M. 2006. Hillslope erosion and small watershed sediment yield following a wildfire on the San Dimas Experimental Forest, southern California. Pages 41–48 *in* J. M. Bernard, J. W. Webb, and M. M. Garsjo, editors. Proceedings of the 8th Federal Interagency Sedimentation Conference, Reno, Nevada, April 2–6, 2006, Washington, D.C., USA, Interagency Advisory Committee on Water Information, Subcommittee on Sedimentation.

Wohlgemuth, P. M. 2015. Hillslope erosion and small watershed sediment yield before and after fire in southern California. Pages 977–986 *in* J. M. Bernard, M. M. Garsjo, and C. L. Hoeft, editors. Proceedings of the 3rd Joint Federal Interagency Conference (10th Federal Interagency Sedimentation Conference and 5th Federal Interagency Hydrologic Modeling Conference), Reno, Nevada, April 19–23, 2015, Washington, D.C., USA, Interagency Advisory Committee on Water Information, Subcommittee on Sedimentation.

Wohlgemuth, P. M., J. L. Beyers, C. D. Wakeman, and S. G. Conard. 1998. Effects of fire and grass seeding on soil erosion in southern California chaparral. Pages 41–51 *in* Grey, S., chair. Proceedings of the 19th Forest Vegetation Management conference, Redding, California, USA.

Wohlgemuth, P. M., J. L. Beyers, and K. R. Hubbert. 2009. Rehabilitation strategies after fire: the California, USA experience. Pages 511–535 *in* A. Cerda, and P. R. Robichaud, editors. Fire effects on soils and restoration strategies. Science Publishers, Enfield, New Jersey, USA.

Chapter 8
Water Provision in Chaparral Landscapes: Water Quality and Water Quantity

Christopher W. Solek and Vince H. Resh

Abstract Rivers and streams in chaparral landscapes provide both direct and indirect critical services to humans. Water provision services can be broadly parsed into five categories: improvement of extractive water supply, improvement of in-stream water supply, water damage mitigation (e.g., flood control), water-related cultural services such as recreation, and water associated supporting services, such as enhancement of aquatic species biodiversity. Each of these services is influenced by the quantity and quality of water, location, and timing of flow. Water quantity and quality in California's chaparral landscapes are affected by sequential flooding and drying, particularly in small seasonal streams, resulting from the highly seasonal precipitation patterns in Mediterranean-type climate regions. Fire is also a key factor affecting water quality and quantity. In these systems, water quantity is limited and quality often degraded, especially during the dry season. This is further exaggerated by diversions and withdrawals for urban, agriculture, and industrial uses, while future climate change could be particularly severe in these highly seasonal climate regions. Arguably, streams and rivers in chaparral landscapes are among the most vulnerable ecosystems to human activities, and are regularly subjected to various influences that may have deleterious effects on surface waters, such as groundwater pumping, conversion of natural lands to agriculture, cattle grazing, waste disposal, and urban encroachment. Because of high human population and agricultural demands in southern California, water security is essential. Reservoirs and storage facilities help provide this, although these alter the natural hydrographs of streams and rivers. Key management priorities to protect water provision services include the reduction of contaminants, eutrophication, and alteration of biogeochemical processes to reduce nutrient loads, along with establishing water-quality goals and aiding watershed protection. To best retain water provision services supplied by chaparral landscapes, coordination and

C.W. Solek (✉)
Council for Watershed Health, Los Angeles, CA, USA
e-mail: csolek@sbcglobal.net

V.H. Resh
University of California, Berkeley, CA, USA

E. C. Underwood et al. (eds.), *Valuing Chaparral*, Springer Series on Environmental Management, https://doi.org/10.1007/978-3-319-68303-4_8

efficiency of management practices and interventions across land jurisdictions, property lines, and watershed boundaries must be improved.

Keywords Beneficial uses · Flooding · Hydrological services · Hydroperiod · Recreation · Rivers · Stormwater · Streams · Water provision · Water quality

8.1 Introduction

Water encompasses and connects terrestrial, freshwater, and coastal ecosystems providing a range of ecosystem services that are critical for human survival. In particular, the hydrological functions of rivers and streams provide human societies some of the most valued, recognized, and diverse ecosystem services on the planet. These include *direct* market services, such as the provision of drinking water, electricity generation, pollution disposal, irrigation, fisheries, recreation, and transportation, as well as *indirect*, non-market, supporting services, such as nutrient re-cycling and renewal, carbon storage, biodiversity, support for terrestrial and estuarine ecosystems, and habitat for plant and animal life (Gleick 1993; Naiman et al. 1995; Costanza et al. 1997; Postel and Carpenter 1997; Grizzetti et al. 2016). The spatial scale of these services varies. Some supporting services, such as nutrient cycling, soil stabilization, and climate regulation sustain human life on a global scale. Other regulatory services, such as water purification by vegetated watersheds and flood damage mitigation provided by riparian wetlands, directly benefit human health and safety at a more regional scale. Rivers and streams can also confer profound aesthetic and cultural services to humans. They provide recreation opportunities and, at an even more essential and basic level, the innate satisfaction derived from knowing that a river ecosystem exists (Costanza et al. 1997). This attribute can also result in higher residential property values nearby scenic rivers and riparian areas (Purcell et al. 2002; Weber and Berrens 2006).

Some of the indirect services provided by rivers and streams can be difficult to quantify from the traditional market perspective, but are important to people nonetheless. For example, environmental economists have attempted to assess the monetary value of protecting non-use services of water, such as in-stream flows and associated riparian areas, for at least several decades (Loomis 1987, 1998). Particularly in the western United States, the worth of in-stream flows and stream- or river-based recreational access (a cultural service) has motivated a conservationist perspective to market and cultural forces that are more historically based on resource extraction (a provisioning service). Stream-based recreational values documented across the western US that have spurred the protection of environmental flows and associated endangered fish habitat include whitewater rafting (Ward 1987; Leones et al. 1997), angling (Duffield et al. 1992; Loomis and Creel 1992), and bird-watching (Eubanks et al. 1993; Berrens et al. 1996, 2000). Some of these benefits have been developed using non-market valuation techniques such as the contingent valuation method, hedonic price index method, and travel-cost method. In fact, non-use values of river systems, such as water quality enhancement or contributions to the larger habitat of critical fish and wildlife species, have been reported in some cases to be much larger than use values (Brown 1992).

Most often, valuation studies of freshwater-related services measure the environmental amenities of water provision in terms of its quantity and quality (Leggett and Bockstael 2000; Carson et al. 2003), and can be broadly parsed out into five categories: (1) improvement of extractive (diverted) water supply, (2) improvement of in-situ (in-stream) water supply, (3) water damage mitigation, (4) provision of water-related cultural services, and (5) water associated supporting services (Fig. 8.1). Each of these services has associated attributes of quantity, quality, location, and timing of flow. The categories of diverted water supply, in-situ water supply, and water damage mitigation refer to the quantity of water, whereas the provision of water-related cultural and supporting services relate to water quality aspects. The attributes of location and timing describe where and when water is available, respectively. A number of ecosystem processes affect each attribute (Fig. 8.1). Some services, such as municipal water supply, require not just an adequate quantity of water, but also that it be of acceptable quality and in the right place at the right time (Brauman et al. 2007). Human impacts on landscapes often diminish the capabilities for ecosystems to provide these essential services for people (Foley et al. 2005; Huber-Sannwald et al. 2006).

There are multiple, clear linkages among waterways, habitat, and ecosystem services in chaparral landscapes. Water quantity and delivery can particularly affect the establishment of riparian vegetation in chaparral dominated landscapes, and, consequently, the habitat-related services provided by the streams themselves in a number of ways, with all processes inter-related in a feedback loop. For example, flood magnitude affects propagule dispersal, seedling survival and growth, and, to a lesser extent, sapling and adult survival and/or growth. Flood timing has major effects on propagule dispersal, but less on seedling survival. Conversely, streamflow variability (and depth to water table) affects seedling survival and, to a lesser extent, germination, sapling survival, and adult survival, whereas sediment deposition affects germination and seedling survival, and to a lesser extent, adult plant survival.

In this chapter, we provide an overview of the water provision services provided by chaparral landscapes, including the factors that influence these water provision services, management implications, the economic valuation of water, regulations affecting water provision services, fire effects on water provision, and recommendations of future needs.

8.2 Overview of Water Provision Services

Water provision is arguably the highest value ecosystem service associated with the ecosystems of California (including shrubland ecosystems like chaparral), providing both hydropower and water supply to downstream users. In 2010, California used an estimated 45.6 billion m^3 of water for public supplies, irrigation, and livestock (Maupin et al. 2014), more than any other state in the

Ecohydrologic Process (what the ecosystem does)	Hydrologic Attribute (direct effect on ecosystem)	Ecosystem Service (beneficial use)	Hydrologic system change or perturbation to provisioning service
■ Local climate (temperature, precipitation) ■ Water use/uptake by plants	Water quantity (surface and groundwater storage and flow)	Provisioning (diverted water supply for municipal, agricultural, commercial, industrial uses)	Change in evapotranspiration Change in amount of water runoff
■ Filtration by soil ■ Soil stabilization ■ Chemical/biological additions and subtractions	Water quality (pathogens, nutrients, salinity, sediment)	Provisioning (in-situ water supply for hydropower, transportation, supply of fish and other freshwater products)	Change in soil moisture or condition Decreased porosity of native soils
■ Soil development ■ Ground surface modification ■ Surface flow path alteration ■ River bank development	Location (ground/surface, up/ down stream in/out channel)	Regulating (water damage mitigation, reduction of flood damage, salt water intrusion, sedimentation)	Loss of permeable soil surfaces (upland, floodplain, and in-channel) Depletion of groundwater stores
■ Control of flow speed ■ Water storage (short and long-term) ■ Seasonality of water use	Timing (peak flow, base flow, velocity)	Cultural (spiritual, aesthetic provision religious, educational, and recreational values) Supporting (water and nutrients for aquatic habitats and downstream estuaries)	Change in timing of water runoff

Fig. 8.1 Relationship of hydrological ecosystem processes to the ecosystem services they provide (modified from Brauman et al. 2007 and Chang and Bonnette 2016)

US. Surface water rights are approximately five time the state's mean annual runoff while substantial uncertainty surrounds actual use estimates, especially for groundwater (Grantham and Viers 2014). California's complex, intensively developed water storage and delivery system depends almost entirely on the collection and redistribution of winter precipitation, surface runoff, and groundwater. In southern California, the demand for water now far exceeds local supplies and the region now imports the majority of its water which may be as high as 85% of its total water use (Carter and Resh 2005). This water is pumped through the 4800 km (2983 miles) of pipelines, tunnels and canals (Stokes and Horvath 2009) of the Central Valley Project, the State Water Project, the Colorado River Aqueduct, and others (WIW 2013).

Chaparral ecosystems, in particular, have long been recognized as an important resource and for their intrinsic watershed values (Kinney 1900; Clar 1959). This is exemplified by the chaparral landscapes of coastal California, a region of many small rivers draining steep, mountainous catchments with varied land cover characterized by large seasonal and inter-annual variations in runoff (Bonada and Resh 2013) (Fig. 8.2). In these landscapes, vegetated slopes commonly receive underground water subsidies from upslope drainages (Rowe et al. 1954). The numerous streams that originate in the mountains, like those fringing the northern coast of the Santa Barbara Channel, transport nutrients and sediments into coastal estuaries and the near-shore oceanic environment (Goodridge and Melack 2012).

In chaparral dominated watersheds, physical and hydrological processes (e.g., infiltration, runoff, and erosion) are heavily influenced by fire. Steep hillslopes in these landscapes typically respond during postfire winter rains with increased runoff and accelerated erosion, resulting in debris flows, landslides, and floods, completing what has been dubbed the "fire-flood" sequence (Keeley and Zedler 2009) (see Chap. 7). These processes have profound effects on the water provision services these landscapes are able to provide, especially in terms of water quantity and quality.

Water quality and quantity can also be influenced by climatic (e.g., wet-dry season) and anthropogenic perturbations (e.g., water impoundments and recreation) in chaparral landscapes. In a broad sense, hydrological regimes are strongly determined by seasonal, inter-annual, and spatial dimensions that reflect these Mediterranean-type climate (MTC) regions (Cooper et al. 1986), with sequential and predictable flooding and drying over an annual cycle (Gallart et al. 2012; Bonada and Resh 2013) (see Chap. 7). These characteristics profoundly influence water availability in these systems and, in turn, their associated services. Often these services are manifested and demonstrated through the (often detrimental) effect that human activities have on them, such as water diversions made during the dry season or water augmentation that turns seasonal streams into perennial ones. The hydrological extremes of streams in these regions can also affect nutrient dynamics and pollution events by diluting these features in the wet season and concentrating them in the dry periods (Gasith and Resh 1999).

8.2.1 Water Quantity

Water quantity is a measure of the amount (e.g., volume or discharge) of water supplied by a contributing watershed. The relative amount of water available in support of ecosystem services depends on the quantity of water delivered to a landscape and how it is partitioned within the landscape for urban, agricultural, industrial, conservation, and other uses. In terms of water as a provisioning service, it has been estimated that 13% of the global terrestrial land surface could be managed for urban water use alone (Reid 2001). The partitioning of water among hydrological processes of surface water flow, groundwater recharge, and evaporation from soils and plants depends primarily on climate and relatively static landscape properties, including topography, soil texture, and the underlying geology (Havstad et al. 2007). Natural landscapes, including chaparral and other shublands, promote the transfer of surface water to groundwater by infiltration, which reduces flood peaks while increasing baseflow, generally increasing the predictability of water availability (Smakhtin 2001; Stella et al. 2013). One key supporting service provided by natural ecosystems is maintaining an area's hydroperiod, the characteristic seasonal fluctuations of wet and dry conditions. A predictable hydroperiod allows for the continued supply of services such as habitat provision for native fish valued by fishermen and migratory birds enjoyed by bird-watchers.

In California, water quantity is one of the most pressing management and environmental issues. A continually increasing population along with ever-intensive, irrigated agriculture continues to increase demands for water (Grantham et al. 2013), and subsequently affect natural water conveyance hydroperiods. Large volumes of water are made available annually to meet these demands through summer groundwater pumping and water conveyance systems (the networks of canals, pipes, and pumps that carry water from one place to another). One of the fundamental determinants of the energy use of water conveyance systems is the relationship between the elevation of where water is sourced and where it is used. Water volume and the distance travelled are other key factors that influence the amount of energy required to transport and distribute water needed for various industrial, agriculture, and residential uses (WIW 2013). As population expands into places where water must be imported from distant locations or pumped to get water up and over hills and mountains (as is the case for southern California), water supplies become more energy intensive. Water-related energy uses consume roughly 19% of all electricity used in the state and 32% of natural gas (CEC 2005; GEI Consultants/Navigant Consulting 2010).

With 1128 km (701 miles) of canals and pipelines, the California State Water Project is one of the most extensive water conveyance systems in the world, linking high precipitation regions in the north and east with high population regions in the south and west, and mid-state agriculture. Pumps are the most energy-intensive devices in most conveyance systems (CEC 2005), rendering the California State Water Project alone the largest single user of energy in California for agriculture and municipal uses. In the process of delivering water from the San Francisco Bay-Delta to southern California, the project uses 2–3% of all electricity consumed in the state (Wolff et al. 2004). When the supply of surface water is unable to meet demand, groundwater pumped from aquifers provide up to 30% of freshwater in California (Wolff et al. 2004), and can account for nearly 60% of the state's water supply during high-demand, dry years.

This issue has been exacerbated over the last 5 years owing to California's drought, but represents a problem that is reflective of the region's Mediterranean-type climate as the predictability of natural water supply varies greatly between wet and dry years (Ball et al. 2013). As the climate changes, altered precipitation patterns could further affect water conveyance and storage infrastructure, as their original locations may no longer be where the needs are (WIW 2013).

8.2.2 Water Quality

Water quality is a measure of the chemicals, pathogens, nutrients, salts, and sediments in surface and groundwater. Not only is water quality important for drinking water supplies, but quality is an important attribute of all other water provision services, such as production of fish and other freshwater organisms that are consumed by humans. Clean water sources are important for agriculture and industry for many communities, and thus can be integral to supporting regional economies (CBD 2016). Water quality can also influence supporting services such as the delivery of water and nutrients to estuaries (via natural hydrological processes) that impact downstream water quality. This, in turn, can provide cultural services such as recreation and water play at beaches and coastal areas.

As water moves through a landscape, ecosystem processes can improve or degrade the supply of hydrological services and uses through their effect on multiple attributes of streams (Fig. 8.1, Brauman et al. 2007). Water quality can be improved as it is filtered through natural vegetation, particularly forests, through a process known as eutrophication reduction (Stella et al. 2013; Kreye et al. 2014). For example, headwaters can buffer downstream aquatic ecosystems from nutrient export effects, such as nitrogen and phosphorous, that impair water quality by filtering, retaining, and storing water (McNeil et al. 2008; Kreye et al. 2014). Likewise, within an ecosystem, different ecohydrological processes may have competing effects on the same attribute or have simultaneously positive and negative effects on different attributes of a particular service. For example, an intact chaparral upland with stable soils can increase infiltration while decreasing total runoff volume. Concentrating the focus on the way ecosystems affect hydrological attributes provides a way to translate traditional hydrological science into an ecosystem services context that is useful to decision makers.

8.2.2.1 Water Quality Relationships to Aquatic Habitat and Species

In southern California, where the climate is characterized by extreme seasonal flow variability and water scarcity, periods of low flow or drought tend to be associated with increased water temperatures and concentrations of total dissolved solids and specific conductivity. Together with changes in streamflow, changes in stream temperature and concentrations of ions and solids can have negative consequences on aquatic habitats and species. As stream temperature increases, many cold-water species are exposed to stream temperatures well above survival temperature thresholds (Null et al. 2012). Species living in water-limited, small headwater streams are particularly vulnerable to changes in streamflow and stream temperature, while species living in estuaries will be vulnerable to changes in salinity as it affects the solubility of dissolved oxygen. Higher salinity is associated with lower dissolved oxygen content in water. For example, record high levels of salinity in the Sacramento-San Joaquin Delta in November 2015 corresponded directly to record low inflow resulting from low rainfall in the autumn of 2015 (CDWR 2016). After a prolonged drought, the first autumn storm event is likely to flush pollutants that have been accumulated during the dry period, increasing the concentration of pollutants and thus impairing water bodies (Chen and Chang 2014).

8.2.2.2 Water Quality Protection and Treatment

There are various regulations and activities at the federal and state level to protect water quality. These include enforcing federal clean water and safe drinking water laws, providing support for municipal water infrastructure improvements, and pollution prevention efforts aimed at protecting watersheds and sources of drinking water (USEPA 2016). One of the ways water quality protection can be achieved is via artificial water purification processes, including surface/groundwater treatment prior to use, wastewater treatment after use, and desalination processes. However, the energy requirements and associated costs with water purification vary greatly and can offset the benefits (i.e., services provided) of the treatment process. According to the California Energy Commission, wastewater treatment uses one percent of California's electricity (CEC 2005). To provide some perspective of these energy requirements, water reuse-related energy requirements of reclaimed water treatment and conveyance beyond that required for wastewater discharge range from 0.4 to 1.2 kWh/kilogallon (kgal) (or 0.38–1.1 megajoule [MJ]/m^3), compared to as low as 0.1 kWh/kgal (0.095 MJ/m^3) for traditional raw water treatment. In contrast, GEI Consultants/Navigant Consulting (2010) estimate the energy requirements of seawater desalination at 12.2 kWh/kgal (MJ/m^3) and inland brackish water desalination at 4.0–5.5 kWh/kgal (MJ/m^3) (NRC 2012).

In southern California, protecting water quality via artificial water purification and reuse is a high priority due to the region's overall water scarcity and high consumer demand. For example, Los Angeles County operates one of the largest, engineered wastewater recycling programs in the world (LACSD 2017), and the City of Los Angeles possesses a growing number of distributed facilities for stormwater capture and treatment (Geosyntec Consultants 2015) (see Box 8.1). Estimates of the combined energy use for water reclamation and non-potable reuse for utilities in Los Angeles were 1.84 kWh/kgal (1.75 MJ/m^3) for treatment and distribution, with an estimated energy cost of \$0.46/kgal (\$0.12/m^3) (NRC 2012).

Other water treatment options to increase the amount of available water in southern California include seawater desalination. However, several local comparisons of energy requirements for water reuse scenarios in California have estimated that desalination requires substantially more energy than potable and non-potable reuse (Equinox Center 2010). Stokes and Horvath (2009) calculated comparative total energy use for a hypothetical water treatment facility in southern California and found that desalination used significantly more energy than other processes, such as reclaimed water. Based on their analysis, the cost of desalination was comparable to the costs of water importation. However, both seawater desalination and water importing are very expensive, and currently impractical options. For example, a seawater desalination plant built in northern San Diego County, California (the largest such facility in the United States), cost $1 billion to construct, with an estimated $50 million required annually to generate the power to run the facility. Similarly, the estimated cost of importing water to San Diego is about $2300 dollars an acre-foot (1233 m^3), which equates to more than double the cost that most southern California cities pay for water (Gorn 2016).

Building or improving treatment plants is not the only way to deliver potable water. Natural ecosystems, including chaparral dominated landscapes, clean water without using any energy and are therefore by far the most energy-efficient "treatment" process. These systems can provide net energy gains provided that distribution systems are comparable. There is a fairly large body of literature establishing that protected areas can be maintained to avoid significant costs and associated energy demands of traditional treatment works (White et al. 2006; Matamoros et al. 2007). The Cleveland, San Bernardino, Angeles, and Los Padres national forests in southern California preserve some of the most pristine chaparral shrubland ecosystems known to exist on the planet (CCI 2010). In contrast to many other national forests, these were originally

Box 8.1 Management Actions and Stormwater Capture in Los Angeles

With increased pressure on traditional water resources, many Mediterranean-type climate regions worldwide are undertaking significant efforts to augment local water supplies through increased and more efficient capture of the runoff (both dry and wet weather) that is not infiltrated to soil and potentially contributes to the process of perennialization. In southern California, the City and County of Los Angeles are actively investing in urban stormwater capture to augment its local water supply portfolio. Los Angeles has large portions of its coastal draining watersheds comprised of chaparral plant communities, including coastal sage scrub (soft chaparral) (Fig. 8.2, see Chaps. 1 and 2).

Stormwater has historically contributed a significant amount of water for Los Angeles. Currently, municipalities are actively recharging the local groundwater aquifers with approximately 35,800,000 m^3 (~29,000 acre-ft) per year, while another 43,000,000 m^3 (~35,000 acre-ft) per year is recharged into those same aquifers by incidental infiltration through mountain front zones and unpaved surfaces

Fig. 8.2 Typical coastal draining stream in southern California chaparral, East Fork San Gabriel River, San Gabriel Mountains. Photo by Chris Solek

(Geosyntec Consultants 2015). It has been demonstrated that an additional 83 to 140 million m^3 (~68,000 to 114,000 acre-ft) per year could be realistically captured over the next 20 years via infrastructure projects and distributed programmatic approaches (Geosyntec Consultants 2015). This capture of local stormwater runoff creates usable supplies that would not otherwise exist. The approximate value of this water to the City over the same 20 year time period is $1100 per acre-foot (1233 m^3) for recharged water compared to $1550 per acre-foot for imported surface water, which represents a sound investment in Los Angeles' future water supply portfolio and an example of the economic translation of the value of stormwater for urban uses (Geosyntec Consultants 2015). In this case, stormwater capture projects benefit a region like Los Angeles because by providing a local, reliable water source in addition to the value of avoiding purchases of imported water.

The value of captured stormwater accrues as water infrastructure projects increase in the region and provide water for other uses. Therefore, water agencies can monetize this value by avoiding expenses of purchased water. In addition, given the limited availability of purchased imported water, there is value in developing local supplies that are in excess of the value of the purchased water itself. Monetizing the resource values is a critical component to understanding how much investment in a given water infrastructure project is required for the project to ultimately pay for itself.

Challenges in Assessing the Economic Values Associated with Water Supplies Created by Stormwater Capture Projects

- Water supplies can be variable and estimating the monetary value of a particular water supply can differ based on where in the delivery chain the value of this supply is calculated (e.g., at the source or with the end user).
- The capture of urban runoff and stormwater may not be immediate but realized over time.
- The vast majority of economically attractive stormwater capture projects are groundwater recharge projects, some of which will provide ancillary benefit to surface water quality (e.g., open space, peak flow reduction), all of which provide ecological benefits and services.
- There may be additional or hidden costs to accessing some sources of water (e.g., mechanisms to allow groundwater to be pumped safely).

This is an important consideration for water supply in Mediterranean-type climate regions and chaparral dominated landscapes, like the greater Los Angeles area. For example, projects that supply water for groundwater recharge, a source that must be pumped and treated (depending on water quality conditions), are less valuable than direct use projects that provide water sufficiently treated for their intended use. Moreover, water that is provided during off-peak use periods (i.e., winter, autumn, and spring) are less valuable than projects that provide water during high demand periods and potential shortage conditions (i.e., summer). When viewed in the guise of avoided purchases of imported water, the value of stormwater would vary depending upon the type of stormwater project and the period when the water is produced. For example, any value assigned to water recharged into a groundwater basin depends upon the ability to pump and safely use this groundwater (a use and provisioning service in terms of drinking water). Costs for groundwater recovery in the San Fernando Basin in the City of Los Angeles, for example, are abnormally high because of the widespread contamination of the basin by legacy pollutants (Geosyntec Consultants 2015). If these costs were included in the value of the recharge water, then it would suggest that the value of recharge water is exceptionally low. However, it is assumed that costs for groundwater treatment are being addressed independently of any stormwater implementation to be carried out by the City (e.g., from stormwater capture projects) and therefore these costs do not impact the value of captured water (Geosyntec Consultants 2015).

set aside to protect watershed values and upper portions (headwaters) of coastal draining watersheds (CCI 2010) (see Chaps. 5 and 16). In addition, other major metropolitan areas, such as New York City (Catskills region), San Francisco (Hetch Hetchy), and Portland, Oregon (Bull Run), all rely almost exclusively on watershed protection and management for their potable supply treatment (USEPA 2002).

8.2.2.3 Water Quality and Recreation

When valuing land conservation efforts from a water provision and regulating perspective, some economic studies focus exclusively on the water quality protection benefits (Holmes et al. 2004; Shrestha and Alavalapati 2004). However, other studies present water quality protection as part of a bundle of co-benefits along with conservation of green spaces, recreation opportunities, wildlife habitat preservation, and environmental education (Whitehead 1990; Blaine and Lichtkoppler 2004; Johnston et al. 2005).

In particular, surface water quality improvements can create improved conditions for water-based recreational opportunities (USFS 2012). Lakes, rivers, and other water bodies have been identified as magnets for recreation, with water-based activities being one of the top ten activities in all national forests. Fishing and swimming were ranked seventh and eighth, respectively out of a distribution of the primary activities of nearly 90 million annual activity days on federal lands (BBC Research and Consulting 2010). Furthermore, water-based recreation has long been recognized as having a positive influence on human health and well-being. In California, direct expenditures for outdoor recreation are over $20 billion, with the largest expenditure total (approximately $5.1 billion) occurring in the southern California region, followed by the San Francisco Bay Area region and the Los Angeles region (over $4 billion in each region) (BBC Research and Consulting 2010).

In the case of southern California, the four national forests encompassed in the area (the Cleveland, San Bernardino, Angeles, and Los Padres) are different from others in the forest system because they are predominantly comprised of chaparral shrublands and almost exclusively used for recreation by a growing urban population (CCI 2010) (see Chap. 10). Although in close proximity to dense urban areas, these lands harbor many ecologically significant rivers and creeks which are immensely popular swimming spots for urbanites seeking relief from the summer heat. Recognizing this value, they have been proposed for designation as National Chaparral Recreation Areas (NCRAs) within these national forests (CCI 2010).

8.3 Influences on Water Quantity in Chaparral Landscapes

8.3.1 Seasonal Climatic influences

In California, highly seasonal precipitation and temperature patterns occur annually, with hot, dry summers, and cool, wet winters predominating. Inter-annual variability in precipitation is a key characteristic in the five Mediterranean-type climate regions worldwide. The amount of precipitation or stream runoff typically dictates whether the year is classified as a dry, normal, or wet year (Resh et al. 2013), although defining a "typical" wet or dry year based on fixed thresholds is problematic, as these indices have been arbitrarily determined based on historical climate data, and have typically focused more on runoff volume rather than on the timing of flows

(Vicuna et al. 2007). Given the naturally variable hydroclimates of California, coupled with changing climate scenarios, accurately describing (and hence, characterizing) the differences between year types, especially from an ecological perspective, is difficult (Null and Viers 2013).

Streams in chaparral landscapes in California are often characterized by a single annual flow peak in winter, although those with snowmelt influence (e.g., those draining the Sierra Nevada mountain range) may have a second flow peak in spring (Erman et al. 1988; Carter and Resh 2005). Although the timing of drying and flooding is very predictable, the intensity of these events is not. Some years have longer dry periods than others and this has a strong effect on the biological communities in these streams (Gasith and Resh 1999). Similarly, the frequency of floods can be highly variable inter-annually, which is reflected in the high inter-annual variability in physicochemical and biological features, another characteristic of all Mediterranean-type climate regions (Resh et al. 2013).

Highly variable flow regimes with large, periodic floods shaping the geomorphology of the (often-braided) channels are common in natural flowing (i.e., non-engineered) streams in climate regions such as southern California, which is in contrast to stable, temperate-climate rivers (Kondolf et al. 2013). Flow regimes in the region are further influenced by the El Niño/Southern Oscillation (ENSO) phenomenon when higher flows than normal occur during El Niño years and lower flows during La Niña years (Dettinger et al. 2000). Fluvial systems in chaparral regions not only have high flow variability but are often episodic (ephemeral and seasonal stream types are most common). The annual shift between these two extremes of flow-flood and flow-cessation often leads to mosaics of dry channels and lentic waters interspersed among flowing reaches and floodplains. Connectivity of habitats during these dry periods create conservation challenges (Merenlender and Matella 2013) as do the scouring effects of floods on the biota (Resh et al. 2013).

8.3.2 Spatial and Temporal Influences

Streams in chaparral landscapes are also highly variable spatially, with a mosaic of flow conditions, even at the reach scale. This condition results in a unique freshwater ecosystem from a hydrological and, consequently, biological perspective (Gasith and Resh 1999; Bonada and Resh 2013). Because hydrological connectivity varies between seasons, an expansion phase in the wet period (i.e., autumn to winter) and a contraction phase in the dry period (i.e., spring to summer) occurs (Bernal et al. 2013). During the wet period, precipitation restores longitudinal-, lateral-, and vertical-flow connectivity. At this time, disconnected pools reconnect, and the river functions again as a sequence of pools connected to riffles. In small and steep basins, this flow expansion can occur within a very short time lag because precipitation usually falls as intense storms that often lead to sudden flash floods from late-summer to autumn (Camarasa-Belmonte and Segura-Beltrán 2001; Llasat et al. 2010) (see Chap. 14). In contrast, during the dry period, the lack of precipitation and the high evapotranspiration rate of these rivers result in a steady reduction of the longitudinal-, lateral-, and vertical-flow connectivity. This reduction process can be very extreme in certain circumstances (Bonada et al. 2007) and lead to a sequence of disconnected pools that may ultimately lack any surface water, leaving riverbeds dry. In some chaparral streams, perennial flows are maintained in summer due to the presence of groundwater near or at the surface of the riverbed, although even these streams experience seasonal reductions in discharge from winter to summer (Vidal-Abarca 1990; Bonada et al. 2007). Therefore, chaparral landscape streams are those with sequential seasonal flooding and drying periods, with increasing loss of habitat connectivity over an annual cycle that can result in temporary habitats, especially during severe droughts.

The attribute of "timing" describes when water is available in streams and rivers and also encompasses the predictability of flows. The water supplied by rivers and streams is "valued" by humans only when users have access to it (e.g., in downstream diversion ditches, in wells located at a distance from a surface water source, or as accessible groundwater that can be pumped). However, water supplies can be harmful and destructive to humans when water ends up in the wrong place, such as when flooding inundates crops, homes, or leads to loss of human life. In chaparral landscapes precipitation is not spread evenly over the course of the year, so sudden influxes can render much of the mean annual runoff from a catchment unusable or even hazardous. One of the challenges in MTC regions such as southern California is that low flows and flood peaks are poorly described by average or annual water volumes within a typical discharge/flow year (a time period of 12 months for which precipitation totals are measured). As a result, anticipation and management of the supply of hydrological services requires information about the duration, seasonality, and predictability of absolute and relative changes in flood peaks and low flows (Jones and Post 2004). Seasonal variations in water use and the time required for a new hydrological regime to be established in the wake of land use alterations can greatly influence the services that rivers and streams provide.

8.3.3 *Climate Change Influences*

Mediterranean-type climate regions like southern California, are considered to be most vulnerable to the effects of climate change due to their highly seasonal precipitation, high water demand for irrigation, and expanding human populations. Due to these factors, climate change is projected to have profound effects on water-related provisioning, regulating, and cultural services (Chang and Bonnette 2016). As climate change modifies the water cycle via changes in the spatial and temporal distribution of precipitation and the form of precipitation (e.g., snow versus rain), water-related ecosystem services are projected to change in coming decades with the amplified seasonality of water availability and increased frequency of extreme hydrological events (Chang and Bonnette 2016). For example, rising air temperatures are projected to deliver more rain than snow in winter in high-elevation areas of watersheds and result in less snow accumulation and earlier snowmelt (Barnett et al. 2005). This would limit summer water supply when humans and ecosystems need water most.

Climate change may also shift the distribution of rainfall with more extreme weather events, which may lead to more frequent floods and droughts. Flood probability is projected to rise in mid-latitudes where the majority of the world's population resides (Jiménez Cisneros et al. 2014). Such extreme weather events may exceed a drainage basin's capacity to regulate floods, thus damaging human communities that are prone to floods. At the other extreme, climate change-induced water shortages in the form of drought could increase groundwater overdraft throughout the drought period and slowly deplete groundwater reserves. Climate change-induced drought has been implicated in leading to the fallowing of productive farm lands in California and subsequent seasonal job loss, with a total economic impact of $2.74 billion when other effects are considered (Howitt et al. 2015).

In addition to these annual and decadal climatic influences and resulting modifications in the spatial and temporal distribution of water storage, the water provision services of rivers and streams, especially in water-limited ecosystems like chaparral landscapes, may be particularly influenced by climate change and face continued and rapid alteration (Tague et al. 2009). Hydroecologic modeling indicates that vegetation acts as an important control on streamflow and responds to soil moisture availability (Tague et al. 2009). This suggests that future hydrological behavior and ecosystem productivity will depend on the balance between carbon dioxide controls on vegetation water use efficiency and vegetation responses to increasing temperatures. Recent climate scenarios predict that the average temperature in southern California will increase by 1.5 °C to 5 °C in the next century (Goodrich et al. 2000; Wilkinson 2002; Cayan et al. 2008) (see Chap. 14). Results suggest that as biomass increases in chaparral dominated landscapes, reductions in summer streamflow will follow. Changes due to increases in fire frequency will also impact summer streamflow, but these will be small relative to changes in vegetation productivity.

From a water quality perspective, high-elevation areas of watersheds are projected to experience the most climate-related stresses on water-related ecosystem services, including substantial declines in suitable habitats for cold-water organisms with the projected rise in stream temperature. Low-elevation areas may experience other water quality problems such as non-point source pollution loads which could either increase or decrease, depending on flow changes projected by different climate change scenarios (Chang and Bonnette 2016). Increases in temperature and decreases in precipitation could also increase the risk of deoxygenation in rivers (Whitehead et al. 2009). These effects are expected to be most pronounced in chaparral dominated MTC regions.

Faunal changes in stream biota due to perturbations in water quantity and quality from climate change likely include displacement of species and populations to higher elevations and upper latitudes, loss of species diversity and homogenization of community composition, and changes that may result in faunal shifts (Filipe et al. 2013). While the direct effects of climate change on fish populations are difficult to discern, decreased dissolved oxygen, changing sediment and nutrient loads, and changes to hydrographs could lead to decreased productivity of native fish populations and eventually extinction of some fish species around the world (Ficke et al.

2007). Projections of habitat suitability for the large brown trout in the Cabriel River in the Iberian Peninsula, another MTC region, estimate a decreased flow (of 20–29%) and increased water temperature (of 4 °C to 4.2 °C) will cause habitat suitability to decline in the near future (2011–2040) (Muñoz-Mas et al. 2016). Together with other human stressors (e.g., watershed degradation, recreational overuse of river water), climate change is projected to decrease most of California's native salmon species in the twenty-first century (Katz et al. 2013).

Changes in the amount and timing of freshwater inputs as a result of climatic perturbations can also impact estuaries, by changing salinity and negatively affecting fish species that are adapted to specific salinity regimes. These types of changes could have significant economic impacts, particularly in the coastal regions of Mediterranean-type climate regions where commercial and recreational fishing activities are abundant. In addition to over-fishing, ocean warming and acidifications will have negative consequences on shellfish fisheries, aquaculture and corals, leading to shifts in tourism flows and thus revenues associated with the industry (Weatherdon et al. 2016).

8.3.4 Anthropogenic Influences

Mediterranean-type climate regions worldwide have been profoundly influenced by human induced land use changes which have greatly affected terrestrial and aquatic environments in these regions (Mount 1995; Cooper et al. 2013). MTC regions are characterized by water scarcity together with a growing demand for water associated with increasing human populations and irrigated agriculture (Moyle 2014; Howitt et al. 2015). Agriculture in MTC regions, for example, has been shown to use five to ten times more water to irrigate crops than agricultural regions of more temperate Atlantic climates (Kenny et al. 2009).

As a result of this condition, streams and rivers in chaparral landscapes are perhaps among some of the most vulnerable ecosystems to human activities (Gasith and Resh 1999; Bonada and Resh 2013). Impacts to surface waters include groundwater pumping, conversion of natural lands to agriculture, cattle grazing, waste disposal, and urban encroachment, all of which can have implications for the functioning of these systems and the associated water provision services. Not only have these actions dramatically modified hydrological flows in the region, but they have also come at great environmental expense in some cases. For example, river impoundments, dam operations, and other water infrastructure interfere with fundamental hydrological processes that control riverine habitat structure, maintain natural patterns of longitudinal and lateral connectivity, trigger behavioral responses in native organisms, and influence water quality conditions (Prat and Ward 1994; Bunn and Arthington 2002; Pringle 2003).

Water provision in any landscape is in part determined by the topography and morphology of the landscape through which it flows. However, urban development can modify the permeability of the landscape and consequently, the amount of runoff contributed to streams. Riverbeds have a wide variety of substrates that determine their porosity and consequently, the amount of water that drains into groundwater.

Urban related modifications can lead to changes in the patterns of energy and matter distribution in watersheds, including evapotranspiration, surface runoff, discharge, nutrient availability (nitrogen and phosphorus), soil erosion, and sedimentation (He et al. 2000) (see Chap. 7). Furthermore, as a consequence of modifications to the rate, volume, and timing of streamflow, ecosystems can be altered, both spatially and temporally, and influence the structure and composition of lotic communities (organisms that live in flowing water habitats) (Miltner et al. 2004; Konrad and Booth 2005).

At the landscape scale, water withdrawal, flow regulation, surface water diversions, water storage, and irrigation projects represent some of the largest potential perturbations to these systems that influence the provision of water and associated services (Gasith and Resh 1999; Kondolf and Batalla 2005; Kondolf et al. 2013). One of the principal driving factors for flow regulation through large scale water diversions and reservoir construction is the seasonal availability of water in the region. At a finer scale, smaller water impoundments, such as the effects of seasonal rock dams created for recreational water-play, are less studied (Benstead et al. 1999), even though they may have many of the same ecological consequences of larger dams by affecting in-stream flow, deteriorating water quality (Singh 1995), impacting fish passage, and reducing the habitat diversity for various aquatic organisms (Solek 2008). In California, for example, water management is the primary factor responsible for declines in fish biodiversity, where 83% of native freshwater fish taxa are extinct or at risk of becoming so (Moyle et al. 2011).

Reservoirs are commonly constructed to store water to compensate for seasonal and inter-annual variability in water supply, in addition to protecting against flooding downstream (see Chap. 7). In areas with high variability in precipitation and consequently water availability, controls tend to be more rigid to ensure water security, and this is certainly evident in the management of water in California. For example, rainfall-runoff dominated rivers characteristic of chaparral landscapes run high in the winter, when water demands for agriculture are lowest, but these can dry out in the summer, when water need is greatest. For this reason, reservoirs are commonly constructed to store water to compensate for the seasonal and inter-annual variability in water supplies. Indeed, streams and rivers in chaparral landscapes tend to be more heavily regulated by reservoirs than their temperate-climate counterparts (Kondolf and Batalla 2005; Grantham et al. 2010).

8.3.5 Management Influences and Impacts on Hydrology

Management practices relating to soil and vegetation at a particular location within a watershed can have significant effects on hydrological processes. The properties most sensitive to management include soil structure and vegetation cover, and the spatial pattern and composition of vegetation types (Thurow 1991). Soil and vegetation properties also affect how much of the remaining water is available for plant production, and how much is lost to evaporation from the soil or its entry into groundwater. For example, some studies have reported that shrublands lose more water to runoff than grasslands (Schlesinger et al. 2000; Wilcox and Thurow 2006), while other studies from southern California have shown the opposite trend (e.g., Rowe et al. 1954;

Hibbert 1971). Modification of topography, particularly through road construction, can both redirect runoff and accelerate it by facilitating gully or rill formation. In terms of land use change, it is generally assumed that groundwater recharge is higher under herbaceous cover than in areas dominated by woody vegetation because more water is accessed and transpired by the deep-rooted shrubs and trees. While this pattern is generally true for non-arid systems, the effects of vegetation in arid and semi-arid systems are highly variable and context dependent (Havstad et al. 2007).

When water is removed from rivers and aquifers by humans through extraction, retention, or diversion, the hydraulics and hydrology of rivers are altered, regardless of whether they (naturally) flow perennially, intermittently, or ephemerally. Because streams and rivers in chaparral landscapes tend to be more extensively dammed, downstream reaches are commonly deprived of high flows that mobilize and transport sediments, modify channel morphology, and maintain habitat complexity (Kondolf et al. 2013). Restoration to the entire pre-dam hydrograph without losing the benefits of the dam is usually impossible, but restoration of specific components of the natural hydrograph (to which native species are adapted) can restore some ecosystem components, such as native fish species, to rivers and streams in MTC regions (Kondolf et al. 2013) (see Box 8.2). For example, the Trinity River, a major tributary of the Klamath River in Northern California, offers an example of how restoration flows can be incorporated in dam operations (USFWS and Hoopa Valley Tribe 1999; Krause 2011). In this case, minimum flows below the dam were specified to facilitate downstream migration of juvenile salmonids and periodic high flows were made to mobilize the river bed in all but dry years. In addition, gravels were added to the channel below the dam to compensate for the lack of coarse sediment supply.

The effects of artificially induced water loss on riverine habitat and biota, via water diversion or withdrawal, appear to mimic those of hydrological drought. Aquatic habitats contract and fragment corresponding to changes in water quality and availability of food resources, with the survival, reproduction, and movement of water-dependent species becoming modified and often more limited (Gasith and Resh 1999). With complete loss of surface water, aquatic biota are eliminated from surface habitat, but they often can recolonize following the resumption of flow. While the responses of river and stream communities to human alterations of land use, vegetation, hydrological, and hydrochemical conditions are similar in MTC regions to other climate regions, the high variation in hydrological regimes in these semi-arid regions tends to exacerbate the magnitude of these responses (Mount 1995). For example, land use changes promote longer dry season flows, concentrating contaminants, allowing the accumulation of detritus, algae, and plants, and fostering higher temperatures and lower dissolved oxygen levels, all of which may extirpate sensitive native species (Cooper et al. 2013).

The addition of water (e.g., from wastewater or dam releases, ruptured water mains, recharging of springs from watering golf courses) can also induce perennial flow in naturally intermittent or ephemeral systems, a process referred to as perennialization (Chiu et al. 2017). There are, however, few estimates of the extent of perennialized streams in the Unites States or other parts of the world. Stream surveys in arid and semi-arid regions of California found that hydrological systems in urban and agricultural areas have higher proportions of perennial stream length than systems in natural areas, a difference attributed to augmented flows from discharge and urban runoff (Mazor 2015).

Release of water from diversions used for livestock, irrigation, and municipal and industrial use to and from seasonal streams are common and may alter the flow regime by augmenting (i.e., perennializing) flow. In urban catchments, impervious surfaces, such as sidewalks and streets, prevent water from infiltrating into the ground and result in direct runoff into streams. Perennialization may also result from the removal of riparian vegetation, which reduces evapotranspiration and can increase water yield, potentially converting seasonal streams to perennial systems. Restoration to a more natural hydrograph in perennialized reaches of seasonal streams may, therefore, facilitate the recovery of native aquatic biota assemblages adapted to intermittent conditions.

Box 8.2 The Los Angeles River: Managing Water Provision in an Urban Watershed

The Los Angeles River watershed in southern California illustrates the complexities of balancing water provision services within the context of urban watershed management needs. Although much of the lower drainage area of this watershed is highly developed, comprising much of the City of Los Angeles as it flows to the Port of Los Angeles in Long Beach, its headwaters and upper tributaries drain a predominantly chaparral landscape of open space in the San Gabriel Mountains, within the Angeles National Forest. Many of these upper tributary streams are intensely used for water-based recreation such as swimming, wading, and fishing, especially during the summer months. The US Forest Service continues to strive to meet these growing recreational demands while preserving and protecting the other important services (e.g., biodiversity) that these public lands provide.

Outside of the Angeles National Forest, most of the watershed exists as a network of concrete flood control channels, which historically have been considered to have no ecological value. However, federal approval of a $1 billion proposal (at the urging of the City and local watershed advocates) to restore riparian habitat, widen the river corridor, create wetlands, and provide access points for river recreation along an 18 km (11 mile) stretch of the river has increased its profile (Jamison 2015). City of Los Angeles officials estimate that this large scale restoration effort will not only create numerous water-based recreation opportunities such as kayaking, biking, and fishing (Fig. 8.3), but attract more than $5 billion in investment over the next 10–15 years, generating up to 18,000 jobs (Sahgun 2014).

Much of the water that flows through this revitalized portion of the river that is used for water-based activities, nearly 16 million gallons a day (~60.5 million liters), is a result of inputs of treated wastewater from the city's water reclamation plants. Increasing (augmenting) streamflow, such as with treated wastewater, has been shown to have a positive effect on some ecosystem services, all of which are germane to streams in chaparral landscapes. This includes increases or improvements in recreational amenities, natural aesthetics, and habitat for threatened species (Bischel et al. 2013). Readily observed benefits to humans documented from flow augmentation studies include creating water-based recreation opportunities where they did not exist before, increasing the aesthetic values of a stream, promoting better ecological status in river "rating" systems, and enhancing the biodiversity of an area.

Fig. 8.3 Opportunities for river-based recreation now exist in the Los Angeles River where they were not permitted before due to changing perceptions and values associated with the region's urban waterways. Kayaking a reach of the lower Los Angeles River, north of downtown Los Angeles, known as the "Glendale Narrows". Photo by Chris Solek

8.4 Influences on Water Quality in Chaparral Landscapes

While not unique to chaparral landscapes, impacts to water quality in the form of water pollution can affect underlying ecosystem functions and processes and consequently, also affect a range of ecosystem services in a complex causal chain (Mills and Harmens 2011; Harmens and Mills 2012; Jones et al. 2012, 2014; Mills et al. 2013). Water pollution from nutrient enrichment, fecal coliform, heavy metals, saline intrusion, agricultural pesticides from crop runoff, and pharmaceutical and disinfection by-products, particularly in urban areas, can greatly influence river and stream ecosystems (Grantham et al. 2013). These impacts can alter basic ecosystem functions such as primary production (plant growth) and biogeochemical cycling which in turn affect the ecosystem services and the benefits that humans derive from these environments.

The main mechanisms of impact of current pollutant levels are eutrophication (nitrogen), acidification (nitrogen and sulphur), and direct toxicity (ozone, ammonia and nitrogen oxides), all of which affect a wide range of services. Pathogens and contaminants that are harmful to humans or other organisms can be difficult to remove from water and can incur costs. For example, the intestinal parasite Giardia is difficult and expensive to remove from drinking water sources (USEPA 1999). It occurs in higher concentrations in water receiving urban pollution than in water flowing through watersheds with intact natural vegetation.

Eutrophication, a major cause of fish kills, accounts for 60% of the impaired rivers in the United States (Smith 2003). In southern California, it is a major concern for regional streams as local climatic conditions and changes to land cover influence the chemical characteristics of runoff and can negatively impact stream organisms (Fetscher et al. 2013). For instance, high nutrient concentrations may lead to water quality impairments such as hypoxia or elevated pH, harming sensitive fauna (Dodds and Welch 2000; Heiskary and Bouchard 2015). Recent analyses have indicated that small rivers in mountainous chaparral catchments can make substantial contributions to the eutrophication of downstream areas (Smith et al. 2003). In contrast, however, some studies have shown that small rivers can play a role in offsetting these effects. For example, denitrification, a pathway for the permanent loss of nitrate by conversion to nitrogen gas (Alexander et al. 2000) within streams can help offset the total nitrogen (N) load from runoff and groundwater to N-sensitive coastal marine environments (Howarth et al. 1996; Alexander et al. 2000).

The seasonal streams that are characteristic of chaparral dominated landscapes can be equally impacted by direct inputs of contaminants which affect the biota, just like perennial streams. The biota of seasonal rivers are particularly vulnerable to contaminant and excess nutrient inputs during dry phases, because of a lack of dilution and the subsequent flushing of contaminants that have built up on and within their dry beds. Such impacts are likely to occur in other seasonal systems that have been converted to perennial rivers as a result of waste-water discharge. The response of seasonal streams to eutrophication differ to those of perennial stream ecosystems in several ways (Chiu et al. 2017). For example, non-perennial streams typically exhibit slow or stagnant water, warm temperatures, and small volumes of water during dry seasons, with these factors associated with increased eutrophic responses. In chaparral areas, these systems may have lower canopy cover, and therefore, more nutrient loading and often more solar exposure than perennial streams, further enhancing the conditions that lead to eutrophication and algal blooms. Therefore, eutrophic responses may be more common in chaparral landscape seasonal streams than in perennial streams receiving equivalent nutrient loads. In addition, seasonal stream biota may have life history traits (e.g., respiratory pigments, insensitivity to pH, and ability to disperse) that allow them to be more adaptive to eutrophic conditions than the biota of perennial streams that are not adapted to extreme seasonality (episodicity) in flow regimes.

Biogeochemical cycles operating in rivers provide organic matter and nutrient transformations to adjacent ecosystems such as floodplains, groundwater, and riverbanks. Biogeochemical processes that reduce nutrient loads to downstream reaches are more accelerated and pulsed in seasonal versus perennial streams. As a result, this may play a disproportionately important role in reducing nutrient loads to downstream waterbodies during the dry phase, resulting in deposition of transported material. Many of the factors that enhance biogeochemical processing of nutrients, such as the presence of shallow water, low velocity, warm temperature, and extensive contact between surface water and the streambed substrate, are far more evident in seasonal streams. Studies that facilitate an understanding of the biogeochemical processes that reduce nutrient loads to downstream reaches should be a focal point of nutrient management research and practice for seasonal streams and waterbodies.

Soil health and chemistry are affected by the short-term drying and wetting pulses that characterize the transition from summer drought to winter rains in many MTC regions. These pulses may have a disproportionate effect on long-term carbon and nitrogen fluxes, the magnitude of which may be strongly controlled by soil carbon pools (see Chap. 6). These seasonal transitions mark a period of potentially large losses of carbon and nitrogen from these environments that are already nutrient-limited and have little capacity for nutrient retention (Miller et al. 2005). Repeated soil drying and re-wetting causes losses of up to 18% and 10% of total soil C and N, respectively, which indicates the potential importance of these pulse events to annual nutrient budgets (Miller et al. 2005). Elucidation of the mechanism(s) promoting carbon turnover in these soils is therefore critical in understanding the biogeochemical interactions that occur during seasonal transitions, and the processes that control nutrient retention and loss over the long-term. Any perturbations to the natural soil wetting processes will likely have implications for water quality in chaparral landscapes.

8.4.1 Regulations Influencing Water Provision Services

Annually, the United States spends more than $2 billion for clean water initiatives (ESA 2017) for maintaining and improving water quality for the well-being of people and the environment, and it is a high priority for most municipalities. From a regulatory standpoint, water provision services form the basis of watershed protection in the US. Termed "beneficial uses", these provide the foundation for establishing water quality goals and setting standards. Beneficial uses for surface waters originally were designated under the Clean Water Act in 1972 and are defined as "the uses of water necessary for the survival or well-being of man, plants and wildlife". These uses serve to promote the tangible and intangible economic, social and environmental goals of mankind (Barker et al. 1994). Once beneficial uses are designated, appropriate water quality objectives can be established and programs that maintain or enhance water quality can be implemented to ensure the protection of beneficial uses.

Beneficial use designation of surface waters must take into consideration the "use and value" of water for direct services. These include water supply for drinking, recreation in and on the water, industrial and agricultural uses, and indirect services such as the support of freshwater (e.g., cold- and warm-water habitat) and saline aquatic habitats for the protection and propagation of fish, shellfish and wildlife, or the ability to support rare, threatened and endangered species. The water quality requirements of wildlife pertain to the water directly ingested, the aquatic habitat itself, and the effect of water quality on the production of food materials. The US Fish and Wildlife Service estimates that up to 43% of threatened and endangered species rely directly or indirectly on riparian areas associated with surface water for their survival (Klein et al. 2013; USEPA 2017).

At a practical level, states, counties, and cities can take an ecosystems services approach to try to address the impacts of water pollution. For example, policy questions such as "What are the costs and benefits associated with controlling water pollution (e.g., mitigating urban stormwater and runoff sources) versus relying on natural purification processes via infiltration to keep pollutants from even reaching source water?" (DEFRA 2007) can be used as a risk assessment method. One classic example of when preventing pollution is easier than cleaning contaminated water is the decision by New York State to spend $1 billion to restore the watershed that provided the City's drinking water, rather than spend $8 billion on a water treatment facility in New York City (Sagoff 2002). In southern California, the City and County of Los Angeles continue to look for innovative ways (e.g., "green streets", "green alleys", and other types of green infrastructure) to more effectively capture stormwater so that natural filtration processes can clean it before it is delivered to groundwater supplies (Fig. 8.4). Implementation of a suite of centralized water infrastructure projects, and the adoption of distributed programmatic approaches that employ a holistic, watershed-based approach appear promising (Geosyntec Consultants 2015).

8.4.2 Areas of Future Water Treatment Research

Future research could examine the multiple cascading effects, potential synergies, and trade-offs of the various water treatment practices on water quality in MTC regions and chaparral landscapes. For example, although it is apparent that wastewater discharge affects river organisms through toxin and nutrient inputs, there have been few studies on how differences in sewage treatment levels (primary, secondary, and tertiary treatment) or stormwater treatment (e.g., the use of bioswales or other managed wetlands) affect receiving water bodies in MTC regions (Cooper et al. 2013). Similarly, research on bacterial communities in waterways in Mediterranean-type climate regions is generally limited to indicators of fecal contamination (Cao et al. 2011a, b). Fecal indicators of bacterial concentrations increase with increasing urban development and usually peak during and after storms, leading to the posting of health advisories. Using molecular methods, Sercu et al. (2009) observed that

Fig. 8.4 An example of "green infrastructure" in the City of Sun Valley, California. Surface runoff from an adjacent residential street is directed into this constructed bioswale to improve water quality by naturally removing contaminants in the water as it percolates into the soil. Photo by Eileen Alduenda

bacterial community composition in a southern California stream shifted from the wet to the dry season, with fecal indicator bacteria apparently derived from catchment runoff during rains but from point sources of human waste in the dry season (Hagedorn et al. 2011). Future research could investigate the presence of other types of pathogens that cause water-borne diseases, such as Giardia or Cryptosporidium, and the long-term health impacts on recreation users of chaparral landscape streams.

In addition, further investigation is needed to evaluate the effectiveness of various land management practices on the whole suite of water provision, regulating, supporting, and cultural services that are specific to chaparral landscapes. Land use decisions and water infrastructure (dams, reservoirs) could either enhance or reduce these services, especially given climate projections. Therefore, future studies should investigate the specific and combined effects of different infrastructure investment and management strategies on the provisioning of water-related ecosystem services.

8.5 Recreational Use Effects on Water Provision

The water-based recreational services provided by waterways in southern California are well recognized and streams in the regions are commonly used for various forms of water activities (Fig. 8.3, see Chap. 10). These activities, unfortunately, can expose people to a variety of health risks linked to physical, microbial, and chemical hazards, which vary according to the type of water body (e.g., natural lake, reservoir, stream, artificially created plunge pool), geographic location within the watershed (e.g., headwater versus estuary), and local use patterns (e.g., intensity of visitor use). In turn, recreational activities can adversely affect water provision in streams in the form of physical alterations to waterways and contribute to water quality impairment, such as human-caused fecal contamination. These effects can be more pronounced in chaparral landscapes, where opportunities for water-based recreation are more limited and, given the close proximity to large human populations, recreation can be concentrated in a small number of sites.

Of particular management concern are the small seasonal "summer" dams that are constructed annually by day visitors in the coastal watersheds of California. The largest impact posed by these recreational dams probably lies in the cumulative effect of their sheer abundance (Chase et al. 2000). In southern California, over 300 handmade dams have been counted at one time within a 5.6 km (3.5 mile) reach of the East Fork San Gabriel River in 2003 (Ally 2004) and over 250 counted within a 5 km (3.1 mile) reach in October 2005 (Solek 2008).

Summer dams have been documented to diminish the quality of summer rearing habitat for juvenile salmonid species in California coastal streams by changing streamflow patterns, reducing habitat diversity, diminishing water quality, and creating barriers to natural instream movements of juvenile stages. Conversely, these dams can also enhance the quality of habitats for species that are predators of juvenile salmon and steelhead (Pejchar and Warner 2001). Other evidence suggests that construction of these dams may have significant negative impacts on fish spawning and movement, water temperature, and the benthic substrate (Ally 2004). The accumulation of sediment behind these dams and increased algal growth are other potential issues that could become environmental or public health problems. The reduction in available salmonid habitat, changes in water quality, and increases in predator populations in these streams could impact fish stocks in streams that are also used for recreational fishing.

Anthropogenic modifications to a stream channel can sometimes create opportunities for one service at the expense of another. For example, the damming of the Montego River, a catchment in Coimbra, Portugal, eliminated the opportunity for traditional summer wading during low flows, but created additional opportunities for other water-based activities, such as boating and swimming (Kondolf et al. 2013). However, some of the resulting lakes created behind the dams began to be plagued with algal blooms (a water quality issue), creating an undesirable aesthetic and impacting their recreational value.

8.6 Fire Effects on Water Provision

One of the most significant impacts to river and stream ecosystems in chaparral landscapes is fire and the associated loss of vegetative cover (see Chap. 1). Chaparral and other MTC regions are among the most fire prone areas of the world (Cooper et al. 2013). Fire can impact specific ecosystem services of chaparral streams including water delivery, nutrient cycling, and biodiversity. High intensity fires can also change soil physical properties such as water infiltration as a result of hydrophobic soil surfaces (Neary et al. 1999), particularly for peak flows in the first year following fire (DeBano 2000; Hubbert and Oriol 2005) (see Chap. 7). For low flows, however, fast recovery of chaparral in the context of high year-to-year variation in climate suggest that the effect of fire frequency on decadal streamflow behavior is small (Tague et al. 2009). Although short fire-return intervals have been implicated in replacement of southern California chaparral with non-native grasses (Keeley and Brennan 2012; Lippitt et al. 2013), the long-term consequences for landscape hydrology remain poorly understood.

8.6.1 Fire Effects on Water Quantity

Fires can further alter hydrological response by reducing infiltration (DeBano 1981; Wells 1987; DeBano et al. 1998; Gabet 2003), evapotranspiration, and interception (Tiedemann et al. 1979, Wells et al. 1979). The magnitude of responses depend on the severity of the fire, type, amount of vegetation burned, topography, and geologic setting (DeBano and Conrad 1976; Wells et al. 1979; Earles et al. 2004). Fire-induced changes in hydrological response contributes to increased soil erosion (Wells 1987), sediment mobilization and transport (Rice 1974), and nutrient export (DeBano et al. 1998). Scouring floods after fires can also reduce habitat for fish, invertebrates, and algae. Fire also can influence adjacent riparian zones by affecting seed germination, seedling and sapling survival and, to a lesser extent, adult plant survival. However, chaparral stream vegetation is more resistant to fire than other vegetation types because of the rapid recovery of riparian vegetation in these ecosystems. Although drought and mass sediment movements can prolong fire effects, the return to prefire conditions is typically associated with vegetation recovery over time (Verkaik et al. 2013).

Postfire there is a reduction in evapotranspiration from reduced amounts of vegetation and a decrease in infiltration associated with vegetation prefire. Consequently, runoff increases water quantity that can lead to higher streamflow (Stoof et al. 2014). This results in greater and flashier discharge at burned sites compared to unburned sites (DeBano et al. 1998) (see Chap. 7). Hubbert et al. (2006) tracked the changes in water repellency at the soil surface in a chaparral watershed in southern California and documented (during a period without rain) a return to prefire repellency after approximately 2 months postfire. They also observed similar variability in water repellency at both the 10-cm scale and watershed scale. Although flashier discharge also occurred at burned sites during the wet season, especially in early season storms, closely spaced storms and above average precipitation diminished the fire-related impact on total discharge as the wet season progressed. Storm runoff and peak discharge in burned compared with unburned sites increased and were greatest during early season storms when enhanced runoff generation occurred. As the winter progressed, closely spaced storms and above average precipitation reduced the fire-related impacts that resulted in significant increases in annual postfire runoff. Other studies conducted in southern California chaparral have found similar patterns (Riggan et al. 1994; Loaiciga et al. 2001). Finally, modifications of hydrological inputs can have long-term impacts on the recovery of native ecosystems. For example, altered water delivery and nitrogen inputs have been shown to influence the recovery of coastal sage scrub plant communities (soft chaparral) following a fire (Kimball et al. 2014).

8.6.2 *Fire Effects on Water Quality*

The impact of fire not only result in decreased ecosystem services through loss of natural stormwater benefits (e.g., decreased water storage capacity, added cost of retaining additional storage), but can decrease water quality. Water quality can be impacted by increased contaminant loadings due to land cover change, in terms of increased erosion and inputs of organic matter, sediments, and solutes can produce contaminant fluxes, including increased metals, polyaromatic hydrocarbons, and nutrients. Fires in coastal watersheds of California have been shown to result in episodic alterations in hydrology and export of solutes and suspended sediment (Coombs and Melack 2013). These increased contaminant loads can increase the costs of additional water treatment (Geiling 2015).

Although there have been numerous studies of the effects of fire on sediment erosion, transport, and deposition (Shakesby and Doerr 2006; Shakesby 2011), few watershed scale studies have determined nutrient export to coastal waters in California or other regions with similar climate and topography (Verkaik et al. 2013). Although impacts generally last less than a year, rainfall can influence the persistence of effects. Because concentrations of nutrients, metals, and certain organic pollutants can be elevated in postfire runoff, the receiving waters downstream of burned areas can also be affected. In southern California, watersheds affected by fire often drain to waterbodies that support sensitive resources or that have been designated as impaired under Section 303(d) of the Clean Water Act. Moreover, the contaminants elevated in postfire runoff are often the same constituents already elevated in the receiving water (Stein and Brown 2009).

Recovery of soil nutrient levels after fire can be fairly slow in some ecosystems, particularly those with limited nitrogen, and in semi-arid regions where decomposition rates are slow (Neary 2004). While more mobile forms of nitrogen (e.g., nitrates and ammonium) are more accessible for plant uptake, they are also more prone to off-site movement through surface runoff and leaching

(Neary et al. 1999). Although this is assumed to have implications for long-term water quality during the recovery process, relatively little is known about the effect of postfire runoff on water quality in chaparral landscapes. This would be a rewarding topic for future investigation.

8.7 Conclusions and Recommendations

The many ecosystem services associated with water provision intimately link human populations and their needs and values to a particular landscape. As human populations increase over the coming decades in chaparral landscapes across the world, managing these ecosystems for water provision services will become increasingly important to prevent both shortages of water and degradation of natural habitats (Millennium Ecosystem Assessment 2003). The variability and unpredictable nature of the annual precipitation and consequently the uncertain supply of fresh water are common features of chaparral dominated landscapes in MTC regions. Although climate change projections from global climate models predict the timing and quantity of precipitation is likely to increase (Shaw et al. 2011), chaparral landscapes will continue to serve an important role in supplying water to human populations. However, these changes suggest that management issues will become even more critical in the future.

Many other provisioning and regulating ecosystem services in chaparral landscapes are strongly linked to the quantity of water delivered by watersheds and how it is allocated among processes of surface water flow, groundwater recharge, evaporation, and transpiration. In these systems, a series of relatively static factors (topography, soils, and geology) and relatively dynamic factors (climate, land management, and land use) interact to determine how this water is partitioned and how it will be delivered (Havstad et al. 2007). Although water yield may be difficult to measure and value, the changing availability of water from chaparral landscapes, such as through water production and filtering, must be evaluated and quantified across multiple spatial scales, including production at the landscape, watershed, and basin level, as well as over long time horizons (Huntsinger and Oviedo 2014).

The value of the unique biological components of river and stream ecosystems cannot be overestimated. With their variable flow regimes and episodic channels, watersheds in chaparral landscapes support a diverse and unique assemblage of native species that evolved under conditions of sequential floods and droughts. However, changes to their fundamental characteristics through incidental or deliberate actions has severely degraded river ecosystems and facilitated both the extirpation of native fauna and the establishment of non-native species (Kondolf et al. 2013). Streams in California and other MTC regions are among the aquatic habitats most altered by human actions, invasions of non-native species, and typically support novel ecosystems, defined as ecosystems dominated by new combinations of organisms in highly altered habitats (Moyle 2014).

To retain delivery of the provisioning services supplied by chaparral landscapes, it will be increasingly necessary to coordinate management practices and interventions across land jurisdictions, property lines, and watershed boundaries (Goldman et al. 2007; Huntsinger et al. 2010; Plieninger et al. 2012; Ferranto et al. 2013). Increased cooperation among local communities, local governance structures, landowners, agencies, and land trusts that own or control access to chaparral habitats to develop collaborative management approaches is a way to accomplish this goal. Moreover, research into the decision making of landowners and managers about land management practices in chaparral landscapes is as important to ecosystem service production, as is further ecological research about the relationships between ecosystem services and land use practices within these landscapes. Building spatial databases and tools with information about ecosystem services, land use change, vegetation, soils, topography, and political and social boundaries, will also be important to the process. Finally, from a policy analysis perspective, chaparral landscapes could benefit from scale appropriate cost-sharing programs, a form of Payment for Ecosystem Services. These include the Environmental Quality Improvement Program that offers incentives for certain management practices by landowners and managers, including managing watersheds and water developments to conserve aquatic species, improve wildlife habitat, and protect water quality. Through these approaches, the sustainability of water provision ecosystem services generated from chaparral landscapes will hopefully be ensured for future generations.

References

Alexander, R. B., R. A. Smith, and G. E. Schwarz. 2000. Effect of stream channel size on the delivery of nitrogen to the Gulf of Mexico. Nature 403:758-761.

Ally, J. R. R. 2004. Survey of selected stream parameters in the East Fork San Gabriel River and its tributary Cattle Canyon, and in the North Fork San Gabriel River. Department of Fish and Game, South Coast Region, Inland Fisheries Unit. Los Alamitos, California, USA.

Ball, J. E., L. A. Bêche, P. K. Mendez, and V. H. Resh. 2013. Biodiversity of Mediterranean-climate streams of California. Hydrobiologia 719:187-213.

Barker, D. T., K. K. Schwall, and L. L. Pardy. 1994. Water quality control plan for the San Diego Basin (9), California Regional Water Quality Control Board San Diego Region, San Diego, California, USA.

Barnett, T. P., J. C. Adam, and D. P. Lettenmaier. 2005. Potential impacts of a warming climate on water availability in snow-dominated regions. Nature 438:303-309.

BBC Research and Consulting. 2010. California outdoor recreation economic study: statewide contributions and benefits. Denver, Colorado, USA.

Benstead, J. P., J. G. March, C. M. Pringle, and F. N Scatena. 1999. Effects of a low-head dam and water abstraction on migratory tropical stream biota. Ecological Applications 9:656-668.

Bernal, S., D. von Schiller, F. Sabater, and E. Martí, 2013. Hydrological extremes modulate nutrient dynamics in Mediterranean climate streams across different spatial scales. Hydrobiologia 719:31-42.

Berrens, R., A. Bohara, C. Silva, M. McKee, and D. Brookshire. 2000. Contingent valuation of instream flows in New Mexico: with tests of scope, group-size reminder and temporal reliability. Journal of Environmental Management 58:73-90.

Berrens, R. P., P. Ganderton, and C. L. Silva. 1996. Valuing protection of minimum instream flows in New Mexico. Journal of Agriculture and Resource Economics 21:294-309.

Bischel, H. N., J. E. Lawrence, B. J. Halaburka, M. H. Plumlee, A. S. Bawazir, J. P. King, J. E. McCray, V. H. Resh, and R. G. Luthy. 2013. Renewing urban streams with recycled water for streamflow augmentation: hydrologic, water quality, and ecosystem services management. Environmental Engineering Science 30:455-479.

Blaine, T. W. and F. R. Lichtkoppler. 2004. Willingness to pay for green space preservation: a comparison of soil and water conservation district clientele and the general public using the contingent valuation method. Journal of Soil and Water Conservation 59:203-208.

Bonada, N. and V. H. Resh. 2013. Mediterranean-climate streams and rivers: geographically separated but ecologically comparable freshwater systems. Hydrobiologia 719:1-29.

Bonada, N., M. Rieradevall, and N. Prat. 2007. Macroinvertebrate community structure and biological traits related to flow permanence in a Mediterranean river network. Hydrobiologia 589:91-106.

Brauman, K. A., G. C. Daily, T. K. Duarte, and H. A. Mooney. 2007. The nature and value of ecosystem services: an overview highlighting hydrologic services. Annual Review of Environmental Resources 32:67-98.

Brown, T. C. 1992. Water for wilderness areas: instream flow needs, protection, and economic value. Rivers 2:311-325.

Bunn, S. E. and A. H. Arthington, 2002. Basic principles and ecological consequences of altered flow regimes for aquatic biodiversity. Environmental Management 30:492-507.

Camarasa-Belmonte, A.M., and F. Segura-Beltrán, 2001. Flood events in Mediterranean ephemeral streams (ramblas) in Valencia region, Spain. Catena 45:229-249.

Cao Y., L. C. van de Werfhorst, B. Sercu, J. L. S. Murray, and P. A. Holden. 2011a. Application of an integrated community analysis approach for microbial source tracking in a coastal creek. Environmental Sciences and Technology 45:7195-7201.

Cao Y., C. H. Wu, G. L Anderson, and P. A. Holden. 2011b. Community analysis-based methods. Pages 251-282 *in* C. Hagedorn, A. R. Blanch, and V. J. Harwood, Microbial source tracking: methods, applications, and case studies. Springer, New York, New York, USA.

Carson, R. T., R. C. Mitchell, M. Hanemann, R. J. Kopp, S. Presser, and P. A. Ruud. 2003. Contingent valuation and lost passive use: damages from the Exxon Valdez oil spill. Environmental and Resource Economics 25:257-286.

Carter, J. L. and V. H. Resh. 2005. Pacific coast rivers of the coterminous United States. Pages 541-590 *in* A. Benke and C. Cushing, editors. The rivers of North America. Elsevier Academic Press, Burlington, Massachusetts, USA.

Cayan, D. R., Maurer, E. P., Dettinger, M. D., M. Tyree, and K. HayHoe. 2008. Climate change scenarios for the California region. Climatic Change 87:21-42.

CBD [Center for Biological Diversity]. 2016. Introduction to the four southern California national forests: Los Padres, Angeles, San Bernardino, Cleveland. Tucson, Arizona, USA. http://www.biologicaldiversity.org/programs/public_lands/forests/southern_california_forests/pdfs/Intro-4-S-CA-National-Forests.pdf

CCI [California Chaparral Institute]. 2010. The California chaparral preservation plan. Escondido, California, USA. http://www.californiachaparral.org/images/NEW_Chaparral_Preservation_Plan.pdf

CDWR [California Department of Water Resources]. 2016. Drought contingency plan for water project operations. Sacramento, California, USA. http://www.water.ca.gov/waterconditions/docs/2016-DroughtContingencyPlan-CVP-SWPOperations-Feb-Nov_1.19.16-FINAL.pdf

CEC [California Energy Commission]. 2005. California's water-energy relationship. CEC-700-2005-011-SF. Sacramento, California, USA. http://www.energy.ca.gov/2005publications/CEC-700-2005-011/CEC-700-2005-011-SF.PDF

Chang, H., and M. R. Bonnette. 2016 Climate change and water-related ecosystem services: impacts of drought in California, USA. Ecosystem Health and Sustainability 2(12):e01254.
Chase, S., R. Benkert, D. Manning, S. White, and S. Brady. 2000. Results of the Sonoma County Water Agency's Mirabel Rubber Dam/Wohler pool reconnaissance fish sampling program. Sonoma County Water Agency, Santa Rosa, California, USA.
Chen, H. and H. Chang. 2014. Response of discharge, TSS, and E. coli to rainfall events in urban, suburban, and rural watersheds. Environmental Science: Processes and Impacts 16:2313-2324.
Chiu, M. C., C. Leigh, R. Mazor, N. Cid, and V. Resh. 2017. Anthropogenic threats to intermittent rivers and ephemeral streams. Pages 433-454 *in* T. Datry, N. Bonada, and A. Boulton, editors. Intermittent rivers: ecology and management. Academic Press, London, UK.
Clar, C. R. 1959. California government and forestry from Spanish days until the creation of the Department of Natural Resources in 1927. California Division of Forestry, Sacramento, California.
Coombs, J. S. and J. M. Melack. 2013. The initial impacts of a wildfire on hydrology and suspended sediment and nutrient export in California chaparral watersheds. Hydrological Processes 26:3842-3851.
Cooper, S. D., T. L. Dudley, and N. Hemphil. 1986. The biology of chaparral streams in southern California. Pages 139-151 *in* J. DeVries, editor. Proceedings of the Chaparral Ecosystem Research Conference, Report No. 62. California Water Resources Center, Davis, California, USA.
Cooper, S. D., P. S. Lake, S. Sabater, J. M. Melack, and J. L. Sabo. 2013. The effects of land use changes on streams and rivers in Mediterranean climates. Hydrobiologia 719:383-425
Costanza, R., R. d'Arge, R. de Groot, S. Faber, M. Grasso, B. Hannon, K. Limburg, S. Naeem, R. V. O'Neill, J. Paruelo, R. G. Raskin, P. Sutton, and M. van den Belt. 1997. The value of the world's ecosystem services and natural capital. Nature 387:253-260.
DeBano, L. F. 1981. Water repellent soils: a state-of-the-art. USDA Forest Service General Technical Report GTR-PSW-46. Pacific Southwest Forest and Range Experimental Station, Berkeley, California, USA.
DeBano, L. F. 2000. The role of fire and soil heating on water repellency in wildland environments: a review. Journal of Hydrology 231-232:195-206.
DeBano, L. F., and C.E. Conrad. 1976. Nutrients lost in debris and runoff water from a burned chaparral watershed. Pages 13-27 *in* Proceedings of the Third Federal Inter-Agency Sedimentation Conference. Water Resource Council, Denver, Colorado, USA.
DeBano, L. F., D. G. Neary, and P. F. Ffolliott. 1998. Fire's effect on ecosystems. John Wiley & Sons, New York, New York, USA.
DEFRA [Department for Environment, Food and Rural Affairs]. 2007. An introductory guide to valuing ecosystem services. Defra Publications, London, UK.
Dettinger, M. D., D. R. Cayan, G. J. McCabe, and J. A. Marengo. 2000. Multiscale streamflow variability associated with El Nino/Southern oscillation. Pages 113-148 *in* H. F. Diaz and V. Markgrarf, editors. El Niño and the Southern Oscillation: multiscale variability and global and regional impacts. Cambridge University Press, Cambridge, UK.
Dodds, W. K., and E. B. Welch. 2000. Establishing nutrient criteria in streams. Journal of the North American Benthological Society 19:186-196.
Duffield, J. W., C. J. Neher, and T. C. Brown. 1992. Recreation benefits of instream flow: application to Montana's Big Hole and Bitterroot rivers. Water Resources Research 28:2169-2181.
Earles, T. A., K. R. Wright, C. Brown, and T. E. Langan. 2004. Los Alamos forest fire impact modeling. Journal of the American Water Resources Association 40:371-384.
ESA [Ecological Society of America]. 2017. Water purification fact sheet for communicating ecosystem services: tools for scientists to engage the public. The Ecological Society of America and the Union of Concerned Scientists. http://www.esa.org/ecoservices/
Equinox Center. 2010. The potential of purified recycled water. H2Overview Project, San Diego, California, USA. https://www.scribd.com/document/34901564/Equinox-Potential-of-Recycled-Water-July-Final-Rev

Erman, D. C., E. D. Andrews, M. Yoder-Williams. 1988. Effects of winter floods on fishes in the Sierra Nevada. Canadian Journal of Fisheries and Aquatic Sciences 45:2195-2200.

Eubanks, T., P. Kerlinger, and R.H. Payne. 1993. High Island, Texas: a case study in avitourism. Birding 25:415-420.

Ferranto, S., L. Huntsinger, C. Getz, M. Lahiff, W. Stewart, G. Nakamura, and M. Kelly. 2013. Management without borders? a survey of landowner practices and attitudes toward cross-boundary cooperation. Society and Natural Resources 26:1086-1100.

Fetscher, A. E., M. A. Sutula, L. B. Busse, and E. D. Stein. 2013. Condition of California perennial, wadeable streams based on algal indicators. Final Technical Report 2007-11 prepared for the California State Water Board. Southern California Coastal Water Research Project and San Diego Regional Water Quality Control Board, Costa Mesa, California, USA.

Ficke, A. D., C. A. Myrick, and L. J. Hansen. 2007. Potential impacts of global climate change on freshwater fisheries. Reviews in Fish Biology and Fisheries 17:581-613.

Filipe, A. F., N. Bonada, and J. E. Lawrence. 2013. Vulnerability of stream biota to climate change in Mediterranean climate regions: a synthesis of ecological responses and conservation challenges. Hydrobiologia 719:331-351.

Foley, J. A., R. DeFries, G. P. Asner, C. Barford, G. Bonan, S. R. Carpenter, F. S. Chapin, M. T. Coe, G. C. Daily, H. K. Gibbs, J. H. Helkowski, T. Holloway, E. A. Howard, C. J. Kucharik, C. Monfreda, J. A. Patz, C. Prentice, N. Ramankutty, and P. K. Snyder. 2005. Global Consequences of Land Use. Science 309:570-574.

Gabet, E. J. 2003. Sediment transport by dry ravel. Journal of Geophysical Research 108:2049.

Gallart, G., F. Gallego, J. Latrón, P. Llorens, G. G. Barberá, J. Froebrich. 2012. A novel approach to analysing the regimes of temporary streams in relation to their controls on the composition and structure of aquatic biota. Hydrology and Earth System Sciences 16:3165-3182.

Gasith, A., and V. H. Resh. 1999. Streams in Mediterranean climate regions: abiotic influences and biotic responses to predictable seasonal events. Annual Review of Ecology and Systematics 30:51-81.

GEI Consultants/Navigant Consulting. 2010. Embedded energy in water studies-study 1: statewide and regional water-energy relationship, San Francisco, California, USA, Prepared for the California Public Utilities Commission, Energy Division.

Geiling, S. 2015. The west is literally on fire, and the impacts could be widespread. ThinkProgress. https://thinkprogress.org/the-west-is-literally-on-fire-and-the-impacts-could-be-widespread-b93b4c9c38a

Geosyntec Consultants. 2015. Stormwater capture master plan. Prepared for the Los Angeles Department of Water and Power in partnership with TreePeople, Los Angeles, California, USA. https://www.ladwp.com/cs/idcplg?IdcService=GET_FILE&dDocName=OPLADWPCCB421767&RevisionSelectionMethod=LatestReleased

Gleick, P. H. 1993. Water in crisis: a guide to the world's fresh water resources. Oxford University Press, New York, New York, USA.

Goldman, R. L., B. H. Thompson, and G. C. Daily. 2007. Institutional incentives for managing the landscape: inducing cooperation for the production of ecosystem services. Ecological Economics 64:333-343.

Goodrich, D. C., R. L. Scott, J. Qi, B. Goff, C. L. Unkrich, M. S. Moran, D. Williams, S. Schaeffer, K. Snydery, R. MacNish, T. Maddock, D. Pool, A. Chehbouni, D. I. Cooper, W. E. Eichinger, W. J. Shuttleworth, Y. Kerr, R. Marsett, and W. Ni. 2000. Seasonal estimates of riparian evapotranspiration using remote and *in situ* measurements. Agricultural and Forest Meteorology 105:281-309.

Goodridge, B. M. and J. M. Melack. 2012. Land use control of stream nitrate concentrations in mountainous coastal California watersheds. Journal of Geophysical Research 117:G02005.

Gorn, D. 2016. Desalination's future in California is clouded by cost and controversy. Drought Watch, KQED Science, Public Media for Northern California, San Francisco, California, USA. https://ww2.kqed.org/science/2016/10/31/desalination-why-tapping-sea-water-has-slowed-to-a-trickle-in-california/

Grantham T. E., R. Figueroa, and N. Prat. 2013. Water management in Mediterranean river basins: a comparison of management frameworks, physical impacts, and ecological responses. Hydrobiologia 719:451-482.

Grantham, T. E., A. M. Merenlender, and V. H. Resh, 2010. Climatic influences and anthropogenic stressors: an integrated framework for streamflow management in Mediterranean-climate California, USA. Freshwater Biology 55:188-204.

Grantham T. E., and J. H. Viers. 2014. 100 years of California's water rights system: patterns, trends and uncertainty. Environmental Research Letters 9:084012.

Grizzetti, B., D. Lanzanova, C. Liquete, A. Reynaud, and A. C. Cardoso. 2016. Assessing water ecosystem services for water resource management. Environmental Science and Policy 61:194-203.

Hagedorn C., A. R. Blanch, and V. J. Harwood. 2011. Microbial source tracking: methods, applications and case studies. Springer, New York, New York, USA.

Harmens, H., and G. Mills. 2012. Ozone pollution: impacts on carbon sequestration in Europe. ICP Vegetation Programme Coordination Centre. Centre for Ecology and Hydrology. Bangor, UK.

Havstad, K. M., D. P. C. Peters, R. Skaggs, J. Brown, B. Bestelmeyer, E. Fredrickson, J. Herrick, and J. Wright. 2007. Ecological services to and from rangelands of the United States. Ecological Economics 64:261-268.

He, C., S. B. Malcom, K. S. Dahlberg, and B. Fu. 2000. A conceptual framework for integrating hydrological and biological indicators into watershed management. Landscape and Urban Planning 49:25-34.

Heiskary, S. A., and R. W. Bouchard Jr. 2015. Development of eutrophication criteria for Minnesota streams and rivers using multiple lines of evidence Freshwater Science 34:574-592.

Hibbert, A. R. 1971. Increases in streamflow after converting chaparral to grass. Water Resources Research 7:71-80.

Holmes, T. P., J. C. Bergstrom, E. Huszar, S. B. Kask, F. Orr III.2004. Contingent valuation, net marginal benefits, and the scale of riparian ecosystem restoration. Ecological Economics 49: 19-30.

Howarth, R. W., G. Billen, D. Swaney, A. Townsend, N. Jaworski, K, Lajtha, A. Downing, R. Elmgreen, N. Caraco, T. Jordan, F. Berendse, J. Freney, V. Kudeyarov, P. Murdoch, and Z. Zhao-Liang. 1996. Regional nitrogen budgets and riverine N and P fluxes for the drainages to the North Atlantic Ocean: Natural and human influences. Biogeochemistry 35:181-226.

Howitt, R., D. MacEwan, J. Medellin-Azuara, J. R. Lund, D. A. Sumner. 2015. Preliminary analysis: 2015 drought economic impact study. Center for Watershed Sciences, University of California, Davis, California, USA.

Hubbert, K. R. and V. Oriol. 2005. Temporal fluctuations in soil water repellency following wildfire in chaparral steeplands, southern California. International Journal of Wildland Fire 14: 439-447.

Hubbert, K. R., H. K. Preisler, P. M. Wohlgemuth, R. C. Graham, and M. G. Narog. 2006. Prescribed burning effects on soil physical properties and soil water repellency in a steep chaparral watershed, southern California, USA. Geoderma 130:284-298

Huber-Sannwald, E., F. T. Maestre, J. E. Herrick, J. F. Reynolds. 2006. Ecohydrological feedbacks and linkages associated with land degradation: a case study from Mexico. Hydrological Processes 20:3395-3411.

Huntsinger, L., and J. L. Oviedo. 2014. Ecosystem services are social-ecological services in a traditional pastoral system: the case in California's Mediterranean rangelands. Ecology and Society 19:8.

Huntsinger, L., M. Johnson, M. Stafford, and J. Fried. 2010. Hardwood rangeland landowners in California from 1985 to 2004: production, ecosystem services, and permanence. Rangeland Ecology and Management 63:324-334.

Jamison, P. 2015. Army Corps Board Approves $1.3 billion LA River Restoration Proposal. Los Angeles Times, Los Angeles, California, USA. http://www.latimes.com/local/lanow/la-me-ln-army-corps-l-a-river-restoration-20150716-story.html

Jiménez Cisneros, B. E., T. Oki, N.W. Arnell, G. Benito, J. G. Cogley, P. Döll, T. Jiang, and S. S. Mwakalila. 2014. Freshwater resources. Pages 229-269 *in* C. B. Field, V. R. Barros, D. J.

Dokken, K. J. Mach, M. D. Mastrandrea, T. E. Bilir, M. Chatterjee, K. L. Ebi, Y. O. Estrada, R. C. Genova, B. Girma, E. S. Kissel, A. N. Levy, S. MacCracken, P. R. Mastrandrea, and L. White, editors. Climate change 2014: impacts, adaptation, and vulnerability. Part A: Global and sectoral aspects. Contribution of Working Group II to the Fifth Assessment Report of the Intergovernmental Panel on Climate Change, Cambridge, UK and New York, New York, USA, Cambridge University Press.

Johnston, R. J., E. Y. Besedin, R. Iovanna, C. J. Miller, R. F. Wardwell, and M. H. Ranson. 2005. Systematic variation in willingness to pay for aquatic resource improvements and implications for benefit transfer: a meta-analysis. Canadian Journal of Agricultural Economics 53:221-248.

Jones, J. A., and D. A. Post. 2004. Seasonal and successional streamflow response to forest cutting and regrowth in the northwest and eastern United States. Water Resources Research 40:W05203.

Jones, M. L. M., A. Provins, L. Harper-Simmonds, M. Holland, G. Mills, F. Hayes, B. A. Emmett, J. Hall, L. G. Sheppard, R. Smith, M. Sutton, K. Hicks, M. Ashmore, R. Haines-Young. 2012. Using the ecosystems services approach to value air quality. NE0117. Department for Environment, Food and Rural Affairs, Bangor, UK.

Jones, L., A. Provins, L. Harper-Simmonds, M. Holland, G. Mills, F. Hayes, B. A. Emmett, J. Hall, L. G. Sheppard, R. Smith, M. Sutton, K. Hicks, M. Ashmore, R. Haines-Young. 2014. A review and application of the evidence for nitrogen impacts on ecosystem services. Ecosystem Services 7:76-88.

Katz, J., P. B. Moyle, R. M. Quinones, J. Israel, and S. Purdy. 2013. Impending extinction of salmon, steelhead, and trout (Salmonidae) in California. Environmental Biology of Fishes 96:1169-1186.

Keeley, J. E., and T. J. Brennan. 2012. Fire-driven alien invasion in a fire-adapted ecosystem. Oecologia 169:1043-1052.

Keeley, J. E., and P. H. Zedler. 2009. Large, high-intensity fire events in southern California shrublands: debunking the fine-grain age patch model. Ecological Applications 19:69-94.

Kenny, J. F., N. L. Barber, S. S. Hutson, K. S. Linsey, J. K. Lovelace, and M. A. Maupin. 2009. Estimated use of water in the United States in 2005. US Geological Survey Circular 1344. Reston, Virginia, USA.

Kimball, S., M. L. Goulden, K. N. Suding, and S. Parker. 2014. Altered water and nitrogen input shifts succession in a southern California coastal sage community. Ecological Applications 24:1390-1404.

Kinney, A. 1900. Forest and water. Post Publishing, Los Angeles, California, USA.

Klein, C., F. Cheever, B. C. Birdsong. 2013. Natural resources law: a place-based book of problems and cases, third edition (Aspen Casebook), New York, New York, USA, Wolters Kluwer Law and Business.

Kondolf, G. M. and R. J. Batalla. 2005. Hydrological effects of dams and water diversions on rivers of Mediterranean-climate regions: examples from California. Developments in Earth Surface Processes 7:197-211.

Kondolf, G. M., K. Podolak and T. E. Grantham. 2013. Restoring mediterranean-climate rivers. Hydrobiologia 719:527-545.

Konrad, C. P. and D. B. Booth. 2005. Hydrologic changes in urban streams and their ecological significance. American Fisheries Society Symposium 47:157-177.

Krause, A., 2011. Mechanical extraction and augmentation of coarse and fine sediment on the Trinity River, California, 1950-2011. TR-TRRP-2011-3, Weaverville, California, USA, Trinity River Restoration Program.

Kreye, M. M., D. C. Adams, and F. J. Escobedo. 2014. The value of forest conservation for water quality protection. Forests 5:862-884.

LACSD [Sanitations Districts of Los Angeles County]. 2017. Wastewater treatment and water reclamation. Whittier, California, USA. http://www.lacsd.org/wastewater/wwfacilities/moresanj.asp

Leggett, C. G., and N. E. Bockstael. 2000. Evidence of the effects of water quality on residential land prices. Journal of Environmental Economics and Management 39:121-144.

Leones, J., B. Colby, D. Cory, and L. Ryan. 1997. Measuring regional economic impacts of streamflow depletions. Water Resources Research 33:831-838.

Lippitt, C. L., D. A Stow, J. F. O'Leary, J. Franklin. 2013. Influence of short-interval fire occurrence on post-fire recovery of fire-prone shrublands in California, USA. International Journal of Wildland Fire 22:184-193.

Llasat, M. C., M. Llasat-Botija, A. Rodriguez, and S. Lindbergh, 2010. Flash floods in Catalonia: a recurrent situation. Advances in Geosciences 26:105-111.

Loaiciga, H. A., D. Pedreros, and D. Roberts. 2001. Wildfire-streamflow interaction in a chaparral watershed. Advances in Environmental Research 5:295-305.

Loomis, J. 1987. The economic value of instream flow: methodology and benefit estimates of optimum flow. Journal of Environmental Management 24:169-179.

Loomis, J. 1998. Estimating the public's values for instream flow: economic techniques and dollar values. Journal of the American Water Resources Association 34:1007-1014.

Loomis, J. and M. D. Creel. 1992. Recreation benefits of increased flows in California's San Joaquin and Stanislaus Rivers. Rivers 3:1-13.

Matamoros, V., C. Arias, H. Brix and J. Bayona. 2007. Removal of pharmaceuticals and personal care products (PPCPs) from urban wastewater in a pilot vertical flow constructed wetland and a sand filter. Environmental Science and Technology 41:8171-8177.

Maupin, M. A., J. F Kenny, S. S. Hutson, J. K. Lovelace, N. L. Barber, and K. S. Linsey. 2014. Estimated use of water in the United States in 2010. US Geological Survey Circular 1405. Reston, Virginia, USA.

Mazor, R. D. 2015. Bioassessment of perennial streams in southern California: a report on the first five years of the Stormwater Monitoring Coalition's regional stream survey. Technical Report 844. Southern California Coastal Water Research Project, Costa Mesa, California, USA.

McNeil, B. E., J. M. Read, T. J. Sullivan, T. C. McDonnell, I. J. Fernandez, and C. T. Driscoll. 2008. The spatial pattern of nitrogen cycling in the Adirondack Park, New York, Ecological Applications 18:438-452.

Millennium Ecosystem Assessment. 2003 Ecosystems and human well-being: a framework for assessment. Island Press, Washington, D.C., USA.

Merenlender, A. M., and M. K. Matella 2013. Maintaining and restoring hydrologic habitat connectivity in Mediterranean streams: an integrated modeling framework. Hydrobiologia 719:509-525.

Miller, E. K., A. VanArsdale, G. J. Keeler, A. Chalmers, L. Poissant, N. Kamman, and R. Brulotte. 2005. Estimation and mapping of wet and dry mercury deposition across Northeastern North America. Ecotoxicology 14:53-70.

Mills, G. and H. Harmens. 2011. Ozone pollution: a hidden threat to food security. ICP Vegetation Programme Coordination Centre, Centre for Ecology and Hydrology. Bangor, UK.

Mills, G., S. Wagg, and H. Harmens. 2013. Ozone pollution: impacts on ecosystem services and biodiversity. ICP Vegetation Programme Coordination Centre, Centre for Ecology and Hydrology, Bangor, UK.

Miltner, R. J., D. White, and C. Yoder. 2004. The biotic integrity of streams in urban and suburbanizing landscapes Landscape and Urban Planning 69:87-100.

Mount, J. F., 1995. California rivers and streams: the conflict between fluvial process and land use. University of California Press, Berkeley, California, USA.

Moyle, P. B. 2014. Novel aquatic ecosystems: the new reality for streams in California and other Mediterranean climate regions. River Research and Applications. 30:1335-1344.

Moyle, P. B., J. V. E. Katz and R. M. Quiñones, 2011. Rapid decline of California's native inland fishes: a status assessment. Biological Conservation 144:2414-2423.

Muñoz-Mas, R., A. Lopez-Nicolas, F. Martinez-Capel, and M. Pulido-Velazquez. 2016. Shifts in the suitable habitat available for brown trout (*Salmo trutta* L.) under short-term climate change scenarios. Science of the Total Environment 544:686-700.

Naiman, R. J., J. J. Magnuson, D. M. McKnight, J. A. Stanford, and J. R. Karr. 1995. Freshwater ecosystems and their management: a national initiative. Science 270:584-585.

Neary, D. G. 2004. An overview of fire effects on soils. Southwest Hydrology 3:18-19.
Neary, D. G., C. C. Klopatek, L. F. DeBano, and P. F. Ffolliott. 1999. Fire effects on belowground sustainability: a review and synthesis. Forest Ecology and Management 122:51-71.
NRC [National Research Council]. 2012. Water reuse: potential for expanding the nation's water supply through reuse of municipal wastewater, The National Academic Press, Washington, D.C., USA.
Null, S. E., J. H. Viers, M. L. Deas, S. K. Tanaka, and J. F. Mount. 2012. Stream temperature sensitivity to climate warming in California's Sierra Nevada: impacts to cold water habitat. Climatic Change 116:149-170.
Null, S. E., and J. H. Viers. 2013. In bad waters: water year classification in nonstationary climates. Water Resources Research 49:1137-1148.
Pejchar, L., and K. Warner. 2001. A river might run through it again: criteria for consideration of dam removal and interim lessons from California. Environmental Management 28:561-575.
Plieninger T., S. Ferranto. L. Huntsinger. M. Kelly, and C. Getz. 2012. Appreciation, use, and management of biodiversity and ecosystem services in California's working landscapes. Environmental Management 50:427-440.
Postel, S., and S. Carpenter. 1997. Freshwater ecosystem services. Pages 195-214 *in* G. C. Daily, editor. Nature's services. Island Press, Washington, D.C., USA.
Prat, N., and J. V. Ward. 1994. The tamed river. Pages 219-236 *in* R. Margalef, editor. Limnology now: a paradigm of planetary problems. Elsevier, London, UK.
Pringle, C. 2003. What is hydrologic connectivity and why is it ecologically important? Hydrological Processes 17:2685-2689.
Purcell, A. H., C. Friedrich, and V. H. Resh. 2002. An assessment of a small, urban stream restoration project in Northern California. Restoration Ecology 10:685-694.
Reid, W. V. 2001. Capturing the value of ecosystem services to protect biodiversity. Pages 197-225 *in* V. C. Hollowell, editor. Managing human dominated ecosystems. Missouri Botanical Garden, St. Louis, Missouri, USA.
Resh, V. H., L. A. Bêche, J. E. Lawrence, R. P. Mazor, E. P. McElravy, A. P. O'Dowd, D. Rudnick, and S. M. Carlson. 2013. Long-term population and community patterns of benthic macroinvertebrates and fishes in Northern California Mediterranean-climate streams. Hydrobiologia 719:93-118.
Rice, R. M. 1974. The hydrology of chaparral watersheds. Pages 27-34 *in* M. Rosenthal, editor. Proceedings of the symposium on Living with the Chaparral. Sierra Club, San Francisco, California, USA.
Riggan, P. J., S. E. Franklin, J. A. Brass, and F. E. Brooks. 1994. Perspectives on fire management in Mediterranean ecosystems of southern California. Pages 140-162 *in* J. Moreno and W. Oechel, editors. Fire and global change in Mediterranean ecosystems. Ecological studies 107. Springer-Verlag, New York, New York, USA.
Rowe, P. B., O. M. Countryman, and H. C. Storey. 1954. Hydrologic analysis used to determine effects of fire on peak discharge and erosion rates in southern California watersheds. US Forest Service, California Forest and Range Experiment Station, Berkeley, California, USA.
Sagoff, M. 2002. On the value of natural ecosystems: the Catskills parable. Politics and the Life Sciences 21:16-21.
Sahgun, L. 2014. Army Corps to recommend $1-billion LA River project. Los Angeles Times, Los Angeles, California, USA. http://www.latimes.com/science/la-me-la-river-approval-20140529-story.html
Schlesinger, W. H., T. J. Ward, and J. Anderson. 2000. Nutrient losses in runoff from grassland and shrubland habitats in southern New Mexico: II. Field plots. Biogeochemistry 49:69-86.
Sercu, B., L. C. van de Werfhorst, J. Murray, and P. A. Holden. 2009. Storm drains are sources of human fecal pollution during dry weather in three urban southern California watersheds. Environmental Science and Technology 43:293-298.
Shakesby, R. A. 2011. Post-wildfire soil erosion in the Mediterranean: review and future research directions Earth Science Reviews 105:71-100.

Shakesby, R. A. and S. H. Doerr. 2006. Wildfire as a hydrological and geomorphological agent. Earth Science Reviews 74:269-307.

Shaw, M. R., L. Pendleton, D. Cameron, B. Morris, G. Bratman, D. Bachelet, K. Klausmeyer, J. MacKenzie, D. Conklin, J. Lenihan, E. Haunreiter, C. Daly, and R. Roehrdanz. 2011. The impact of climate change on California's ecosystem services. Climatic Change 109:465-484.

Shrestha, R. K., and J. R. R Alavalapati. 2004. Valuing environmental benefits of silvopasture practice: a case study of the Lake Okeechobee watershed in Florida. Ecological Economics 49:349-359.

Singh, V. P. 1995. Computer models of watershed hydrology. Pages 1-22 *in* V. P. Singh, editor. Watershed modeling. Water Resources Publications, Highlands Ranch, Colorado, USA.

Smakhtin, V. U. 2001. Low flow hydrology: a review. Journal of Hydrology 240:147-186.

Smith, V. H. 2003. Eutrophication of freshwater and coastal marine ecosystems: a global problem. Environmental Science and Pollution Research 10:126-39.

Smith, H., P. J. Wood, and J. Gunn. 2003. The influence of habitat structure and flow permanence on invertebrate communities in karst spring systems. Hydrobiologia 510:53-66.

Solek, C. W. 2008. Ecology of the San Gabriel River catchment: anthropogenic effects on black flies (Diptera: Simuliidae) and other benthic macroinvertebrates. Dissertation. University of California, Berkeley, California, USA.

Stein, E. D. and J. Brown. 2009. Effects of post-fire runoff on surface water quality: Development of a southern California regional monitoring program with management questions and implementation recommendations. Technical Report 598. Southern California Coastal Water Research Project, Costa Mesa, California, USA.

Stella, J. C., P. M. Rodríguez-González, S. Dufour, and J. Bendix. 2013. Riparian vegetation research in Mediterranean-climate regions: common patterns, ecological processes, and considerations for management. Hydrobiologia 719:291-315.

Stokes, J. R. and A. Horvath. 2009. Energy and air emission effects of water supply. Environmental Science and Technology 43:2680-2687.

Stoof, C. R., E. C. Slingerland, W. Mol, J. van den Berg, P. J. Vermeulen, A. J. D. Ferreira, C. J. Ritsema, J. Y. Parlange, and T. S. Steenhuis. 2014. Preferential flow as a potential mechanism for fire-induced increase in streamflow. Water Resources Research 50:1840-1845.

Tague, C., L. Seaby, and A. Hope. 2009. Modeling the eco-hydrologic response of a Mediterranean type ecosystem to the combined impacts of projected climate change and altered fire frequencies. Climatic Change 93:137-155.

Thurow, T.L. 1991. Hydrology and erosion. Pages 141-159 *in* R. K. Heitschmidt and J. W. Stuth, editors. Grazing management: an ecological perspective. Timber Press, Portland, Oregon, USA.

Tiedemann, A. R., C. E. Conrad, J. H. Dieterich, J. W. Hornbeck, W. F. Megahan, L. A. Viereck, and D. D. Wade. 1979. Effects of fire on water: a state-of-knowledge review. USDA Forest Service General Technical Report GTR-WO-10. US Government Printing Office, Washington, D.C., USA.

USEPA [United States Environmental Protection Agency]. 1999. Giardia: drinking water health advisory. EPA-822-R-99-008. Office of Science and Technology, Office of Water, Washington D.C., USA.

USEPA [United States Environmental Protection Agency]. 2002. Evaluation of impacts to underground sources of drinking water by hydraulic fracturing of coalbed methane reservoirs. EPA 816-R-04-003. Office of Water, Office of Ground Water and Drinking Water, Drinking Water Protection Division, Prevention Branch, Washington D.C., USA.

USEPA [United States Environmental Protection Agency]. 2016. Water quality. Pacific Southwest, Region 9, Los Angeles, California, USA. https://archive.epa.gov/region9/socal/web/html/index-9.html

USEPA [United States Environmental Protection Agency]. 2017. Wetland functions and values. Online training module. Washington D.C, USA. https://cfpub.epa.gov/watertrain/moduleFrame.cfm?parent_object_id=285

USFWS [United States Fish and Wildlife Service] and Hoopa Valley Tribe. 1999. Trinity River flow evaluation final report. Arcata, California, USA.
USFS [United States Forest Service]. 2012. Improving recreation opportunities. http://www.funoutdoors.com/files/USFS%20Recreation%20Strategy%20Final%20PDF.pdf
Verkaik, I., M. Rieradevall, S. D. Cooper, J. M. Melack, T. L. Dudley, and N. Prat. 2013. Fire as a disturbance in Mediterranean climate streams. Hydrobiologia 719:353-382.
Vicuna, S., E. P. Maurer, B. Joyce, J. A. Dracup, and D. Purkey. 2007. The sensitivity of California water resources to climate change scenarios. Journal of the American Water Resources Association 43:482-498.
Vidal-Abarca, M. R. 1990. Los ríos de las cuencas áridas y semiáridas: una perspectiva ecológica comparativa y de síntesis. Scientia gerundensis 16:219-228.
Ward, F. A. 1987. Economics of water allocation to instream uses in a fully appropriated river basin: evidence from a New Mexico wild river. Water Resources Research 23:381-392.
Weatherdon, L. V., A. K. Magnan, A. D. Rogers, U. Rashid Sumaila, and W. W. L. Cheung. 2016. Observed and projected impacts of climate change on marine fisheries, aquaculture, coastal tourism, and human health: an update. Frontiers in Marine Science 3:48.
Weber, M. A., and R. P. Berrens. 2006. Value of instream recreation in the Sonoran Desert. Journal of Water Resources Planning and Management 132:53-60.
Wells, W. G. 1987. The effects of fire on the generation of debris flows in southern California. Geology 7:105-114.
Wells, C. G., R. E. Campbell, L. F. DeBano, C. E. Lewis, R. L. Fredriksen, E. C. Franklin, R. C. Froelich, and P. H. Dunn. 1979. Effects of fire on soil: a state of knowledge review. USDA Forest Service General Technical Report GTR-WO-7. Washington, D.C., USA, US Government Printing Office.
White, J., M. Belmont, and C. Metcalfe. 2006. Pharmaceutical compounds in wastewater: wetland treatment as a potential solution. The Scientific World Journal 6:1731-1736.
Whitehead, J. 1990. Measuring willingness-to-pay for wetlands preservation with the contingent valuation method. Wetlands 10:187-201.
Whitehead, P. G., R. L. Wilby, R. W. Batterbee, M. Kernan, and A. J. Wade. 2009. A review of the potential impacts of climate change on surface water quality. Hydrological Sciences Journal 54:101-123.
Wilcox, B. P., and T. L. Thurow. 2006. Emerging issues in rangeland ecohydrology: vegetation change and the water cycle. Rangeland Ecology and Management 59:220-224.
Wilkinson, R. 2002. The potential consequences of climate variability and change for California. A report of the California Regional Assessment for the United States Global Change Research Program. Washington D.C., USA. http://www.ncgia.ucsb.edu/pubs/CA_Report.pdf
WIW [Water in the West]. 2013. Water and energy nexus: a literature review, Stanford, California, USA, Stanford Woods Institute for the Environment and Bill Lane Center for the American West.
Wolff, G., R. Cohen, and B. Nelson. 2004. Energy down the drain: the hidden costs of California's Water Supply. Natural Resources Defense Council. Pacific Institute. Oakland, California, USA. https://www.nrdc.org/sites/default/files/edrain.pdf

Chapter 9
Mapping the Value of National Forest Landscapes for Ecosystem Service Provision

Emma C. Underwood, Allan D. Hollander, Patrick R. Huber, and Charlie Schrader-Patton

Abstract Natural landscapes provide ecosystem services that are critical to human health and society. However, as landscapes are threatened by urban development, climate change, intensive agriculture, and altered fire regimes, this negatively affects the condition of natural ecosystems and reduces the provision of these services for which there may not be viable alternatives. We report on a project to map the value of national forest lands in southern California for ecosystem service provision. Our focus is on quantifying five types of ecosystem services, water runoff, groundwater recharge, sediment erosion retention, carbon storage, and biodiversity, across an area that encompasses the four southern national forests—the Angeles, Los Padres, San Bernardino, and Cleveland. We first develop environmentally and climatically driven ecological units as a practical way to summarize information on services for resource managers. Second, we map the spatial distribution of the five services under current climate conditions and assess the spatial concordance between the five services. Third, using a conceptually straightforward approach, we prioritize the ecological units that provide the highest amount of each service and identify hotspots of ecosystem services where ecological units contain multiple services. By providing results to resource managers and the means to access and query the data, information can be used to guide decision making (e.g., prioritizing areas for conservation and restoration activities), assess the impacts of proposed activities (e.g., the impact of fuel management on ecosystem service provision), provide estimates

E. C. Underwood (✉)
University of California, Davis, CA, USA

University of Southampton, UK
e-mail: eunderwoodrussell@ucdavis.edu

A. D. Hollander · P. R. Huber
University of California, Davis, CA, USA

C. Schrader-Patton
RedCastle Resources, Inc. and USDA Forest Service, Western Wildland Environmental Threat Assessment Center, Pineville, OR, USA

E. C. Underwood et al. (eds.), *Valuing Chaparral*, Springer Series on Environmental Management, https://doi.org/10.1007/978-3-319-68303-4_9

of the value of chaparral shrublands for fire damage assessments, and help secure the long-term provision of ecosystem services across the landscape. The maps and data generated in this project provide the foundation for the next step that is calculating the economic value of these services.

Keywords Biodiversity · Carbon storage · Groundwater recharge · Hotspots · Landsat EVI · Sediment erosion retention · Water runoff

9.1 Introduction

Functioning ecosystems provide a suite of natural assets, or ecosystem services, that are critical to human health and society and have a distinct set of beneficiaries. The Millennium Ecosystem Assessment underscores the importance of these services and highlights the threats posed to their supply from unsustainable anthropogenic activities (Millennium Ecosystem Assessment 2005). As natural landscapes experience urbanization and intensive agricultural development, climate change, altered fire regimes, and invasion of non-native plants this negatively affects the condition of the ecosystems they support. This leads to reduced provision of the services for which substitutes are costly or completely unavailable (Benayas et al. 2009; Bullock et al. 2011; Shaw et al. 2011). Understanding the quantity and spatial distribution of ecosystem services, along with their associated economic value, is a fundamental first step to developing effective means of ensuring the long-term provision of services by natural ecosystems.

Attention on ecosystem services from a research and policy perspective has increased substantially over the last two decades (Daily 1997; Sachs and Reid 2006), as evidenced by the numerous studies that evaluate ecosystem services at regional (Chan et al. 2006; Nelson et al. 2009; Polasky et al. 2011), national (Egoh et al. 2008), and global scales (Costanza et al. 1997; Naidoo et al. 2008). Examples of ecosystem service evaluations on federal lands in the USA that focus on quantifying the benefits of forested landscapes include a collaborative project with multiple stakeholders in the Deschutes National Forest, Oregon (Smith et al. 2011) and a monetary valuation of the Tolt River watershed in Washington (Batker 2005). Accompanying the proliferation of ecosystem services studies is the development of software and computing tools to systematize spatial evaluations and produce maps of ecosystem services across the landscape, including InVEST (Integrated Valuation of Ecosystem Services and Trade-Offs) (Tallis et al. 2014), the US Geological Survey's SolVES (Social Values for Ecosystem Services, https://solves.cr.usgs.gov/), and ARIES (Artificial Intelligence for Ecosystem Services, http://aries.integratedmodelling.org/).

Given the available techniques and tools for assessing the spatial distribution of ecosystem services, there is great potential for providing important inputs for resource management. In terms of planning activities, data on ecosystem services can allow short-term costs of specific management actions to be weighed against long-term benefits provided by ecosystem services (Benayas et al. 2009). For example, the costs of restoration activities are viewable in conjunction with the economic value of services such as carbon storage and sediment erosion retention as they recover in parallel with restored shrubland vegetation. Data on ecosystem services can also highlight any unintended negative (or positive) consequences of management actions, such as quantifying the loss of aesthetic or recreational value associated with creating fire breaks. Information on the provision of ecosystem services and their value can help prioritize the spatial location of management actions, e.g., native revegetation in areas that will both restore habitat for sensitive species and reduce sediment erosion. Finally, data on ecosystem services can assist in providing science-based estimates of the broader impact of wildfires on ecosystem services, which is particularly valuable in non-forested landscapes where timber values cannot be used to estimate fire damage.

In mapping ecosystem services and using this information in resource management, one challenge is the variation in spatial distribution and abundance of different services across the landscape (Egoh et al. 2008; Raudsepp-Hearne et al. 2010). Many recent studies routinely include an examination of the spatial concordance between ecosystem services. A particular focus from an academic and practical interest is evaluating the overlap between ecosystem services and biodiversity, although evidence of overlap is inconsistent. For example in the Willamette Basin, Oregon, Nelson et al. (2009) evaluated the impacts of different land use scenarios on biodiversity and ecosystem services, finding scenarios that enhanced biodiversity conservation also benefit the production of services. In contrast, Chan et al. (2006) in the central coast of California, and Naidoo et al. (2008) using global ecoregions, both find weak relationships between ecosystem services and biodiversity. Another technique often integrated into mapping studies is to identify hotspots or bundles of ecosystem services across the landscape. Identifying bundles of multiple ecosystem services may be important for identifying and prioritizing conservation areas (Gos and Lavorel 2012; Schroter and Remme 2016). One caveat, however, when combining maps and values of multiple ecosystem services is the need to recognize that the benefits associated with each ecosystem service span different spatial scales, which may be challenging to convey in the final integrated map. For example, biodiversity benefits are global as well as local, while sediment erosion retention benefits relatively localized populations in downstream floodplains.

9.2 The Relevance of an Ecosystem Services Project in Southern California

Developing a framework for ecosystem services in southern California is of key importance given the extensive area that Mediterranean-type shrubland covers and the close proximity of urban centers and the people who benefit from these ecosystem services (22 million people are estimated to live in the seven southern counties). Some of the ecosystem services provided by intact shrubland include high levels of plant richness and endemism (Olson and Dinerstein 2002; Burge et al. 2016; see Chaps. 1 and 2), the retention of sediment thereby preserving soil productivity and protecting communities downstream (see Chap. 7), water provision for surrounding populations, and potentially significant contributions to global carbon storage (see Chaps. 1 and 6). However, southern California is a region experiencing enormous threats. These include rapidly growing urban centers, high levels of air pollution, agricultural intensification, and drastic alteration of pre-EuroAmerican settlement fire regimes that is causing type-conversion from shrub to non-native grasses (Haidinger and Keeley 1993; Keeley 2005; Safford 2007; see Chap. 12). Fire activity is also generally expected to increase under future climates (see Chap. 14), in part due to increased growth of fuels under higher CO_2 and decreased fuel moisture from warmer temperatures (Oechel et al. 1995; Westerling and Bryant 2006; Lenihan et al. 2008), but see Batllori et al. (2013). Climate change will also affect the timing and distribution of water runoff and recharge (see Chap. 14), rates of carbon sequestration, and the distribution of species. Developing a suite of data layers on the provision of ecosystem services across the landscape provides additional information for planning, prioritizing, and decision making in relation to resource management under these existing and impending threats.

This chapter reports on work undertaken in conjunction with the US Forest Service (USFS) Pacific Southwest Region to develop a framework to quantify and economically value ecosystem services in southern California's chaparral dominated ecosystems. At the outset of the project, discussions with USFS staff and partners identified priority ecosystem services in the region. Six of these, achievable in the project timeframe of 3 years, became the focal services—carbon storage, water runoff, groundwater recharge, sediment erosion retention, biodiversity, and recreation.

Here we describe the process for mapping, quantifying, and assessing patterns in five of these six services (see Chap. 10 for details relating to recreation services). We first create environmentally- and climatically-based ecological units across the landscape as practical reporting units for resource managers. Second, we map and quantify the distribution of the ecosystem services by these ecological units and examine the spatial concordance between services (e.g., are areas that are important for water runoff also important for carbon storage?). Third, we used a straightforward approach to identify priority ecological units within each service, and then prioritize units across multiple services to identify hotspots of ecosystem service provision. A longer term component of this project is to develop an online mapping tool that will integrate spatial data on ecosystem services for resource managers to utilize in routine decision making and planning, thereby allowing users to query and prioritize different values of services depending on the context.

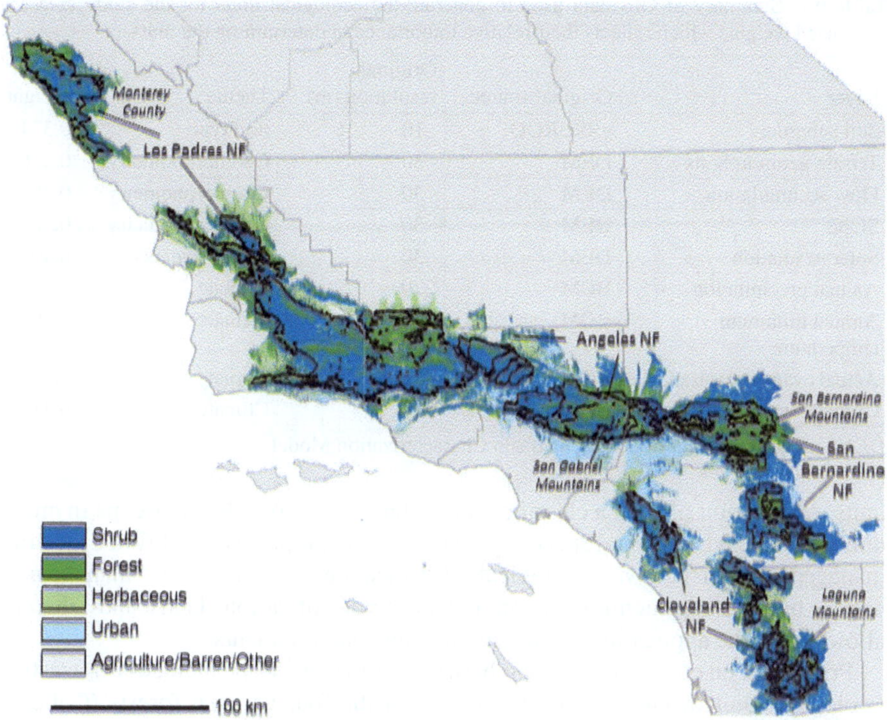

Fig. 9.1 Study area encompassing four national forests in southern California for which five ecosystem services were assessed. Major vegetation types are derived from FVEG (FRAP 2015). Key features highlighted in black text appear in descriptions below

9.3 Framework Description

9.3.1 Developing Ecological Units for the Study Area

The project area encompasses the Los Padres, Angeles, San Bernardino, and Cleveland national forests that account for 39% of the total study area footprint (Fig. 9.1). The predominant vegetation type in the study area is chaparral shrubland (54%) with conifer and hardwood forest comprising less than 20% (Fig. 9.1). There are a total of 375 watersheds at the HUC12 scale (USGS Hydrologic Unit Code) that intersect the four national forests, with the addition of 11 other HUC12 units to provide a practical and continuous boundary to the study area encompassing a total of 386 watersheds. The total footprint is 3,515,805 ha (8,687,731 acres). Within this boundary the first step was to create a physiographically-based classification and mapping of the region into a relatively limited number of ecologically relevant landscape types, hereafter referred to as ecological units. Such a classification can improve geographical understanding of the magnitude of these ecological services and allows for simple statistical summaries of their values to be generated by these

Table 9.1 Summary of GIS data used to generate the ecological units for the study area and associated weights which reflects their relative importance in determining the units

Layer	Original source	Original resolution (m)	Theme	Weight
Soil suborders	gSSURGO	10	Soil type	0.374
Terrain geomorphons	DEM	30	Geomorphometry	0.226
Flow accumulation	DEM	30	Geomorphometry	0.355
Slope	DEM	30	Geomorphometry	0.507
Solar irradiation	DEM	30	Energy balance	0.404
Annual precipitation	BCM	270	Climate	1.00
Annual minimum temperature	BCM	270	Climate	0.600
Actual evapotranspiration	BCM	270	Climate	0.367
Climatic water deficit	BCM	270	Climate	0.413

DEM Digital Elevation Model, *BCM* Basin Characterization Model

units. A key point of such a classification is that it partitions the landscape in multidimensional environmental space, as opposed to geographic space. From a resource management perspective, ecological classifications can facilitate management actions by focusing attention on common properties of various landscapes, thereby allowing for the application of unified planning and treatments.

Prior work on classifying the southern California landscape includes a Terrestrial Ecological Unit Inventory (TEUI) analysis of the four national forests (Gallegos et al. 2001). We opted to develop our own classification for a number of reasons. First, the 2001 TEUI did not incorporate climate data, and climate is the most fundamental factor driving land use capability, both directly and indirectly through its influence on soils and vegetation. Second, much better data are available now for building this classification, including digital elevation models and climate rasters. Third, the existing TEUI units were confined to federal lands, hindering analyses on the lands buffering the forests. Finally, in contrast to some approaches (e.g., McMahon et al. 2004; Cullum et al. 2016), we decided not to include vegetation as an input data layer as the aim was to capture biophysical characteristics and vegetation often reflects historical land use and patterns (e.g., fire history, type-conversion from shrubland to grassland, or agricultural use).

We first compiled Geographic Information System (GIS) layers from a variety of sources and generated a number of input layers from a Digital Elevation Model (30 m [0.2 acres]). Resolutions ranged from 10 to 270 m (0.02–18 acres), which were resampled to 30 m raster resolution where necessary (GIS modeling used a combination of GRASS 7.0.0 and R). Input data for generating the ecological units organized nine variables into four themes (Table 9.1). First, soil type reflected 22 soil suborders derived from the county scale gridded Soil Survey Geographic Database (gSSURGO). Second, geomorphometry was captured by creating terrain geomorphons derived from the DEM (30 m) which classified the landscape into ten discrete landform types (e.g., ridges, slopes, hollows, and valleys) (Jasiewicz and Stepinski 2013). Other geomorphometry type inputs included flow accumulation which provides a measure of the upslope area that conceivably drains into a given pixel and also slope measured in degrees. Third, energy balance was calculated based on direct, diffuse, and reflected solar irradiation for a given day, location, topography, and atmospheric conditions (assuming clear-sky conditions). Fourth, the Basin Characterization Model (Flint et al. 2013) provided a suite of climate variables for the period 1981–2010 at 270-m resolution; average annual precipitation, average annual minimum temperature (to represent montane winter conditions), average actual evapotranspiration, and climatic water deficit which are known to be strong drivers of vegetation distribution (Stephenson 1998).

Onto these GIS data layers we overlaid 10,000 random points across the study area and extracted the values. To sort the random points into a limited number of clustered types we calculated the distance of each random point to every other point with respect to these environmental variables in multivariate space, to create a dissimilarity matrix.

To overcome the issue of the nine environmental variables using different metrics and mixing numerical and categorical data types we assigned weightings to each variable to sum up their relative distances. These were generated from a sub-analysis focusing on a subset of the study area, the Santa Clara River watershed, which used three proxy variables to represent biomass (from MODIS derived Enhanced Vegetation Index), hydrological response (water recharge, Flint et al. 2013), and biodiversity (California Wildlife Habitat Relationships classification from the FVEG vegetation data, FRAP 2015). To create an overall set of weightings reflecting the relative importance of these variables to the ecological units, we averaged the ranking of importance values from two random forest regressions (for EVI and recharge) and a random forest classification (for habitat type) (Liaw and Wiener 2002; Hastie et al. 2009). The random forest regression model for recharge explained 72.5% of the variance in recharge, the regression model for EVI explained 54.9% of the variance, and the classification model for habitat type had an out-of-bound error rate of 50.1%. We applied these weightings in the dissimilarity matrix (Gower 1971; Maechler et al. 2017).

To group these random points into discrete cluster types we used a technique called Partitioning Around Medoids (PAM) (Kaufman and Rousseeuw 2005; Hollander 2012). PAM is similar to the more commonly used k-means algorithm but differs in that cluster centroids are assigned to actual data observations rather than means of data, it accepts categorical as well as numerical data, and the number of clusters is specified in advance. To determine the optimum number of clusters to portray the landscape variability we tested clusters ranging in number from 23 to 50 and evaluated performance using a combination of internal and external homogeneity metrics. We measured internal homogeneity using the average silhouette width of the clustering (Rousseeuw 1987) while external homogeneity used a measure of within cluster variability for the three proxy variables. Based on these measures we selected the clustering with 37 classes to best group the ecological variation across the study area.

To map these clusters we constructed a random forest predictor model which took the cluster assignment for the 10,000 random points and applied these assignments across the study area to create ecological units using the nine environmental variables from the raster stack. Finally, we smoothed the GIS cluster map by running a 3 × 3 majority filter over it. Patches of clusters in the final raster map ranged in size from 0.5 to 12,500 ha (1.2 to 30,888 acres) with 747,000 patches total, and a minimum mapping unit of 90 m by 90 m (0.8 ha).

9.3.2 Quantifying Carbon Storage

Carbon stored in above- and below-ground biomass of natural landscapes provides an ecosystem service to humans at local to global scales by securing carbon, thereby helping to regulate climate change. When natural landscapes are lost through conversion to urban use or agriculture, CO_2 is released and exacerbates global warming (Foley et al. 2005). In this project we use above-ground live biomass as captured in the Enhanced Vegetation Index (EVI) as a proxy for carbon stored on the landscape. It is generally concluded that about half (0.45) of vegetation dry matter is carbon (Schlesinger 1997). Although using EVI will indicate levels of above-ground live biomass, it does not account for above-ground dead biomass, nor the substantial amounts of carbon stored below-ground (live or dead), and would be less reliable in recently burned areas with substantial vegetation resprouting. From a resource management perspective, characterizing patterns of biomass is not only important for understanding ecosystem function, but also for understanding the impacts of fire or other management activities on carbon storage.

In this project, we used EVI derived from remote sensing imagery as a proxy for the amount of above-ground living biomass on the landscape. Vegetation indices such as the Normalized Difference Vegetation Index (NDVI) and the Enhanced Vegetation Index (EVI) measure the photosynthetic activity and the chlorophyll content of vegetation present in each image pixel. NDVI has been used substantially for vegetation mapping since the 1970s (Rouse et al. 1974) and has since been applied to a variety of different geographic settings and ecosystems. The development of the Enhanced Vegetation Index intended to address issues associated with atmospheric influences, such as aerosols, and also variable background soil reflectance by adjusting reflectance in the red spectral band by incorporating the blue band (Liu and Huete 1995; Huete et al. 1997). The EVI, in contrast to NDVI, is also sensitive to high levels of biomass (Huete et al. 2002) and studies have demonstrated its successful application for determining seasonal patterns of vegetation and detecting differences in plant characteristics such as leaf area index, canopy, and structure (Wittenberg et al. 2007; Kinoshita and Hogue 2011). Applications of EVI in Mediterranean-type climates include Kinoshita and Hogue (2011) in the San Bernardino Mountains to assess ecosystem recovery postfire and the corresponding response of seasonal and annual hydrology. A study by Clark et al. (2013) of six fires in chaparral and conifer habitats in southern California used EVI to assess postfire vegetation recovery, finding EVI to be well correlated with photo-interpreted ground cover (Clark et al. 2013). In another Mediterranean-type climate region,

Wittenberg et al. (2007) used EVI to assess vegetation recovery after fire on Mount Carmel in Israel.

To estimate mean biomass for southern California, we generated EVI from Landsat imagery for 2015. We obtained Landsat 8 Operational Land Imager data (OLI) from the Google Earth Engine data catalog (https://earthengine.google.com/datasets/) for each month in 2015. To create the EVI product we used standard USGS surface reflectance Landsat data (http://landsat.usgs.gov/CDR_LSR.php) and applied the following equation:

$2.5 * (\mathrm{NIR} - \mathrm{red})/(\mathrm{NIR} + 6 * \mathrm{red} - 7.5 * \mathrm{blue} + 1)$

Where NIR = near-infrared band (OLI band 5), red = red band (OLI band 4), blue = blue band (OLI band 2). EVI raster layers were created from the images available in each month, where images overlapped the maximum pixel value was selected for inclusion in the final EVI layer for the month. To remove anomalous short-term data spikes that are outside the range of general trends (e.g., owing to atmospheric conditions or variations in the angle of the sensor on the satellite) we applied a Savitzky-Golay least squares polynomial smoothing filter to the 2015 monthly data (TIMESAT software, Jönsson and Eklundh 2004; Eklundh and Jönsson 2015). We calculated the annual mean EVI value for each pixel and summarized the data using the ecological units.

9.3.3 *Quantifying Sediment Erosion Retention*

Native chaparral vegetation plays an important role in preventing sediment erosion. It stabilizes soils through its deep roots, intercepts rainfall, reduces overland flow of water, and traps sediment transported by water (Wohlgemuth et al. 1999, 2009). Southern California is particularly prone to sediment erosion because of steep topography, non-cohesive soils, and intense rainfall events (see Chap. 7). Fire occurrence in the region poses a particular problem for sediment erosion. First, incinerated vegetation and litter means reduced rainfall interception and so denuded hillsides are subjected to unimpeded raindrop impacts (Rice 1974). Second, fire and the combustion of organic material in the soil can create a subsurface water-repellent layer that restricts infiltration and promotes overland flow (DeBano 1981). As a consequence, surface runoff, increased sheet wash, and rill erosion intensify which leads to increased sediment yield, particularly after infrequent high magnitude storms (Scott and Williams 1978). These impacts are most noticeable in the first year after fire. In the San Dimas Experimental Forest for example, 85% of the total sediment delivered over four years resulted from the first year after fire (Wohlgemuth et al. 2009). Ultimately, sediment is transported from ephemeral channels to watershed outlets and deposited into debris basins at the mouths of steep canyons, which must be cleared regularly at high cost by the county.

For resource managers, sediment erosion is a severe problem causing damage to downstream human communities, threatening lives, damaging infrastructure such as roads, bridges and utility lines, and negatively impacting wildlife and aquatic habitats and water quality (Wohlgemuth et al. 2009). Consequently, identifying those areas on the landscape that offer the greatest ability to retain sediment is of management interest, particularly given the uncertainties associated with future climates when altered precipitation regimes may further increase fire frequency and increase the potential for heavy rainfall events (Flannigan et al. 2000; Westerling et al. 2006).

To estimate sediment erosion retention for southern California we utilized the sediment delivery ratio model (SDR) from InVEST software (version 3.3.0) (Hamel et al. 2015). It builds off the widely-used Revised Universal Soil Loss Equation (RUSLE) (Renard et al. 1997) for computing soil loss on a point basis but extends it by accounting for linkages between upper and lower watershed areas. Data inputs for running the model (at 30 m resolution) included a DEM, a rainfall erosivity index (Renard et al. 1997) (provided the California State Water Resource Control Board), and soil erodibility (K factor) developed from county scale SSURGO soils maps. This used a weighted average of the whole soil K factor value of the surface horizon across different soils components within each soil map polygon. A land use/land cover map was specified in conjunction with a practice factor of 1 (P), which refers to landscape modifications that reduce erosion. A value of one is typically assigned for applying RUSLE at large landscape scales (see Perović et al. 2013). We also specified a cover factor (C), which represents the influence of vegetation cover on soil erosion, based on the maximum NDVI from Landsat 8 imagery in the rainy season when most erosion occurs (October 2014 to April 2015). An equation of the form:

$$C = \exp(-2(\mathrm{NDVI})/(1 - \mathrm{NDVI}))$$

has been found to correspond well to plot-measured C factors in Mediterranean-type climate landscapes (van der Knijff et al. 2000; Perović et al. 2013). We specified the HUC12 watershed boundaries as the subunits for the SDR model and used model default values for other parameters (e.g., threshold flow and accumulation). Outputs from the model included an index that represents the avoided soil loss under vegetative cover relative to bare soil, weighted by the sediment delivery ratio.

9.3.4 *Quantifying Water Runoff and Groundwater Recharge*

In this project we focus on the benefits that water provides as a provisioning service, the supply of water for consumptive uses such as drinking, domestic use, agriculture, and industrial use and also non-consumptive purposes, such as generating power or freshwater for aquatic organisms. In addition, water provides cultural services such as river recreation and other tourist activities, as well as regulating services by maintaining water quality through natural filtration (Millennium Ecosystem Assessment 2005). There is an increasing awareness that the secure provision of future water supplies requires the protection and management of natural landscapes in the source watershed (Barten and Ernst 2004). In southern California, water is a critical issue. In the city of Los Angeles, for example, 89% of the water supply is imported from more than 322 km (200 miles) away, although efforts are now underway to rely on more local water resources (Gold et al. 2015). Future climates are estimated to further increase the pressures on water supply. For example, Hayhoe et al. (2004) find warmer temperatures and more precipitation falling as rain rather than snow will cause snowmelt runoff to shift earlier. As a result, spring and summer streamflows are reduced while evaporation increases due to warmer temperatures, leading to increasing reliance on groundwater resources. Understanding the spatial patterns of water runoff and groundwater recharge across the landscape is an important first step to identifying where water conservation measures can be implemented.

We extracted data from the Basin Characterization Model (BCM) on water runoff and groundwater recharge (Flint et al. 2013). This statewide raster model (original resolution of 270 m) combines empirical data on topography, soils, and geology with rainfall and temperature data. Unique to BCM is that hydrologic responses are spatially distributed based on bedrock permeability. We calculated the average, standard deviation, and range of current (1981–2010) runoff and recharge for each ecological unit in the study area (mm/year).

9.3.5 Quantifying Biodiversity

The Millennium Ecosystem Assessment (2005) describes biodiversity as providing the foundation of ecosystem services to which human well-being is intimately linked. On federal lands, resource managers have to manage for multiple objectives (see Chap. 15) which, in southern California, includes fire and fuel management, recreation, resources, restoration, the conservation of habitat for wildlife, and threatened, endangered, and sensitive species.

In this project, we define biodiversity services as a combination of natural features on the landscape including rare and native species, vegetation, aquatic biota, and areas of important landscape connectivity. Although these biodiversity features fall within the national forest boundaries and have some degree of protection, we wanted to generate a synoptic biodiversity layer for prioritization purposes. We used the conservation planning software Marxan to generate this layer (Ball et al. 2009), rather than alternatives such as taking a weighted sum of the input layers. Marxan works by using a simulated annealing algorithm to explore many configurations of planning units, incrementally moving towards solutions that meet inputted conservation objectives in “low cost” ways. “Cost” is defined by the user; it can refer to monetary cost but more generally refers to the suitability of a given planning unit for inclusion in a final conservation network. Unlike focusing on a single species or proxy such as intact habitat, the irreplaceability score takes account of multiple biodiversity data inputs and identifies multiple sets of planning units that comprise relatively low cost—or most suitable—solutions to meet a user’s conservation objectives, thereby prioritizing the landscape into a single value.

To undertake this component we minimized the areal extent of the original raster ecological units to make them practical from a conservation management perspective, setting a minimum size of 4.04 ha (10 acres) and maximum size of 6475 ha (1600 acres) for each ecological (planning) unit (minimum mapping unit 200 m). Water bodies such as lakes and reservoirs were omitted, along with any units with greater than 50% urban land cover. For each ecological unit we developed a cost based on the area of each unit modified by its native species richness and rare species richness scores from the statewide Areas of Conservation Emphasis (ACEII) dataset (California Department of Fish and Wildlife) using:

$$\left(1-\left(\frac{\left(\frac{\text{nat}}{\text{nat max}}+\frac{\text{rare}}{\text{rare max}}\right)}{2}\right)\right) * \text{acres}$$

Ecological units with high native species diversity and/or high rare species richness are given lower cost scores than units with equivalent areas and consequently, are more likely to be selected in the Marxan analysis.

We next identified conservation targets within the study area and assigned a corresponding conservation goal for each target. Conservation targets included land cover type, sensitive species, landscape connectivity, steelhead trout, and Watershed Condition Class (Table 9.2), which were selected owing to their region-wide extent and well-recognized status. The conservation goal for each target was determined based on experience conducting Marxan analyses in other geographic locations (Smith et al. 2008; Huber et al. 2010). Goals were higher for targets where there were fewer records or they covered a smaller spatial area. For example, our conservation goal for the least tern (*Sternula antillarum*) was 100% as there was only a single point record in the study area, compared to the more widely distributed Lawrence's goldfinch (*Spinus lawrencei*) with 581 records which received a conservation goal of 25%.

One further requirement for running the Marxan software was to specify the boundary file which identifies the length of every adjacent border shared between each ecological unit. Increasing or decreasing this alters the number of output priorities generated by the software. We conducted a sensitivity analysis to determine the optimal value and found a boundary length modifier of 1.0 resulted in outputs containing high value areas that were of an appropriate size for practical management actions. Marxan was set to output 100 runs at two million iterations each, enough to overcome the effects of the random seeding at the beginning of each run. Each ecological unit was given a value (0–100) representing the number of runs in which it was selected as part of an optimal solution. Higher scores represent units with a greater irreplaceability for meeting the inputted conservation goals (as specified in Table 9.2).

Table 9.2 Summary of conservation targets and associated conservation goals used in the mapping of biodiversity services in southern California

Conservation target	Source	Description	Target by ecological unit	Conservation goal assigned (%)
Land cover	FRAP 2015[a]	Native vegetation types (with exception of annual grasslands)	Area of each land cover type calculated for each ecological unit	0–40.4 ha = 100% 40.4–404.7 ha = 75% 404.7–4046.9 ha = 50% >4046.9 ha = 25%
Sensitive species	CNDDB and NRIS[b]	Selected 203 plant and animal species listed in the USFS Southern California Forest Plan Revision	Point data for species were summarized for each unit Polygon data were used to assign areal extent within each unit	0–4.04 ha = 1 4.04–40.5 ha = 75% 40.5–202.3 ha = 50% >202.3 ha = 25%
Landscape connectivity	Central Coast, South Coast, and Central Valley critical linkage studies[c]	Species-specific least cost corridors calculated for linking large core areas	If the centroid of the unit intersected with connectivity area, the unit was given a 1, else 0	Any unit with connectivity present = 25%
Steelhead trout	National Marine Fisheries Service[d]	Occurrence data from the National Marine Fisheries Service (NMFS)	Total length of steelhead habitat reaches within each unit	Any unit with steelhead habitat present = 25%
Watershed Condition Class	USFS national assessment of HUC12 watersheds that contain federal lands[e]	Used Aquatic Biota Index which reflects life-forms presence, native species, and non-native or invasive species	Identified HUC12 watersheds classed as 'good' quality and selected units whose centroids were located in these watershed	Any unit with 'good' aquatic biota present = 25%

Reference and data sources:

[a](FRAP 2015) http://frap.cdf.ca.gov/data/frapgisdata-sw-fveg_download

[b](USDA Forest Service 2004)

[c](Huber et al. 2010; Spencer et al. 2010)

[d]http://www.nmfs.noaa.gov/pr/species/criticalhabitat.htm

[e]http://www.fs.fed.us/biology/resources/pubs/watershed/maps/watershed_classification_guide2011FS978.pdf. Note: 88% of total study footprint encompassed

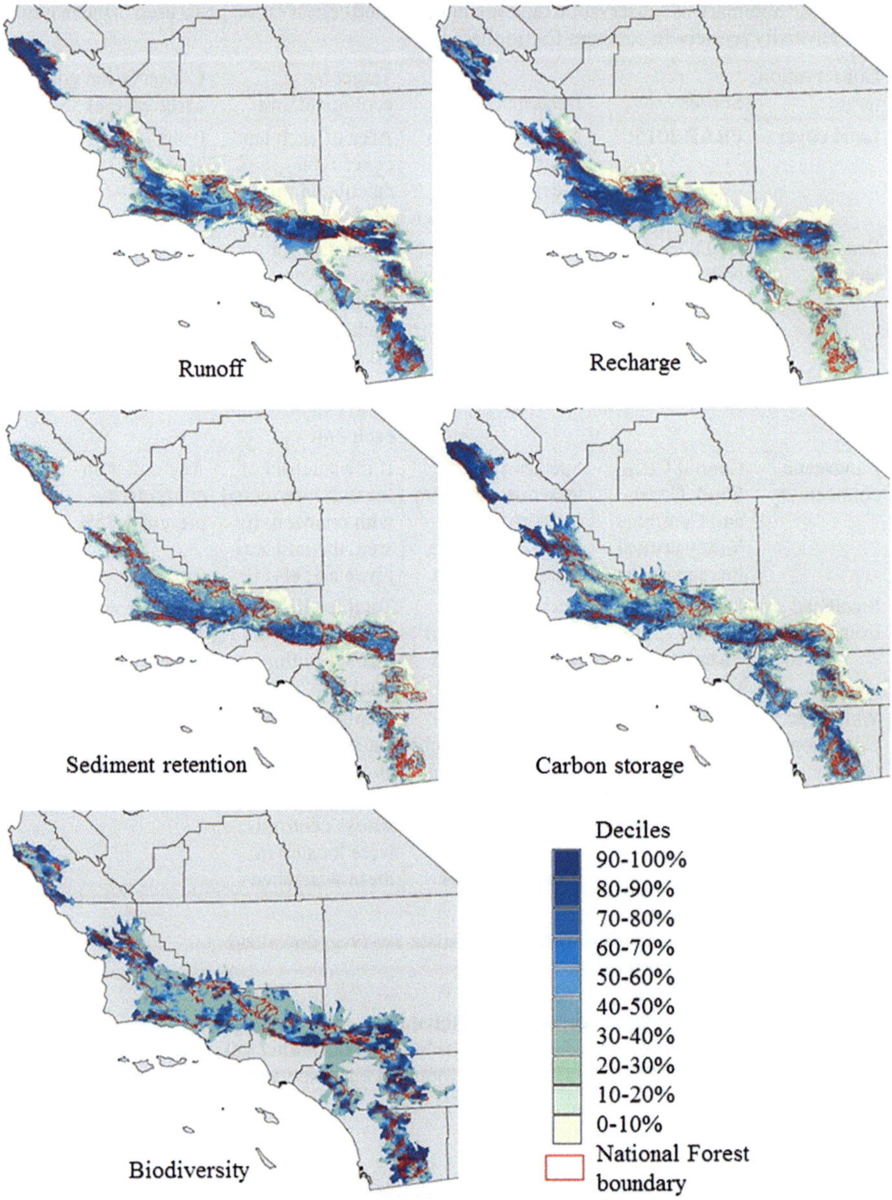

Fig. 9.2 Maps of five ecosystem services across the southern California study area. The legend reflects deciles based on the ranked raw values of each service in its original units (higher deciles indicate higher provision of ecosystem services)

Table 9.3 Correlation coefficients (r_s) between the five ecosystem services using the mean value of each service per unit area (n = 37)

	Recharge		Runoff		Sediment Ret.		Biodiversity	
Runoff	0.930	<0.0001						
Sediment Ret.	0.801	<0.0001	0.818	<0.0001				
Biodiversity	0.497	0.002	0.442	0.006	0.414	0.011		
Carbon storage	0.843	<0.0001	0.854	<0.0001	0.697	<0.0001	0.351	0.033

9.3.6 Summarizing Ecosystem Services by Ecological Units

For each ecological unit we present the value for each of the five ecosystem services individually using 30 m resolution raster grids with an accompanying legend which divides the original units into deciles to facilitate comparisons between services (Fig. 9.2). We next calculate the mean, standard deviation and minimum and maximum value of each service by the ecological units and analyze the spatial concordance between them (Table 9.3). We then present two prioritizations of the ecosystem services data. Discussions with USFS resource managers and partners identified whether low or high values of each ecosystem service are prioritized. For example, sediment erosion retention could be a negative service in that sediment produces and replenishes beach sand, or it could be a positive service, with areas of greater retention valued more highly as it prevents damage to property and infrastructure downstream. Similarly, higher values of biodiversity, water runoff, groundwater recharge, and carbon storage in this project are all assigned a higher priority for management (although we recognize that in some contexts resource managers may want to identify areas with low values to mitigate negative effects).

We use a conceptually straightforward method to identify priority ecological units. The first prioritization approach filters the ecological units to identify the top quintile (20% or the top eight units) for each ecosystem service. The second approach sums these priority units to provide a map of ecosystem service hotspots, ranging from units which contain no services to ones which contain four (none contained all five). For these hotspots we describe their spatial location along with summary statistics for each of the five services (Tables 9.4 and 9.5).

9.4 Spatial Patterns of Ecosystem Services Across Southern California

The spatial distribution of the five ecosystem services shows a variety of patterns across the north-south extent of the study area (Fig. 9.2). Groundwater recharge values were notably lower in the southern portion of the study area around the Cleveland National Forest. Higher values for carbon storage are distributed in patchy fashion throughout the study area and mostly align with higher elevation areas, encompassing conifer and hardwood vegetation. Higher value areas

Table 9.4 Description of ecological units in the southern California study area supporting high numbers of priority ecosystem services

ECL unit	Description	No. of Services
3	Steep slopes and ridges in the Central Coast and Transverse ranges with relatively high precipitation, dominated by mixed chaparral and coastal oak woodlands	4
13	Steep south-facing mid-elevation slopes in the Central Coast and Transverse ranges, dominated by mixed chaparral	4
15	Steep north-facing mid-elevation slopes with relatively high precipitation, dominated by mixed chaparral and montane hardwood-conifer	4
25	Higher elevation, very steep slopes of the Transverse Ranges, heavily dominated by mixed chaparral	4
26	Mid-elevation steeper slopes of the Transverse Ranges and San Bernardino Mountains, dominated by mixed chaparral	4
22	Higher elevation steep, south-facing slopes with colder temperatures of the Transverse Ranges and Southern California mountains, dominated by mixed chaparral and montane hardwood-conifer	3
30	High-elevation mountains with steep slopes, dominated by Sierran mixed conifer and subalpine conifer forest	3
20	Steep, wet and cool south-facing mid-elevation slopes of the Big Sur region and San Bernardino Mountains, dominated by mixed chaparral	2
35	High-elevation relatively dry, cooler slopes in Transverse Ranges and the San Bernardino range, dominated by pinyon-juniper woodlands and Jeffrey pine	2
1	Moderately high interior ridges and slopes across Central and Southern California mountains, dominated by mixed chaparral and grasslands. Moderately high precipitation	1
10	Steep north-facing mid-elevation slopes in the Tranverse Ranges and interior Central Coast Ranges, dominated by mixed chaparral and coastal oak woodlands	1
12	Interior Central and Southern California valleys and gentle slopes with moderate precipitation, dominated by annual grasslands and mixed chaparral	1
16	Mid-elevation moderate slopes on mountains of Central and Southern California with moderately low precipitation, dominated by mixed chaparral and coastal oak woodland	1
21	Higher elevation south-facing slopes of Central and Southern California mountains, relatively dry, dominated by mixed chaparral and chamise-redshanks chaparral	1
23	Mid-elevation steep slopes of the Transverse Ranges, dominated by mixed chaparral and coastal scrub	1
29	Mid-elevation, moderately warm, moderate slopes and ridges mostly in the Transverse Ranges, dominated by mixed chaparral and coastal scrub	1

Table 9.5 Summary of ecosystem services contained in tier one (*) (unit contained 4 priority services) and tier two (α) (unit contained 3 priority services) priority ecological units based on an analysis of water recharge, runoff, sediment erosion retention, carbon storage, and biodiversity

ECL unit	Water recharge (mm/year)			Water runoff (mm/year)			Sediment retention index			Carbon storage			Biodiversity		
	Mean	Std Dev	Range	Mean	Std Dev	Range	Mean	Std Dev	Range	Mean	Std Dev	Range	Mean	Std Dev	Range
3*	150.2	91.3	0–573.4	287.5	122.2	0–573.4	9.3	33.3	0–3964.7	0.3	0.1	0.04–0.7	36.6	38.2	0–100
13*	163.2	93.2	0–573.4	294.0	121.8	0–573.4	13.0	40.7	0–3480.7	0.3	0.1	0.05–0.7	25.9	35.2	0–100
15*	179.6	114.1	0–687.6	299.6	147.3	0–687.6	8.8	26.7	0–2682.7	0.3	0.1	0.04–0.7	22.2	35.0	0–100
25*	168.7	102.6	0–633.7	335.6	169.7	0–633.7	7.3	19.9	0–2215.8	0.3	0.1	0.06–0.7	19.2	32.8	0–100
26*	277.8	138.7	0–696.6	301.2	164.7	0–696.6	10.0	30.0	0–1921.4	0.3	0.1	0.05–0.7	12.6	29.8	0–100
22α	163.8	99.0	0–644.1	195.4	103.0	0–644.1	6.0	18.7	0–1274.7	0.3	0.1	0.05–0.7	28.6	39.2	0–100
30α	174.6	110.7	0–604.0	385.6	169.6	0–604.0	5.3	13.4	0–847.6	0.2	0.1	0.01–0.6	40.1	46.7	0–100

Water recharge and runoff = mm/year; sediment retention = index that represents the avoided soil loss under vegetative cover relative to bare soil weighted by the sediment delivery ratio; carbon storage = Enhanced Vegetation Index averaged across a whole year; biodiversity = the number of times the particular ecological unit was identified as a needed element in 100 different random Marxan runs (i.e., an irreplaceability value) identified as irreplaceable for biodiversity are scattered throughout the study area (Fig. 9.2).

To assess the spatial correlation between the five ecosystem services across the 37 ecological units, we calculated the mean value of each service per unit area. A subsequent non-parametric Spearman rank correlation analysis showed the correlation between six pairs of ecosystem services to be highly significant ($p < 0.001$, Spearman rank correlation) and the correlation between the remaining four was significant ($p < 0.05$) (Table 9.3).

All correlations between variables within the ecological units are positive (Table 9.3). One interpretation is that some of the models used to map services across the landscape utilize the same key data layers, such as precipitation. For example, recharge and runoff are highly correlated, and similarly sediment erosion retention with both recharge and runoff ($r_s = 0.93$, $r_s = 0.80$ and $r_s = 0.82$ respectively, $p < 0.0001$, Spearman rank correlation) as precipitation is a key factor in both the Basin Characterization Model and the erosivity index of the InVEST sediment delivery ratio model. Correlations between carbon storage and recharge, runoff, and sediment retention were also all highly significant ($r_s = 0.84$, 0.86, and 0.70 respectively, $p < 0.0001$). Precipitation, again, is a key determinant of vegetation patterns and the hydrology and sediment delivery ratio models include topographical variables such as a DEM which helps explain these patterns. Another factor is that correlations are conducted using the average value of each service for each ecological unit, with these units already reflecting environmentally and climatically similar areas of the landscape. Correlations with biodiversity are less easy to interpret, however, since the biodiversity value represents an irreplaceability score assigned based on conservation targets and associated goals, and is less directly tied to the underlying physical characteristics of the landscape.

9.5 Identifying Hotspots of Ecosystem Services

For each ecosystem service we identified the top tier of units, those where service values are in the highest quintile (Figs. 9.1 and 9.3). High value areas for water runoff and groundwater recharge are concentrated along the Big Sur coastline in Monterey county and ranging inland up to ~1800 m (~6000 ft). Other high values for both runoff and recharge are scattered through the high-elevation areas of the southern portion of the Los Padres National Forest, the San Gabriel Mountains of the Angeles National Forest, and the conifer and shrub dominated San Bernardino Mountains. The spatial pattern of priority ecological units for carbon storage is similar to runoff and recharge, with the exception that the San Bernardino Mountain area does not meet the priority threshold. Priority sediment erosion retention areas include the western edge of the Los Padres National Forest, the southern edge of the Angeles, and the northern tip of the Cleveland National Forest.

Priority ecological units based on high irreplaceability values for biodiversity are clustered throughout the study area owing to the distribution of species with very

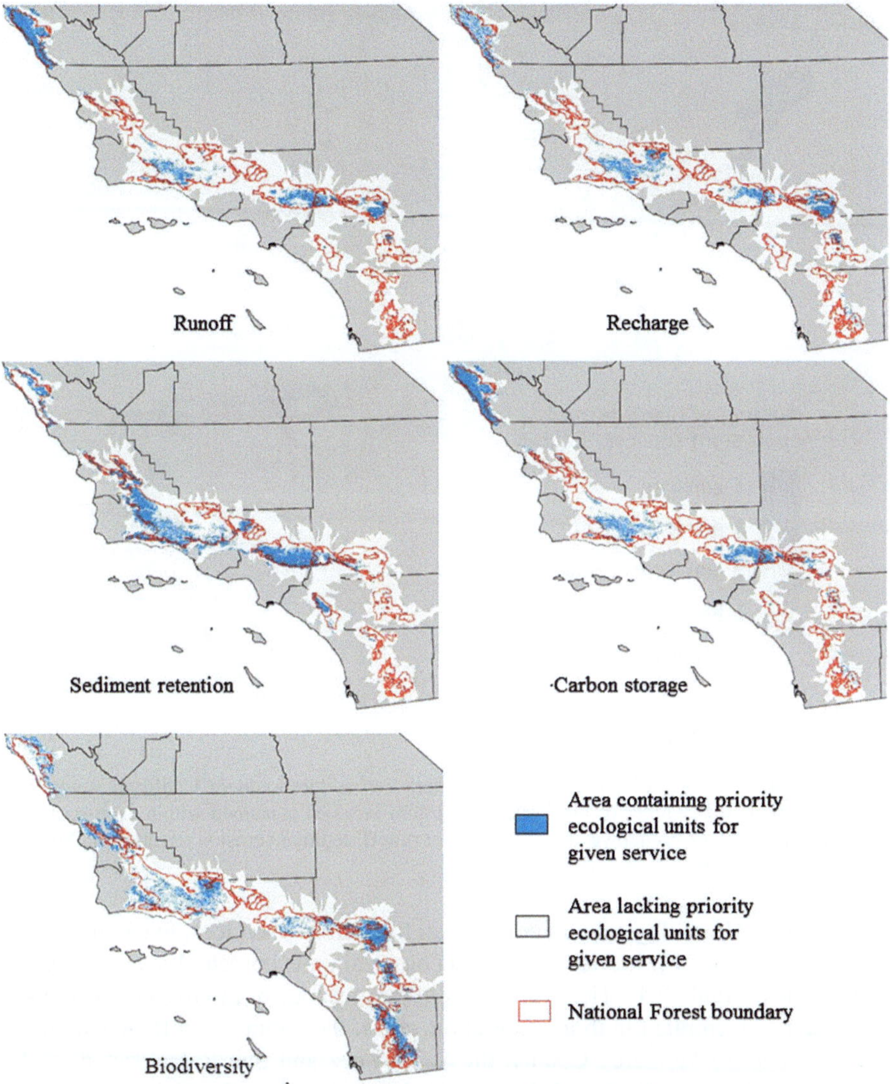

Fig. 9.3 Priority ecological units for each ecosystem service across the southern California study area. Priorities are defined as those ecological units in the top quintile when ranked based on the mean value per unit area of each service

few documented occurrences, but with a notable concentration in the San Bernardino Mountains and along the eastern edge of the Cleveland National Forest encompassing the Laguna Mountains.

The second prioritization we conducted involved adding the highest value ecological units for each service (i.e., the sum of the maps shown in Fig. 9.3). Less than half of the ecological units contain one or more priority ecosystem services (16 out

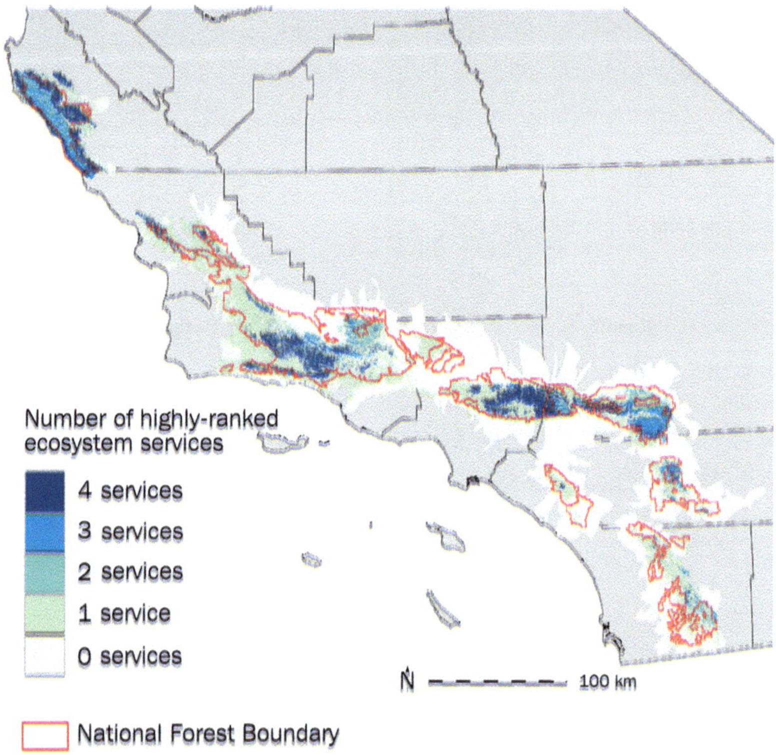

Fig. 9.4 Hotspots of multiple ecosystem services across the southern California study area. Legend reflects the number of high value ecosystem services contained within each ecological unit. Services include groundwater recharge, water runoff, sediment erosion retention, carbon storage, and biodiversity

of 37 units). No ecological unit contains all five priority services, five units contain four services (tier 1 priorities), and two ecological units contain three services (tier 2 priorities) (Table 9.4). Tier 1 priorities harboring four of the five ecosystem services are concentrated in three geographic areas: the Santa Lucia Range along the central coast in Monterey County, the Santa Ynez and San Rafael ranges of the Transverse Ranges in the Los Padres National Forest, and the San Gabriel Mountains in the Angeles National Forest (Fig. 9.4). More specific descriptions of priority ecological units include the steep south-facing mid-elevation slopes in the central Coast and Transverse ranges, dominated by mixed chaparral (ecological unit 13, Tables 9.4 and 9.5).

Second tier priority geographic areas containing three of the five ecosystem services include the San Bernardino Mountains in the Transverse Ranges and the adjacent Mount San Jacinto State Park in the Peninsular Ranges, higher elevations in the Monterey County portion of the Los Padres National Forest, and small, scattered areas running north-south along the Laguna Mountains and Cuyamaca Mountains of Cleveland National Forest's eastern edge (Table 9.5). Specific ecological units include the steep, wet and cool south-facing mid-elevation slopes of the Big Sur area and San Bernardino Mountains, dominated by mixed chaparral (ecological unit 20, Tables 9.4 and 9.5), and the high-elevation steep, south-facing slopes with colder temperatures of the Transverse Ranges and Southern California mountains, dominated by mixed chaparral and montane hardwood-conifer (ecological unit 22, Tables 9.4 and 9.5).

9.6 Implications for Natural Resource Management

In this project, we use ecological units as a basis to map the value of national forest landscapes in southern California for ecosystem service provision. We map and quantify five types of ecosystem services across the study landscape, setting the stage for the next step of calculating the associated economic value of these services. Based on the data compiled for each type of ecosystem service, we identified two categories of priorities: the highest value ecological units for each of the five services separately, and ecological units containing multiple high value ecosystem services, i.e., hotspots of service provision.

The type of prioritization (and there are many others besides the two methods we used) which is most useful depends on the management objectives. For specific resource questions, understanding how a single service is distributed could be most informative, e.g., investigating opportunities for Payment for Ecosystem Services schemes for carbon credits on private lands, which can generate income for conservation efforts (Jack et al. 2008). On the other hand, focusing attention on hotspots containing multiple services has the potential for reducing the resources and effort required for managing them (Egoh et al. 2008). In hotspots, policies and actions to protect or restore the provision of one ecosystem service may yield ancillary benefits for others. In contrast, if there is little overlap in services then trade-offs must be confronted (Anderson et al. 2009). Compared to a number of other ecosystem services studies (e.g., Chan et al. 2006; Naidoo et al. 2008; Anderson et al. 2009) which found weak correlations between ecosystem services, our assessment showed significant overlap between the different ecosystem services summarized by ecological units in southern California. In addition, although the value of ecosystem services can be mapped and assessed within the study area boundaries, the benefits of some of these services, such as carbon storage and biodiversity, reach far beyond these borders and are global in extent.

In Chap. 15, Safford et al. identify areas of management focus and the relationship between these foci and ecosystem services for federal and state conservation units in southern California (Table 15.1). There is a strong positive relationship between conservation and restoration and ecosystem services, consequently there is great scope for the information generated in this project to support decision making and planning relating to these broad management areas on federal and state lands (see Box 15.3). The longer term goal of the project is for the ecosystem services maps and digital spatial data to be accessible to resource managers through an online mapping tool, which can then be employed, for example, to select areas for postfire restoration (see Box 15.3).

In undertaking this project and interpreting findings for resource management there are a number of caveats that need to be noted. For example, the biodiversity priorities are contingent on the input data layers we selected, which may not necessarily reflect the biodiversity objectives of all resource managers across the four national forests. However, access to the digital data and maps can help resolve this problem as the ecosystem service data can be viewed in conjunction with other biodiversity data layers. Second, the sediment erosion retention model is ultimately derived from the Universal Soil Loss Equation developed for farmland in the Midwest of the USA, which is unlikely to be the most appropriate for the steep topography in southern California (we are exploring ways to account for this in the sediment delivery ratio model). Third, we use EVI as a proxy of above-ground live biomass to indicate carbon storage, about half of which is carbon, however, this does not account for the substantial amounts of carbon stored in extensive underground root systems of chaparral (Kummerow et al. 1977). Fourth, a key ecosystem service that needs to be included in the future is recreation, given the proximity of chaparral dominated national forests to high-density urban areas in southern California. Finally, as illustrated in other chapters in this book (see Chap. 11 for example), many ecosystem services, such as the intrinsic value of chaparral and other cultural services, cannot be quantified.

By classifying the landscape into ecological units which reflect the climatic and environmental variability and quantifying ecosystem services within them, we provide a ranked set of priority units (Table 9.4). These data, particularly when accessible to resource managers in a non-technical tool with a straightforward interface, can help inform management of the impact of activities on ecosystem services (e.g., fuel management, see Chap. 15), and identify priority areas where activities can secure the continued provision of services in the future. Furthermore, providing estimates of the amount and economic value of ecosystem services provided by national forests can highlight the contribution of national forest lands to providing benefits to the public, as well as heighten awareness of often under-valued and under-appreciated chaparral landscapes.

References

Anderson, B. J., P. R. Armsworth, F. Eigenbrod, C. D. Thomas, S. Gillings, A. Heinemeyer, D. B. Roy, and K. J. Gaston. 2009. Spatial covariance between biodiversity and other ecosystem service priorities. Journal of Applied Ecology 46:888-896.

Ball, I. R., H. P. Possingham, and M. Watts. 2009. Marxan and relatives: software for spatial conservation prioritisation. Pages 185-195 *in* A. Moilanen, K. A. Wilson, and H. P. Possingham, editors. Spatial conservation prioritisation: quantitative methods and computational tools. Oxford University Press, Oxford, UK.

Barten, P. K., and C. Ernst. 2004. Land conservation and watershed management for source protection. Journal of the American Water Works Association 96:121-135.

Batker, D. K. 2005. Supplemental ecological services study: Tolt River watershed asset management plan. Report for Earth Economics, Seattle, Washington, USA.

Batllori, E., M. A. Parisien, M. A. Krawchuk, and M. A. Moritz. 2013. Climate change-induced shifts in fire for Mediterranean ecosystems. Global Ecology and Biogeography 22:1118-1129.

Benayas, J. M. R., A. C. Newton, A. Diaz, and J. M. Bullock. 2009. Enhancement of biodiversity and ecosystem services by ecological restoration: a meta-analysis. Science 325:1121-1124.

Bullock, J. M., J. Aronson, A. C. Newton, R. F. Pywell, and J. M. R. Benayas. 2011. Restoration of ecosystem services and biodiversity: conflicts and opportunities. Trends in Ecology and Evolution 26:541-549.

Burge, D. O., J. H. Thorne, S. P. Harrison, B. C. O'Brien, J. P. Rebman, J. R. Shevock, E. R. Alverson, L. K. Hardison, J. Delgadillo Rodríguez, S. A. Junak, T. A. Oberbauer, H. Riemann, S. E. Vanderplank, and T. Barry. 2016. Plant diversity and endemism in the California Floristic Province. Madroño 63:3-206.

Chan, K. M. A., M. R. Shaw, D. R. Cameron, E. C. Underwood, and G. C. Daily. 2006. Conservation planning for ecosystem services. PLoS Biology 4:379.

Clark, J., M. Stamer, K. Cooper, C. Napper, T. S. Hogue, and A. M. Kinoshita. 2013. Using remote sensing to monitor post-fire watershed recovery as a tool for management. Pages 370-375 *in* G. J. Gottfried, P. Folliott, B. S. Gebow, L. G. Eskew, and L. C. Collins, editors. Merging science and management in a rapidly changing world: biodiversity and management of the Madrean Archipelago III. Proceedings RMRS-P-67. USDA Forest Service, Rocky Mountain Research Station, Fort Collins, Colorado, USA.

Costanza, R., R. D'Arge, R. de Groots, S. Farber, M. Grasso, B. Hannon, K. Limbrug, R. V. O'Neill, J. Paruelo, R. G. Raskin, P. Sutton, and M. van den Belt. 1997. The value of the world's ecosystem services and natural capital. Nature 387:253-260.

Cullum, C., K. H. Rogers, G. Brierley, and E. T. F. Witkowski. 2016. Ecological classification and mapping for landscape management and science. Progress in Physical Geography 40:38-65.

Daily, G. C. 1997. Nature's services: societal dependence on natural ecosystems. Island Press, Washington, D.C., USA.

DeBano, L. F. 1981. Water repellent soils: a state-of-the-art. General Technical Report PSW-GTR-46. USDA Forest Service, Pacific Southwest Forest and Range Experiment Station, Berkeley, California, USA.

Egoh, B., B. Reyers, M. Rouget, D. M. Richardson, D. C. Le Maitre, and A. S. van Jaarsveld. 2008. Mapping ecosystem services for planning and management. Agriculture Ecosystems and Environment 127:135-140.

Eklundh, L., and P. Jönsson. 2015. TIMESAT 3.2 software manual. Lund and Malmö University, Lund, Sweden. http://web.nateko.lu.se/timesat/timesat.asp

Flannigan, M. D., B. J. Stocks, and B. M. Wotton. 2000. Climate change and forest fires. Science of the Total Environment 262:221-229.

Flint, L. E., A. L. Flint, J. H. Thorne, and R. Boynton. 2013. Fine-scale hydrologic modeling for regional landscape applications: the California Basin Characterization Model development and performance. Ecological Processes 2:25.

Foley, J. A., R. DeFries, G. P. Asner, C. Barford, G. Bonan, S. R. Carpenter, F. S. Chapin, M. T. Coe, G. C. Daily, H. K. Gibbs, J. H. Helkowski, T. Holloway, E. A. Howard, C. J. Kucharik, C. Monfreda, J. A. Patz, C. Prentice, N. Ramankutty, and P. K. Snyder. 2005. Global consequences of land use. Science 309:570-574.

FRAP [Fire and Resource Assessment Program]. 2015. California Department of Forestry and Fire Protection's CALFIRE Fire and Resource Assessment Program (FRAP). Fveg15_1 vegetation data. http://frap.fire.ca.gov/data/frapgisdata-sw-fveg_download

Gallegos, A., F. Levitan, C. Phillips, and B. Roath. 2001. Southern California Province land type association ecological unit inventory. Unpublished report prepared for the USDA Forest Service for the Los Padres, Angeles, San Bernardino, and Cleveland national forests.

Gold, M., S. Pincetl, and F. Federico. 2015. Environmental report card for Los Angeles County. UCLA Institute of the Environment and Sustainability, Los Angeles, California, USA. http://www.environment.ucla.edu/perch/resources/files/report-card-2015-2.pdf

Gos, P., and S. Lavorel. 2012. Stakeholders' expectations on ecosystem services affect the assessment of ecosystem services hotspots and their congruence with biodiversity. International Journal of Biodiversity Science Ecosystem Services Management 18:93-106.

Gower, J. C. 1971. A general coefficient of similarity and some of its properties. Biometrics 27:623-637.

Haidinger, T. L., and J. E. Keeley. 1993. Role of high fire frequency in destruction of mixed chaparral. Madroño 40:141-147.

Hamel, P., R. Chaplin-Kramer, S. Sim, and C. Mueller. 2015. A new approach to modeling the sediment retention service (InVEST 3.0): case study of the Cape Fear catchment, North Carolina, USA. Science of the Total Environment 524-525:166-177.

Hastie, T., R. Tibshirani, and J. Friedman. 2009. The elements of statistical learning: data mining, inference, and prediction. Springer, Berlin, Germany.

Hayhoe, K., D. Cayan, C. B. Field, P. C. Frumhoff, E. P. Maurer, N. L. Miller, S. C. Moser, S. H. Schneider, K. N. Cahill, E. E. Cleland, L. Dale, R. Drapek, R. M. Hanemann, L. S. Kalkstein, J. Lenihan, C. K. Lunch, R. P. Neilson, S. C. Sheridan, and J. H. Verville. 2004. Emissions pathways, climate change, and impacts on California. Proceedings of the National Academy of Sciences 101:12422-12427.

Hollander, A. D. 2012. Using GRASS and R for Landscape Regionalization through PAM Cluster Analysis. OSGeo Journal 10:6.

Huber, P. R., S. E. Greco, and J. H. Thorne. 2010. Spatial scale effects on conservation network design: trade-offs and omissions in regional versus local scale planning. Landscape Ecology 25:683-695.

Huete, A., K. Didan, T. Miura, E. P. Rodriguez, X. Gao, and L. G. Ferreira. 2002. Overview of the radiometric and biophysical performance of the MODIS vegetation indices. Remote Sensing of Environment 83:195-213.

Huete, A. R., H. Q. Liu, K. Batchily, and W. Van Leeuwen. 1997. A comparison of vegetation indices over a global set of TM image for EOS-MODIS. Remote Sensing of Environment 59:440-451.

Jack, B. K., C. Kousky, and K. R. E. Sims. 2008. Designing payments for ecosystem services: lessons from previous experience with incentive-based mechanisms. Proceedings of the National Academy of Sciences 105:9465-9470.

Jasiewicz, J., and T. F. Stepinski. 2013. Geomorphons - a pattern recognition approach to classification and mapping of landforms. Geomorphology 182:147-156.

Jönsson, P., and L. Eklundh. 2004. TIMESAT – a program for analyzing time-series of satellite sensor data. Computer and Geosciences 30:833-845.

Kaufman, L., and P. Rousseeuw. 2005. Finding groups in data: an introduction to cluster analysis. Wiley, Hoboken, New Jersey, USA.

Keeley, J. E. 2005. Fire as a threat to biodiversity in fire-type shrublands. Pages 97-106 *in* B. E. Kus, and J. L. Beyers, technical coordinators. Planning for biodiversity: bringing research and management together. General Technical Report PSW-GTR-195. USDA Forest Service, Pacific Southwest Research Station, Albany, California, USA.

Kinoshita, A. M., and T. S. Hogue. 2011. Spatial and temporal controls on post-fire hydrologic recovery in Southern California watersheds. Catena 87:240-252.

Kummerow, J., D. Krause, and W. Jow. 1977. Root systems of chaparral shrubs. Oecologia 29:163-177.

Lenihan, J. M., D. Bachelet, R. P. Neilson, and R. Drapek. 2008. Response of vegetation distribution, ecosystem productivity, and fire to climate change scenarios for California. Climate Change 87:S215-S230.

Liaw, A., and M. Wiener. 2002. Classification and regression by random forest. R News 2:18-22.

Liu, H. Q., and A. R. Huete. 1995. A feedback based modification of the NDVI to minimize canopy background and atmospheric noise. IEEE Transactions on Geoscience and Remote Sensing 33:457-465.

Maechler, M., P. Rousseeuw, A. Struyf, M. Hubert, and K. Hornik. 2017. Cluster: cluster analysis basics and extensions. R package version 2.0.6.

McMahon, G., E. B. Wiken, and D. A. Gauthier. 2004. Toward a scientifically rigorous basis developing mapped ecological regions. Environmental Management 34:S111–24.

Millennium Ecosystem Assessment. 2005. Millennium ecosystem assessment synthesis report. Island Press, Washington, D.C., USA.

Naidoo, R., A. Balmford, R. Costanza, B. Fisher, R. E. Green, B. Lehner, T. R. Malcolm, and T. H. Ricketts. 2008. Global mapping of ecosystem services and conservation priorities. Proceedings of the National Academy of Sciences 105:9495-9500.

Nelson, E., G. Mendoza, J. Regetz, S. Polasky, H. Tallis, D. Cameron, K. M. A. Chan, G. C. Daily, J. Goldstein, P. M. Kareiva, E. Lonsdorf, R. Naidoo, T. H. Ricketts, and M. R. Shaw. 2009. Modeling multiple ecosystem services, biodiversity conservation, commodity production, and tradeoffs at landscape scales. Frontiers in Ecology and the Environment 7:4-11.

Oechel, W. C., S. J. Hastings, G. L. Vourlitis, M. A. Jenkins, and C. L. Hinkson. 1995. Direct Effects of Elevated CO_2 in chaparral and Mediterranean-type ecosystems. Pages 58-75 *in* J. M. Moreno and W. C. Oechel, editors. Global change and Mediterranean-type ecosystems. Springer, New York, New York, USA.

Olson, D. M., and E. Dinerstein. 2002. The Global 200: Priority ecoregions for global conservation. Annals of the Missouri Botanical Garden 89:199-224.

Perović, V., L. Životić, R. Kadović, A. Đorđević, D. Jaramaz, V. Mrvić, and M. Todorović. 2013. Spatial modelling of soil erosion potential in a mountainous watershed of South-eastern Serbia. Environmental Earth Sciences 68:115-128.

Polasky S., Nelson E., Pennington D., and Johnson K. A. 2011. The impact of land-use change on ecosystem services, biodiversity and returns to landowners: a case study in the state of Minnesota. Environmental Resource Economics 48:219-242.

Raudsepp-Hearne, C., G. D. Peterson, and E. M. Bennett. 2010. Ecosystem service bundles for analyzing tradeoffs in diverse landscapes. Proceedings of the National Academy of Sciences 107:5242-5247

Renard, K. G., G. R. Foster, G. A. Weesies, D. K. McCool, and D. C. Yoder. 1997. Predicting soil erosion by water: a guide to conservation planning with the revised Universal Soil Loss Equation (RUSLE). Agricultural Handbook No. 703. USDA Agricultural Research Service, Washington D.C., USA.

Rice, R. M. 1974. The hydrology of chaparral watersheds. Pages 19-26 *in* M. Rosenthal, editor. Proceedings of the Symposium on Living with the Chaparral, Riverside, California, March 30-31, 1973. Sierra Club Special Publication, San Francisco, California, USA.

Rouse, J. W., R. H. Haas, J. A. Scheel, and D. W. Deering. 1974. Monitoring vegetation systems in the Great Plains with ERTS. Pages 309-317 *in* S. C. Freden, E. P. Mercanti, and M. A. Becker, editors. Proceedings, third Earth Resource Technology Satellite (ERTS) symposium, volume 1, Washington, D.C., December 10-14, 1973. NASA, Washington, D.C., USA.

Rousseeuw, P. J. 1987. Silhouettes: a graphical aid to the interpretation and validation of cluster analysis. Journal of Computational and Applied Mathematics 20:53-65.

Sachs, J. D., and W. V. Reid. 2006. Environment: investments toward sustainable development. Science 312:1002.

Safford, H. D. 2007. Man and fire in southern California: doing the math. Fremontia 35:25-29.

Schlesinger, W. H. 1997. Biogeochemistry: an analysis of global change. Academic Press, San Diego, California, USA.

Schroter, M., and R. P. Remme. 2016. Spatial prioritisation for conserving ecosystem services: comparing hotspots with heuristic optimisation. Landscape Ecology 31:431-450.

Scott, K. M., and R. P. Williams. 1978. Erosion and sediment yields in the Transverse Ranges, southern California USA. USGS Professional Paper 1030. US Geological Survey, Washington D.C., USA.

Shaw, M. R., L. Pendleton, D. R. Cameron, B. Morris, D. Bachelet, K. Klausmeyer, J. MacKenzie, D. R. Conklin, G. N. Bratman, J. Lenihan, E. Haunreiter, C. Daly, and P. R. Roehrdanz. 2011. The impact of climate change on California's ecosystem services. Climatic Change 109:S465-S484.

Smith, N., R. Deal, J. D. Kline, D. Blahna, T. Patterson, T. A. Spies, and K. Bennett. 2011. Ecosystem services as a framework for forest stewardship: Deschutes National Forest overview. General Technical Report PNW-GTR-852. USDA Forest Service, Pacific Northwest Research Station. Portland, Oregon, USA.

Smith, R. J., J. Easton, B. A. Nhancale, A. J. Armstrong, J. Culverwell, S. D. Dlamini, P. S. Goodman, L. Loffler, W. S. Matthews, A. Monadjem, C. M. Mulqueeny, P. Ngwenya, C. P. Ntumi, S. Soto, and N. Leader-Williams. 2008. Designing a transfrontier conservation landscape for the Maputaland centre of endemism using biodiversity, economic and threat data. Biological Conservation 141:2127-2138.

Spencer, W. D., P. Beier, K. Penrod, K. Winters, C. Paulman, H. Rustigian-Romsos, J. Strittholt, M. Parisi, and A. Pettler. 2010. California essential habitat connectivity project: a strategy for conserving a connected California. Prepared for California Department of Transportation, California Department of Fish and Game, and Federal Highways Administration, http://www.wildcalifornia.org/wp-content/uploads/2014/04/CEHC_Plan_MASTER_030210_3-reduced.pdf

Stephenson, N. 1998. Actual evapotranspiration and deficit: biologically meaningful correlates of vegetation distribution across spatial scales. Journal of Biogeography 25:855-870.

Tallis, H. T., T. H. Ricketts, D. Ennaanay, E. Nelson, K. Vigerstol, G. Mendoza, and D. Cameron. 2014. InVEST 2.5.6 beta User's Guide. The Natural Capital Project. http://data.naturalcapitalproject.org/invest-releases/documentation/InVEST_2.5.6_Documentation.pdf

USDA [US Department of Agriculture]. 2004. Southern California National Forests land management plan. US Department of Agriculture, Forest Service, Vallejo, California, USA. http://www.fs.usda.gov/Internet/FSE_DOCUMENTS/fsbdev7_007721.pdf

van der Knijff, J. M., R. J. A. Jones, and L. Montanarella. 2000. Soil erosion risk assessment in Europe. EUR 19044 EN. European Commission Directorate General. Joint Research Centre, Space Applications Institute, European Soil Bureau, Ispra, Italy.

Westerling, A. L., and B. Bryant. 2006. Climate change and wildfire in and around California: fire modeling and loss modeling. CEC-500-2006-190-SF. Report from the California Climate Change Center to the California Energy Commission. http://www.energy.ca.gov/2005publications/CEC-500-2005-190/CEC-500-2005-190-SF.PDF

Westerling, A. L., H. G. Hidalgo, D. R. Cayan, and T. W. Swetnam. 2006. Warming and earlier spring increase western US forest wildfire activity. Science 313:940-943.

Wittenberg, L., D. Malkinson, O. Beeric, A. Halutzya, and N. Teslera. 2007. Spatial and temporal patterns of vegetation recovery following sequences of forest fires in a Mediterranean landscape, Mt. Carmel Israel. Catena 71:76-83.

Wohlgemuth, P. M., J. L. Beyers, and S. G. Conard. 1999. Postfire hillslope erosion in southern California chaparral: a case study of prescribed fire as a sediment management tool. General Technical Report PSW-GTR-173. USDA Forest Service, Pacific Southwest Research Station, Albany, California, USA.

Wohlgemuth, P. M., K. R. Hubbert, J. L. Beyers, and M. G. Narog. 2009. Post-fire watershed response at the Wildland/Urban Interface, southern California. Pages 137-142 *in* R. M. T. Webb, and D. J. Semmens, editors. Planning for an uncertain future - monitoring, integration, and adaptation. Proceedings of the Third Interagency Conference on Research in the Watersheds, Estes Park, Colorado, September 8-11, 2008. USGS Scientific Investigations Report 2009-5049. US Geological Survey, Reston, Virginia, USA.

Chapter 10
Recreation Ecosystem Services from Chaparral Dominated Landscapes: A Baseline Assessment from National Forests in Southern California

Cloé Garnache, Lorie Srivastava, José J. Sánchez, and Frank Lupi

Abstract This chapter examines recreation ecosystem services provided by chaparral dominated landscapes. Such areas are popular around the world amongst recreation users, including hikers, mountain bikers, campers, and nature enthusiasts. Yet, relatively few studies have documented the recreation services provided by chaparral landscapes such as national forests. For policy makers to manage these areas effectively, baseline information on the provision of recreation services and the populations who benefit is important, especially given current stressors such as overuse and projected climate change effects. To this end, this chapter examines four chaparral dominated national forests surrounding the Los Angeles and San Diego metropolitan areas, namely the Angeles, Cleveland, Los Padres, and San Bernardino. Using data from the USDA Forest Service's National Visitor Use Monitoring (NVUM) survey, we discuss the types of visitors using these public lands and their recreation use patterns. Our analyses suggest recreation in chaparral dominated national forests is especially important for minorities. Yet, these landscapes are facing altered human and natural disturbance regimes that may affect the recreation services they provide.

Keywords Chaparral · National forests · National Visitor Use Monitoring survey · Recreation demand · Visitor survey

C. Garnache (✉) · F. Lupi
Michigan State University, East Lansing, MI, USA
e-mail: garnache@msu.edu

L. Srivastava
University of California, Davis, CA, USA

J. J. Sánchez
USDA Forest Service, Pacific Southwest Research Station, Riverside, CA, USA

E. C. Underwood et al. (eds.), *Valuing Chaparral*, Springer Series on Environmental Management, https://doi.org/10.1007/978-3-319-68303-4_10

10.1 Introduction

Chaparral landscapes are enjoyed by outdoor recreationalists in each of the world's five Mediterranean-type climate regions—in the United States (US) and northern Baja of Mexico, the Mediterranean Basin, Chile, South Africa, and Australia. Quantification of these uses and their associated benefits, however, has been limited, both on private and public lands. Policy makers wishing to manage landscapes such as national forests and national parks for present and future recreation use can potentially benefit from an understanding of the provision of recreation services provided and the stressors these services face. Before embarking on a more detailed study of future challenges, baseline information and insight on recreation services is needed.

The United Nation's Millennium Ecosystem Assessment (MEA) provides an assessment framework and defines ecosystem services as "the benefits people obtain from ecosystems [including] provisioning services such as food, water, timber, and fiber; regulating services that affect climate, floods, disease, wastes, and water quality; cultural services that provide recreational, aesthetic, and spiritual benefits; and supporting services such as soil formation, photosynthesis, and nutrient cycling" (Millenium Ecosystem Assessment 2005). Thus, recreation is a cultural service within an ecosystem service framework. The United States Forest Service (USFS) has long recognized that ecosystems from public lands provide multiple benefits to users of national forests. The multiple uses acknowledged by national forest policy include outdoor recreation, timber, watersheds, and wildlife and fish, which were made explicit in the Multiple-Use Sustained-Yield Act of 1960. As such, in the United States, recreation has long been an element of the multiple uses management paradigm (Fedkiw 1998).

Recreation activities in natural landscapes appeal to outdoor enthusiasts, in particular when the forests are in close proximity to heavily urbanized areas where substitute outdoor activities may be scarce. In the US, national forests, managed by the USFS, are examples of such landscapes. One of the mandates of the USFS is to manage its land to provide recreation opportunities for the general public (Vincent et al. 2014; USDA Forest Service 2015a). In addition, managers of national forests are increasingly mindful of such pursuits for physical and mental health as they design and implement policies to sustainably manage such landscapes (Bell et al. 2009) both for the present as well as the future. As a first step, it could be helpful for decision makers to understand who visits the national forests for recreation and the types of activities they participate in, so as to prioritize investments of effort and money in national forest maintenance and enhancement.

This chapter reviews studies of recreation services provided by chaparral landscapes around the world, although the existing literature is relatively scant. To develop a baseline assessment that may aid management for present and future generations, we undertake a case study of the recreation services provided by four southern California national forests dominated by chaparral vegetation, namely, the Angeles, Cleveland, Los Padres, and San Bernardino. The proximity of these chaparral dominated national forests to large urban centers makes them unique as they provide recreation opportunities to millions of visitors every year (USDA Forest Service 2015b).

We use the National Visitor Use Monitoring (NVUM) survey to determine the types of visitors and the activities they engage in while in the four national forests. Organized by the USFS, the NVUM surveys—started in 2004 and continuing to the present—collect data from visitors in each national forest using a detailed questionnaire that typically takes five minutes or less to complete. The results provide information on a broad range of recreation activities from day use at picnic areas to overnight visitation. They encompass hundreds of individual sites at each national forest, including trailheads, picnic areas, visitor centers, parking lots, as well as road stops set up throughout the forests. Given the length and depth of data collection, the NVUM data are an important information source. For example, in conjunction with other surveys, these data suggest that multiple factors determine the type and frequency of recreation activities in which visitors participate, including the characteristics of the landscape, the recreation infrastructure, and distance from urban centers. As a result, the NVUM data are potentially helpful to the USFS for forest planning and decision making. For instance, the description of visitor characteristics can help forest staff identify recreation niches for each national forest such as recreating near water sources, provide staff with a profile of the visitors they serve, and help determine their needs.

Regardless of the location of chaparral landscapes, decision makers and managers of these lands are faced with the challenge of how to best manage the natural resources successfully for present and future recreationalists. Many resource-based recreation activities will likely be overused by growing populations seeking natural landscapes for recreation and by climate change as it can potentially affect both weather patterns and landscape characteristics. In the study area of southern California, projected climate effects include changes in precipitation, temperature, drought, and wildfire, along with changes in the distribution of chaparral vegetation (see Chap. 14).

This chapter addresses gaps in our knowledge of visitor characteristics, recreation activities, overuse and degradation, activities by under-represented groups, and the potential effects of climate change by providing a baseline assessment of the study area. An understanding of these issues may help inform appropriate and timely policies to improve natural resource management of chaparral for recreation, especially in the face of human and natural stressors. A baseline assessment of visitor numbers and uses of chaparral dominated forests may be foundational for managers of these public spaces as they plan and budget for initiatives to successfully deal with these challenges.

10.2 Literature Review of Recreation in Chaparral and Similar Landscapes

Despite chaparral landscapes being used for recreation in many parts of the world, few studies have attempted to measure the recreation services provided by these landscapes, much less address future challenges or use by minority groups. Where studies have been undertaken, differences in study designs and their respective findings make generalizations difficult. For example, multiple environmental studies show the negative effects of human disturbance (e.g., recreation and trail use) within chaparral landscapes (Sauvajot et al. 1998; Witztum and Stow 2004; George and Crooks 2006), but these studies do not document recreation ecosystem benefits provided to visitors. In another example, a study of visitor preferences conducted in central Spain investigates the preferences of cultural tourists for recreation activities and landscape types (Schmitz et al. 2007). Chaparral landscapes, one of the four landscape types examined, was overall less preferred than other landscapes with different vegetation types, such as woodland mosaics.

Koniak et al. (2011) investigated the preferences of hikers and picnickers for different types of natural vegetation formations in Israel. They found that visitors preferred to hike in the open garrigue (a type of chaparral) more than other nearby vegetation types, namely dense scrub or pine forest. Alternatively, picnickers preferred either open garrigue or planted pine forest to dense scrub. Their study suggests chaparral provides a valuable service to recreation users. In contrast, in a classic study of global ecosystem services, Costanza et al. (1997) assign very small recreation values to chaparral, grassland, and rangeland compared to other vegetation types.

We can infer the range of welfare losses due to climate change or overuse and degradation by examining studies of recreation site closures in chaparral and other areas. For example, in the Kassandra peninsula in northeastern Greece, Papaspyropoulos et al. (2015) estimate the loss of a chaparral site for the hunting season increased travel expenditures (a rough proxy for value) by about $67,000 in 2007 dollars, largely due to greater travel distances required to get to alternate sites. In comparison, the closure of trailheads in the Alpine Lakes Wilderness in Washington State in the US is associated with an annual loss ranging from $5,500 to $2,500 (Hilger and Englin 2009). Baerenklau et al. (2010) and Sánchez et al. (2016) show that the closure of popular hiking sites in the San Bernardino National Forest—one of our case study national forests—in southern California results in losses in the range of $0.5 million to $3 million per season per site. These differences in economic value across sites are likely due to the differences in the number of users and the availability of nearby substitute recreation sites, highlighting the need to have a baseline understanding of recreation use.

In our study area of southern California, wildfires have increased in frequency and size as a consequence of fuel accumulation, drought effects, and type-conversion to non-native grasses (see Chaps. 12 and 15). While some research exists on the implications of climate change on forest-based recreation in some parts of the US (e.g., Irland et al. 2001; Richardson and Loomis 2004), a larger literature has explored the effect of wildfire on demand for forest recreation. Duffield et al. (2013) estimate that about 35,000 trips were lost per year due to wildfires in the states of Montana, Idaho, and Wyoming over the period 1986–2011. Nevertheless, the effect of fire on recreation often depends upon the type of activity in which the visitors engage, on landscape and fire characteristics, and on the recovery time after a fire. The extent of the effect has been found to vary by activity, for example, the number of hiking trips reduced by 5%, but mountain biking trips declined by more than double (−12%) in national forests in Montana (Hesseln et al. 2004). Similarly, demand for camping was affected differently in Montana, Idaho, Washington, and Alberta (Rausch et al. 2010; Duffield et al. 2013).

In terms of landscape and fire characteristics, crown fires (i.e., forest fires that burn through the vegetation canopy) and the percentage of area burned visible from the trail negatively affect the number of recreation trips relative to less intense fires (Loomis et al. 2001). It is noteworthy that almost all fires in chaparral dominated landscapes are crown fires. Studies have also found that the lag time since a fire also affects trip demand. Several studies indicate that trip demand is negatively affected immediately after a fire, but does eventually return to prefire levels (Loomis et al. 2001; Hesseln et al. 2003; Englin et al. 2006; Boxall and Englin 2008), with the lag period dependent upon the activity type. Studies record a spike in visits in response to well-managed prescribed burns and recent fires (Englin et al. 2001; Boxall and Englin 2008; Sánchez et al. 2016). In particular, Hilger and Englin (2009) record a hiking trip demand that surpasses prefire levels just 4 years after a burn in the Alpine Lakes Wilderness in Washington State. Englin et al. (2001) describe the demand dynamics for hiking trips in response to fire for the US Rocky Mountain region as initially resulting in an increase in the number of trips immediately after a fire, followed by a decrease after 3 years, and then the number of trips slowly returned to prefire levels. Several reasons may drive this initial spike in visits to areas impacted by fire, including wildflower blooms following a fire (Loomis et al. 2001) and an abundance of various animal species (Englin et al. 2001; Hilger and Englin 2009). In addition, Starbuck et al. (2006) suggest that since low-intensity fires thin the wood biomass while leaving large trees unharmed, they may be seen as opening and enhancing the viewshed. Indeed, given this finding it could be assumed that a reduction of hazardous fuels and forest restoration activities may be viewed by recreation users as increasing the quality at a given recreation site.

How might wildfires affect the value of lost recreation opportunities in chaparral landscapes? Studies by Shechter and colleagues have addressed recreation use values in Mount Carmel National Park in Israel, which includes notable areas of Mediterranean maquis landscape (a type of chaparral). Using revealed and stated preference approaches, they find values for recreation use in chaparral that parallel the rental value of converting the land to agriculture (Shechter et al. 1998). Freeman (2012) used values from Shechter et al. (1998), along with other values, to develop a model to estimate the economic consequences of a road expansion infrastructure project on woodland succession, a maquis landscape in the Carmel range of northern Israel. The construction of the road was predicted to alter the natural fire regime resulting in an increase of open maquis and loss of moderate and dense maquis. Freeman's findings suggest that the road expansion project would reduce natural maquis cover thereby affecting both economic use values (including recreation) and non-use ecological values (although a breakdown of the effect on recreation was not provided). In summary, although the literature indicates that chaparral landscapes provide valuable services to the public, studies on recreation use in chaparral landscapes remain scant and more studies are needed to improve our ability to meet present and future challenges to these landscapes.

10.3 Case Study: Southern California

We explore relationships between recreation use and chaparral dominated national forests in the study area using National Visitor Use Monitoring data. The information collected by these surveys is now required for national forest plans as part of Executive Order 12862 (Setting Customer Service Standards). Using the NVUM data, we examine visitors' demographics and the activities in which they engage while visiting the four southern California national forests. In addition, by linking NVUM visitor use data to site characteristics such as vegetation type and distance to water, we are able to provide information for managing the threats faced by these national forests. For example, the visitation estimates can be helpful in considering visitor capacity issues to prevent site overcrowding, overuse, and degradation.

10.3.1 Study Area and the National Visitor Use Monitoring Survey

The study area encompasses the four national forests in southern California that border the Los Angeles and San Diego metropolitan areas: the Angeles, Cleveland, Los Padres, and San Bernardino (Fig. 10.1). Collectively, these four national forests cover over 1.4 million ha (3.5 million acres). The most northern one is the Los Padres National Forest, north of Los Angeles and stretching nearly to Monterey, and the most southern is the Cleveland, which runs to the international border with Mexico. The area is heavily urbanized, with a population of over 23 million people across the ten different counties within which at least one of the national forests is situated (US Census Bureau 2014a), with eight million people living within a 1 h drive of these forests (USDA Forest Service 2005a). The primary vegetation type across these national forests is chaparral shrubland, but other vegetation types exist including hardwoods, conifers, and grasslands.

Although the NVUM surveys started in 2004, the initial round of data collection (Round 1) was part of a pilot program to test the methodology. The methodology was subsequently updated and finalized as a result of the pilot. We therefore only use the subsequent Rounds 2 and 3. Each national forest is surveyed once every 5 years to provide representative information on visitors and use, including seasonal variations throughout the year. The Angeles National Forest was surveyed in 2006 (Round 2) and 2011 (Round 3), while the Cleveland, Los Padres, and San Bernardino were surveyed in 2009 (Round 2) and 2014 (Round 3) (USDA Forest Service 2015b). Over the two NVUM rounds, a total of 126 interview sites were surveyed in the Angeles, 136 in the San Bernardino, 126 in the Los Padres, and 92 in the Cleveland. A total of 9614 people were interviewed in Rounds 2 and 3 of NVUM (Table 10.1). Of the visitors intercepted and questioned, 89% stated their primary purpose was to recreate in the national forest (in contrast to simply passing through or being on a work-related visit). Hereafter we refer to these people as "visitors" or "recreation users" of the national forests.

Fig. 10.1 Location of the four national forests in the southern California study area: the Angeles, San Bernardino, Los Padres, and Cleveland. Source: US Forest Service Pacific Southwest Region

Table 10.1 Number of people interviewed in the NVUM survey in Rounds 2 and 3

	Round 2	Round 3	Total
Angeles	1303	2150	3453
San Bernardino	682	558	1240
Los Padres	922	1090	2012
Cleveland	1452	1457	2909

Source: NVUM surveys, rounds 2 and 3 (USDA Forest Service 2015b)

The NVUM survey uses a random stratified sample, ensuring it is representative of total use at the national forest level (English et al. 2002). The sampling frame is defined using a combination of four basic "site type" categories: Day Use Developed Sites (DUDS), Overnight Use Developed Sites (OUDS), Designated Wilderness Areas (Wilderness), and General Forest Areas (GFA). It defines "use level" based on the expected level of recreation visitors for each site and day of the year: very high, high, medium, low, or no use. Days and sites are randomly sampled within each stratum of site type and use level. On average, DUDS represents 25%, OUDS 28%, GFA 34%, and Wilderness 14% of the interview sites across the four forests.

To extrapolate the total estimated number of annual visits to each national forest using the NVUM interviews collected, we used the weights from the stratified sampling scheme to aggregate data from individual interviews to the level of each forest following the procedures developed by English et al. (2002). All the numbers we report regarding users and uses of the national forest take into account total estimated annual visitation. Furthermore, given that two rounds of the NVUM survey are available, our analysis focuses on the average users and uses across the two rounds.

The NVUM survey collects information on age, gender, race, ethnicity, home zip code, date of visit, name of the site visited, primary activity of the user, and activities in which the user engages. Detailed activity information is collected using a list of 31 potential activities divided into six general categories; fishing/hunting, viewing, learning about nature and culture, non-motorized activities, motorized activities, camping or other overnight activities, and other activities. Using this detailed information, we define water-related activities as fishing and non-motorized and motorized water travel. We also define snow-related activities as downhill skiing and snowboarding, cross-country skiing, snowshoeing, and snowmobile travel.

10.3.2 Additional Data Sources

We combine the NVUM data with other data sources to yield further insights with respect to preferences, income level, and allow spatial analysis. To analyze the distance visitors travel from home to the sites they visit in the national forests, we use the software PC*Miler (http://www.pcmiler.com/). This calculated the travel distance and travel time through a road network from the national forest sites where users are interviewed to the centroid of the visitor's home zip code that was recorded in the surveys.

In addition, to better grasp users' economic background, we matched the visitors' reported zip code from NVUM with Census data (US Census Bureau 2014b). The Census provides information on mean and median socio-economic characteristics at the zip code level. For our purposes, we report information on median income for the home zip code of each visitor.

To indicate the extent that water features may attract visitors, we calculate the distance of each national forest site in the NVUM survey to lakes, and perennial and intermittent streams (CDFW 2015) using ArcGIS. Visitors access streams and lakes using paths and trails; however, since we do not have accurate trail information for most sites, we calculate access to water as the shortest distance to the closest stream or lake (i.e., in a straight line). Note that we do not know whether users actually use the streams and lakes.

Finally, we link each recreation site in the NVUM survey to vegetation data (FRAP 2015) in ArcGIS. We are thus able to examine which vegetation types are most visited in the national forests. While some sites are surrounded by a single vegetation type, others are surrounded by a mix. To facilitate the analysis, we focus on the dominant vegetation type within a 3.22 km (2 mile) radius from the site's parking lot.

10.4 Results: Patterns of Visitation and Activities

Analyzing the NVUM data provides a foundational assessment of recreation within chaparral dominated public forests. We are able to quantify the number of visits across the four national forests, distance travelled by visitors, their socio-demographic characteristics, and recreation activities. In addition, the analysis can

Table 10.2 Total annual visitation estimates (millions) based on two rounds of NVUM data collected across the four southern California national forests

	Round 2		Round 3	
	Total site visits	Total national forest visits	Total site visits	Total national forest visits
Angeles	3.4	3.0	4.4	3.6
San Bernardino	2.7	2.4	2.2	1.9
Los Padres	1.1	0.9	1.3	0.9
Cleveland	0.5	0.5	0.7	0.6

Note: a single national forest visit may include multiple site visits
Source: NVUM Surveys

potentially help to determine overuse of recreation sites, which may have implications for future management decisions of these and other recreation facilities.

10.4.1 Estimated Number of Total Visits to the National Forests

Of the four national forests in southern California, annual visitation is highest at Angeles National Forest with over three million visitors, resulting in four million site visits (as visitors stop at multiple sites), followed by the San Bernardino with around two million visits annually. The Los Padres and Cleveland, which are more spatially dispersed and farther from the city of Los Angeles, attract about 1.6 million visitors annually (Table 10.2). Despite slight variations across the two NVUM rounds (which span 5 years), visitation numbers remain strong and tend to increase over the time period of 2006–2011, with the exception of the San Bernardino. Note, however, that given the extent of snow-driven activities in the San Bernardino, a poor snow year may dramatically lower the number of visitors.

10.4.2 Visitors to the National Forests

Users' demographics reveal that females typically make up less than one-third of all the visitors to the four national forests, with 69% of the visitors in the Angeles being male, 64% in the San Bernardino, 68% in the Los Padres, and 74% in the Cleveland. In addition, Whites generally account for over three quarters of national forest visitors. Hispanics comprise the second largest group of visitors, averaging about 17% of visitors across the four forests (note that some Hispanics self-identify as Whites in addition to Hispanic). Asian-Americans are a distant third place, accounting for an average of about 7% of visitors (again, with some self-identifying as Whites in

Table 10.3 Average percentage of national forest visits by race and ethnicity based on two rounds of NVUM data collected across the four southern California national forests

Race/Ethnicity	Angeles (%)	San Bernardino (%)	Los Padres (%)	Cleveland (%)
White	76	84	87	89
Asian-American	12	7	5	5
African-American	5	3	2	2
Native-American	4	4	4	3
Pacific and Islander	3	2	1	0
Hispanic	21	19	17	12

Note: race was asked separately from ethnicity. Percentages are conditional upon respondents agreeing to answer about the race with which they most closely identify. Sums can be greater than 100% since people could self-identify with more than one group and, in some cases, race and ethnicity may overlap
Source: NVUM surveys

Box 10.1 Minority Outdoor Participation and Perceived Barriers

The Latino Outdoors non-profit organization (http://latinooutdoors.org/) conducted a national survey that focuses on Hispanics who have never (or only a few times) visited national forests, national parks, or other public lands. Individuals and families were recruited by Latino Outdoors to participate in outdoor excursions throughout the US (California, Florida, New York, Texas, and Washington, D.C.) during the 2015 and 2016 calendar years. For this example, we focus on excursions in southern California. Latino Outdoors surveyed participants both before and after the excursions. The non-profit was responsible for recruitment and the administration of the survey. They provided completed surveys to the USFS for analysis (survey response rate is unknown). The voluntary surveys were designed to provide information on outdoor recreation activities, potential barriers, and determine the effectiveness of the Latino Outdoors program in promoting outdoor participation by under-represented visitors.

The pre-trip survey was administered prior to the visit to the local recreation site. The survey asked whether they had ever visited a national or state park or national forest. If they had not, they were asked why. If they had visited one, they were asked questions about the number of those trips in the past 12 months, the travel distance, travel hours, and travel cost. In addition, several questions were asked about the characteristics of national or state parks or national forests that were important to them. The survey also included six open ended questions for survey participants to write qualitative information concerning their sense of responsibility toward public lands, meaning of the outdoors, expectations of participating in outdoor activities, perceptions of Hispanics in the outdoors, and memorable outdoor experiences.

The individuals were then taken on a one-day trip to a local outdoor recreation site. The post-trip survey was composed of three types of questions: (a)

questions about the trip they took, including an open-ended willingness to pay question; (b) similar questions to the pre-trip survey to see if the Latino Outdoors experience had a substantial effect on their responses; and (c) demographic questions (e.g., age, education level, gender, home zip code, and household income). A total of 28 participants completed both the pre- and post-trip surveys in either English or Spanish. The survey respondents were relatively young, averaging 31 years of age, almost half (49%) had at least a bachelor's degree, and the average reported annual household income was $46,000. A significant portion were recent immigrants, as 39% responded that they are the first generation of their family to live in the US. Regarding recreation trips to the outdoors, respondents were willing to pay an average of $27 to visit a local day trip recreation site. Results show that the majority (84%) of respondents have visited a national forest, national park, or other public land in the past 12 months and they travel approximately 61 miles (~100 km) to reach their recreation destination. When visiting a national forest or national park, 82% of respondents felt welcomed by the staff and 77% felt welcomed by other races/ethnicities. The most common activities in which they engaged during their visits are hiking (80%), camping (34%), picnicking (25%), and bird-watching (16%).

Survey respondents believe that public lands have important functions and value them. For example, they rated national forests as important or very important in providing opportunities for recreation, viewing natural scenery, environmental education, spending time with family, a source of inspiration, and habitat for wildlife. Despite these valuable benefits, 65% of respondents believe that their family and friends do not spend enough time visiting national forests. Some reasons provided include lack of financial resources and limited knowledge of recreation opportunities available to the public.

addition to Asian-American). The highest percentages of Hispanic and Asian-American visitors were both found on the Angeles National Forest (Table 10.3).

Most visitors to national forests are over 20 years of age. Nevertheless, our findings show that the number of children under the age of 16 is substantial in all four national forests, especially in the Angeles and San Bernardino (Fig. 10.2, see final bar in each plot).

The vast majority of visitors to these four national forests are local residents, with over half the visits to the Angeles, Los Padres, and Cleveland made by people who live within 80 km (50 miles) of the forests. The San Bernardino attracts people from farther away, in part related to opportunities provided for winter sports (Table 10.4).

The annual household median income for visitor's home zip code, based on the 2015 American Community Survey 1-Year Estimates (US Census Bureau 2014b),

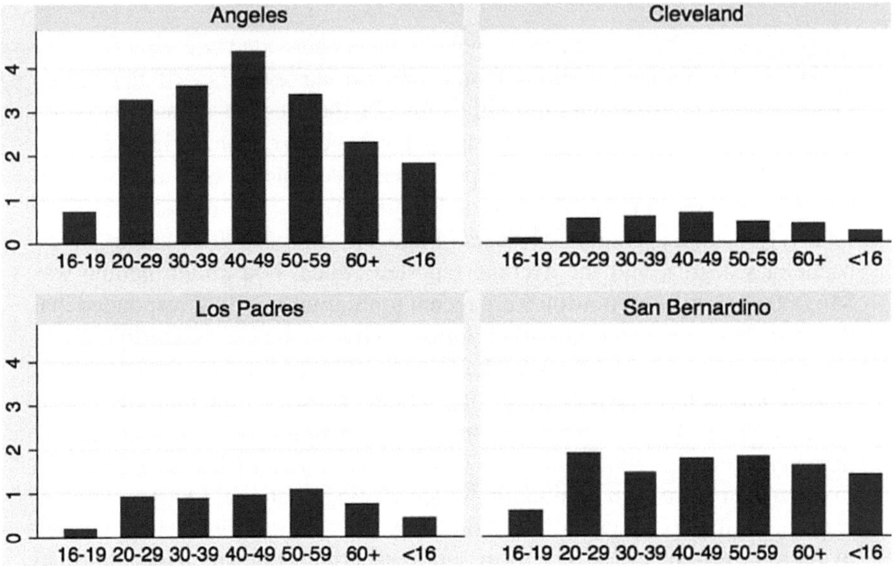

Fig. 10.2 Average self-reported annual visits to the national forests by age. Source: NVUM surveys

Table 10.4 Proportion of visits by average distance to national forests based on two rounds of NVUM data collected across the four southern California national forests

Distance (miles)	Angeles (%)	San Bernardino (%)	Los Padres (%)	Cleveland (%)
<25	40	24	46	33
25–50	24	14	12	29
50–75	18	14	9	29
75–100	14	18	7	4
100–250	3	26	17	3
>250	10	4	10	2

Source: NVUM surveys

provides a partial picture of the neighborhoods from which visitors are drawn (Table 10.5). The annual household median income from visitor's home zip codes appears to be higher than both the national median income (around $54,000) as well as the southern California median income (around $64,000).[1] Over half of visitors come from zip codes where annual household median income is between $50,000–$100,000 (Table 10.5).

[1]The ten counties in which at least one of the four National Forests is situated were used in the median income calculation, which are: Kern, Los Angeles, Monterey, Orange, Riverside, San Bernardino, San Diego, San Luis Obispo, Santa Barbara, and Ventura. The values are in 2015 dollars.

Table 10.5 Annual household median income for the zip code of origin of visitors to national forests in southern California based on two rounds of NVUM data collected across the four southern California national forests

Income range	Angeles (%)	San Bernardino (%)	Los Padres (%)	Cleveland (%)
<$50,000	14	13	9	28
$50,000–75,000	43	32	48	37
$75,000–100,000	35	37	28	26
$100,000–150,000	6	16	13	8
>$150,000	2	1	2	2

Source: NVUM surveys

Table 10.6 Proportion of national forest visits by dominant vegetation type

Vegetation type	Angeles (%)	San Bernardino (%)	Los Padres (%)	Cleveland (%)
Chaparral (shrubs)	70 (71)	11 (38)	97 (83)	100 (100)
Conifers	30 (29)	89 (61)	2 (16)	0 (0)
Hardwood trees	0 (0)	0 (1)	1 (1)	0 (0)

Note: the proportion of NVUM sites by their dominant vegetation type is reported in parenthesis (e.g., 38% of the NVUM sites in San Bernardino have chaparral shrubland as the dominant vegetation within a 2-mile radius of the parking lot). Our analysis excludes NVUM data from the Monterey District of the Los Padres and from the Palomar District of the Cleveland National Forest, which contain areas with heavily-used conifer and hardwood forests. Source: NVUM surveys

10.4.3 Recreation Within the National Forests

To determine what types of outdoor recreation activities respondents pursued in the national forests we analyzed the NVUM data for relationships between visits and vegetation type, month, type of recreation activity, and recreation features. The analysis of NVUM site location data showed the majority of annual visits to the four national forests were to sites dominated by chaparral vegetation (70% in the Angeles, 97% in the Los Padres, and virtually 100% in the Cleveland). The exception is the San Bernardino where visits to chaparral sites accounted for only 11% of total visits, while conifer vegetation accounted for the remaining visits (Table 10.6). When contrasted with the dominant vegetation type of the sites where the NVUM interviews were conducted, visitors disproportionally targeted chaparral vegetation in the Los Padres, while they disproportionally targeted sites with conifers in the San Bernardino—which can be partially explained by higher elevation, winter sports recreation. For example, in the San Bernardino, 61% of the NVUM sites are conifer but 89% of the visits were to these sites.

The number of visits to these four national forests is generally highest in the summer between June and August, except for the San Bernardino which is most heavily used in the winter between December and April for downhill skiing and other snow-related activities, and the Cleveland which is most popular in the non-summer months (Fig. 10.3).

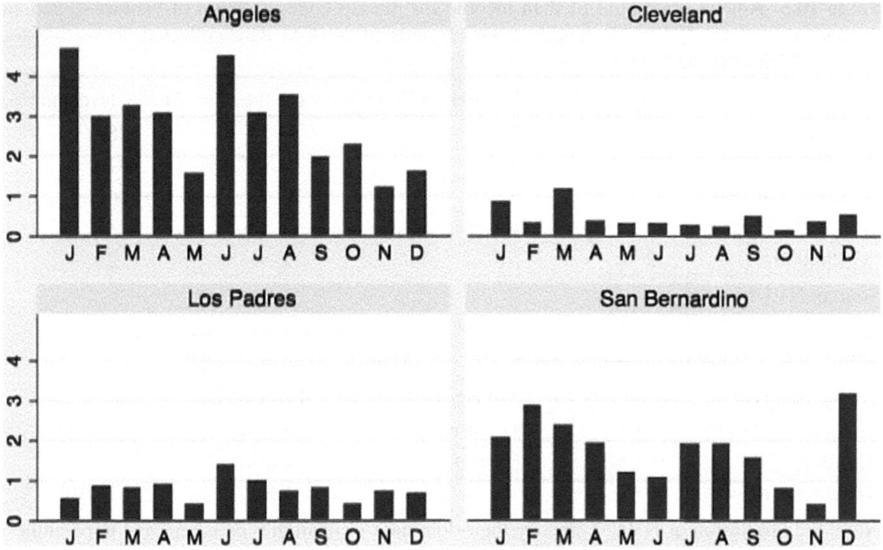

Fig. 10.3 Average monthly visits to the national forests in southern California (millions). Source: NVUM survey

Table 10.7 Average percentage of visits to national forests by primary recreation activity

Activity	Angeles (%)	San Bernardino (%)	Los Padres (%)	Cleveland (%)
Hiking	47	27	39	46
Biking	4	4	9	13
Relaxing, picnic, nature	12	13	21	10
Water activities	9	3	2	0
Snow activities	11	30	1	0
OHV use	3	5	2	2

Source: NVUM survey

In terms of different activities, the most common primary activity in the four national forests is hiking, with the exception of snow-related activities in the San Bernardino. Many visitors also report relaxing, picnicking, or nature viewing as their primary activity. Biking is relatively more popular in the Los Padres and Cleveland than in the Angeles and San Bernardino, where water-related activities and Off-Highway Vehicle (OHV) use are slightly more important (Table 10.7).

Visitors often participate in more than one activity, i.e., other activities in addition to their primary activity. For example, although relaxing, picnicking, and being in nature were not the primary activity of the majority of visitors, virtually all visitors reported participating in these pursuits (Table 10.8). The same applies to hiking, albeit to a lesser extent. In terms of patterns for different ethnicities (not shown in the tables), the distribution of activities that Hispanics engage in is relatively similar to that of the average visitor to these national forests, with slightly higher

Table 10.8 Average percentage of national forest visits by recreation activity

Activity	Angeles (%)	San Bernardino (%)	Los Padres (%)	Cleveland (%)
Hiking	63	47	66	64
Biking	5	6	11	15
Relaxing, picnic, nature	94	78	100	100
Water activities	11	6	4	1
Snow activities	11	30	1	0
OHV use	3	9	4	4

Note: totals for each national forest can exceed 100% as visitors may engage in multiple activities during each visit. Source: NVUM surveys

Table 10.9 Proportion of national forest visits by distance to water

Distance (m)	Angeles (%)	San Bernardino (%)	Los Padres (%)	Cleveland (%)
<100	29	47	34	9
100–250	11	27	2	14
250–500	32	1	3	3
>500	28	25	61	73

Source: NVUM surveys

participation levels in water-related activities and lower participation in snow activities, especially in the San Bernardino.

Our analysis of site survey location to lakes and streams shows most visits take place very close to water sources (Table 10.9). In the San Bernardino National Forest, for example, almost 50% of visits were <100 m from water. Two caveats, however, are important to note. First, the distance measured reflects the shortest distance which may dramatically under-estimate the actual distance a visitor may have to walk to access a stream or lake, especially in areas characterized by steep topography. Second, the presence of a stream or lake nearby does not imply that the visitor takes advantage of this nearby access to water. Nevertheless, this information may be used in conjunction with data on water activities to hypothesize that in addition to water activities, visitors engaged in hiking, picnicking, and other activities may also be drawn to sites near water, for instance, hikers may prefer trails near streams or lakes (Tables 10.7 and 10.8).

10.4.4 Recreation Site Degradation from Overuse

The NVUM survey provides USFS decision makers with baseline data to assess current use patterns that can help inform future initiatives and strategic directions. Overuse and degradation of recreation sites are of concern as they could potentially prevent the agency from fulfilling its mandate to protect public lands for present and future generations. Our analysis of the NVUM data on total visitation by site reveal that although many of the recreation sites within the four southern California national forests receive low or moderate use levels, some sites are heavily used.

Fig. 10.4 Heavy use and user-made recreation dams at the East Fork of the San Gabriel River in the Angeles National Forest. Photo by Charles White

Two sites which are heavily used in the Angeles National Forest, especially during the weekends, include the West Fork of the San Gabriel River on San Gabriel Canyon Road (Highway 39) and the Oaks Picnic Area on East Fork road of the San Gabriel River. Both sites are dominated by chaparral vegetation with oak trees near the parking area and cottonwood and other riparian trees along the stream. The West Fork of the San Gabriel River site is visited by an average of 100,000 visitors per year (representing about 3% of the total annual visits to the Angeles), while the Oaks Picnic Area, which has a much smaller parking area, sees an average of 17,000 visits a year.

These sites are highly used by Hispanics who account for 30% of all visitors at the West Fork of the San Gabriel site, and the majority of visits at Oaks Picnic Area. Access to water is an important characteristic of these sites: 17% of visitors at the West Fork report water activities as their primary activity and 28% report participating in water activities. These activities may involve wading in the stream but also more ecologically disruptive activities such as building dams to form shallow pools in the stream (Fig. 10.4).

10.5 Discussion: Climate Impacts, Overuse, and Creating Opportunities

Chaparral landscapes are sought by outdoor enthusiasts in many parts of the world for their recreation opportunities. Our analysis of NVUM data along with a supplementary community survey conducted by Latino Outdoors (see Box 10.1) provides baseline information that could help land managers better manage public lands when faced with present and future challenges like climate change and resource degradation.

10.5.1 Climate Change and Its Effect on Recreation

Many ecosystem services may be adversely affected by climate change (Millennium Ecosystem Assessment 2005). In the western US, wildfires are predicted to continue to increase in size and frequency (Barbero et al. 2015; Restaino and Safford 2018). Wildfires have been shown to negatively affect the demand for recreation (Duffield et al. 2013) and wildfires often lead to site closures until the site is deemed safe for the public. The NVUM data indicate that the top three recreation activities for visitors are hiking, relaxing/picnic/nature activities, and snow-related pursuits. Unfortunately, the literature indicates that these pursuits can potentially be affected—if not curtailed—by changes in climate. Keeley et al. (2004) suggest the need for better tactical planning to address the destructive fires in southern California, for example, in the form of engineering solutions akin to that of earthquakes and other natural disasters. Whatever the approach chosen to combat wildfires or mitigate their effect, managers may consider the heterogeneous preferences of different groups when weighing their options.

In California, ecosystem services such as water provisioning and snow recreation are predicted to be significantly reduced as a result of climate change (Shaw et al. 2009). Based on our analysis of current use patterns in southern California's national forests using the NVUM data, we can hypothesize that changes in vegetation and access to water can potentially affect visitation. In particular, this might affect the San Bernardino, popular for winter sports, and the Angeles, heavily used in the summer months for water-related activities and consequently the user groups associated with these activities, e.g., Hispanics who prefer picnicking and relaxing in the shade, and those who place a high value on access to water, open views (shrubs), and shade around streams and/or picnic areas.

10.5.2 Recreation Site Degradation from Overuse

The low to moderate use levels of many sites within the four national forests in southern California suggests human disturbance may not be as problematic as it has been at some of the more heavily used sites studied in the literature for California

(Sauvajot et al. 1998; George and Crooks 2006). Nevertheless, the total visitation numbers in the NVUM data indicate potential overuse and resource degradation at several sites, including in the Angeles, likely resulting in ecosystem degradation.

The USFS recognizes increased demand and overuse of some of its recreation sites and facilities (USDA Forest Service 2005a). For example, the San Bernardino National Forest management plan explicitly focuses on relieving "concentrated demand within existing high use areas and…accommodate[ing] future growth and new uses elsewhere" (USDA Forest Service 2005b). One response on the ground by USFS staff has been to post signs at the parking lots of heavily used areas to educate visitors about the ecological damages of certain activities such as building recreation dams in streams (Fig. 10.4), although these practices seem to persist. Nevertheless, funding for providing information and public outreach to address degradation and overuse may be limited in the future.

For instance, during the period 2012–2015 the recreation budget for all four national forests declined both in nominal and real terms by 22% and 25%, respectively. In 2016, a year after part of the Angeles was declared a national monument, it was the only national forest out of the four to have an increase in its recreation budget over the previous year, and in fact higher than 2012, from $1.6 million to $2.2 million (2016 dollars).[2] Excluding the Angeles, the recreation budget for the remaining southern California's national forests declined by 26% over the 2012–2016 period (G. Macias, US Forest Service, pers. comm.).[3]

10.5.3 Recreation Opportunities for Under-represented Visitors: Ethnic and Racial Minorities

A NVUM study (USDA Forest Service 2013) reports that national forest visitors are comprised of 95.3% White, 5.4% Hispanic, 3.2% Asian/Pacific Islander, and 1.1% African American.[4] The NVUM data in our study area align with these findings from the literature and the national NVUM findings that people who did not self-identify as "white" comprised on average 16% of all visitors to the four national forests in southern California (Table 10.3). It is important to note that some respondents identified themselves both as White and other, such as Hispanic or Asian-American. Other studies have found similar responses. For example, in a national

[2] As part of President Obama's initiative to get "Every Kid in a Park," fourth grade classes became eligible to receive reimbursement for transportation costs when visiting the Angeles National Forest and San Gabriel Mountains National Monument, as well as outdoor classroom programming. See: https://www.fs.usda.gov/angeles

[3] Over the years 2012–2016, the US Forest Service allocated an average annual budget of $1.5 billion (in 2015 dollars) for its National Forest System (Hoover 2016). Total budgets have declined by 9% over this period, with an 11% decline in the budget allocated to recreation facilities (USDA Forest Service 2016).

[4] Respondents could choose to self-identify with more than one group.

study, Outdoor Foundation (2015) established that Whites (73%) participate most in outdoor recreation, while only 7% of participants are Asian/Pacific Islander, 8% are Hispanics, and 9% are African-American. Similar to our results, they identified that these under-represented groups do not participate in outdoor recreation activities in national forests, national parks, and other public lands at the same level as Whites (Tierney et al. 1998; Crano et al. 2008; Outdoor Foundation 2015). Without national efforts to connect under-represented groups to the outdoors, this trend is likely to continue.

Management concerns with providing recreation opportunities for under-represented populations may be monitored using future data collected by NVUM. There is evidence to suggest that minority groups may suffer disproportionately from changes in recreation sites due to climate change, e.g., wildfires and droughts, due to their limited financial resources and transportation access to recreation sites further away. Moreover, the scarcity or crowding of substitutes such as parks and other green spaces in the heavily urban environment of southern California may decrease their enjoyment of outdoor pursuits (Tierney et al. 1998).

The barriers to participation found by Latino Outdoors (see Box 10.1), namely financial constraints and limited knowledge of recreation opportunities, mirrors what others have found. Extensive research throughout the decades has focused on constraints faced by under-represented groups to explain their limited participation in outdoor recreation activities. Roberts and Chitewere (2011) found that outdoor participation barriers for minorities are associated with perceived or real discrimination, accessibility issues, and having no prior outdoor experience. In particular, Green et al. (2009) found that some of the perceived constraints for African-Americans and Asian/Pacific Islanders with respect to outdoor recreation were: lack of transportation, inadequate facilities and information, crowded activity areas, and pollution problems. Other studies reported personal safety as a perceived barrier to recreation (Johnson et al. 2001; Scott et al. 2004; Green et al. 2009; Byrne 2012). Relevant for our study area, Crano et al. (2008) surveyed Los Angeles County residents and found that Hispanics tend not to visit national forests due to the lack of transportation means, financial constraints, and physical limitations.

The NVUM data indicate that visitors belonging to minority groups tend to prefer different recreation activities; for example, within the study area Hispanics have a slight preference for water-based activities. Across other minority groups, the literature implies that recreation preferences of under-represented groups differ from those of Whites (Dwyer 1994). For instance, Washburne and Wall (1980) conducted a survey of the continental US population and found that African-Americans prefer urban recreation experiences, while Whites prefer wildland activities. Similarly, African-Americans are less likely than Whites to travel to wildland recreation areas (Washburne 1978; Dwyer 1994; Johnson 1998).

Hispanics also prefer to recreate in a non-traditional way (in the United States), picnicking or barbecuing in large groups for the entire day (Carr and Chavez 1993; Chavez 2002; Chavez and Olson 2009). A recent study of the immigrant Hispanic community of Cache Valley in Utah found they prefer recreation areas characterized by gathering spaces for large groups, facilities, scenery, and close to home (Madsen et al. 2014). Research on outdoor recreation activities of Hispanic and Asians use of southern California national forests also supports these findings, indicating that they tend to participate in outdoor recreation activities in large groups (Chavez 1992, 1993).

The US Forest Service's mission as summarized by Gifford Pinchot, the first Forest Service Chief, is "to provide the greatest amount of good for the greatest amount of people in the long run." A critical component of this mission is providing recreation opportunities to everyone and, in particular, under-represented communities by reducing barriers to visitation and usage. A better understanding of how minorities use chaparral landscapes and which barriers limit their use may help decision makers better serve these communities.

10.6 Conclusion

There are currently few studies on the recreation services provided by chaparral landscapes across the world. More work is needed to better understand the importance of recreation benefits these environments provide to various populations, especially in the future with stressors such as climate change and resource degradation. As a case study, we focused on the four national forests in southern California—the Angeles, Cleveland, Los Padres, and San Bernardino. Their proximity to highly urbanized counties around Los Angeles and San Diego make these forests a unique and primary outdoor recreation destination, including for under-represented groups.

This chapter discussed several threats faced by chaparral dominated national forests, in particular, climate change and site degradation from overuse. Predicted changes in precipitation, temperature, incidence of droughts, and wildfires can potentially affect chaparral vegetation and the activities in which visitors engage. Water- and snow-based activities may be directly affected. It is currently unclear how visitors participating in those activities might adapt their use of the public lands. For example, in this study we found minorities disproportionately engaged in water-based activities, raising the question of how their participation might change with new conditions—will they concentrate on fewer sites that still offer their preferred activity, engage in other activities, or stop recreating in these national forests?

Aside from climate change, an additional threat to chaparral resources is site degradation, as evidenced by the NVUM data. Though many sites within the forests receive low levels of use, some popular recreation sites are subject to heavy use, often leading to ecosystem degradation by users who seek to alter the natural landscape, e.g., stream modification, erosion, vegetation removal, litter, and facility degradation. These negative effects could be somewhat relieved by providing educational brochures (in multiple languages), adequate oversight of highly-visited sites, and increased maintenance of these sites, although in practice, there might be limited resources within the national forest to do this. Moreover, managers of public lands are faced with the dilemma of serving the public while protecting the natural ecosystem from the very public they serve. Recreation site overuse and degradation highlights the potential trade-off between cultural and other ecosystem services (see

Chap. 8). Recreation is a valuable service to people but increased use may negatively affect the provision and value of other ecosystem services, such as biodiversity or water quality. Policy makers can potentially use information from baseline assessments such as that presented here to understand relationships between vegetation characteristics, visitors, and recreation activities and use this to inform policy planning and management decisions.

Additionally, understanding the social and economic barriers as well as outdoor recreation preferences of under-represented groups can potentially increase outdoor recreation participation for these demographic groups. Based on literature and the case study presented in this chapter of minority outdoor recreation, funding transportation services could be considered to improve participation in outdoor activities. Under the Obama administration, transportation services were being subsidized through "Every Kid in a Park" initiative which targeted all fourth grade children across the country. Under this program, the Angeles National Forest became an eligible destination after part of it was designated as a national monument.[5]

The goal of this chapter is to provide a baseline of information about the recreation services that chaparral provides to southern California residents. Based on the results, we conclude that public lands dominated by chaparral provide valuable recreation services to visitors of the four national forests in southern California. Our analysis reveals minority populations, especially Hispanics, depend on these resources for outdoor recreation. Yet, studies with community groups indicate social and financial barriers exist that prevent some from accessing these national forests.

We also assessed potential threats to these landscapes, including ecosystem degradation from climate change and overuse of recreation sites. For example, as California faces shorter wet seasons and longer dry seasons, this suggests that certain winter and water recreation activities may be limited or eventually non-existent. Consequently, more visitors might visit fewer recreation sites, degrading the sites at a faster rate. In addition, understanding the barriers and the recreation activities in which minorities engage can assist future management planning decisions to reduce these barriers. Finally, this chapter can help decision makers by providing them with information to increase visitation rates by under-represented communities.

References

Baerenklau, K., A. González-Cabán, C. Paez, and E. Chavez. 2010. Spatial allocation of forest recreation value. Journal of Forest Economics 16:113–126.

Barbero, R., J. T. Abatzoglou, N. K. Larkin, C. A. Kolden, and B. Stocks. 2015. Climate change presents increased potential for very large fires in the contiguous United States. International Journal of Wildland Fire 24:892–899.

Bell, S., M. Simpson, L. Tyrväinen, T. Sievänen, and U. Pröbstl. 2009. European forest recreation and tourism: a handbook. Taylor and Francis, New York, New York, USA.

[5] See: https://www.fs.usda.gov/angeles

Boxall, P., and J. Englin. 2008. Fire and recreation values in fire-prone forests: exploring an intertemporal amenity function using pooled RP-SP data. Journal of Agricultural and Resource Economics 33:19–33.

Byrne, J. 2012. When green is white: the cultural politics of race, nature and social exclusion in a Los Angeles urban national park. Geoforum 43:595–611.

CDFW [California Department of Fish and Wildlife]. 2015. GIS Clearinghouse. California Lakes and Streams. https://www.wildlife.ca.gov/Data/GIS/Clearinghouse

Carr, D. S., and D. J. Chavez. 1993. A qualitative approach to understanding recreation experiences: Central American recreation on the National Forests of southern California. Pages 181-194 *in* A. W. Ewert, D. J. Chavez, and A. W. Magill, editors. Culture, conflict, and communication in the wildland-urban interface. Westview Press, Boulder, Colorado, USA.

Chavez, D. J. 1992. Hispanic recreationists in the wildland-urban interface. Trends 29:23–25.

Chavez, D. J. 1993. Visitor perceptions of crowding and discrimination at two National Forests in southern California. Research Paper PSW-RP-216. USDA Forest Service, Pacific Southwest Research Station, Albany, California, USA.

Chavez, D. 2002. Adaptive management in outdoor recreation: serving Hispanics in southern California. Western Journal of Applied Forestry 17:129–133.

Chavez, D., D. Olson. 2009. Opinions of Latino outdoor recreation visitors at four urban National Forests. Environmental Practice 11:263–269.

Costanza, R., R. d'Arge, R. de Groot, S. Farber, M. Grasso, B. Hannon, K. Limburg, S. Naeem, R. Oneill, J. Paruelo, R. Raskin, P. Sutton, and M. van den Belt. 1997. The value of the world's ecosystem services and natural capital. Nature 387:253–260.

Crano, W., R. Quist, and P. L. Winter. 2008. Forest visitation, media consumption, and diverse publics: lessons for outreach. General Technical Report PSW-GTR-210. USDA Forest Service, Pacific Southwest Research Station, Albany, California, USA.

Duffield, J., C. Neher, D. Patterson, and A. Deskins. 2013. Effects of wildfire on National Park visitation and the regional economy: a natural experiment in the northern Rockies. International Journal of Wildland Fire 22:1155–1166.

Dwyer, J. 1994. Customer diversity and the future demand for outdoor recreation. General Technical Report RM-GTR-252. USDA Forest Service, Rocky Mountain Forest and Range Experiment Station, Fort Collins, Colorado, USA.

English, D., S. Kocis, S. Zarnoch, and J. Arnold. 2002. Forest Service national visitor use monitoring process: research method documentation. General Technical Report SRS-GTR-57. USDA Forest Service, Southern Research Station, Asheville, North Carolina, USA.

Englin, J., J. Loomis, and A. González-Cabán. 2001. The dynamic path of recreational values following a forest fire: a comparative analysis of states in the Intermountain West. The Canadian Journal of Forest Research 31:1837–1844.

Englin, J., J. McDonald, K. Moeltner. 2006. Valuing ancient forest ecosystems: an analysis of backcountry hiking in Jasper National Park. Ecological Economics 57:665–678.

Fedkiw, J. 1998. Managing multiple uses on National Forests, 1905–1995: a 90-year learning experience and it isn't finished. Technical report FS-628. USDA Forest Service, Washington, D.C., USA.

FRAP [Fire and Resources Assessment Program]. 2015. California Department of Forestry and Fire Protection's CALFIRE. Fveg15_1 vegetation data. http://frap.fire.ca.gov/data/frapgisdata-sw-fveg_download

Freeman, S. 2012. Ecological and economic dimensions of fire and anthropogenic disturbance in maquis woodlands of the Carmel Range: implications for planning and management. Geography Research Forum 32:62–80.

George, S. L., and K. Crooks. 2006. Recreation and large mammal activity in an urban nature reserve. Biological Conservation 133:107–117.

Green, G., J. Bowker, X. Wang, K. Cordell, and C. Johnson. 2009. An examination of perceived constraints to outdoor recreation. Journal of Public Affairs and Issues 12:28–53.

Hesseln, H., J. Loomis, A. González-Cabán, and S. Alexander. 2003. Wildfire effects on hiking and biking demand in New Mexico: a travel cost study. Journal of Environmental Management 69:359–368.

Hesseln, H., J. Loomis, and A. González-Cabán. 2004. The effects of fire on recreation demand in Montana. Western Journal of Applied Forestry 19:47–53.

Hilger, J., and J. Englin. 2009. Utility theoretic semi-logarithmic incomplete demand systems in a natural experiment: forest fire impacts on recreational values and use. Resource and Energy Economics 31:287–298.

Hoover, K. 2016. Forest Service appropriations: five-year data and trends and FY2017 budget request (CRS Report No. R43417). Congressional Research Service, Washington, D.C., USA.

Johnson, C. Y. 1998. A consideration of collective memory in African-American attachment to wildland recreation places. Human Ecology Review 5:5–15.

Johnson, C., J. Bowker, and K. Cordell. 2001. Outdoor recreation constraints: an examination of race, gender, and rural dwelling. Southern Rural Sociology 17:111–133.

Irland, L., D. Adams, R. Alig, C. Betz, C. Chen, M. Hutchins, and B. Sohngen. 2001. Assessing socioeconomic impacts of climate change on US forests, wood-product markets, and forest recreation. Bioscience 51:753–764.

Keeley J., C. Fotheringham, and M. Moritz. 2004. Lessons from the October 2003 wildfires in southern California. Journal of Forestry 102:26–31.

Koniak, G., E. Sheffer, and I. Noy-Meir. 2011. Recreation as an ecosystem service in open landscapes in the Mediterranean region in Israel: public preferences. Israel Journal of Ecology and Evolution 57:151–171.

Loomis J., A. González-Cabán, and J. Englin. 2001. Testing for differential effects of forest fires on hiking and mountain biking demand and benefits. Journal of Agricultural and Resource Economics 26:508–522.

Madsen, J., C. Radel, and J. Endter-Wada. 2014. Justice and immigrant latino recreation geography in Cache Valley, Utah. Journal of Leisure Research 46:291–312.

Millennium Ecosystem Assessment (MEA). 2005. Millennium ecosystem assessment synthesis report. Island Press, Washington, D.C., USA.

Outdoor Foundation. 2015. Outdoor recreation participation report 2015. http://www.outdoorfoundation.org/research.participation.html

Papaspyropoulos, K. G., C. Sokos, and P. Birtsas. 2015. The impacts of a wildfire on hunting demand: a case study of a Mediterranean ecosystem. iForest 8:95–100.

Rausch, M., P. Boxall, and A. Verbyla. 2010. The development of fire-induced damage functions for forest recreation activity in Alberta, Canada. International Journal of Wildland Fire 19:63–74.

Restaino, C. R., and H. D. Safford. 2018. Fire and climate change. Pages 493-505 *in* J. W. van Wagtendonk, N. G. Sugihara, S. L. Stephens, A. E. Thode, K. E. Shaffer, and J. A. Fites-Kaufman, editors. Fire in California's ecosystems. Second edition. University of California Press, Berkeley, California, USA.

Richardson, R. B., and J. Loomis. 2004. Adaptive recreation planning and climate change: a contingent visitation approach. Ecological Economics 50:83–99.

Roberts, N. S., and T. Chitewere. 2011. Speaking of justice: exploring ethnic minority perspectives of the Golden Gate National Recreation Area. Environmental Practice 13:1–16.

Sánchez, J., K. Baerenklau, and A. González-Cabán. 2016. Valuing hypothetical wildfire impacts with a Kuhn-Tucker model of recreation demand. Forest Policy and Economics 71:63–70.

Sauvajot, R. M., M. Buechner, and D. Kamradt. 1998. Patterns of human disturbance and response by small mammals and birds in chaparral near urban development. Urban Ecosystems 2:279–297.

Schmitz, M. F., I. Aranzabal, and F. Pineda. 2007. Spatial analysis of visitor preferences in the outdoor recreational niche of Mediterranean cultural landscapes. Environmental Conservation 34:300–312.

Scott, D., S. Herrera, and K. Hunt 2004. Constraints to outdoor recreation among ethnic and racial groups. Pages 17-20 *in* P. T. Tierney, and D. J. Chavez, technical coordinators. Proceedings of the 4th Social Aspects and Recreation Research Symposium, San Francisco, California, February 4-6, 2004. San Francisco State University, San Francisco, California, USA.

Shaw, M. R., L. Pendleton, D. Cameron, B. Morris, G. Bratman, D. Bachelet, K. Klausmeyer, J. MacKenzie, D. Conklin, J. Lenihan, E. Haunreiter, and C. Daly. 2009. The impact of climate change on California's ecosystem services. Report CEC-500-2009-025-F, California Climate Change Center, Sacramento, California, USA.

Shechter, M., B. Reiser, and N. Zaitsev. 1998. Measuring passive use value: pledges, donations and CV responses in connection with an important natural resource. Environmental and Resource Economics 12:457–478.

Starbuck, C., R. Berrens, and M. McKee. 2006. Simulating changes in forest recreation demand and associated economic impacts due to fire and fuels management activities. Forest Policy and Economics 8:52–66.

Tierney, P. T., R. Dahl, and D. J. Chavez. 1998. Cultural diversity of Los Angeles County residents using undeveloped natural areas. Research Paper PSW-RP-236. USDA Forest Service, Pacific Southwest Research Station, Albany, California, USA.

US Census Bureau. 2014a. American FactFinder. https://www.census.gov/geo/reference/zctas.html

US Census Bureau. 2014b. American Community Survey, 1-Year Estimates. https://factfinder.census.gov/faces/nav/jsf/pages/searchresults.xhtml?refresh=t#none

USDA Forest Service. 2005a. Final environment impact statement, volume 1 land management plans: Angeles National Forest, Cleveland National Forest, Los Padres National Forest, San Bernardino National Forest. R5-MB-074-A. Washington, D.C., USA.

USDA Forest Service. 2005b. Land management plan part 2 San Bernardino National Forest strategy. R5-MB-079. Washington, D.C., USA.

USDA Forest Service. 2013. National visitor use monitoring results national summary report. http://www.fs.fed.us/recreation/programs/nvum/

USDA Forest Service. 2015a. Fiscal year 2016 budget overview. Washington, D.C., USA.

USDA Forest Service. 2015b. National visitor use monitoring: visitor use report. Database queried for: Angeles National Forest (2006, 2011), Cleveland National Forest (2009, 2014), Los Padres National Forest (2009, 2014), and San Bernardino National Forest (2009, 2014). https://apps.fs.usda.gov/nfs/nrm/nvum/results/

USDA Forest Service. 2016. Fiscal year 2017 budget overview. Washington, D.C., USA.

Vincent, C. H., L. Hanson, and J. Bjelopera. 2014. Federal land ownership: overview and data. CRS Report No. R42346. Congressional Research Service, Washington, D.C., USA.

Washburne, R. F. 1978. Black underparticipation in wildland recreation: alternative explanations. Leisure Sciences 1:175–189.

Washburne, R., P. Wall. 1980. Black-White ethnic differences in outdoor recreation. Research Paper INT-249. USDA Forest Service, Intermountain Forest and Range Experiment Station, Ogden, Utah, USA.

Witztum, E. R., and D. A. Stow. 2004. Analysing direct impacts of recreation activity on coastal sage scrub habitat with very high resolution multispectral imagery. International Journal of Remote Sensing 25:3477–3496.

Chapter 11
Connecting Californians with the Chaparral

Richard W. Halsey, Victoria W. Halsey, and Rochelle Gaudette

Abstract Chaparral is California's most extensive, native ecosystem. We examined nature centers, publications, curricula, and volunteer naturalist programs in southern California to determine how the chaparral is being presented to the public. We found that a number of centers do an excellent job presenting accurate content. However, the majority need updates to reflect current science and the major contribution chaparral makes to the state's biodiversity. Easily accessible publications and curricula with accurate information about the chaparral are lacking. More than half of the nature centers reviewed offer extensive naturalist training courses. Passion and enthusiasm of staff and volunteers are as important as content in creating and maintaining successful volunteer naturalist programs as well as the nature centers themselves. Utilizing active learning methodology versus lecturing can be a key factor in a program's success. We offer an approach to nature education that combines active learning where students participate in the teaching process, and meaningful interpretation that establishes personal connections with nature. The greater understanding resulting from this approach can inspire a diverse, new generation of long-term nature advocates and help create an informed public, facilitating an appreciation for the value of the chaparral ecosystem.

Keywords Active learning · California native plants · Chaparral · Naturalist · Nature center · Nature education

R. W. Halsey (✉)
California Chaparral Institute, Escondido, CA, USA
e-mail: rwh@californiachaparral.org

V. W. Halsey
Ken Blanchard Companies, Escondido, CA, USA

R. Gaudette
Canyoneers, San Diego Natural History Museum, San Diego, CA, USA

E. C. Underwood et al. (eds.), *Valuing Chaparral*, Springer Series on Environmental Management, https://doi.org/10.1007/978-3-319-68303-4_11

11.1 Chaparral Education—For the Love of Nature

"We shall not cease from exploration, and the end of all our exploring will be to arrive where we started and know the place for the first time."
—T.S. Eliot

For many Californians, the most accessible place to experience nature is in the chaparral, a shrub dominated ecosystem rich in biodiversity that can be found in every county in the state (Keeley and Davis 2007). As a consequence, chaparral provides one of the best places for Californians to learn about and connect with the natural environment (Fig. 11.1).

Connecting and learning about nature, and chaparral specifically, can be greatly enhanced by information and interpretation provided in nature centers and the natural history programs they offer. The success of these facilities and their efforts to help people better understand nature and encourage a broader appreciation of chaparral depends on not just the quality of the exhibits, but the enthusiasm of the rangers and volunteers who support these centers. One of the first nature centers we visited during our research highlighted the role of such enthusiasm.

Devil's Punchbowl, a protected natural area administered by the Los Angeles County Department of Parks and Recreation, is a deep canyon carved through huge, uptilted slabs of sedimentary rock on the northern side of the San Gabriel Mountains. Surrounded by numerous earthquake faults, including the San Andreas, the

Fig. 11.1 Mixed chaparral in the Santa Monica Mountains National Recreation Area. Photo by Richard Halsey

Fig. 11.2 Ranger Dave Numer at the Devil's Punchbowl Natural Area, Los Angeles County Parks. Photo by Richard Halsey

Punchbowl is a testament to California's dynamic geological history. Hardy desert chaparral decorates much of the landscape with pinyon pines scattered throughout.

When we first entered the park's wood-paneled nature center, a small building with an open, uncluttered display hall, it appeared as if we were on our own. Along two walls were cabinets filled with an array of bird and mammal taxidermy, artifacts, and photos. Embedded within the other two walls were ten terrariums holding an assortment of live reptiles and insects. In the far corner was a tall counter with a doorway behind that led into a small office containing an old book case with additional specimens from the park, all neatly arranged on shelves. On the counter sat a dissection scope and a two-foot tall glass cylinder filled with soil. Attached to the wall was a two-foot diameter section of a cut pine log. It was hollowed out and faced with a plate of glass to reveal an active beehive within.

Dave Numer, ranger and superintendent of Devil's Punchbowl, emerged from the back office. "So, how do you think the bees get in and out of there?" he asked. Wearing a crisp uniform and a wide brimmed hat, he leaned against the desk with a broad smile and he let us consider the question for a brief moment. Tapping the back of the log with a small, metal pointer that he extended with a snap, he provided the answer. "It's connected to the outside by a PVC pipe right back there" (Fig. 11.2).

Numer then turned our attention to the glass cylinder. "And take a look at this!" He pulled out a thumb-sized flash light and shined it into a small, excavated chamber. Translucent, marble-sized golden globes hung from the chamber's ceiling. The globes were moving slightly from side to side, glistening in the beam of light. The scene was right out of a nature documentary.

"Honeypot ants!"

"Honeypot ants, yes!" Numer answered back. "And they're native here in the park." Numer added that the globes were the enlarged abdomens of specialized worker ants of the species (*Myrmecocystus mexicanus*) capable of storing a honey-like substance used to provide food for the colony during lean times. Then he pointed out the glass cylinder's partially open lid and a little pile of dirt on his desk. "How do you think that got there?"

"The ants?"

"Yes, the ants! They leave the nest at the end of the day and roam the nature center all night." He enjoyed waiting for our reaction. "They always return by morning though. One day when I came in early, I caught them hauling up a dead moth they had found on the floor somewhere."

Numer shared more of his knowledge about the ants, the bees, and the center's other displays, then offered to take us outside. Joined by his assistant Olivia, they discussed their favorite points of interest until being interrupted by a raucous noise from above. An old raven was calling out from his perch in the large pinyon pine near a small building across from the center. Numer laughed then nodded toward the building. "That was my home for five years when I lived on site." He looked back at us. "I've been here for nearly 43 years now."

Numer continued sharing his knowledge about the park's desert chaparral, scattered between the pines and junipers, and the recent drought's impact on the manzanita. The gnarled, gray stems of several dead individuals were still pointing skyward. Looking off into the distance he said, "A fire burned over that farthest ridge in 1953." The area was still relatively bare. The pines and junipers had not come back well. Turning our attention into the canyon below, Numer encouraged us to explore it. "There's a little turn off along the trail by a large log. You can climb over a few boulders there and have a great view."

When finalizing our review of the Punchbowl's nature center, we were surprised to realize that references to the chaparral, and nearly everything else we were measuring in our evaluation including the Mediterranean-type climate, fire ecology, and the significance of watersheds, were missing. Yet the place inspired us. It was the learning environment Numer had created over the past four decades, his questions, and more importantly how he asked them, that caught our attention. The most compelling part of our experience was Numer's enthusiasm and personal warmth, not the content of the exhibits. We wanted to come back.

After completing our evaluation of southern California's chaparral connected nature centers and speaking with dozens of naturalists, our Devil's Punchbowl conclusions were affirmed. The state of chaparral education cannot be discovered by merely reviewing content. The people behind the desk, the outdoor educators, and the volunteer naturalists on the trail play a critical role in whether or not the content is meaningful to visitors. These people create relationships that establish the foundation required to encourage lasting change within the minds of those they inspire. This observation offers an alternative to how nature education is often approached.

Rather than focusing *what* we want to teach, a naturalist's goals might be better served by focusing on *how* we can teach to better facilitate the change we hope to achieve.

What change do we seek as nature educators? A general consensus emerged during our research—to inspire a love for nature that will foster curiosity to learn more and the desire to care for and protect the natural environment. For us, Dave Numer's approach to interpretation, one that involved us and allowed him to share his passion for nature, achieved exactly that. A nature center's content and design becomes more effective when a naturalist facilitates the center's learning experience with interactivity and sincere enthusiasm.

11.2 The Benefits of Chaparral Education

Although the positive attributes of experiencing nature have received a significant amount of attention since the publication of *Last Child in the Woods* (Louv 2008), and more recently, *How to Raise a Wild Child* (Sampson 2015), strong societal pressures continue to prevent people from getting outside. According to several studies, the average American young person spends less than 7 min a day in unstructured outdoor play, but more than 7 to 10 hr in front of some kind of electronic screen (Hofferth and Sandberg 1999; Juster et al. 2004; Rideout et al. 2010). "That outdoor time is 90% less than most of their parents had," Scott Sampson (2016) said in a recent interview. "So in one generation, we've flipped this around."

It is reasonable to speculate that many adults suffer from the same deficit. Such lack of activity led the former US Surgeon General Richard Carmona (2004) to speculate that, "… we may see the first generation that will be less healthy and have shorter life expectancy than their parents."

Enhancement of mental and physical health through recreation and relaxation is one of the most direct, obvious contributions the chaparral can provide. These benefits have been elucidated in a number of interesting studies examining how people respond while being in nature (Suttie 2016) including reduced stress (Lee et al. 2014; Tyrvainen et al. 2014), less worrying (Bratman et al. 2015), increased creativity (Atchley et al. 2012; Aspinall et al. 2013), increased generosity (Zhang et al. 2014; Piff et al. 2015), feeling more "alive" (Ryan et al. 2010), and improved immunity (Li et al. 2008).

In addition to empirical research, a number of writers and philosophers have suggested that nature can play a role in increasing confidence and self-respect. Abrams (2014) writes,

"Nature teaches you that there is nothing wrong with you. When you're in nature, you don't have to look at advertising that tries to convince you there's something wrong with you, in order to sell a product. Nor do you have to look in mirrors. Instead, you're either focused on the setting around you, or on what you are doing, like climbing, setting up a tent, or gardening".

Fostering an awareness and greater appreciation for the chaparral will likely help increase the public's interest in exploring their local, natural environment and hence, acquire the physical and mental benefits such activity can provide. Multiple studies have also shown that personal experience and direct knowledge can reduce negative biases (Tadmor et al. 2012). When nature becomes familiar, fear of nature can be replaced by enjoyment. As knowledge and appreciation for the environment builds, local nature can become incorporated into a person's place attachment, creating a "sense of place"—an intimate connection that combines physical setting with what we bring to the place and how we interact with it (Steele 1981).

A sense of place is extremely personal. It does not exist independent of us. This is why nature education programs in chaparral environments should emphasize creating enjoyable, engaging experiences that help develop personal meanings about the chaparral and the organisms that live there. Establishing a sense of place with the chaparral can form a basis for community cooperation and action such as participation in local volunteer programs and political activism (Manzo and Perkins 2006). This community involvement can then be mobilized to expand participation in nature education programs, ultimately leading to efforts to protect local nature because people have learned to care about the natural environment in which they live. Teaching content is important, but without personal connections and meaning (why do I care about this place?), content usually fails to inspire the behavioral changes we seek as naturalists.

This is what conservationist Baba Dioum (1968) was referring to in his speech to the General Assembly of the International Union for Conservation of Nature and Natural Resources when he said, "In the end, we will conserve only what we love, we will love only what we understand, and we will understand only what we are taught."

11.3 The Consequences of Anonymity

Despite the chaparral's dominant presence and its significant contribution to biodiversity in California (Halsey and Keeley 2016), a large portion of the public remains unaware of its presence or that it is a viable community of living things on the same level as more familiar ecosystems like forests and grasslands. This lack of awareness has been observed anecdotally during hundreds of educational presentations presented by the non-profit California Chaparral Institute to thousands of people over the past 10 years. When asked to explain what chaparral is, many audiences are unable to do so. In fact, many chaparral ecological processes remain a mystery to the scientific community, especially those relating to animal species (van Mantgem et al. 2015).

Huell Howser, the well-known host of the long-running public television series, *California's Gold*, highlighted the chaparral's anonymity several months after producing an episode about the ecosystem. With his characteristic enthusiasm and Tennessee twang he said, "I've been up and down the state filming our show, and you know what I've seen? I've seen chaparral! It's everywhere! I had no idea!" (pers. comm. 2005).

The public's lack of awareness of chaparral is not particularly surprising since the landscapes it dominates are not identified as such. In fact, chaparral is frequently referred to as a forest or is ignored all together in deference to other features present. The four largest public chaparral preserves in California are the Cleveland, San Bernardino, Angeles, and Los Padres national "forests." Chaparral and other native shrublands cover 52–88% of these federally managed landscapes. Similarly, one of the last remaining stands of old-growth southern maritime chaparral is named for the trees present—the Torrey Pines State Natural Reserve. The University of California's Emerson Oaks Reserve in Riverside County is primarily chaparral. Even where chaparral is the exclusive feature, it is identified as a miniature "forest" such as the El Moro Elfin Forest Natural Area in Los Osos and the Elfin Forest Recreational Reserve in Escondido.

The chaparral ecosystem's anonymity likely reflects a historical bias favoring more economically valued forests and rangelands over native shrublands (Halsey 2011a, b). This bias led to derogatory references to chaparral as "worthless brush" on late nineteenth century survey maps (Ward 1984), as a dangerous and unsightly "menace" (as reported in the local Daily San Diegan newspaper on September 26, 1889), and as troublesome "brushfields" in more recent documents from the Stanislaus National Forest that claim chaparral will prevent the establishment of forest habitat and "impede wildlife movement" (USFS 2016). In the popular press, chaparral is often disparaged as merely "over-grown brush" (Rocha and Kelley 2017). Some have even expressed the view that large stands of chaparral are unnatural because early Spanish explorers made "little mention of brush" in their written accounts (Santa Barbara County Fish and Game Commission 2008). What is ignored by all these perspectives is that chaparral has come to dominate much of California's natural landscape over 15 million years of evolution (Rundel et al. 2016) and that large, impenetrable, old-growth stands of chaparral represent the ecosystem's natural condition (Keeley and Davis 2007).

So chaparral remains nameless and misunderstood for many, and hence unknown as a viable ecosystem. Under such conditions, an entity can be defined by incorrect perceptions and biases. Such definitions can lead to eventual loss because the entity's value is not properly recognized. The California grizzly bear, the now extinct denizen of the chaparral, but central on the state's flag, serves as a poignant example. While the grizzly bear was certainly well known, its important ecological contribution was unrecognized until it was too late to protect the species.

Ignorance of a group has also been directly tied to prejudice and stereotypes (Stephan and Stephan 1984). While prejudice most often refers to unsubstantiated judgments and opinions between people, it can also influence one's view of nature. Such views can impact public policy and lead to damaging land management practices. Ignorance and prejudice can also cause individuals unfamiliar with the chaparral's sometimes counter-intuitive ecology to be susceptible to accepting misconceptions as fact. The misconceptions are then promulgated frequently enough to form stereotypes that are difficult to dislodge. Most chaparral misconceptions relate to fire suppression impacts and vague notions about ecological health, all of which have been used to justify various forms of chaparral management for "ecological reasons" by mechanical clearing (mastication), the application of herbicides, or prescribed burning. The most commonly encountered misconceptions include the following:

- Past fire suppression is claimed to have created large stands of unnaturally dense chaparral leading to large, high-intensity wildfires. This is false. Chaparral naturally creates large stands of dense vegetation that are historically subject to large, high-intensity fires (Keeley et al. 2004; Keeley and Zedler 2009; Lombardo et al. 2009). In fact, fire suppression has provided important ecological benefits by preventing excessive fire in chaparral that can lead to a loss of biodiversity and eventually type-conversion, the transformation of native shrubland to non-native grassland (Keeley 2001).
- It is frequently claimed that chaparral "needs" frequent fire to stay "healthy." This is a misnomer because old-growth chaparral is a viable ecosystem that is capable of regenerating after long fire-free periods (Keeley et al. 2005). The chaparral's natural fire-return interval of 30 to 150 years or more allows shrub species to restock the soil with seeds and restore starch supplies in under-ground tubers or roots between fires. If fire burns chaparral more than once per 10- to 15-year period (depending on location), native species can be extirpated. The void is usually filled with non-native grasses, leading to increased flammability and eventual type-conversion (Keeley and Brennan 2012).
- It has also been hypothesized that fire is needed in chaparral to remove germination-inhibiting chemicals in the soil that were produced by mature plants (allelopathy), resulting in reduced competition. This hypothesis has been rejected because the seeds of most chaparral species are innately dormant, with germination stimulated by fire cues, not the denaturing of soil chemicals by fire or other means (Halsey 2004).

One way to help the public recognize the chaparral's value is by publicizing the benefits the ecosystem offers human communities. Recently, there has been a renewed interest in quantifying these "ecosystem services." For example, the chaparral's dense cover offers protection to a region's watershed, allowing water to slowly recharge aquifers rather than racing across the landscape and causing flash floods, large debris flows, and significant damage to public infrastructure (Gartner et al. 2009) (see Chap. 7). Supplying water is one of the most important provisioning ecosystem services that chaparral provides. Other benefits include "regulating" services (e.g., carbon sequestration), "supporting" services (e.g., nutrient cycling), and "cultural" services (e.g., recreation).

The ecosystem services concept is limited, however, because of its utilitarian, anthropocentric focus—it fails to recognize that nature has value regardless of the benefits it provides humans (Vucetich et al. 2015). Strictly focusing on ecosystem services to value nature is similar to valuing one's friends purely on the material "friendship services" they provide. Clearly, friends and nature are more than the goods and services we can gain from them. In addition to providing a significant amount of habitat for a multitude of species, chaparral has intrinsic value, a value that is neither conferred nor revocable, but springs from its evolutionary heritage and potential (Soulé 1985). Chaparral has value for what it is, not only for what it does for human society.

11.4 The Current State of Chaparral Education

To examine how chaparral is presented to the public, we visited nature/interpretive centers in southern California associated with native shrublands, reviewed volunteer naturalist training programs, and evaluated available educational curricula and publications.

We did not conduct an extensive review of nature programs in individual schools because many of these are connected to the local nature centers and parks we visited. We were also most interested in young adults and families because this is where we believe the greatest need for "nature literacy" education exists. We define nature literacy as being able to identify several local, native species (plant and animal), familiarity with the community in which those species exist, participating in several excursions per year to local wild areas for the purpose of enjoying nature, and having a clear understanding of the need to protect nature for its intrinsic value as well as for the benefits it provides us.

There are a number of programs exposing elementary-aged children to nature including sixth grade camps, school field trips, and opportunities supported by organizations like the Children & Nature Network. However, based on interactions with many young adults over the last 10 years we have found a surprising lack of nature literacy. Equally concerning is the current children-in-nature-movement, while booming, lacks needed diversity as it largely operates "at the fringes of affluent white society" (Sampson 2015).

11.4.1 Nature/Interpretive Centers

All of the nature/interpretive centers we visited in Los Angeles, Orange, and Riverside Counties are either deeply embedded in chaparral or have a significant amount of chaparral or California sage scrub nearby (Fig. 11.3). There are also a number of nature centers located within urban areas, such as the El Dorado Nature Center in Long Beach and the Turtle Rock Nature Center in Irvine. Although we did not evaluate these centers, they can also play an important role in helping the public become familiar with nature, and are particularly important given their strategic position in developed areas (Halsey 2009).

We evaluated four key components at each nature center: the level of focus on chaparral (ranging from 0 to 4, with 4 being the greatest, most accurate focus), the presence and accuracy of the description of chaparral fire ecology (again, ranging

Fig. 11.3 Location of nature centers/museums. Image from Google Earth

from 0 to 4), the availability of naturalist training, and finally, the date the exhibit(s) were last updated (Table 11.1). Each center was visited once. The levels were based on the representation of the most current science and the absence or presence of misconceptions.

An encouraging finding was that the most reliable predictor for the amount and accuracy of information provided about chaparral was the age of the nature center exhibits. Seven of the ten centers updated or built over the past decade offered the most thorough information about chaparral, especially those at Placerita Canyon, Chino Hills, and the San Diego Natural History Museum. Their displays provide extensive details about the natural history of species within the context of the chaparral (rather than providing information about the species in isolation), accurately explain the ecological role of fire and the threat of type-conversion from too frequent fire, and how the ecosystem is shaped by the Mediterranean-type climate. One of our top-rated centers, San Diego Natural History Museum's exhibit *Coast to Cactus* in which the chaparral is significant component, won the prestigious Overall Excellence for an Exhibition Award from the American Alliance of Museums in 2016 (Fig. 11.4).

Some newer nature centers also offer specialized features or expanded content. For example, the Santa Rosa Plateau center in Riverside County is unique because

Table 11.1 Chaparral exhibits in nature centers/museums (by last update)

Name of center	Location	Admin agency	Exhibit update	Chap focus	Fire ecology	Naturalist training
Mission Trails Regional Park	San Diego	San Diego City	2016	1	2	Yes
Placerita Canyon Natural Area	Newhall	LA County	2015	4	4	Yes
San Diego Nat. Hist. Museum	San Diego	Museum	2015	4	4	Yes
Topanga Canyon State Park	Topanga	CA State	2015	3	2	Yes
Chino Hills State Park	Chino Hills	CA State	2014	3	4	
Santa Monica Mountains	Calabasas	National Park Serv.	2012	2	4	
Santa Rosa Plateau	Murrieta	Riverside County	2010	3	3	Yes
Elfin Forest Rec. Reserve	Escondido	Olivenhain Water Dist.	2009	1	0	Yes
Nix Nature Center	Laguna Beach	Orange County	2007	1	0	—[a]
Tucker Wildlife Sanctuary	Silverado	Cal State Fullerton	2007[b]	2	m	
Tecolote Canyon Natural Park	San Diego	San Diego City	2003	1	0	—[c]
Stough Canyon	Burbank	Burbank	2003	1	0	
Eaton Canyon	Pasadena	LA County	1998	0	m	Yes
San Dimas Canyon	San Dimas	LA County	1993	2	m	Yes
Torrey Pines State Reserve	San Diego	CA State	1988	1	1	Yes
Devils Punchbowl	Pearblossom	LA County	1988	0	0	
Oak Canyon	Anaheim	Anaheim	1980[d]	0	0	
Malibu Creek State Park	Calabasas	CA State	1976[e]	1	m	Yes
Cabrillo National Monument	San Diego	National Park Serv.	—[f]	0	0	—[f]

Chaparral Focus score key: (4) an entire exhibit area focused on chaparral, (3) separate set of panels/displays describing chaparral, (2) a distinct panel/display describes chaparral in context to the nature center, (1) mention of chaparral only in context of a species or general habitat reference, and (0) no mention of chaparral

Fire Ecology score key: (4) accurate explanations with recognition of type-conversion due to increased fire frequency, (3) two correct references to fire ecology, (2) one correct reference to fire ecology, (1) causal reference to fire, (0) no mention of fire, (m) misconceptions concerning fire

[a]1 day public interaction workshop

[b]Updates in progress

[c]Uses Mission Trails training

[d]Exhibit update is approximate

[e]Date represents when the park was created, but center updates/rotates exhibits on an ongoing basis

[f]No permanent chaparral exhibits, but has an extensive intertidal education program

Fig. 11.4 The Coast to Cactus exhibit at the San Diego Natural History Museum which has separate displays for all the region's major habitat types. The chaparral portion is accessed through the opening to the left of the central mural. Photo by Richard Halsey

it extended its displays into the field with detailed trailside signs along a short circular path. The adjoining Granite Loop Trail has fifteen numbered signposts that correspond to content in a brochure. Also noteworthy was the Santa Monica Mountains National Recreation Area center which did an excellent job describing the chaparral's relationship to fire.

Four of the nine nature centers with older exhibits did not mention chaparral at all. The five that did, casually reference it in connection with fire or a particular species. Concerning fire, four of the centers with older exhibits contained misconceptions in the display text, mostly implying that dense chaparral is either unnatural or undesirable. Specifically, the exhibits mistakenly claim that: (1) fire suppression has created overly dense chaparral, or (2) fire "prunes tightly packed vegetation," or (3) fire "thins thick brush" to create vast new stands of "open grasslands." The natural growth pattern of chaparral shrubs, their physical structure, and the dynamics of fire contradict such claims. And while chaparral can be converted to open grasslands, such a landscape is typically composed of non-native annual grasses (e.g., *Bromus* spp.) and is caused by unnatural disturbances such as overly frequent, low-intensity fire or soil disturbance. One of the centers with newer exhibits contained some of these errors, but staff indicated the display is scheduled for an update.

We encountered passion and enthusiasm from staff and volunteers at most of the nature centers we visited. Four in particular stood out for us: Placerita Canyon, Tucker Wildlife Sanctuary, Santa Rosa Plateau, and of course, Devil's Punchbowl.

Like Devil's Punchbowl, Placerita Canyon is a Los Angeles County Park. It has its own supporting non-profit as well as being part of the greater Nature Center Associates of Los Angeles County. The development of the nature center's new exhibits was a collaborative effort between the center's staff, volunteers, and the community, all successfully coordinated by Jennifer Rigby, the director of the interpretive planning and design firm, The Acorn Group.

Recreation Supervisor Frank Hoffman is the dynamic public face of Placerita Canyon and has served as a ranger there on three different occasions over the past 20 years. We found his enthusiasm infectious. He summarized his approach to nature education by saying, "Mother Nature is talking all the time. You just have to know what she is saying. You need to execute all your senses. You need to know the audience. It's not what you say, it's how you say it." Hoffman credits a supportive staff and dedicated group of volunteers for the nature center's positive environment.

Tucker Wildlife Sanctuary is unique from the other nature centers we visited as it is administered and funded by California State University, Fullerton. About 9 years ago when site manager Marcella Gilchrist arrived, the facility was in disrepair. Her first task was moving the center to another building that was larger and more accessible. Through her vision and dedication, the nature center has become a bustling educational opportunity for local schools and the surrounding community. Kurt Miethke, the center's grounds manager and landscaper, has been working closely with Gilchrist to improve and maintain the facility's well-marked trails, interpretive panels, and study areas. Between the two of them, they have turned the Tucker Wildlife Sanctuary into a compelling natural history experience.

The Santa Rosa Plateau Nature Center has a successful education program for local schools led by volunteer naturalists and interpretive ranger Rob Hicks. All of the volunteers have their own unique way of sharing information. Volunteer Jim Lockyer has developed a method to inspire students through what he calls "a simple formula for nature discovery." Printed out on a twofold business card that Lockyer gives to students is a curious photo of a cork with the word "Inspire" underneath. Inside, a "simple formula for nature discovery" is revealed: "COR = K", meaning Curiosity, Observation, Research, equals Knowledge. Lockyer uses the formula to provide his students an easily remembered approach to exploring the natural environment.

Although passionate individuals are essential in shaping successful nature center experiences, centers need adequate financial support. The one facility we found lacking such support was the Oak Canyon Nature Center, operated by the city of Anaheim. During its heyday, from 1972 through the mid-1990s, the nature center was a bustling place, open every day and supported by a full-time staff and dedicated volunteers. Three large notebooks at the center document this period, filled with newspaper clippings, photos, and art. The public's ability to experience the center now is limited to weekends. A part-time staff does its best to operate the center and accommodate school groups during weekdays.

In summary, we found 7 out of 19 nature centers reviewed accurately represented chaparral in a manner commensurate with its presence. All seven had exhibits installed within the last decade.

11.4.2 Naturalist Programs

As well as evaluating individual nature centers, we also assessed the extent they provided naturalist programs. We noted the number of training sessions provided to volunteers on a number of topics including interpretation, ecology, botany, zoology, history, and geology (Table 11.2).

Nine of the 19 nature centers have independent volunteer naturalist programs designed to lead public nature hikes and guide school groups through the connected shrubland preserves. Each training program has its own unique blend of topics, but all are organized by individuals who are passionate about helping the public understand, appreciate, and value the local natural environment. Interestingly, more than half of the naturalist training programs began in the 1970s at the height of the environmental movement and many were integral components of preservation battles to protect open space.

For example, a community action group, Citizens to Save Open Space (SOS), was formed in 1971 to stop the development of San Diego's Tecolote Canyon. After years of effort, including letter writing campaigns and a march on city hall, SOS was successful in achieving its goal. The city acquired the canyon and dedicated it as open space in 1978. The nature center was built 5 years later and a small group of volunteers began sharing the canyon's natural history with the public. Two of the current volunteers, Eloise Battle and Sherlie Miller, have been there since the beginning. Both women were instrumental in leading the preservation battle. Commenting on what the preserve means to her, Miller expressed a perspective that aligns with one of the primary goals of chaparral education. "People ask, Chaparral? What does it mean? Sage scrub? What is that? People need to know!"

The San Diego Natural History Museum's Canyoneers was founded by Helen Chamlee, an associate botanist at the museum, after she helped defeat a plan by the San Diego Zoo to turn the nearby Florida Canyon into a parking lot. The Canyoneer's mission was to educate the public to appreciate the chaparral filled open space so they would support its protection. The Canyoneers led their first hikes into the canyon in 1973. Chamlee's continued efforts eventually lead to a master plan for the canyon in 1975. The plan's landscape architect, Stephen L. Halsey, received the American Society of Landscape Architect's National Honor Award for the design. Although efforts to properly manage the canyon have yet to be formally adopted, its wild space remains undeveloped. The Canyoneers eventually expanded beyond Florida Canyon and now lead public hikes in 70 locations throughout San Diego County. Descriptions of the hikes are available in a comprehensive publication co-written by numerous Canyoneers (Lindsay et al. 2016).

The Canyoneer model has spread to other locations in San Diego County as well. Bill Howell, the Canyoneer's lead trainer since 1988, founded the Trail Guides naturalist program at San Diego's Mission Trails Regional Park in 1994. Mission Trails is one of the largest municipally owned parks in the United States. The volunteer Chaparral Naturalist Certification program at the Elfin Forest Recreational Reserve in Escondido was started in 2015 by one of Howell's long-time Canyoneer colleagues, Richard Halsey.

Table 11.2 Naturalist training programs

Naturalist Program (Date Originated)	Vol #	Int	Hab	Fire	Eco	Bot	Vert	Arth	Hist	Geo	Total hours
Placerita Nature Center (1971)	35	x			xx	x	xxx	x	xx	x	90[a]
Mission Trails (1994)	100		Ch	p		xx	xx	x	xx	x	63
Tecolote Canyon (1994)	10										—[b]
Topanga Canyon State Park (1974)	35	x			x	xx	xxx	x	x		51[c]
San Diego Natural Hist. Mus. (1973)	100 [d]		Ch	p		xx	xx	x	x	x	47
CA Naturalist Ojai (2016)	—[d]		p		x	x	x	p		x	43[e]
Torrey Pines (1975)	250	x			x	x	x	x	xx	x	42[f]
Elfin Forest Rec. Reserve (2015)	—[d]	x	Ch	x		x	p	p	x	x	36
Eaton Canyon (1978)	30				x	x	xxx	x	x	x	35
Irvine Ranch (2011)	400		Ch p	p	p	p	p	p	p	p	32[g]
Malibu Creek State Park (1976)	60	x	x			x	x		xx		30[h]
Cold Creek Docents (1977)	8 [d]	x	Ch		x	x		x	xx	x	30
Santa Rosa Plateau (1991)	87										30[i]
San Dimas Canyon (1997) (1) Jr. Naturalist (2) Jr. Ranger	30 Age: 12–17	x									25[j]
	180 Age: 7–11				x		x	x			12[k]
Nix Nature Center (2007)	125[d]	x									6[l]
Cabrillo National Mon. (1996)	70	x									—[m]

Topic key: Vol # = Active Volunteers; Int = Interpretation methods: Hab = Habitats (Ch indicates class focuses on chaparral and the Mediterranean-type climate), Fire = Chaparral fire ecology, Eco = Ecology, Bot = Botany training sessions, Vert = Vertebrates (reptiles, birds, and mammals), Arth = Arthropods, Hist = History/Native American, Geo = Geology

Training sessions key: x = number of full in-class sessions dedicated to the topic (x = 1, xx = 2, xxx = 3), with "p" indicating partial sessions dedicated to the topic, based on class syllabi. Class sessions typically run between 2–3 h. Both in-class sessions and field trips included in total hours

[a]Ongoing in-service opportunities

[b]Uses Mission Trails training

[c]18 h additional workshops

[d]Naturalists not tied to a central location

[e]40 additional volunteer hours required

[f]Final project required

[g]Plus 1½ days for introduction/interpretation

(continued)

Table 11.2 (continued)

[h]3 additional classes per year, 7 h habitat class per year
[i]No formal training, 30 h volunteering required, occasional lectures
[j]Volunteer hours, learn to care and feed animals on site, lead tours
[k]Fall and spring sessions offered (60–90 students each), 8 classes per session
[l]1 day public interaction workshop
[m]Tide pool training only with 2 day conference plus supplemental training

A strong social fabric fostered by some of the volunteer programs is also an important part of the volunteer naturalist experience. Two programs stood out in particular: the Torrey Pines Docent Society and the Placerita Canyon docents. Both organize frequent social and advanced educational events.

The Torrey Pines organization, with 250 active volunteers, is the largest of the groups we reviewed. Members keep in touch with each other through the *Torreyana* newsletter, published regularly since 1975. Their monthly meetings and other social gatherings usually attract up to 100 participants. In addition, there are various subgroups that have their own get-togethers. Reflecting on the personal nature of the Docent Society, Jeannie Smith, one of the longest serving members said, "You should see the trouble the president has every month in getting everyone into the meeting and shutting us up before the poor speaker can take the floor."

The origin of the Placerita docents is a testament to volunteer action. Prior to the opening of the nature center in 1970, several members of the American Association of University Women would show up on weekends at the entrance of the canyon and share natural history and scientific knowledge with the public. Operating out of the trunk of their cars, the women would pull out animal pelts and other specimens to inspire the visitors. One of these early docents, Shirley Morano, still volunteers at the nature center. Morano is 94 years old and drives herself the 30 miles (~50 km) from her home to center on weekends.

Orange County has a well coordinated effort to expose the public to their parks, although not chaparral in particular. The Orange County *Wild Passport* program provides a way for members of the public to earn a "Wilderness Lover" patch by visiting five different parks and preserves. Orange County Docent Day is held each year, attracting about 100 volunteer naturalists from various parks and organizations who attend a full day of workshops and lectures concerning the region's natural history.

The California Chaparral Institute is initiating a similar Passport program in San Diego County with its *Passport to the Chaparral*. Participants earn a pocket naturalist guide and a certificate if they complete 10 hikes or activities within the Escondido Creek and San Dieguito River watersheds.

Although the Irvine Ranch Conservancy in Orange County does not have a nature center, it does manage about 8000 ha (20,000 acres) of open space and has taken a lead role in training the region's naturalists and other volunteers. The trained volunteers not only assist with the Conservancy's public education programs, but also the recreational and stewardship activities on behalf of landowners who contract with the Conservancy. Their training emphasizes not only natural history, but also interpretive skills modeled by the National Association for Interpretation.

Brian Hughes, the co-facilitator of the Conservancy's training program clarified their educational philosophy. "What we do is more about sharing than teaching." He added that the best strategy is to, "think of yourself as a host for the site you are interpreting, treating visitors as invited guests."

Although there is not a chaparral based program at the National Park Service's Cabrillo National Monument, there is one that focuses on the intertidal zone that can provide a model for a future effort. Facilitated by approximately 70 trained volunteers, the program engaged 37,010 interpretive contacts in 2015. The program's success can be partially attributed to what the monument's scientific program coordinator Alexandria Warneke called the "three touch approach." Monument educators go to the participating schools and prepare the students for what they will see, interpret what the students encounter during their visit, then return to the classrooms for a follow up. The program is all based on an engaging/interactive model whereby the students are not lectured to, but rather actively participate in the learning process.

The naturalist program at the San Dimas Canyon Nature Center is unique in that it focuses exclusively on youngsters. Founded in 1997 by Park Supervisor Roddy Gregory, who still plans and coordinates the effort every year, the program has two components. The first inspires as many as 180 7- to 11-year-olds per year to learn about nature as Junior Rangers through nine, 90-min engaging, activity-based classes at the center. About 10–20% of these students go on to participate in the Junior Naturalist program for 12- to 17-year-olds. These naturalists learn the skills of interpretation and care for the animals at the center's zoo which houses a number of native species.

Gregory provides the vision and inspiration for the entire program and organizes an annual Hawaiian Ho'olaulea Festival in June to raise the needed funds to support his volunteers. When asked about training adults, Gregory laughed and suggested kids are much more fun and do not have set ideas that get in the way of learning. "Adults are harder to control," he added with another laugh. Discussing the program's future, he grew pensive when considering what would happen to it all after he retired. Programs usually fade when the organizer leaves, he said with a sense of resignation in his voice. Although it will be difficult to replace Gregory, we are hopeful Los Angeles County Parks will endeavor to continue the valuable program he has inspired and has worked so hard to maintain.

At the state level, the California Naturalist Program is an ambitious effort sponsored by the University of California, Division of Agriculture and Natural Resources. It endeavors to create a working group of naturalists who receive certification after completing a 40+ h course that involves classroom and field experience. Once certified, the naturalists are encouraged to complete another 40 h of community service focusing on nature-based volunteer opportunities.

The program is relatively decentralized, allowing local providers to tailor about half of the content to what exists in their surrounding natural environment. The courses are generally organized around the program's official text, *The California Naturalist Handbook* (de Nevers et al. 2013). However, the text presents a content problem for chaparral education because of its forest/rangeland centric bias.

Chaparral is only given two paragraphs within a chapter on plants. Forests on the other hand are given 20 pages in a chapter dedicated to forest, woodland, and range resources (the only chapter dedicated to plant communities). The forest/rangeland centric influence may partially explain the presence of a common misconception in the text that chaparral is merely a successional phase in forest development. The text claims that, "Mature stands of chaparral provide a shaded seedbed for their successors, oak woodland and conifer forest." While this can be true for some montane chaparral stands which arise after forest fires, the vast majority of chaparral is a climax community and not a successional stage to forest.

Fortunately, revisions and supplemental documents to The California Naturalist Handbook text are being discussed. The program is a vision in progress, so the text is a working document. Creating a state-wide group of certified naturalists who will be interacting with the public in a coordinated manner has tremendous potential in helping citizens become better connected with the natural environment. Hopefully, future editions of the program's main text and supplemental materials will reflect the chaparral's importance in the state.

Our general finding regarding the inclusion of chaparral ecology in naturalist training programs is that most mention the ecosystem at various times during their courses. Only half of the programs, however, dedicate an entire class session or a specified amount of time focusing on chaparral as a functioning ecosystem.

11.4.3 Publications

To assess the extent of published material about chaparral that is easily available to the general public and schools, we conducted a review of books on the subject. There are relatively few publications that focus exclusively on the ecosystem. Unfortunately, it is not unusual to find misconceptions in these materials, especially regarding the chaparral's fire regime. We have only reviewed those publications that avoid such errors.

The most thorough treatments of chaparral are contained within books that review all of the vegetation communities in California (e.g., Keeley and Davis 2007; Parker et al. 2016). Halsey (2008) offers an overview of chaparral ecology, a detailed discussion of the chaparral's fragile relationship to fire, and suggestions on how human communities can adapt to California's fire-prone environment. Keeley et al. (2012) provides an excellent analysis of fire in Mediterranean-type climate regions.

An old publication that is still available in the used book market is *The Elfin Forest*, by Francis M. Fultz (1927), the first popular book written about California's chaparral. Fultz provides a description of the chaparral community that most chaparral explorers can appreciate when he wrote, "The Chaparral is very dear to me now, but when I first "hit the trail" that led me into it, it did not strike me at all favorably. And everything about it was so new and strange that I almost felt as if I were in another world."

There are several plant identification publications that offer excellent information on the most common chaparral species. Kauffmann et al. (2015) is a well-organized field guide to the chaparral's most characteristic shrub, manzanita (*Artcostaphylos* spp.). McLean and McLean (2003) produced a CD containing an easy to use guide to the common plants of the San Gabriel foothills (obtained at the Eaton Canyon Nature Center in Pasadena, California). A guide to the flowering plants of the Santa Monica Mountains written by Dale (1986) is out of print but can be purchased used online. It remains one of the best chaparral plant books available because in addition to the excellent photos, it provides interesting anecdotes about each species. Fillius (2005) offers a comprehensive guide to native plants at Torrey Pines State Reserve. The classic work on chaparral shrubs that should be on every chaparral enthusiast's bookshelf is Lester Rowntree's (1939) *Flowering Shrubs of California*.

11.4.4 Curricula

In terms of materials for schools, we found limited chaparral related curricula available to Kindergarten through 12th grade teachers. High school biology texts, mostly those for higher level courses such as Advanced Placement (AP) Biology, often include chaparral in the section on ecology as one of the world's major biomes. For example, the AP edition of Campbell Biology (Reece et al. 2011) provides an accurate, one page summary of chaparral, including the negative impacts of human-caused fires.

A popular elementary science text series in California by Macmillian/McGraw-Hill mentions chaparral in the third and fourth grade versions (Hackett et al. 2008). However, when it comes to listing plant communities, chaparral is missing. Only deserts, grasslands, forests, and Arctic tundra are discussed. Ignoring chaparral in deference to other communities is replicated in the new national Framework for K-12 Science Education which forms the foundation for California's Next Generation Science Standards. The Framework mentions deserts, grasslands, rainforests, coral reefs, and wetlands, but not Mediterranean-type climate region plant communities (NRC 2016). The absence is unfortunate because of the significant contribution chaparral and other Mediterranean-type climate ecosystems provide to the world's biodiversity.

In an effort to improve environmental education, California Assemblywoman Fran Pavley authored legislation in 2003 to develop an environmentally-based curriculum called the California Education and the Environment Initiative (EEI). It produced an environmental curriculum in 85 separate units for kindergarten through twelfth grade. The program was approved by the California State Board of Education in 2010.

Chaparral makes its first complete appearance in the EEI curriculum in the third grade. It is given a four page description in a booklet on California's natural regions that provides good descriptions of some of the ecosystem's species. However, the description of chaparral includes problematic language that fosters the anthropocentric notion that fire is needed because it "cleans out old brush and brings about new growth." On the other hand, a creative activity in the student workbook does accurately portray issues such as the threat of too much fire and development. In addition, a map of the major habitats in California including chaparral is provided as a supplement. Chaparral does not appear again in the EEI curriculum until the sixth grade with a brief mention of it being one of the nine biomes of the world. At the high school level, a unit examining biodiversity incorrectly lists chaparral as existing in only three of ten regions in California, the central coast, south coast, and the Sierra Nevada (EEI 2011). In contrast to the chaparral's three appearances in the EEI curriculum, forests and deserts are discussed six times.

There are a large number of implementation partners in the EEI Program including the California Regional Environmental Education Community Network (CREEC) and the non-profit organization Ten Strands. When using the CREEC website search engine (accessed 5/2016) to find programs focusing on chaparral, only two appeared. One was a five-lesson packet about chaparral shrublands for fourth grade provided by the San Diego Child and Nature Collaborative. The other was a field program from the Ocean Institute for fourth, fifth, and sixth graders who visit six California plant communities, one being chaparral.

Overall, we found the coverage of chaparral in available curricula to be lacking in both accuracy and depth.

11.5 Recommendations

Chaparral occurs in every county in California and dominates the southern part of the state where more than 60% of the population lives. Furthermore, since chaparral is especially concentrated near many of the most populated areas of the state, it also represents one of the best places to explore nature, especially for individuals who do not have the resources to travel to more distant locations. As a consequence, chaparral should be a significant topic in environmental education programs throughout California. Otherwise, neither the public nor the government agencies that serve them are likely to grasp the significance of chaparral to us and the native species that live there.

All naturalist programs should identify chaparral as an important ecosystem by providing a full session, or at least a major part of one, dedicated to the subject. Only five of the 17 programs we evaluated did so. The basic themes that should be discussed in context of the chaparral ecosystem should include the following:

- **Mediterranean-type climate**. How it shapes the chaparral ecosystem and how it is expected to change in the future
- **Biodiversity**. The distinction between different types of chaparral (e.g., manzanita chaparral, red shanks chaparral, etc.), the high level of biodiversity in Mediterranean-type climate regions compared to the rest of the USA, and the chaparral's most common native species

- **Botany**. Chaparral plant families, plant structure, and reproductive strategies
- **Evolution and ecology**. The origin and evolution of chaparral, speciation, and the ecological relationships between chaparral species
- **Fire ecology**. Chaparral plant and animal responses to fire, understanding that chaparral is adapted to a particular fire regime (not fire per se), that too much fire can destroy the chaparral ecosystem, and the role fire plays in the spread of non-native grass species
- **Chaparral physical environment**. The role of geology (e.g., soils) and geography in the distribution and diversity of chaparral, chaparral species (e.g., manzanitas), and other Mediterranean-type climate region shrublands such as California sage scrub
- **Chaparral ethics**. The importance of protecting an intact, healthy chaparral ecosystem for its intrinsic value and for human needs

Most nature education programs (and the grants that fund them) focus on elementary aged school children. The lack of opportunities for young adults to experience nature can negatively impact the transfer of these childhood experiences into adulthood. In California, some schools offer sixth grade camp experiences where students spend a week in a wildland area during the school year. Support for such camps needs to be increased and more nature education programs need to reach out to high school and college students, as well as families.

There is also a growing population of retired citizens who have a wealth of environmental knowledge and time to pass along the wisdom they gained when outdoor activity was the norm. Programs need to take advantage of this resource before the wisdom is lost. These volunteers can play an important role in initiating and supporting nature education programs for young adults. This means more than assigning older volunteers as trail guides. Older adults should serve as mentors through individualized programs that inspire younger adults to rediscover nature and motivate their peers to do the same.

The lack of diversity in the naturalist community is a critical issue that needs to be addressed. The United States Census estimates that by 2044, more than half of all Americans are projected to be people of color (any group other than non-Hispanic white) (Colby and Ortman 2015). In contrast, people of color compose only 12.4% of the staff in non-profit environmental groups (Taylor 2014). We have found even less diversity in the naturalist training programs we evaluated. If we intend to expand, much less maintain a population interested in the natural environment, naturalist education organizations must make a major effort to diversify their populations and approaches to learning. Lanham (2015), Professor of Wildlife Ecology at Clemson University, addresses the diversity issue in his short video, *Rules for a Black Birdwatcher* when he says, "When I meet another black birder, it's like encountering an Ivory-billed Woodpecker, an endangered species. Extinction looms. We have to do something to make birding, to make nature study in general, more interesting to people of color."

How do we increase diversity? How do we inspire more people to care about, and ultimately incorporate the chaparral into their lives? One place to start is by examining the way we teach natural history.

The most common method by which natural history information is shared with naturalist trainees and the public in the programs we reviewed is through lecture. While there is definitely a place for lectures, the method has been proven to be an ineffective way to develop meaningful, long-lasting learning (Bligh 2000). Lecturing also does not allow an individual to develop the personal meanings necessary to connect to a place and the species that live there. On the other hand, active learning techniques requiring students to engage in the learning process (e.g., discussions, presentations, interactive projects) result in better retention, deeper understandings (Prince 2004), and higher achievement (Freeman et al. 2014). In addition, active learning differentially benefits students of color (Haak et al. 2011). In other words, lecturing can discriminate against under-represented communities.

The question arises then, why do we keep lecturing when we want to pass along information we feel is important? One possible answer is that most of us have experienced lectures as the dominant teaching strategy, and as such we replicate what is familiar. Unfortunately, lectures fail to take advantage of the multiple ways people receive and process information (Gardner 1993).

Hughes, the interpretive specialist at the Irvine Ranch Conservancy suggested that, "99% of the people you talk to are going to remember you, not what you told them." Therefore, he limits the number of new ideas or topics in any presentation to five, plus or minus two, and actively engages the listeners in the process. "Otherwise, people's eyes just start to gloss over."

11.5.1 The ENGAGE Model

The brain often wanders while being subjected to a lecture, and it frequently does (Halsey 2016). To engage the mind and inspire the passionate naturalists we endeavor to create, we recommend that naturalist education programs employ active learning strategies. One such strategy is the ENGAGE model (Halsey 2011b).

The ENGAGE model takes a six step approach to teaching content by using active learning techniques combined with utilizing meaningful interpretation. The ENGAGE model is based on the observation that single modality teaching (i.e., lecturing) is not effective because it is passive. The model can be used by a single interpreter working with a group on the trail or while teaching content in a traditional classroom. The point to remember is that those who do the teaching do the learning.

It is important, however, that practitioners of this model be prepared for moments of doubt and failure. Helping people to become engaged in their own learning is not easy. It takes a confident, often audacious coach to keep participants involved and willing to do the kinds of activities suggested by the model because most of us are accustomed to passive "sit and get" lectures. But learner involvement is critical if we actually want people to remember what we teach and to use that knowledge to change behaviors.

Work through the following six steps to help others learn, remember, and care about the content you wish to share. Be aware that without coaching, most who are asked to learn and follow this model will at first spend most of their time on step two because it is closest to the teaching style we are accustomed. The percentages listed indicate the approximate amount of time that should be spent on each step.

- **Step 1—Energize Learners (15%)**

Optimal learning begins when you energize people *before* the interpretation experience/class begins. This primes the learning process. Giving visitors/students interesting items to look at and analyze will introduce them to the subject and allow the brain to access similar information, increasing receptivity to learning. Provide a thoughtful question that can be answered individually or by a group. If you have contact information before the class/event, send out a short article on the subject.

- **Step 2—Navigate Content (35%)**

Provide a summary of the content through a brief discussion, supplemented with written documents, outlines, photos, and field specimens as appropriate. Then utilize a wide variety of activities to allow the participants to "teach back" the material to others (e.g., demonstrations, skits, stories, games, interactive discussions, jigsaw learning). Your role as a facilitator is to coach participants through the learning process and create enough diversity of activities to give everyone a chance to leverage their strengths as learners and maximize learning. Finally, review the content your participants have learned with a few quick questions. This is essential. The point of the questions is not to baffle, but to help participants demonstrate how they are smart.

- **Step 3—Generate Meaning (5%)**

Once the learners have grasped the key content through teach backs, help them develop meaning for that content by employing *meaningful interpretation*, a method often utilized by outdoor educators in national parks. Help the learners care for the subject they are learning about.

In his book, Meaningful Interpretation, David Larsen (2011) provides the answer to why we share our passions about the natural environment—we want people to care, and ultimately help protect the natural treasures we value. The way we do this is through meaningful interpretation—helping others find personal connections and meanings to the plants and animals, the places, and the ecological communities we love. Larsen writes, "Helping visitors connect to meanings is the entire goal. Meaning is more important than knowing! Your job is not to fill their heads with information… Even those people who want information want to connect it to personal meanings. Audiences want to connect to your place intellectually and emotionally."

Once a meaningful connection is made, people care and will more likely *care for* the natural places we value.

Establishing meaning is often challenging because we are usually more focused on passing along content. It can begin by showing a sincere interest in the individual you are communicating with. Ask rather than tell. Develop connections between an individual's interests and the subject being shared. Ask, "How do you think this place, this animal, this plant could be important to you?" Asking leading, follow-up questions can further the process. "Watch your daughter as she listens to the wren singing in the distance. What do you think is going on in her mind? Since a major part of a bat's diet is insects, how might your life change if bats disappeared? How did it feel to sit quietly, with your family, listening to the wind moving through the leaves?" When teaching a group, have participants discuss *why* the learning is important in their world. Have them list the values, benefits, and personal meanings of what they have learned in a journal, on an index card or flipchart. Asking "why" questions open minds to the next stage of learning.

- **Step 4—Apply Learning (35%)**

We often teach people what we want them to know (isolated facts), but do not have them demonstrate their knowledge in the setting where they can use it. If observation techniques are being taught, provide field journals and have participants practice describing sounds, sights, and smells on the trail. By using dots and dashes on a graph, bird calls can be easily documented and remembered. To apply interpretation skills, practice asking questions with other willing visitors that inspire curiosity about the subject, making sure the questions are designed to encourage further thought, not just simple answers. Ask participants to pick up a handful of soil and describe what it looks and feels like rather than asking difficult content questions that require prior knowledge. As participants apply what they have learned, they are creating new memories and deepening retention of content.

- **Step 5—Gauge and Celebrate Learning (10%)**

Checking for understanding is a critical component to ensure the learner has an accurate grasp of the new knowledge learned and meanings established. This was started in step two. During your interactions, continually check with the group you are teaching by asking questions, making sure they have understood what has been shared. Use creative ways to reward participants for learning (e.g., flash card games, crossword puzzles, bingo, interactive reviews, stump the panel, quiz shows, group mind maps). Let people "show how they are smart" as they answer questions and demonstrate their new knowledge.

- **Step 6—Extend Learning to Action**

Do not let the experience stop once the class or interpretation has concluded. Collect contact information if appropriate and offer follow-up extensions to what has been learned. Give everyone a token or talisman as a symbol of their new knowledge and ask the participants to pass it along to someone else they may inspire in the future. This is the principle behind Jim Lockyer's "COR = K" card he gives to students at the Santa Rosa Plateau. The idea is to keep the learning alive by creating a process to foster more curiosity, additional meanings, and a stronger relationship to the subject. This can also be used to establish a community of nature advocates. The Chaparral Naturalist Certification program at the Elfin Forest Recreational Reserve which utilizes the ENGAGE model, awards three green marbles to its graduates—one to keep and two to share with people they may meet who also express an interest in inspiring others to love their local, natural environment.

The basic message of the ENGAGE model is that people learn best when they play an active, critical role in the learning process, apply what they have learned, and are encouraged to discover their own meanings to places and things.

11.6 Teaching to Inspire Life-Long Learning

Although we can identify the most effective way to communicate with people about nature, as Larsen (2011) reminds us, audiences hold power. He wrote, "No matter how enthusiastic, professional, knowledgeable, and creative an interpreter is, it is the audience that will ultimately decide if they had a meaningful experience, connected emotionally and intellectually, and believe the place (or subject) is worth caring about and for."

Nature provides a wonderfully complex story with actors of many kinds that can inspire emotional connections and life-long passions. All we need to do is to show we care, provide the guidance, and then get out of the way. People do not care about how much you know unless they know how much you care. Rachel Carson (1965) addressed the significance of emotional connections when she wrote, "If facts are the seeds that later produce knowledge and wisdom, then the emotions and the impressions of the senses are the fertile soil in which the seeds must grow… Once the emotions have been aroused—a sense of the beautiful, the excitement of the new and the unknown, a feeling of sympathy, pity, admiration, or love—then we wish for knowledge about the object of our emotional response. Once found, it has a lasting meaning."

When designing a program to increase awareness of California's most extensive ecosystem, we need to ask not only *what* we want to teach about the chaparral, but *how* can we teach it to help develop meaning and connections. Our purpose as naturalists is to foster behavioral changes that will create a diverse population of life-long nature advocates—individuals who understand how and why nature, and specifically the chaparral, is such an integral part of our sense of place. Engaging, meaningful interpretation that reflects the significance and value of chaparral offers a compelling educational experience to accomplish this goal. Such an approach is within reach of every nature center, naturalist program, and classroom in California.

References

Abrams, K. 2014. Why do you feel so good in nature? Why does nature heal? The top 10 reasons. Cedar Tree Healing Arts. http://www.cedartreehealing.org/blog/2014/11/24/why-do-we-feel-so-good-in-nature-why-does-nature-heal-the-top-10-reasons

Aspinall, P., P. Mavros, R. Coyne, and J. Roe. 2013. The urban brain: analyzing outdoor physical activity with mobile EEG. British Journal of Sports Medicine. https://doi.org/10.1136/bjsports-2012-0918778

Atchley, R. A., D. L. Strayer, and P. Atchley. 2012. Creativity in the wild: improving creative reasoning through immersion in natural settings. PLoS ONE 7:e51474.

Bligh, D. A. 2000. What's the use of lectures? Jossey-Bass, New York, New York, USA.

Bratman, G. N., J. P. Hamilton, K. S. Hahn, G. C. Daily, and J. J. Gross. 2015. Nature experience reduces rumination and subgenual prefrontal cortex activation. Proceedings of the National Academy of Sciences 112:8567-8572.

Carson, R. 1965. Sense of Wonder. Harper and Row Publishers, New York, New York, USA.

Colby, S. L., and J. M. Ortman. 2015. Projections of the size and composition of the US population: 2014 to 2060. Report P25-1143. US Department of Commerce, US Census Bureau.

Carmona, R. H. 2004. The growing epidemic of childhood obesity, testimony, March 2, 2004, before the US Senate, Subcommittee on Competition, Infrastructure, and Foreign Commerce, Committee on Commerce, Science, and Transportation. Text from: https://www.surgeongeneral.gov/news/testimony/childobesity03022004.html

Dale, N. 1986. Flowering plants: The Santa Monica Mountains, coastal & chaparral regions of southern California. California Native Plant Society, Sacramento, California, USA.

de Nevers, G., D. S. Edelman, and A. M. Merenlender. 2013. The California naturalist handbook. University of California Press, Berkeley, California, USA.

Dioum, B. 1968. Public address to the General Assembly of the International Union for Conservation of Nature and Natural Resources. New Delhi, India.

EEI [California Education and the Environment Initiative]. 2011. Biodiversity: the keystone to life on Earth. Student edition. California Environmental Protection Agency, Sacramento, California, USA.

Fillius, M. L. 2005. Native plants: Torrey Pines State Reserve & nearby San Diego County locations. Found Image Press, San Diego, California, USA.

Freeman, S., S. L. Eddy, M. McDonough, M. K. Smith, N. Okoroafor, H. Jordt, and M. P. Wenderoth. 2014. Active learning increases student performance in science, engineering, and mathematics. PNAS 111:8410–8415.

Fultz, F. M. 1927. The elfin-forest of California. The Times-Mirror Press, Los Angeles, CA, USA.

Gardner, H. 1993. Frames of mind: The theory of multiple intelligences. Basic Books, New York, New York, USA.

Gartner, J. E., S. H. Cannon, D. R. Helsel, and M. Bandurraga. 2009. Multivariate statistical models for predicting sediment yields from southern California watersheds. Open-File Report 2009-1200. US Department of the Interior, US Geological Survey, Washington D.C., USA.

Haak, D. C., J. HilleRisLambers, E. Pitre, and S. Freeman. 2011. Increased structure and active learning reduce the achievement gap in introductory biology. Science 332:1213-1216.

Hackett, J. K., R. H. Moyer, J. Vasquez, M. Teferi, D. Zike, K. LeRoy, D. J. T. Terman, and G. F. Wheeler. 2008. Life science 4. Macmillan/McGraw-Hill, Columbus, Ohio, USA.

Halsey, R. W. 2004. In search of allelopathy: an eco-historical view of the investigation of chemical inhibition in California coastal sage scrub and chamise chaparral. Journal of the Torrey Botanical Society 131:343-367.

Halsey, R. W. 2008. Fire, chaparral, and survival in southern California. Sunbelt Publications, San Diego, California, USA.

Halsey, R. W. 2009. The wisdom of sunflowers. The Chaparralian 6:4-8.

Halsey, R. W. 2011a. Chaparral as a natural resource. Pages 82-86 *in* J. W. Willoughby, B. K. Orr, K.A. Schierenbeck, and N. J. Jensen, editors. Proceedings of the CNPS Conservation Conference, January 17-19, 2009. California Native Plant Society, Sacramento, California, USA.

Halsey, V. 2011b. Brilliance by design. Berrett-Koehler Publishers, Inc., San Francisco, California, USA.

Halsey, R. W. 2016. The first ten things I learned in the wilderness: #4 stop talking at me. http://www.californiachaparral.org/images/The_First_Ten_Things_I_Learned_in_the_Wilderness.pdf

Halsey, R. W., and J. E. Keeley. 2016. Conservation issues: California chaparral. *In* S. A. Elias, D. Alterton, J. K. Cochran, D. A. Dellasala, F. Funiciello, M. I. Goldstein, K. Lajtha,

S. Marshall, and P. Mineau. Reference module in earth systems and environmental sciences. Elsevier Publications Inc, Amsterdam, Netherlands. https://doi.org/10.1016/B978-0-12-409548-9.09584-1.
Halsey, V. 2011b. Brilliance by design. Berrett-Koehler Publishers, Inc., San Francisco, California, USA.
Hofferth, S., and J. Sandberg. 1999. Changes in American children's time, 1981-1997. University of Michigan Institute for Social Research, Ann Arbor, Michigan, USA.
Juster, F. H., H. Ono, and F. P. Stafford. 2004. Changing times of American youth: 1981-2003. Institute for Social Research, University of Michigan, Ann Arbor, Michigan, USA. http://www.umich.edu/news/Releases/2004/Nov04/teen_time_report.pdf
Kauffmann, M., V. T. Parker, and M. C. Vasey. 2015. Field guide to manzanitas. California, North America, and Mexico. Backcountry Press. Kneeland, California, USA.
Keeley, J. E. 2001. We still need Smokey Bear! Fire Management Today 61:21-22.
Keeley, J. E., W. J. Bond, R.A. Bradstock, J. G. Pausas, and P. W. Rundel. 2012. Fire in Mediterranean ecosystems: ecology, evolution and management. Cambridge University Press, Cambridge, UK.
Keeley, J. E., and T. J. Brennan. 2012. Fire-driven alien invasion in a fire-adapted ecosystem. Oecologia 169:1043-1052.
Keeley, J. E., and F. W. Davis. 2007. Chaparral. Pages 339-366 *in* M. G. Barbour, T. Keeler-Wolf, and A. A. Schoenherr, editors. Terrestrial vegetation of California. University of California Press, Berkeley, California, USA.
Keeley, J. E., C. J. Fotheringham, and M. Mortiz. 2004. Lessons from the 2003 wildfires in southern California. Journal of Forestry 102:26-31.
Keeley, J. E., A. H. Pfaff, and H. D. Safford. 2005. Fire suppression impacts on postfire recovery of Sierra Nevada chaparral shrublands. International Journal of Wildland Fire 14:255-265.
Keeley, J. E., and P. H. Zedler. 2009. Large, high-intensity fires events in southern California shrublands: debunking the fine-grain age patch model. Ecological Applications 19:69-94.
Lanham, J. D. 2015. Rules for the Black Birdwatcher. https://www.youtube.com/watch?v=4thb2zGuOnU
Larsen, D. L. 2011. Meaningful Interpretation: how to connect hearts and minds to places, objects, and other resources. Eastern National, National Park Service, Washington D.C., USA.
Lee, J., Y. Tsunetsugu, N. Takayama, B-J. Park, Q. Li, C. Song, M. Komatsu, H. Ikei, L. Trvainen, T. Kagawa, and Y. Miyazaki. 2014. Influence of forest therapy on cardiovascular relaxation in young adults. Evidence-Based Complementary and Alternative Medicine 2014:834360.
Li, Q., K. Morimoto, M. Kobayashui, H. Inagaki, M. Katsumata, Y. Hirata, K. Hirata, T. Shimizu, Y. J. Li, Y. Wakayama, T. Kawada, T. Ohira, N. Takayama, T. Kagawa, and Y. Miyazaki. 2008. A forest bathing trip increases human natural killer activity and expression of anti-cancer proteins in female subjects. Journal of Biological Regulators Homeostatic Agents 22:4-55.
Lindsay, D., P. Knoll, and T. Varnell. 2016. Coast to cactus: the canyoneer trail guide to San Diego outdoors. Sunbelt Publications, Inc., San Diego, California, USA.
Lombardo, K. J., T. W. Swetnam, C. H. Baisan, and M. I. Borchert. 2009. Using bigcone Douglas-fir fire scars and tree rings to reconstruct interior chaparral fire history. Fire Ecology 5:32-53.
Louv, R. 2008. Last child in the woods: saving our children from nature-deficit disorder. Algonquin Books, New York, USA.
Manzo, L. C., and D. D. Perkins. 2006. Finding common ground: the importance of place attachment to community participation and planning. Journal of Planning Literature 20:335-350.
McLean, G., and C. McLean. 2003. [CD-ROM]. Common plants of Eaton Canyon and the San Gabriel Foothills. http://www.natureathand.com/EC_Plants/ECFieldGuideBig.htm
NRC [National Research Council]. 2016. A framework for K-12 science education: practices, crosscutting concepts, and core ideas. The National Academies Press, Washington D.C., USA.
Parker, V. T., R. B. Pratt, and J. E. Keeley. 2016. Chapter 24. Pages 479-507 *in* H. Mooney, and E. Zavaleta, editors. Ecosystems of California. University of California Press, Berkeley, California, USA.

Piff, P. K., P. Dietze, M. Feinberg, D. M. Stancato, and D. Keltner. 2015. Awe, the small self, and proscocial behavior. Journal of Personality and Social Psychology 108:883-899.

Prince, M. 2004. Does active learning work? A review of the research. Journal of Engineering Education 93:223-231.

Reece, J. B., L. A. Urry, M. L. Cain, S. A. Wasserman, P. V. Minorsky, R. B. Jackson, and N. A. Campbell. 2011. Campbell Biology, AP Edition. Pearson Publications, New York, New York, USA.

Rideout, V., U. G. Foehr, and D. F. Roberts. 2010. Generation M: media in the lives of 8-18 year olds. The Henry J. Kaiser Family Foundation, Menlo Park, California, USA.

Rocha, V., and S. Kelley. 2017. June 16. Think it's hot now? Just wait Southern California's heat wave just getting started. Los Angeles Times. Los Angeles, California, USA. http://www.latimes.com/local/lanow/la-me-ln-hot-weather-southern-california-20170616-htmlstory.html

Rowntree, L. 1939. Flowering shrubs of California. Stanford University Press, Stanford, California, USA.

Rundel, P. W., M. T. K. Arroyo, R. M. Cowling, J. E. Keeley, B. B. Lamont, and P. Vargas. 2016. Mediterranean biomes: evolution of their vegetation, floras, and climate. Annual Review of Ecology, Evolution, and Systematics 47:383-407.

Ryan, R. M., N. Weinstein, J. Bernstein, K. W. Brown, L. Mistretta, and M. Gagne. 2010. Vitalizing effect of being outdoors and in nature. Journal of Environmental Psychology 30:159-168.

Sampson, S. D. 2015. How to raise a wild child: the art and science of falling in love with nature. Houghton Mifflin Harcourt Publishing Company, New York, USA.

Sampson, S. D. 2016. Hey kids! Go outside, already. WBUR-FM: On Point. Boston, Massachusetts, USA, July 20, 2016. http://www.wbur.org/onpoint/2016/07/20/children-nature-parenting-outside

Santa Barbara County Fish and Game Commission. 2008. Wild Fire Task Force Report and Recommendation submitted to the Clerk of the Santa Barbara County Board of Supervisors. Agenda Letter, September 2, 2008. Santa Barbara County Board of Supervisors, Santa Barbara, California, USA.

Soulé, M. E. 1985. What is conservation biology? Bioscience 35:727-734.

Steele, F. 1981. The sense of place. CBI Publishing Company, Inc., Boston, Massachusetts, USA.

Stephan, W. G., and C. W. Stephan. 1984. The role of ignorance in intergroup relations. Pages 229-257 *in* N. Miller, and M. B. Brewer, editors. Groups in contact: the psychology of desegregation. Academic Press, Inc., Cambridge, Massachusetts, USA.

Suttie, J. 2016. How nature can make you kinder, happier, and more creative. Greater Good Magazine. http://greatergood.berkeley.edu/article/item/how_nature_makes_you_kinder_happier_more_creative

Tadmor, C. T., Y. Y. Hong, M. M. Chao, F. Wiruchnipawan, and W. Wang. 2012. Multicultural experience reduce intergroup bias through epistemic unfreezing. Journal of Personality and Social Psychology 103:750-772.

Taylor, D. E. 2014. The state of diversity in environmental organizations. University of Michigan, School of Natural Resources & Environment, Ann Arbor, Michigan, USA.

Tyrvainen, L., A. Ojala, K. Korpela, T. Lanki, Y. Tsunetsugu, and T. Kagawa. 2014. The influence of urban green environments on stress relief measures: a field experiment. Journal of Environmental Psychology 38:1-9.

USFS [United States Forest Service]. 2016. Rim fire reforestation environmental impact statement. R5-MB-279. USDA Forest Service, Stanislaus National Forest, Sonora, California, USA.

van Mantgem, E. F., J. E. Keeley, and M. Witter. 2015. Faunal responses to fire in chaparral and sage scrub in California, USA. Fire Ecology 11:128-148.

Vucetich, J. A., J. T. Bruskotter, and M. P. Nelson. 2015. Evaluating whether nature's intrinsic value is an axiom of or anathema to conservation. Conservation Biology 29:321-332.

Ward, M. 1984. Rancho de los Peñasquitos - on the road to Yuma. County of San Diego Department of Parks and Recreation, San Diego, California, USA.

Zhang, J. W., P. K. Piff, R. Iyer, S. Koleva, and D. Keltner. 2014. An occasion for unselfing: Beautiful nature leads to prosociality. Journal of Environmental Psychology 37:61-72.

Chapter 12
Chaparral Landscape Conversion in Southern California

Alexandra D. Syphard, Teresa J. Brennan, and Jon E. Keeley

Southern California, where the American Dream came too true

–Lawrence Ferlinghetti

Abstract The low-elevation chaparral shrublands of southern California have long been occupied and modified by humans, but the magnitude and extent of human impact has dramatically increased since the early 1900s. As population growth started to boom in the 1940s, the primary form of habitat conversion transitioned from agriculture to urban and residential development. Now, urban growth is the primary contributor, directly and indirectly, to loss and fragmentation of chaparral landscapes. Different patterns and arrangements of housing development confer different ecological impacts. We found wide variation in the changing extent and pattern of development across the seven counties in the region. Substantial growth in lower-density exurban development has been associated with high frequency of human-caused ignitions as well as the expansion of highly flammable non-native annual grasses. Combined, increases in fire ignitions and the extent of grassland can lead to a positive feedback cycle in which grass promotes fire and shortens the fire-return interval, ultimately extirpating shrub species that are not adapted to short fire intervals. An overlay of a 1930s vegetation map with maps of contemporary vegetation showed a consistent trend of chaparral decline and conversion to sage scrub or

A. D. Syphard (✉)
Conservation Biology Institute, Corvallis, OR, USA
e-mail: asyphard@consbio.org

T. J. Brennan
US Geological Survey, Three Rivers, CA, USA

J. E. Keeley
U.S. Geological Survey, Three Rivers, CA, USA

University of California, Los Angeles, CA, USA

E. C. Underwood et al. (eds.), *Valuing Chaparral*, Springer Series on Environmental Management, https://doi.org/10.1007/978-3-319-68303-4_12

grassland. In addition, those areas type-converted to grassland had the highest fire frequency over the latter part of the twentieth century. Thus, a continuing trend of population growth and urban expansion may continue to threaten the extent and intactness of remaining shrubland dominated landscapes. Interactions among housing development, fire ignitions, non-native grasses, roads, and vehicle emissions make fire prevention a complex endeavor. However, land use planning that targets the root cause of conversion, exurban sprawl, could address all of these threats simultaneously.

Keywords Chaparral · Fire · Housing development · Land use change · Non-native species · Vegetation change

12.1 Introduction

For thousands of years, humans have occupied the vast shrublands blanketing the foothills and mountains of southern California. Native Americans altered their environment to protect and sustain themselves, particularly via controlled burning to open up shrubland landscapes (see Chap. 4). Subsequently, the arrival of Euro-American settlers in the late eighteenth century brought about a sequence of progressively intense phases of rapid population growth and landscape conversion. The California Gold Rush and statehood brought one of the first population booms in 1850, and shortly thereafter, the region was linked to the railroad, enabling faster and safer immigration to the region from the rest of the country. Transportation via automobile soon became possible in the early 1900s, which facilitated even more immigration; plus, it enabled the beginning of suburban development outside of the region's main urban centers, such as Los Angeles and San Diego.

Throughout the progression of the twentieth century, southern California has continued to offer a wide range of economic opportunities. When coupled with the mild Mediterranean-type climate, these have made the region one of the most desirable places to live in the US. In particular, people flocked to the region with the discovery of oil at the turn of the century, which was then followed by growth in numerous other industries, including military defense production, agriculture, and the film industry. In the middle of the century, human population growth exploded; the accompanying massive change in land use dramatically altered the extent and composition of the native vegetation communities in the region. Although large expanses of native shrublands still exist in many areas, southern California has come to be viewed by the world as the land of freeways, strip malls, and endless housing developments.

In this chapter, we explore the trends and drivers of vegetation change in southern California since the early 1900s. In particular, we focus on the interactions between direct habitat conversion through urban growth and indirect changes brought on by non-native annual grasses, increased frequency of fires, and the resultant loss of native shrublands.

12.2 Habitat Conversion

12.2.1 Overview of Land Use Change

By the middle Holocene, Indian populations dominated much of coastal California, and they had a significant impact on landscape patterns through repeated burning and displacement of chaparral with more productive herbaceous communities (see Chap. 4). In the late eighteenth century, Spanish settlements initiated a new wave of changes with the introduction of a wide selection of non-native annual grasses and forbs (Mack 1989). The economy of these early settlements was based on cattle production, and the Mexican vaqueros would often burn off shrublands to increase grazing lands (Kinney 1887). Ever since then, rangeland management has had a significant component of repeated burning of shrublands to increase forage for livestock (Keeley and Syphard 2018).

One of the most significant changes in plant community composition with Euro-American settlement was the replacement of native vegetation with non-native grassland. As a result, non-native annuals were likely a large component of California grasslands by the 1850s (Burcham 1956). Livestock grazing undoubtedly has contributed substantially to this shift (D'Antonio et al. 1992), often in combination with severe droughts (Burcham 1956). Nevertheless, even in the absence of grazing, non-native annuals introduced by Euro-American settlers likely outcompeted native bunch grasses (Bartolome and Gemmill 1981). Intentional conversion of shrublands to create grassland for grazing was common across California (Burcham 1956; Keeley and Fotheringham 2003). Similar patterns of type-conversion have occurred over the 10,000-year history of human occupation in the Mediterranean Basin, where transitions from woody to herbaceous species have also been caused by human disturbance via livestock grazing and accelerated burning due to anthropogenic ignitions. However, in California, this loss of woody cover degrades natural systems and diminishes their conservation value by displacing native flora with non-native species. In the Mediterranean Basin, type-conversion replaces woody natives with herbaceous natives, and thus, native biodiversity increases.

In the early twentieth century, conversion of natural habitat into agricultural lands was the most dominant form of land use change, and by the 1930s, approximately 20% of the land within the South Coast Ecoregion had become croplands, with citrus and other fruit trees becoming especially extensive. At this time, southern California was considered one of the top agricultural regions in the US. However, with population growth and evolving economic opportunities, farming was largely wiped out in the middle of the century in favor of commercial and residential development, a trend that was common nationwide (Alig and Plantinga 2004). In the South Coast Ecoregion of California, less than five percent of the croplands mapped in the 1930s were still present by the early 2000s (derived from data described in next section).

Urban and residential development is now the top contributor to both direct and indirect habitat conversion in southern California. Not only have the major metropolitan areas become denser, but the freeway system developed in the 1940s initiated what has been an ongoing trend of "sprawl" outward from coastal cities into the inland foothills and mountains. This growth was so rapid and extensive that the San Fernando Valley outside of Los Angeles took on the name of "America's Suburb" (Roderick 2002). Across the world, southern California is still perceived as synonymous with urban sprawl.

12.2.2 Spatial and Temporal Patterns of Housing Growth

The spatial pattern of housing development has important implications for landscape conversion because low-density, sprawling-type development typically consumes more land and wildlife habitat than high-density development (Odell et al. 2003). As a consequence, low-density development may have a more negative impact on biodiversity and ecosystem services (Hansen et al. 2005). On the other hand, higher-density, clustered development may be more ecologically degraded with a larger dominance of non-native species (Lenth et al. 2006). Despite these trade-offs, compact urban development has been shown to minimize ecological disruption relative to sprawling development (Sushinsky et al. 2013).

The term wildland-urban interface (WUI) has emerged in the last couple of decades to describe the characteristics and social-ecological effects of those areas where housing development is adjacent to or interspersed with wildland vegetation (Radeloff et al. 2005). Two types of WUI are typically defined, largely as a function of housing density and the extent to which houses are surrounded by wildland vegetation. The "interface WUI" describes those areas where human settlements are denser and form an edge with wildland vegetation, whereas "intermix WUI" reflects areas where sparser, lower-density housing is interspersed with wildland vegetation. Although the exact definition of intermix or interface WUI may vary slightly with regards to how it is mapped (Stewart et al. 2007), these terms have provided a useful framework for understanding how and where human settlements interact with the natural environment, and how different forms of development may differentially affect habitat change and ecological impacts (Bar-Massada et al. 2014).

The spatial pattern of urban development in any given area can vary dramatically over time, but it typically emerges as a result of different characteristic growth types (Herold et al. 2003; Dahal et al. 2017). At one end, compact and high-density development patterns usually result from infill-type growth, where new structures are built within or expand outward from existing urban areas. At the other end, low-density, fragmented, exurban development patterns result from leapfrog-type growth in which new development occurs outside of urban areas and is typically surrounded by wildland vegetation. This lower-density exurban development, characteristic of the intermix WUI, is often the result of homeowner preferences and behaviors, including a desire to live closer to natural amenities (Netusil 2005) or lower land prices at greater distances from the urban core (Wu and Plantinga 2003).

Given the importance of both spatial extent and pattern of housing growth in terms of natural habitat conversion, we quantified historical housing trends in the South Coast Ecoregion from 1940 to 2010. To do this, we evaluated historical housing density maps (Hammer et al. 2004, available at http://silvis.forest.wisc.edu/maps/housing) within the footprint of a modified South Coast Ecoregion boundary (i.e., as in Syphard et al. 2011) that includes the full extent of the Los Padres National Forest. The maps were developed as part of a national data product in which housing density was mapped within partial census block groups and reported as housing units per square kilometer.

We quantified the extent of both low- and medium- to high-density housing from 1940 to 2010 within the seven counties that are located within the ecoregion. Instead of clipping the counties to the ecoregion boundary, we assessed housing growth for the complete extent of each county. To threshold the continuous housing data into classes of low- and medium-high-density, we selected all areas with a housing density between 6.17 and 49 houses per km^2 and classified them as "low density." The number 6.17 corresponds to the minimum housing density cutoff for defining low-density WUI (Radeloff et al. 2005). The threshold of $\geq$50 houses per square kilometer corresponds to those areas defined as medium- or high-density WUI. For each county in each decade, we summarized the total extent of each housing density type and calculated its proportion of the county area.

In all seven counties, housing development, and hence direct habitat conversion, increased substantially from 1940 to 2010 across the region, but the extent of development and pattern of housing growth varied over time and by county (as can be seen in the widely varying range of the Y axis in Fig. 12.1). Medium- to high-density development has dominated the counties closest to Los Angeles, but low-density housing growth has predominated in San Luis Obispo, Riverside, and San Diego counties (Figs. 12.1 and 12.2). Except for Los Angeles, which exhibited slow, steady growth in both housing-density types over time, a pulse in growth was apparent during and shortly after the 1990s for the other counties, which is consistent with nation-wide trends (Glaeser and Shapiro 2003). Orange County stands out in that, as medium-high density increased over time, low-density development has shown a slight decline across most of the record. This also has been evident in recent decades for Santa Barbara and San Luis Obispo Counties, suggesting that, in addition to urban expansion, existing urban areas in these counties may also be infilling and becoming denser. The two southern-most counties (San Diego and Riverside), on the other hand, show no sign of slowing in the expansion of low-density development.

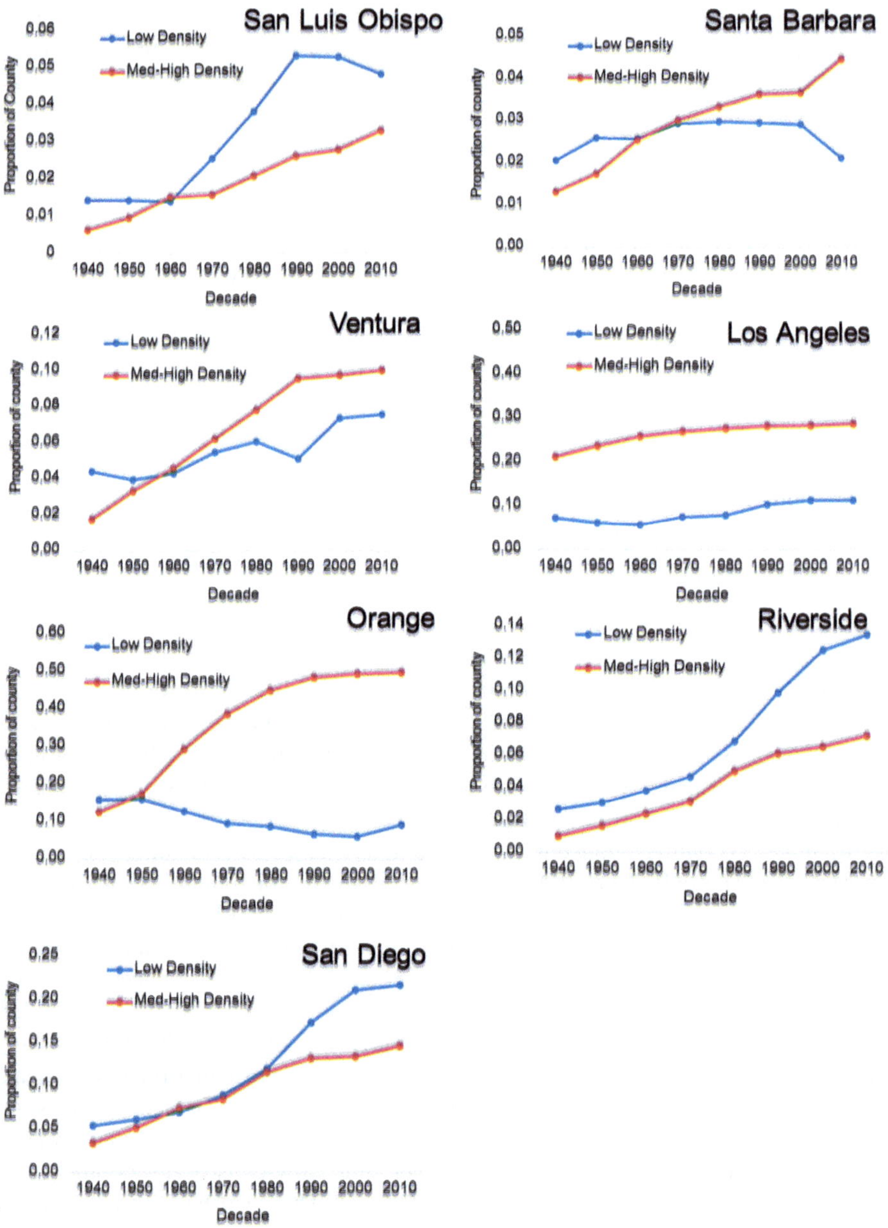

Fig. 12.1 Growth in area (proportion of county) of low and medium-high housing density from 1940 to 2010 within the full boundaries of the seven counties overlapping the South Coast Ecoregion of southern California

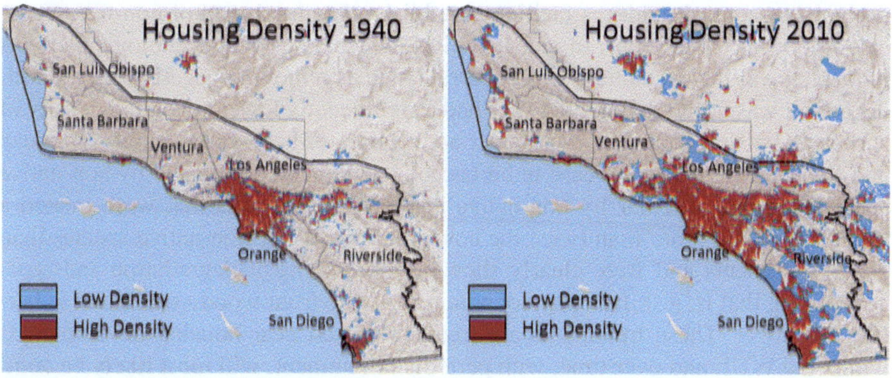

Fig. 12.2 Maps of low and medium-high housing density in 1940 and 2010 in the South Coast Ecoregion of southern California

12.2.3 Indirect Habitat Loss and Conversion

In addition to causing direct conversion of native habitat, housing development in southern California indirectly contributes to chaparral conversion, primarily by facilitating an interaction between increased fire frequency and the expansion of weedy non-native annual grasslands.

Although periodic wildfire is an essential component of chaparral ecosystems, fire frequency has been increasing in southern California to the extent that most of the landscape is burning at fire-return intervals (i.e., the time between fires at a defined area) that are uncharacteristically short relative to pre-EuroAmerican settlement conditions (Safford and Van de Water 2014). In some areas, fires are now so frequent that they outpace the historical baseline by a wide margin; for example, return intervals that once averaged 30 to more than 150 years are now shorter than 10 years in some areas (Keeley and Syphard 2018).

Given that humans cause more than 95% of the fires in the region (Syphard et al. 2007), the trend of increasing fire frequency is primarily explained by population growth and expansion of development into wildland vegetation. Although human-caused fires generally increase with human population, this relationship is mediated by population or housing density. That is, across California and other Mediterranean-type climate regions, studies show that the ignition frequencies tend to peak at low-intermediate population density, such as the WUI intermix areas (Syphard et al. 2007, 2009; Archibald et al. 2010; Syphard and Keeley 2015). The likely explanation for this is that these intermix WUI areas have both enough people to start frequent fires, which wild areas lack, and sufficient wildland areas to facilitate fire spread, which urban areas lack. These are also the areas most difficult to access for fire suppression (Gude et al. 2008).

In addition to increased fire frequency, exurban development provides conduits for non-native species to expand into wildland vegetation, via land disturbance, road networks, and residential landscaping (Gavier-Pizarro et al. 2010). Even fuelbreaks designed to control wildfires facilitate establishment and spread of non-native species (Merriam et al. 2006). In recent studies, which assessed the ecological effects of mechanical fuel treatments on chaparral (such as removing vegetation with bulldozers), it was found that treated sites had a significantly lower cover and density of shrubs and a significantly higher cover and density of herbaceous plants (Brennan and Keeley 2015). The increase in herbaceous plants was dominated by non-native species and in particular by non-native annual grasses. Sites that were treated a second time had more than twice the cover and density of non-native species than single treatments and were clearly showing more signs of degradation and type-conversion, that is, a shift in physiognomic structure from woody shrubland to herbaceous cover. These treatments are frequently used near housing developments within the WUI; and over time, with periodic retreatment, will most likely be completely type-converted to non-native annual grasslands.

The weedy annual grasses that have invaded vast portions of southern California are highly flammable and tolerant of rapidly repeating fires. In the absence of disturbance, chaparral shrublands are relatively resistant to invasion by non-native species, in part due to their dense cover and closed canopy. However, increased human ignitions in these fire-prone grasslands has lengthened the fire season, thereby increasing canopy opening and providing new establishment opportunities for these well-dispersed grasses. This positive feedback process between fires and grass expansion is typically referred to as a grass-fire cycle, and it is recognized as a potential problem in ecosystems across the world (e.g., D'Antonio and Vitousek 1992; Rossiter et al. 2003; Brooks et al. 2004; Bowman et al. 2014), including southern California shrublands (Keeley et al. 2012).

The larger ecological issue is that, despite native shrublands' resilience to periodic wildfire, too-short intervals between fires can lead to their extirpation. This is because many species require a minimum amount of time between fires to recover and regenerate. Non-resprouting species—i.e., obligate seeders—may require up to 25 years to fully establish a seedbank that can effectively recruit new plants after fire (Keeley 1986). Although re-sprouting chaparral species are resilient to shorter intervals between fires than non-re-sprouters, even re-sprouters were reduced when multiple fires occurred within in a six-year interval (Haidinger and Keeley 1993). Thus, as native shrubland species are extirpated, providing opportunities for further grass expansion, the potential exists for large scale vegetation type-conversion.

A number of studies in southern California have provided evidence of vegetation type-conversion from shrubland to grassland. Particularly widespread has been the conversion of coastal sage scrub to non-native grasses (Minnich and Dezzani 1998; Cox et al. 2014). Talluto and Suding (2008) found nearly 50% replacement of sage scrub by annual grasses within a 76-year study period in parts of Orange and Riverside Counties, with a substantial amount being due to fire frequency. Because sage scrub is generally more tolerant of higher fire frequencies than chaparral, chaparral may be even more vulnerable to vegetation type-conversion, depending on species composition and site factors. In some cases, it may even transition to sage scrub vegetation before finally transitioning to herbaceous cover (Syphard et al. 2006).

Chaparral conversion to grasslands after repeated fires has been documented in many localized studies (e.g., Zedler et al. 1983; Haidinger and Keeley 1993; Lippitt et al. 2012; Keeley and Brennan 2012). Given the consistency in these findings across the southern California region, and the fact that large areas across the region have experienced short fire-return intervals, there is reason to suspect that widespread conversion due to repeated fires has already occurred (Keeley 2010). Nevertheless, the empirical evidence for larger landscape scale changes in chaparral has been sparse, with one recent study even questioning the potential for widespread vegetation type change in chaparral to occur (Meng et al. 2014).

12.3 Landscape Scale Vegetation Type-Conversion

As a general means of quantifying historical vegetation change in concert with mean historical fire frequency in southern California, we overlaid contemporary maps of existing vegetation with an historical map delineating broad scale vegetation types and then integrated data on fire frequency. We estimated change using maps from multiple data sources because of the potential for vegetation to be mapped differently. Although variation is much more likely given finer scale vegetation classification schemes, there may even be differences in the way broad vegetation types are mapped due to differences in mapping methods, scales, and definitions.

The historical vegetation type maps (VTM) were developed between the years 1929 and 1934 (Wieslander 1935) as part of an extensive statewide mapping project. In addition to detailed species level plot information, vegetation types and dominant species were mapped on 15-minute topographic quadrangles in the field with a minimum mapping unit of 16 ha (39.5 acres) (Kelly et al. 2005; Kelly 2016). The first contemporary map we evaluated represents existing vegetation and was produced by the US Forest Service using a combination of satellite imagery, field verification, and expert guidance (CalVeg, http://www.fs.fed.us/r5/rsl/projects/classification/system.shtml). The majority of the area in this map was most recently updated in 2002. However, the national forest lands were updated more recently, in 2003, 2009, or 2010. The entire region was mapped at a scale of 1:24,000.

Both the VTM and CalVeg maps provide classification according to the California Wildlife Habitat Relationships System (Mayer and Laudenslayer 1988). Therefore, for both of these maps, we grouped vegetation classes into life-forms, including tree, shrub, coastal sage scrub, and herbaceous. For the other categories, which are mostly unvegetated (e.g., urban/developed land) or wetland, we lumped them into a class named "other."

We also evaluated the 2013 Landfire existing vegetation maps, which were developed based on a combination of decision tree models, field data, Landsat 7 imagery, elevation, and biophysical gradient data (http://landfire.cr.usgs.gov/viewer/ [2013, May 8]). The map comes as a grid at 30 m (0.2 acres) resolution. We developed map classes to match the vegetation types in the other two maps using the map

Table 12.1 Proportion of vegetation types within the historical (VTM) and contemporary (San Diego County, CalVeg, and Landfire) maps

Vegetation Type	VTM	San Diego	CalVeg	Landfire
Grass	0.06	0.09	0.12	0.27
Sage scrub	0.29	0.24	0.10	0.06
Shrubland	0.45	0.35	0.43	0.19
Tree	0.08	0.09	0.12	0.18
Other	0.13	0.23	0.23	0.29
Total	1.0	1.0	1.0	1.0

attribute based on the National Vegetation Classification System Physiognomic Order. Any area that was classified as "sparsely vegetated," "barren," "water," "developed," or "agriculture" in the Landfire vegetation type classification, we converted to the "other" class.

At a finer resolution for a subset of the South Coast Ecoregion, we compared the VTM map to a detailed 2012 vegetation community map that spans part of San Diego County (https://databasin.org/datasets/bcd5db8e6aa540e6b06a371b-de0afde3). This map was developed with a 1 ha (2.5 acre) minimum mapping unit for terrestrial vegetation and has an accuracy of at least 80% as determined through extensive field verification reports. The map was classified according to Sproul et al. (2011), and again, we grouped these into the same life-form vegetation classes and an "other" class.

After re-classifying the vegetation maps into physiognomic types, we quantified the proportion of each vegetation or cover type within each map. We then overlaid the contemporary maps with the VTM map and summarized the mean historical fire frequency that occurred within each change class up to 2013. To estimate the transitions between life-form classes, we assessed changes from shrub to grass, sage scrub to grass, shrub to sage scrub, tree to earlier successional class (shrub, sage scrub, or grass), successional (e.g., grass to sage scrub, sage scrub to shrub, shrub to tree), no change in vegetation, or other (i.e., unvegetated in either map. We used the California Department of Forestry–Fire and Resource Assessment Program (CDF-FRAP 2013) map of overlapping historical fire perimeters (wildfire only) to create a continuous 30 m grid with each cell representing the number of times it had burned since 1878 (http://frap.fire.ca.gov/data/frapgis-data-sw-fireperimeters_download). In this database, any grid cell location may have burned 0–13 times during the time period, although this may under-estimate fire frequency due to the minimum mapping unit of this dataset (Syphard and Keeley 2017).

The contemporary vegetation maps showed consistent trends of increasing grass, tree, and other cover types and decreasing sage scrub and shrubs over time (Tables 12.1 and 12.2, Fig. 12.3). There were substantial areas of agreement in the delineation of all vegetation types that did not change between the VTM map and contemporary maps (Figs. 12.4 and 12.5), particularly in CalVeg and the higher-resolution San Diego County map. The Landfire map, however, delineated a much larger pro-

Table 12.2 Proportion of chaparral in the historical (VTM) map that transitioned to other vegetation types in contemporary (San Diego County, CalVeg, and Landfire) maps

Chaparral Change Class	San Diego	CalVeg	Landfire
Chaparral to chaparral	0.33	0.22	0.24
Chaparral to sage scrub	0.07	0.22	0.26
Chaparral to grass	0.12	0.10	0.26
Chaparral to tree	0.40	0.27	0.17
Chaparral to other	0.09	0.20	0.07
Total	1.0	1.0	1.0

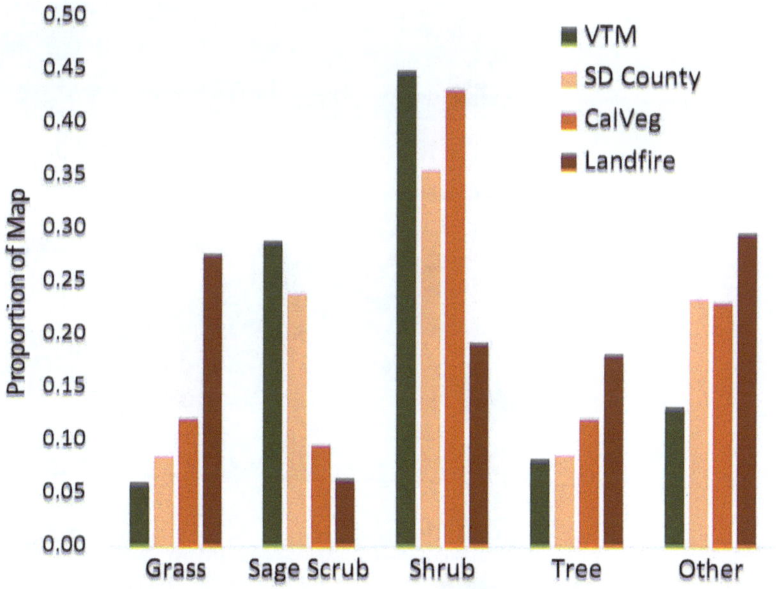

Fig. 12.3 Proportion of vegetation type within four vegetation maps of the South Coast Ecoregion (VTM 1930s; CalVeg 2002; Landfire 2013) and San Diego County (2012)

portion of the landscape as grassland than the other two contemporary maps (Fig. 12.4c). This is reflected in the vast areas of the landscape that were mapped as having changed from sage scrub or shrub to grass (Fig. 12.6a).

In terms of fire frequency, the analysis showed highest mean fire frequencies in classes where either sage scrub or shrub converted to grass, or where shrub converted to sage scrub (Fig. 12.7). The mean number of fires summed across grid cells in each change class ranged from 1.55 to 2.41, but the actual number of times areas burned during the 135-year span of the fire history data ranged from 0 to13.

Fig. 12.4 Vegetation types as mapped in the 1930s (VTM), 2002 (CalVeg maps), and 2013 (Landfire)

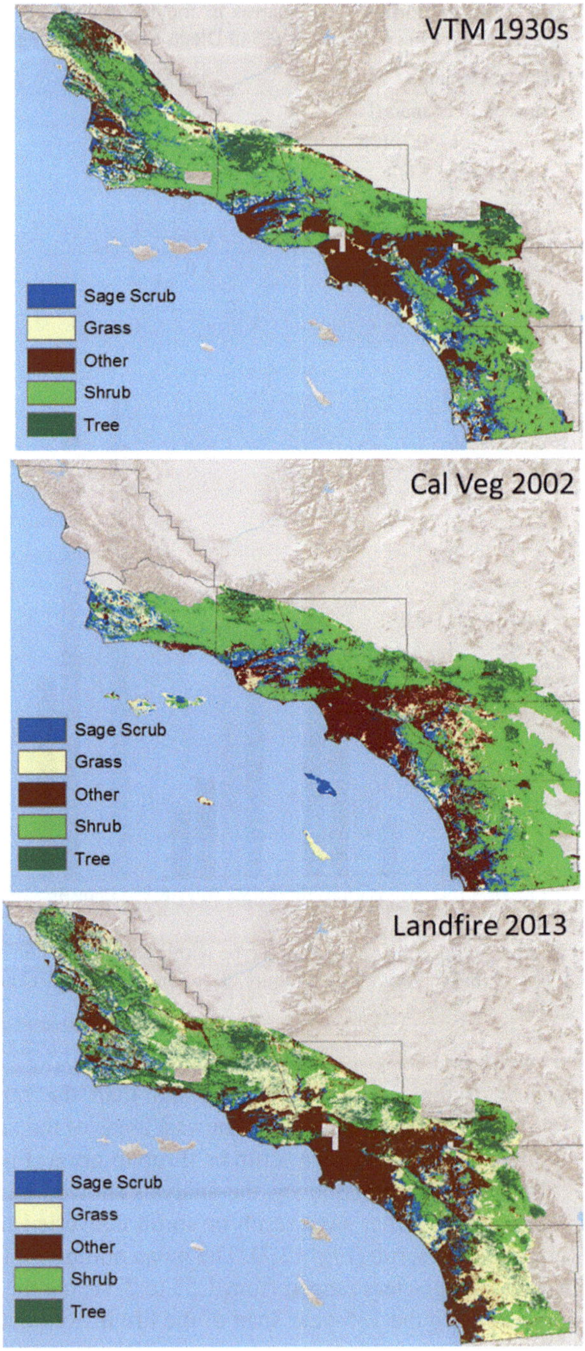

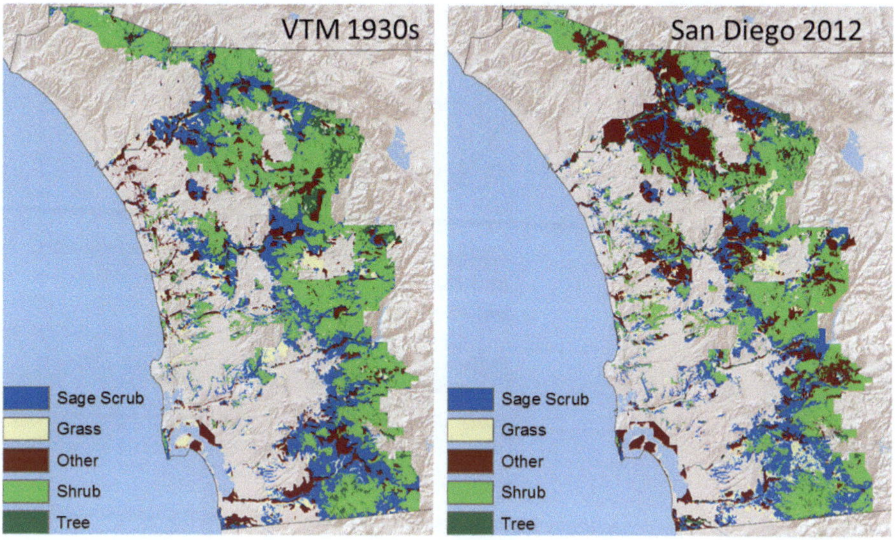

Fig. 12.5 Comparison of vegetation type classes as mapped in the 1930s (VTM) and in 2012 (San Diego County map)

12.3.1 Challenges in Quantifying Vegetation Change

The wide variation apparent among the three contemporary maps illustrates the challenge in overlapping different vegetation maps to accurately delineate and quantify vegetation change, particularly if the objective is to map change at a fine scale. There are multiple sources of uncertainty inherent in any ecological analysis (Regan et al. 2002), and spatial data are particularly susceptible to errors in map boundaries and classification (Goodchild and Gopal 1989). Nevertheless, when vegetation map classes are collapsed into broad categories reflecting vegetation formations, map accuracy can be relatively high (Goodchild et al. 1991).

Clearly, the extent and location of vegetation type-conversion cannot be precisely determined from our analysis, and the vast areas of type change from shrub or scrub to grass mapped using the Landfire data should be interpreted with some caution given that many of these areas were not mapped as grass in the other two contemporary maps. Nevertheless, despite the variation among contemporary maps, the results of all three overlays were remarkably consistent in the kind of change measured. Thus, even using the most conservative estimates, there has been a clear trend of chaparral decline and conversion to either sage scrub or grassland over the last 70–80 years. Furthermore, fire frequency tends to be highest where these changes have been mapped (Fig. 12.7).

In the southern California landscape, the most likely explanation for the differences in maps is the treatment of mixed classes. Depending on the scale of the analysis relative to the heterogeneity of the vegetation, mixed grass and shrub stands must often be lumped into one class or the other. Thus, many of the areas mapped

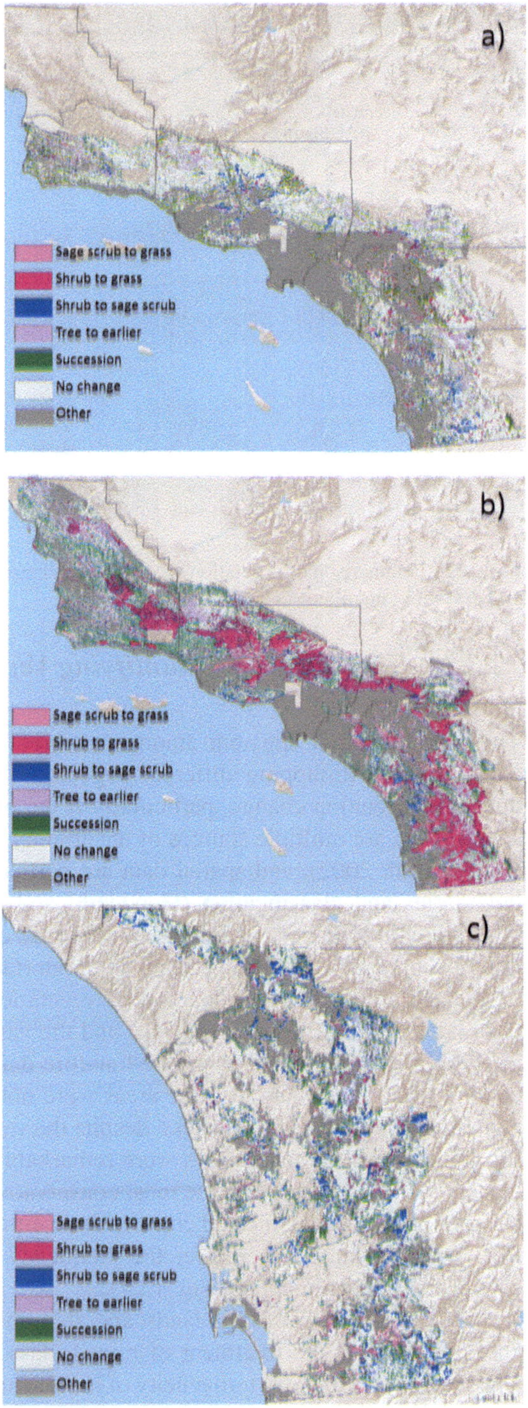

Fig. 12.6 Maps of vegetation type change from (**a**) the 1930s to 2002 (CalVeg), (**b**) the 1930s to 2013 (Landfire), and (**c**) the 1930s to 2012 (San Diego County map)

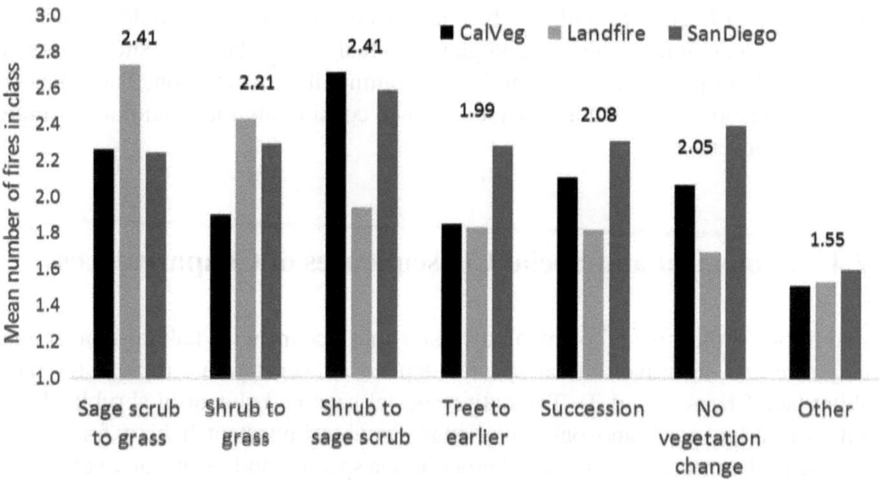

Fig. 12.7 Mean number of fires from 1878–2013 within each vegetation type change class between the 1930s and 2002, using data from CalVeg (2002), Landfire (2013), and San Diego County maps (2012). Numbers above the bars indicate the mean fire frequency averaged across the three maps. "Tree to other" reflects any changes in which trees changed to shrub, sage scrub, or grass. "Succession" reflects any changes in which grass changed to shrub or sage scrub, or sage scrub changed to shrub

as grass in the Landfire map, and mapped as some type of shrubland in the CalVeg or San Diego County map, were probably some mixture of shrub and grass.

Whether these classes were purely grass or represented some mixture with shrubs is one of the central challenges in quantifying landscape scale vegetation change. It also provides one reason for questioning Meng et al.'s (2014) conclusion that widespread vegetation type-conversion is not an immediate threat in southern California, as vegetation type-conversion does not occur as a complete shift at one moment in time. Instead, it occurs as a gradual and cumulative process, which often begins with the elimination of non-resprouting species within mixed stands, habitat simplification, and biodiversity loss (Keeley et al. 2005). In addition, because sage scrub can withstand higher fire frequencies than chaparral, vegetation change may begin with a gradual shift from stands of pure chaparral to mixed stands of chaparral, sage scrub, and grass. This type of transition is suggested in the results here that show substantial change from shrub to sage scrub under higher mean fire frequencies. Given that different species have varying sensitivities to repeat fires, and that overlapping fires exhibit fragmented spatial patterns, multiple repeat fire events are probably necessary for significant vegetation change to be discernable. Thus, one of the methodological challenges in landscape scale analyses like those in Meng et al. (2014) is that type-conversion is only inferred, and the gradual process of vegetation change cannot be documented at a specific location over time the way it can in field studies (Halsey and Syphard 2015). Another challenge is that substantial chaparral conversion had already occurred before vegetation maps became available for modern analysis. There is evidence of chaparral conversion prior to the twentieth century (Cooper 1922), and evidence has also been documented in field studies. In summary, vegetation change is complex, gradual, and related to site factors in addition to long-term fire history and plant community composition. These factors need to be resolved and better understood when considering the potential for future chaparral conversion.

12.4 Ecological and Social Consequences of Chaparral Loss

In addition to the loss of plant biodiversity that occurs with habitat conversion, many rare and sensitive animal species depend on vegetation structure for their habitat (see Chaps. 2 and 3). The native coastal sage and chaparral shrublands, as well as riparian areas and oak woodlands, provide important habitat for a wide range of bird, insect, mammal, and herpetofauna species, and the negative effects of habitat loss and fragmentation have been documented for decades in numerous studies (e.g., Bolger 1991; Soulé et al. 1992; Bolger et al. 2000; Riley et al. 2003; Ruell et al. 2012). Recent studies are also beginning to show how interactions among direct and indirect effects (e.g., fire, climate change, non-native species) of urban development contribute to biodiversity loss (e.g., Franklin et al. 2014; Conlisk et al. 2015; Jennings et al. 2016).

Changes in vegetation structure that occur with the conversion of shrublands to grasslands also impact the physical and hydrological properties of the soil (Martinez-Fernandez et al. 1995; Williamson et al. 2004). The increased density of plants combined with changes in the canopy shape and root distribution of individuals significantly alter how rainfall and organic matter are channeled into and through the soil (Lee and Lauenroth 1994; Martinez-Meza and Whitford 1996). The resultant changes affect the infiltration capacity and water retention of the soil as well as the concentration and dispersal of nutrients and carbon (Gutierrez et al. 1995; Martinez-Fernandez et al. 1995). Shrublands that have been converted to grasslands have more extreme soil temperatures and they tend to develop a thicker, more variable surface (A) horizon with a significantly higher soil bulk density (Williamson et al. 2004). These changes in root distribution decrease the stability of slopes while increasing the potential for hazardous debris flows (Gabet and Dune 2002). External factors such as fire and flooding can further exacerbate the system by increasing runoff and soil erosion, which in turn have the potential to affect water quality and reservoir infilling (Hubbert et al. 2012). Finally, shrublands have substantially better capacity for ecosystem carbon sequestration than grasses (Petrie et al. 2015), which has critical implications in this era of rapid climate change.

Development patterns and chaparral conversion are not only important in terms of ecological effects, but from a social perspective, the intermix WUI areas are also the locations where houses are most likely to be destroyed by wildfire in southern California (Syphard et al. 2012). Large fires at the WUI have been occurring for decades in the region, with an average of 500 houses lost per year in the last 50 years. Furthermore, the rate of destroyed houses and lost lives in the last 10–15 years has been unprecedented (Keeley et al. 2013).

12.5 Discussion and Future Changes

As we march into the twenty-first century, the acceleration of global change is bound to occur, especially given the projections of continued population growth. For example, the San Diego Association of Governments expects a 140% increase in population by 2050 across the county (www.sandag.org/2050forecast). Thus, continuation of direct habitat conversion, particularly in the form of urban development, will continue to reduce and fragment chaparral habitat, as well as increase the length and extent of the WUI (Landis and Reilly 2003; Hammer et al. 2009). Furthermore, these land use changes will likely continue to interact with indirect drivers of conversion, including fire and invasion by non-native grasses.

Climate change will also likely result in chaparral species' range shifts, and possibly type-conversion, through habitat shifts and modifying phenology (Chen et al. 2011; Beltrán et al. 2014, see Chap. 14). However, it is the interaction of climate with the drivers discussed here that may be of most concern (Syphard et al. 2013b; Franklin et al. 2014). For example, future projections suggest that land use change will likely either override or compound the impacts of climate change on shrubland habitat conversion across the state of California (Mann et al. 2014; Riordan and Rundel 2014), and in southern California, loss of chaparral species' suitable habitat may be exacerbated by urban growth, with fire being the most serious threat for obligate seeding chaparral species (Syphard et al. 2013b; Bonebrake et al. 2014). Fire regimes, however, are more likely to be altered due to land use change rather than climate change in chaparral shrublands, as fire activity has not been significantly correlated with historical patterns of temperature and precipitation in these areas (Keeley and Syphard 2015, 2016, 2018). This may be due to the fact that climatic conditions are already suitable for extreme fire activity every year on these landscapes. On the other hand, changing patterns and timing of ignitions may have profound impacts on fire activity and its social and ecological consequences (Syphard and Keeley 2015).

Although the South Coast Ecoregion is relatively homogenous in terms of broad scale climatic and vegetation patterns, questions of scale and geographical context will be important when considering future management needs and priorities. For example, species with similar functional traits and sensitivities to certain threats may be differentially exposed to those threats depending on their distributions (Syphard et al. 2013b). That is, areas with the fastest climate change may not always be the same as the areas of fastest land use change or disturbance regime shifts.

Within the South Coast Ecoregion, different counties have unique histories of development and urban growth, which explains why our data show such variation in the extent and spatial pattern of housing density. Accordingly, habitat loss and fragmentation have and will continue to vary across the region. One of the most serious concerns related to chaparral conversion may be the ongoing expansion of low-density development in counties like San Diego, which still contain substantial areas of intact chaparral. Not only does continued development threaten to reduce shrubland extent and continuity, but intermix WUI is the area most prone to non-native annual grass expansion, increased fire frequency, and corresponding fire risk.

One major concern associated with the increase in fires in the southern California region is that vast areas are now covered with very young chaparral due to the enormous extent of recent wildfires. Also, there have already been extensive areas within southern California that have recently burned at anomalously short intervals (Keeley et al. 2009). These trends greatly increase the risk for future conversion to annual non-native grass. An additional potential factor is increased atmospheric pollution. Non-native grasses respond favorably to elevated atmospheric nitrogen deposition, which will likely accelerate with ongoing development (Cox et al. 2014).

Given the profound recent loss of human lives and property in southern California associated with wildfire, there has been a growing sense of urgency to identify new ways to reduce fire risk and ensure community safety. Aside from active fire suppression to control burning wildfires, the most prevalent form of management has been to burn, modify, or clear wildland vegetation to control fire behavior. While fuelbreaks can be safe and effective tools for firefighter access to chaparral communities, research shows that vegetation management in terms of prescribed fire and fuelbreaks provide little benefit for controlling the most damaging weather-driven fires (Syphard et al. 2011; Price et al. 2012; Penman et al. 2014). Given that vegetation management is a driver of chaparral conversion, trade-offs could be carefully considered in the design and placement of fuelbreaks, which ideally could be strategically placed for firefighter defense of communities.

In addition to strategically placed fuelbreaks, homeowner property preparation in terms of building construction and design and defensible space may significantly reduce the risk of a house being destroyed in a wildfire (Cohen 2004; Quarles et al. 2010; Syphard et al. 2014, 2016a). However, while defensible space does provide significant protection, the effect results primarily from modifying vegetation immediately adjacent to the structure. Research has shown there is no added benefit of treating areas farther than 100 ft. (30 m) from the property, even on steep slopes. In addition, only 40% reduction in woody cover was needed for significant protection (Syphard et al. 2014). This is important with regards to habitat, as there has been a recent push from county governments and insurance companies for homeowners to clear up to 300 ft. (60 m) of defensible space around their houses, which cumulatively could result in substantial areas of habitat loss (Keeley et al. 2013).

Considering house losses from wildfire at both local and landscape scales, the most significant factor that explains whether or not a house is destroyed has been its location and arrangement relative to other houses on the landscape (Syphard et al. 2012; Alexandre et al. 2015). Therefore, land use planning may be the most effective long-term solution for not only preventing house loss to wildfires, but also for maximizing biodiversity. Simulation studies showed that land use planning decisions, either through growth policies or through private land acquisition, could result in mutual benefits for both fire risk reduction and biodiversity conservation (Syphard et al. 2013a, 2016b; Butsic et al. 2017). In particular, both house loss and ecological impacts are likely to be most effectively minimized if future development is designed to be compact and clustered, with development restricted in either high-fire-hazard or species-rich areas, which tend to occur in the same areas (Syphard et al. 2016b). Ignition prevention efforts may also be highly effective as part of a comprehensive fire management program (Prestemon et al. 2010; Syphard and Keeley 2015).

12.6 Conclusion

The sprawling development pattern in southern California has been the primary driver of contemporary chaparral conversion, both through the direct removal and fragmentation of habitat, but also through its indirect role in driving annual grass expansion associated with increased fire frequency. It is also indirectly responsible for other factors such as fuelbreaks to protect communities scattered throughout the wildland, climate change, and perhaps even the increase of nitrogen deposition. For example, the increasing road density and traffic volumes associated with increased population and urban development have and will continue to have numerous effects that threaten chaparral ecosystems. Roads are often the source of fire ignitions (Syphard and Keeley 2015), promote the spread of non-native species (Bar-Massada et al. 2014), contribute to elevated ozone and nitrogen deposition that favors grasses over shrubs (Fenn et al. 2010), and fragment habitat needed for sensitive fauna (Poessel et al. 2014).

Thus, as we move into the future, it may be well worth the effort to seriously consider how developments are designed and arranged across the landscape. Land use planning could systematically address the root causes of fire risk as well as habitat loss (Moritz et al. 2014). It could lower ignitions through reduced human presence in flammable areas, lower non-native species expansion by reducing corridors to invasion, and lower the risk of property loss by arranging houses so that they are less fire-prone (Syphard et al. 2012, 2013a). Land use planning can thus address multiple impacts of global change across California shrublands, and may ultimately be the most powerful tool for a sustainable future.

References

Alexandre, P. M., S. I. Stewart, M. H. Mockrin, N. S. Keuler, A. D. Syphard, A. Bar-Massada, M. K. Clayton, and V. C. Radeloff. 2015. The relative impacts of vegetation, topography and spatial arrangement on building loss to wildfires in case studies of California and Colorado. Landscape Ecology 31:415-430.

Alig, R. J., and A. J. Plantinga. 2004. Future forestland area: Impacts from population growth and other factors that affect land values. Journal of Forestry 102:19-24.

Archibald, S., R. J. Scholes, D. P. Roy, G. Roberts, and L. Boschetti. 2010. Southern African fire regimes as revealed by remote sensing. International Journal of Wildland Fire 19:861-878.

Bartolome, J. W., and B. Gemmill. 1981. The ecological status of *Stipa pulchra* (Poaceae) in California. Madroño 28:172-184.

Bolger, D. T. 1991. Community perturbations: introduced species and habitat fragmentation. University of California, San Diego, California, USA.

Bolger, D. T., A. V. Suarez, K. R. Crooks, S. A. Morrison, and T. J. Case. 2000. Arthropods in urban habitat fragments in southern California: area, age, and edge effects. Ecological Applications 10:1230-1248.

Bonebrake, T. C., A. D. Syphard, J. Franklin, K. E. Anderson, H. R. Akçakaya, T. Mizerek, C. Winchell, and H. M. Regan. 2014. Fire management, managed relocation, and land conservation options for long-lived obligate seeding plants under global changes in climate, urbanization, and fire regime. Conservation Biology 28:1057-1067.

Bar-Massada, A., V. C. Radeloff, and S. I. Stewart. 2014. Biotic and abiotic effects of human settlements in the wildland-urban interface. BioScience 64:429-437.

Beltrán, B., J. Franklin, A. D. Syphard, H. M. Regan, L. E. Flint, and A. L. Flint. 2014. Effects of climate change and urban development on the distribution and conservation of plant functional types in a Mediterranean-type ecosystem. International Journal of Geographic Information Science 28:1561-1589.

Bowman, D. M. J. S., H. J. MacDermott, S. C. Nichol, and B. P. Murphy. 2014. A grass–fire cycle eliminates an obligate-seeding tree in a tropical savanna. Ecology and Evolution 4:4185-4194.

Brennan, T. J., and J. E. Keeley. 2015. Effect of mastication and other mechanical treatments on fuel structure in chaparral. International Journal of Wildland Fire 24:949-963.

Brooks, M. L., C. M. D'Antonio, D. M. Richardson, J. B. Grace, J. E. Keeley, J. M. DiTomaso, R. J. Hobbs, M. Pellant, and D. Pyke. 2004. Effects of invasive alien plants on fire regimes. Bioscience 54:677-688.

Burcham, L. T. 1956. Historical backgrounds of range land use in California. Journal of Range Management 9:81-86.

Butsic, V., A. Syphard, J. E. Keeley, and A. Bar-Massada. 2017. Modeling the impact of private land conservation on wildfire risk in San Diego County, California. Landscape and Urban Planning 157:161-169.

Chen, I.-C., J. K. Hill, R. Ohlemüller, D. B. Roy, and C. D. Thomas. 2011. Rapid range shifts of species associated with high levels of climate warming. Science 333:1024-1026.

Cohen, J. 2004. Relating flame radiation to home ignition using modeling and experimental crown fires. Canadian Journal of Forest Research 34:1616-1626.

Cooper, W. S. 1922. The broad-sclerophyll vegetation of California: an ecological study of the chaparral and its related communities. Carnegie Institution of Washington, Washington D.C., USA.

Conlisk, E., A. D. Syphard, J. Franklin, and H. M. Regan. 2015. Predicting the impact of fire on a vulnerable multi-species community using a dynamic vegetation model. Ecological Modelling 301:27-39.

Cox, R. D., K. L. Preston, R. F. Johnson, R. A. Minnich, and E. B. Allen. 2014. Influence of landscape-scale variables on vegetation conversion to non-native annual grassland in southern California, USA. Global Ecology and Conservation 2:190-203.

Dahal, K. R., S. Benner, and E. Lindquist. 2017. Urban hypotheses and spatiotemporal characterization of urban growth in the Treasure Valley of Idaho, USA. Applied Geography 79:11-25.

D'Antonio, C. M., and P. M. Vitousek. 1992. Biological invasions by non-native grasses, the grass/fire cycle, and global change. Annual Review of Ecology and Systematics 23:63-87.

Fenn, M. E., E. B. Allen, S. B. Weiss, S. Jovan, L. H. Geiser, G. S. Tonnesen, R. F. Johnson, L. E. Rao, B. S. Gimeno, F. Yuan, and T. Meixner. 2010. Nitrogen critical loads and management alternatives for N-impacted ecosystems in California. Journal of Environmental Management 91:2404-2423.

Franklin, J., H. M. Regan, and A. D. Syphard. 2014. Linking spatially explicit species distribution and population models to plan for the persistence of plant species under global change. Environmental Conservation 41:97-109.

Gabet, E. J., and T. Dunne. 2002. Landslides on coastal sage-scrub and grassland hillslopes in a severe El Niño winter: the effects of vegetation conversion on sediment delivery. Bulletin of the Geological Society of America 114:983-990.

Gavier-Pizarro, A. G. I., V. C. Radeloff, S. I. Stewart, D. Cynthia, and N. S. Keuler. 2010. Housing is positively associated with invasive non-native plant species richness in New England, USA. Ecological Applications 20:1913-1925.

Glaeser, E. L., and J. M. Shapiro. 2003. Urban growth in the 1990s: is city living back? Journal of regional science 43:139-165.

Goodchild, M. F., F. W. Davis, M. Painho, and D. M. Stoms. 1991. The use of vegetation maps and Geographic Information Systems for assessing conifer lands in California. Technical Report 91-23. National Center for Geographic Information and Analysis, University of California, Santa Barbara, California, USA.

Goodchild, M., and S. Gopal. 1989. The accuracy of spatial databases. Taylor & Francis, London, UK.

Gude, P. H., R. Rasker, and J. van den Noort. 2008. Potential for future development on fire-prone lands. Journal of Forestry 106:198-205.

Gutierrez, J., R. E. Sosebee, and K. E. Spaeth. 1995. Spatial variation of runoff and erosion under grass and shrub cover on a semiarid rangeland. Pages 11-20 *in* T. J. Ward, editor. Watershed Management-Planning for the 21st Century. American Society of Civil Engineers, San Antonio, Texas, USA.

Haidinger, T. L., and J. E. Keeley. 1993. Role of high fire frequency in destruction of mixed chaparral. Madroño 40:141-147.

Halsey, R. W., and A. D. Syphard. 2015. High-severity fire in chaparral cognitive dissonance in the shrublands. Pages 177-209 *in* The ecological importance of mixed-severity fires: nature's phoenix. First edition. Elsevier, Amsterdam, Netherlands.

Hammer, R. B., S. I. Stewart, R. L. Winkler, V. C. Radeloff, and P. R. Voss. 2004. Characterizing dynamic spatial and temporal residential density patterns from 1940-1990 across the north central United States. Landscape and Urban Planning 69:183-199.

Hammer, R. B., S. I. Stewart, and V. C. Radeloff. 2009. Demographic trends, the wildland-urban interface, and wildfire management. Society & Natural Resources 22:777-782.

Hansen, A. J., R. L. Knight, J. M. Marzluff, S. Powell, K. Brown, P. H. Gude, and K. Jones. 2005. Effects of exurban development on biodiversity: patterns, mechanisms, and research needs. Ecological Applications 15:1893-1905.

Herold, M., N. C. Goldstein, and K. C. Clarke. 2003. The spatiotemporal form of urban growth: measurement, analysis and modeling. Remote Sensing of Environment 86:286-302.

Hubbert, K. R., P. M. Wohlgemuth, J. L. Beyers, M. G. Narog, and R. Gerrard. 2012. Post-fire soil water repellency, hydrologic response, and sediment yield compared between grass-converted and chaparral watersheds. Fire Ecology 8:143-162.

Jennings, M. K., R. L. Lewison, T. W. Vickers, and W. M. Boyce. 2016. Puma response to the effects of fire and urbanization. Journal of Wildlife Management 80:221-234.

Keeley, J. E. 1986. Resilience of Mediterranean shrub communities to fire. Pages 95-112 *in* B. Dell, A. J. M. Hopkins, and B. B. Lamont, editors. Resilience in Mediterranean-type ecosystems. Dr. W. Junk Publishers, Dordrecht, Netherlands.

Keeley, J. E. 2010. Fire on California landscapes. Fremontia 38:2-6

Keeley, J. E., M. Baer-Keeley, and C. J. Fotheringham. 2005. Alien plant dynamics following fire in Mediterranean-climate California shrublands. Ecological Applications 15:2109-2125.

Keeley, J. E., and T. J. Brennan. 2012. Fire-driven alien invasion in a fire-adapted ecosystem. Oecologia 169:1043-1052.

Keeley, J. E., W. J. Bond, R. A. Bradstock, J. G. Pausas, and P. W. Rundel. 2012. Fire in Mediterranean ecosystems: ecology, evolution and management. Cambridge University Press, Cambridge, UK.

Keeley, J. E., and C. J. Fotheringham. 2003. Impact of past, present, and future fire regimes on North American Mediterranean shrublands. Pages 218-262 *in* T. T. Veblen, W. L. Baker, G. Montenegro, and T. W. Swetnam, editors. Fire and climatic change in temperate ecosystems of the western Americas. Springer, New York, USA.

Keeley, J. E., H. D. Safford, C. J. Fotheringham, J. Franklin, and M. A. Moritz. 2009. The 2007 southern California wildfires: lessons in complexity. Journal of Forestry 107:287-296.

Keeley, J. E., A. D. Syphard, and C. J. Fotheringham. 2013. The 2003 and 2007 wildfires in southern California. Pages 42–52 *in* S. Boulter, J. Palutikof, and D. J. Karoly, editors. Natural disasters and adaptation to climate change. Cambridge University Press, Cambridge, UK.

Keeley, J. E., and A. D. Syphard. 2015. Different fire-climate relationships on forested and non-forested landscapes in the Sierra Nevada region. International Journal of Wildland Fire 24:27-36.

Keeley, J. E., and A.D. Syphard. 2016. Climate change and future fire regimes: examples from California. Geosciences 6:37.

Keeley, J. E. and A.D. Syphard. 2018. South Coast bioregion. Pages 319-351 *in* J. W. van Wagtendonk, N. G. Sugihara, S. L. Stephens, A. E. Thode, K. E. Shaffer, and J. A. Fites-Kaufman, editors. Fire in California's ecosystems. Second edition. University of California Press, Berkeley, California, USA.

Kelly, M. 2016. Rescuing and sharing historical vegetation data for ecological analysis: the California Vegetation Type Mapping project. Biodiversity Informatics 11:40-62.

Kelly, M., B. Allen-Diaz, and N. Kobzina. 2005. Digitization of a historic dataset: the Wieslander California vegetation type mapping project. Madroño 52:191-201.

Kinney, A. 1887. Report on the forests of the counties of Los Angeles, San Bernardino, and San Diego, California. First biennial report. California State Board of Forestry, Sacramento, California, USA.

Landis, J. D., and M. Reilly. 2003. How we will grow: baseline projections of California's urban footprint through the year 2100. Pages 55-98 *in* S. Guhathakurta, editor. Integrated land use and environmental models: a survey of current applications and research. Springer Berlin Heidelberg, Heidelberg, Germany.

Lee, C. A., and W. K. Lauenroth. 1994. Spatial distributions of grass and shrub root systems in the shortgrass steppe. American Midland Naturalist 132:117-123.

Lenth, B. A., R. L. Knight, and W. C. Gilgert. 2006. Conservation value of clustered housing developments. Conservation Biology 20:1445-1456.

Lippitt, C. L., D. A. Stow, J. F. O'Leary, and J. Franklin. 2012. Influence of short-interval fire occurrence on post-fire recovery of fire-prone shrublands in California, USA. International Journal of Wildland Fire 22:184-193.

Mack, R. N. 1989. Temperate grasslands vulnerable to plant invasions: characteristics and consequences. Pages 155-179 *in* J. A. Drake, H. A. Mooney, F. D. Castri, R. H. Groves, F. J. Kruger, M. Rejmanek, and M. Williamson, editors. Biological invasions: a global perspective. John Wiley and Sons, New York, New York, USA.

Mann, M. L., P. Berck, M. A. Moritz, E. Batllori, J. G. Baldwin, C. K. Gately, and D. R. Cameron. 2014. Modeling residential development in California from 2000 to 2050: integrating wildfire risk, wildland and agricultural encroachment. Land Use Policy 41:438-452.

Martinez-Fernandez, J., F. Lopez-Bermudez, J. Martinez-Fernandez, and A. Romero-Diaz. 1995. Land use and soil-vegetation relationships in a Mediterranean ecosystem: El Ardal, Murcia, Spain. Catena 25:153-167.

Martinez-Meza, E., and W. G. Whitford. 1996. Stemflow, throughfall and channelization of stemflow by roots in three Chihuahuan desert shrubs. Journal of Arid Environments 32:271-287.

Mayer, K. E., and W. F. Laudenslayer Jr. 1988. A guide to the wildlife habitats of California. California Department of Forestry and Fire Protection, Sacraemento, California, USA.

Meng, R., P. E. Dennison, C. M. D'Antonio, and M. A. Moritz. 2014. Remote sensing analysis of vegetation recovery following short-interval fires in southern California shrublands. PLoS One 9:e110637.

Merriam, K. E., J. E. Keeley, and J. L. Beyers. 2006. Fuel breaks affect nonnative species abundance in Californian plant communities. Ecological Applications 16:515-527.

Minnich, R. A., and R. J. Dezzani. 1998. Historical decline of coastal sage scrub in the Riverside-Perris Plain, California. Western Birds 29:366-391.

Moritz, M. A., E. Batllori, R. A. Bradstock, A. M. Gill, J. Handmer, P. F. Hessburg, J. Leonard, S. McCaffrey, D. C. Odion, T. Schoennagel, and A. D. Syphard. 2014. Learning to coexist with wildfire. Nature 515:58-66.

Netusil, N. R. 2005. The effect of environmental zoning and amenities on property values: Portland, Oregon. Land Economics 81:227-246.

Odell, E. A., D. M. Theobald, and R. L. Knight. 2003. Incorporating ecology into land use planning: the songbirds' case for clustered development. Journal of the American Planning Association 69:72-82.

Petrie, M. D., S. L. Collins, A. M. Swann, P. L. Ford, and M. E. Litvak. 2015. Grassland to shrubland state transitions enhance carbon sequestration in the northern Chihuahuan Desert. Global Change Biology 21:1226-1235.

Penman, T. D., L. Collins, A. D. Syphard, J. E. Keeley, and R. A. Bradstock. 2014. Influence of fuels, weather and the built environment on the exposure of property to wildfire. PLoS ONE 9:e111414.

Poessel, S. A., C. L. Burdett, E. E. Boydston, L. M. Lyren, R. S. Alonso, R. N. Fisher, and K. R. Crooks. 2014. Roads influence movement and home ranges of a fragmentation-sensitive carnivore, the bobcat, in an urban landscape. Biological Conservation 180:224-232.

Prestemon, J. P., D. T. Butry, K. L. Abt, and R. Sutphen. 2010. Net benefits of wildfire prevention education efforts. Forest Science 56:181-192.

Price, O. F., R. A. Bradstock, J. E. Keeley, and A. D. Syphard. 2012. The impact of antecedent fire area on burned area in southern California coastal ecosystems. Journal of Environmental Management 113:301-307.

Quarles, S. L., Y. Valachovic, G. Nakamura, G. Nader, and M. De LaSaux. 2010. Home survival in wildfire-prone areas: building materials and design considerations. Publication 8393. University of California, Agriculture and Natural Resources, Richmond, California, USA.

Radeloff, V. C., R. B. Hammer, S. I. Stewart, J. S. Fried, S. S. Holcomb, and J. F. McKeefry. 2005. The wildland-urban interface in the United States. Ecological Applications 15:799-805.

Regan, H. M., M. Colyvan, and M. A. Burgman. 2002. A taxonomy and treatment of uncertainty for ecology and conservation biology. Ecological Applications 12:618-628.

Riley, S. P. D., R. M. Sauvajot, T. K. Fuller, E. C. York, D. A. Kamradt, C. Bromley, and R. Wayne. 2003. Effects of urbanization and habitat fragmentation on bobcats and coyotes in Southern California. Conservation Biology 17:566-576.

Riordan, E. C., and P. W. Rundel. 2014. Land use compounds habitat losses under projected climate change in a threatened California ecosystem. PloS One 9:e86487.

Roderick, K. 2002. The San Fernando Valley: America's suburb. Los Angeles Times Books, Los Angeles, California, USA.

Rossiter, N. A., S. A. Setterfield, M. M. Douglas, and L. B. Hutley. 2003. Testing the grass-fire cycle: alien grass invasion in the tropical savannas of northern Australia. Diversity and Distributions 9:169-176.

Ruell, E. W., S. P. D. Riley, M. R. Douglas, M. F. Antolin, J. R. Pollinger, J. A. Tracy, L. Lyren, E. E. Boydston, R. N. Fisher, and K. R. Crooks. 2012. Urban habitat fragmentation and genetic population structure of bobcats in coastal Southern California. The American Midland Naturalist 168:265-280.

Safford, H. D., and K. M. Van de Water. 2014. Using fire return interval departure (FRID) analysis to map spatial and temporal changes in fire frequency on national forest lands in California. Research Paper PSW-RP-266. USDA Forest Service, Pacific Southwest Research Station, Albany, California, USA.

Soulé, M. E., A. C. Alberts, and D. T. Bolger. 1992. The effects of habitat fragmentation on chaparral plants and vertebrates. Oikos 63:39-47.

Sproul, F., T. Keeler-Wolf, P. Gordon-Reedy, J. Dunn, A. Klein, K. Harper. 2011. Vegetation classification manual for western San Diego County. Prepared for San Diego Association of Governments, San Diego, California, USA.

Stewart, S. I., V. C. Radeloff, R. B. Hammer, and T. J. Hawbaker. 2007. Defining the wildland – urban interface. Journal of Forestry 105:201-207.

Sushinsky, J. R., J. R. Rhodes, H. P. Possingham, T. K. Gill, and R. A. Fuller. 2013. How should we grow cities to minimize their biodiversity impacts? Global Change Biology 19:401-410.

Syphard, A. D., A. Bar-Massada, V. Butsic, and J. E. Keeley. 2013a. Land use planning and wildfire: development policies influence future probability of housing loss. PLoS One 8:e71708.
Syphard, A. D., T. J. Brennan, and J. E. Keeley. 2014. The role of defensible space for residential structure protection during wildfires. International Journal of Wildland Fire 23:1165-1175.
Syphard, A. D., T. J. Brennan, and J. E. Keeley. 2016a. The importance of building construction materials relative to other factors affecting structure survival during wildfire. International Journal of Disaster Risk Reduction 21:140-147.
Syphard, A. D., V. Butsic, A. Bar-Massada, J. E. Keeley, J. A. Tracey, and R. N. Fisher. 2016b. Setting priorities for private land conservation in fire-prone landscapes: are fire risk reduction and biodiversity conservation competing or compatible objectives? Ecology and Society 21:2.
Syphard, A. D., J. Franklin, and J. E. Keeley. 2006. Simulating the effects of frequent fire on southern California coastal shrublands. Ecological Applications 16:1744-1756.
Syphard, A. D., J. E. Keeley, and T. J. Brennan. 2011. Comparing the role of fuel breaks across southern California national forests. Forest Ecology and Management 261:2038-2048.
Syphard, A. D., J. E. Keeley, A. Bar-Massada, T. J. Brennan, and V. C. Radeloff. 2012. Housing arrangement and location determine the likelihood of housing loss due to wildfire. PLoS ONE 7:e33954.
Syphard, A. D., and J. E. Keeley. 2015. Location, timing, and extent of wildfire varies by cause of ignition. International Journal of Wildland Fire 24:37-47.
Syphard, A. D., and J. E. Keeley. 2017. Historical reconstructions of California wildfires vary by data source. International Journal of Wildland Fire 25:1221-1227.
Syphard, A. D., V. C. Radeloff, J. E. Keeley, T. J. Hawbaker, M. K. Clayton, S. I. Stewart, and R. B. Hammer. 2007. Human influence on California fire regimes. Ecological Applications 17:1388-402.
Syphard, A. D., V. C. Radeloff, T. J. Hawbaker, and S. I. Stewart. 2009. Conservation threats due to human-caused increases in fire frequency in Mediterranean-climate ecosystems. Conservation Biology 23:758-769.
Syphard, A. D., H. M. Regan, J. Franklin, R. M. Swab, and T. C. Bonebrake. 2013b. Does functional type vulnerability to multiple threats depend on spatial context in Mediterranean-climate regions? Diversity and Distributions 19:1263-1274.
Talluto, M., and K. Suding. 2008. Historical change in coastal sage scrub in southern California, USA in relation to fire frequency and air pollution. Landscape Ecology 23:803-815.
Wieslander, A. E. 1935. A vegetation type map of California. Madroño 3:140-144.
Williamson, T. N., R. C. Graham, and P. J. Shouse. 2004. Effects of a chaparral-to-grass conversion on soil physical and hydrologic properties after four decades. Geoderma 123:99-114.
Wu, J., and A. J. Plantinga. 2003. The influence of public open space on urban spatial structure. Journal of Environmental Economics and Management 46:288-309.
Zedler, P. H., C. R. Gautier, and G. S. McMaster. 1983. Vegetation change in response to extreme events: the effect of a short interval between fires in California chaparral and coastal scrub. Ecology 64:809-818.

Chapter 13
Chaparral Restoration

Edith B. Allen, Kimberlyn Williams, Jan L. Beyers, Michala Phillips, Stephanie Ma, and Carla M. D'Antonio

Abstract Chaparral, among the most stable and resilient vegetation types in California, has shown signs of degradation by altered fire frequency, drought, non-native species, recreation, urban development, and possibly anthropogenic nitrogen deposition in southern California. Restoration has been practiced less frequently in chaparral than other, more extensively disturbed vegetation types, but recent degradation suggests that restoration may be important for the maintenance of ecosystem services such as slope stabilization, carbon sequestration, biodiversity conservation, and scenic beauty. Most chaparral restoration has primarily been "passive restoration", the removal of disturbance stressors or management of fire frequency to promote natural successional processes for plant, animal, and soil recovery. However, areas that have suffered severe disturbance, such as topsoil removal and extensive plant invasions, seldom recover passively or at best may be colonized by early successional shrubs. Active restoration, which can include weeding, planting, seeding, treatments to break seed dormancy, and/or stabilizing soil treatments, may be needed in many cases. We review current knowledge of chaparral stressors and dynamics that relate to restoration as well as restoration methods. The limited information on restoration projects to date indicates that early successional, deciduous shrub species, which are common to sage scrub and have low seed dormancy, are most successful in establishment. These may accomplish some restoration objectives, such as soil stabilization and ability to recover from fire, but fall short in biodiversity goals. Application of techniques to establish evergreen chaparral species, such as large-scale dormancy breaking treatments or facilitation by early successional shrubs, is needed. We also discuss plant traits that might be used to guide restoration toward persistent communities under frequent fire. Our aim is to describe knowledge gaps about chaparral restoration and inspire restoration research, planning, and practice.

E.B. Allen (✉) • M. Phillips
University of California, Riverside, CA, USA
e-mail: edith.allen@ucr.edu

K. Williams
California State University, San Bernardino, CA, USA

J.L. Beyers
USDA Forest Service, Pacific Southwest Research Station, Riverside, CA, USA

S. Ma • C.M. D'Antonio
University of California, Santa Barbara, CA, USA

E. C. Underwood et al. (eds.), *Valuing Chaparral*, Springer Series on Environmental Management, https://doi.org/10.1007/978-3-319-68303-4_13

Keywords Active restoration · Chaparral recovery · Chaparral restoration · Passive restoration · Vegetation type-conversion

13.1 Introduction: Why Does Chaparral Need to be Restored?

Ecological restoration is defined as assisting the recovery of a system that has been degraded, damaged or destroyed (SER 2004; Day et al. 2006). For sites with intact soils and seedbanks this can be accomplished passively by removing stressors such as livestock grazing or improving air quality. Sites that are severely disturbed, or where natural rates of succession are too slow for management objectives, may be candidates for active restoration (Allen et al. 2001; Greipsson 2011, see Sect. 13.5). In California, restoration has been practiced extensively in coastal sage scrub, grassland, and wetland ecosystem types that are severely impacted by biotic invasions and anthropogenic disturbance (Mooney and Zavaleta 2016). There are, however, few cases of chaparral restoration because it has been subject to proportionally less anthropogenic disturbance and, until recently, has been considered a resilient vegetation type. Indeed, although there are hundreds of published studies on chaparral species and community responses to fire, much of the information on chaparral restoration is anecdotal (VinZant in prep-b). We highlight situations where chaparral is undergoing disturbances that are reducing its integrity and resilience, and where active restoration may be increasingly needed. We then review approaches to restoration.

13.1.1 Disturbance History and Resilience

Chaparral is one of the most widespread vegetation types in California, and it tends to occur on steep or rocky slopes that are difficult to develop. As a result, proportionally less area has been converted to agricultural or other land uses than other vegetation types. Hence chaparral has not received high priority for conservation and restoration. Instead, it has been considered as one of the most stable ecosystems in terms of constancy of land area occupied, and one of the most resilient ecosystems in its ability to recover from fire (Minnich and Bahre 1995; Keeley et al. 2005b).

Despite its reputation for resilience, chaparral is subject to a range of anthropogenic disturbances that degrade or eliminate it, including purposeful vegetation type-conversion, short fire-return intervals, fire suppression, invasion by non-native species, drought, and fragmentation by urban development, roads, corridors and fuelbreaks. Deliberate type-conversion to grass dominated vegetation has been done to increase forage for livestock (Biswell 1954), enhance water yield (Corbett and

Fig. 13.1 Badger Hill, at California State University San Bernardino, burned in 1980, 1995, and 2003, and has experienced chaparral loss on south-facing slopes. The degree to which various stressors (fire frequency, changing climate, N deposition, and presence of non-native annuals) contribute to poor chaparral recovery at sites such as this may influence decisions about whether to attempt restoration. Photos by the Arthur E. Nelson Archives (1978), held at Pfau Library Special Collections & Archives, California State University San Bernardino, and K. Williams (2006 and 2016)

Rice 1966), or increase wildlife habitat diversity (Rosario and Lathrop 1974). The perception of stability was promulgated in part by the extreme efforts needed to suppress shrub regeneration, which includes combinations of burning, herbicide, and seeding with grasses (Schultz and Biswell 1952; Bentley 1967; Murphy and Leonard 1974). However, chaparral has also been unintentionally degraded by non-native annual grass and forb invasion and reduced shrub density under shortened fire-return intervals (Keeley and Brennan 2012, Fig. 13.1, see Chap. 12) and alternatively by tree encroachment at woodland-chaparral ecotones under fire suppression (Sparling 1994; Van Dyke et al. 2001). Historical repeated burning by Indians may have converted limited areas of chaparral to forb and grassland, especially in southern California (see Chap. 4). Multi-year drought causes mortality of some shrub species with or without fire (Pratt et al. 2014, see Chap. 14), which in turn, increases susceptibility to non-native grasses. Total or partial conversion to non-native grassland can contribute to a positive grass-fire cycle (D'Antonio and Vitousek 1992).

Severely disturbed chaparral, whether by frequent fire or mechanical disturbance, is usually slow, or unable, to recover naturally. Some highly disturbed stands have persisted for decades as non-native forbs and grasses, sometimes with scattered deciduous shrubs that were not constituents of the original chaparral vegetation (Stylinski and Allen 1999). The loss of an extensive area of chaparral has historically been inconceivable given its perceived stability and resilience. To date, there has not been a landscape-scale assessment of the extent of chaparral type-conversion to challenge this perception (but see Chap. 12). Because we know that chaparral degradation is occurring and is likely to accelerate, it is important to develop successful techniques for restoration of chaparral structure and function.

13.1.2 Ecosystem Services of Restored Chaparral

Restoration of chaparral may be justified by the many ecosystem services mature chaparral provides, as described in the preceding chapters in this book. Degraded chaparral results in declines in species diversity (see Chaps. 2 and 3), decreased slope stability from loss of deep-rooted shrubs (see Chap. 7), decreased water infiltration (see Chap. 8), decreased carbon storage caused by a shift from deep-rooted shrubs to shallow-rooted non-native herbs (see Chap. 6), and the loss of recreational opportunities and aesthetic value in damaged lands (see Chap. 10). The extent to which restoration is achieved will be determined by the degree of disturbance and the particular ecosystem services that are prioritized for any site.

13.1.3 Objectives—Steps to Achieve Restoration

Our objectives are to review the steps involved in achieving chaparral restoration. The first consideration is to prioritize sites needing restoration based upon loss of ecosystem services, location in the landscape, and potential for further damage (see Sect. 13.2). The next step is to set restoration goals for provisioning of ecosystem services such as slope stability, ability to recover from fire, reduced fire incidence, enhanced carbon sequestration, or habitat for sensitive species (see Sect. 13.3). Before these goals can be realized, the stressors that caused ecosystem damage in the first place ought to be evaluated, as well as the extent to which pre-disturbance conditions or alternate goals are appropriate to achieve ecosystem services (see Sect. 13.4). Delineation of goals and stressors can assist managers to decide on restoration techniques ranging from passive to active restoration (see Sect. 13.5). Finally, success of restoration activities is assessed in light of the restoration goals, further treatments, adaptive management, and research applied as needed to promote success (see Sect. 13.6).

13.2 Where to Restore

In general, the decision of whether or not to restore chaparral on a specific site will be influenced by: (1) the importance of the benefits of chaparral ecosystem services on that site, (2) the extent of damage, (3) the likelihood of restoration attempts succeeding on that site under current and future conditions (e.g., climate, disturbance), (4) the presence of rare or endangered species, (5) the potential to minimize or mitigate conflicts between chaparral and adjacent landscape elements (e.g., urban areas) at that site, and (6) the cost of restoration. Depending on ownership and available financial resources of affected lands, these decisions are affected by combinations of regulatory constraints and societal input. Spatial information on the value of ecosystem services of chaparral (erosion control, watershed protection, carbon storage, and biodiversity), combined with spatial information on stressors (see Sect. 13.4) that would reduce likelihood of restoration success (e.g., high nitrogen deposition rates and frequent fire), has been used to prioritize areas for postfire restoration on national forest land (see Chap. 15). Explicit consideration of land use around a potential restoration site helps identify both potential benefits of restoration and conflicts that may cause the restoration effort to fail.

Most current restoration projects occur on public lands where agency regulations or permit conditions require revegetation of bare soil after disturbance. Such policies are designed to prevent soil erosion, for example, restoration on road cuts (Allen and Heindl 1993; CalTrans 2008), after pipeline installation (D'Antonio and Howald 1990), or after surface mining (Roy 2009; Wilkin 2009). The US Forest Service in southern California includes permit conditions specifying that restoration is to be carried out after land disturbance by special-use permittees, such as for powerline construction (VinZant in prep-b). Only recently has revegetation of heavily disturbed areas involved pre-disturbance planning and post-disturbance actions to restore chaparral plant communities to a similar pre-disturbance state.

Restoration may also be beneficial where chaparral degradation and type-conversion to annual vegetation has increased threats to adjacent ecosystems or to human infrastructure, such as roads, reservoirs, or urban development. For example, a mudflow through the community of Highland at the base of the San Bernardino Mountains in 2010 may have resulted in part from slope failure on the frequently-burned adjacent foothills during an extreme rain event. Eliminating flashy fuels, typically non-native annual plants and subshrubs of disturbed chaparral, may be most important near ignition sources like urban areas, roads, or areas with moderate housing densities (see Chap. 12). In such areas, heavy human use can lead to fire starts. Even in protected areas such as the Santa Monica Mountains of southern California, chaparral disturbance from trampling, trail creation, off-road vehicle use, purposeful vegetation clearance, and refuse dumping has been extensive and associated with proximity to roads (Sauvajot et al. 1998). The need to replace non-native annual plants with low volume, slow burning shrubs along roads, fuelbreaks, trails, and powerlines through chaparral has been recognized for some time (Nord and Green 1977). The degree to which chaparral restoration, in some form, can contribute to this goal along roadways requires further investigation.

Although benefits of chaparral restoration may be most important at the wildland-urban interface, conflicts between chaparral and other landscape elements are also strongest there, creating challenges to both public acceptance of chaparral restoration and the success of implemented projects. Where mature chaparral abuts urban areas, fuel reduction may be carried out by mastication and other types of mechanical treatments (Brennan and Keeley 2015, see Sect. 13.5.2.1 and Chap. 15). Furthermore, impacts of humans and domesticated animals may threaten the success of restoration projects at the wildland-urban interface (see Sect. 13.4.1.5), and protecting human habitation from wildfire could place restrictions on the design and composition of restored chaparral near human habitation. Therefore, sites that have been type-converted or degraded near urban areas may need thoughtful discussions about the trade-off between reducing ignitions and long-term restoration success (see Chap. 15).

13.3 What to Restore and Restoration Goals

Restoration goals should be specific enough to provide a ready path to measureable criteria to assess success. While many goals in chaparral restoration may focus on ecosystem services, the selection of species to use for the restoration of the site is key for achieving these goals. The choice of species can also be guided by the need to ensure resilience and persistence of the restored community, the ease with which different species establish and spread, and the availability of seed and/or plants. Thus specific goals may have to be adjusted as availability of materials and site constraints are understood. Some chaparral functions may be deemed more critical than others on specific sites and such valuation may influence priorities for re-establishing different types of species. For example, slope stabilization above vulnerable habitats and protection of downslope communities could be enhanced through the establishment of deep-rooted shrubs (see Sect. 13.3.1). Changing climate and landscape conditions (e.g., urban development, fire frequency), as well as practical constraints on seed materials, mean that it may not be possible to restore all the species historically present in all locations. Thus, a useful approach is to focus chaparral restoration projects on recovering ecosystem functions.

13.3.1 Plant Traits to Stabilize Soil and Water Functions

Slope stabilization, water infiltration, and carbon sequestration are typical ecosystem function goals relevant to degraded chaparral. All of these would benefit from the establishment of plants with perennial, deep-root systems that remain intact through the summer drought and are living when the first fall rains arrive. In particular, deep-rooted species may be most effective over long time-scales for both carbon sequestration and protection against slope failure. Deep-rooted shrubs are also, commonly, resprouters (Hellmers et al. 1955; Davis et al. 1999), rendering them relatively resilient to occasional shoot loss from grass fires, grazing, and other disturbances that may be part of the early restoration environment. Once deep-rooted shrubs are well-established they are likely to maintain a well-developed canopy which will deflect the impact of rain on the soil surface. However, they may be slow growing. Thus, to avoid initial runoff and erosion, shorter-lived, fast-growing species such as sage scrub dominants (e.g., California buckwheat [*Eriogonum fasciculatum*], California sagebrush [*Artemisia californica*]) may be employed in combination with slower growing shrubs (Sect. 13.5.2.4).

13.3.2 Habitat Restoration to Maintain Animal Diversity

For the purposes of enhancing animal diversity, vegetation structure and plant species composition are important. Interactions between animal diversity and vegetation structure and composition have been extensively reviewed (Keeley and Swift 1995). Shrub cover attracts small mammals, habitat edges between shrub cover and more open vegetation attract a variety of mammals, and insect pollinators, especially bee species, are abundant in chaparral (Moldenke 1976; Keeley and Swift 1995). The abundance of pollinators can be influenced by plant species diversity and this is particularly important in chaparral because of the large number of specialist pollinator/shrub relationships. Pollinator abundance has also been attributed to vegetation structure and the sheltering effect of the chaparral canopy on ground-nesting bees (Moldenke 1976; Force 1990; Keeley and Swift 1995). If restoration of animal communities is a goal of the restoration project then it may be important to design a project that maximizes vegetation diversity and structure.

13.3.3 Restoring Uncommon and Rare Species

Re-establishing rare plants or habitat for rare animals helps protect global biodiversity. Restoration projects may have permit requirements for development activities that adversely affect rare species. Some of the rare or unusual plant species of chaparral are soil specialists or island endemics (see Chap. 2), and will, therefore, be appropriately re-introduced only at specific sites. In general, restoration of rare plant species can be challenging because material for restoration (seeds or other propagules) is usually limited, little is known about germination and growth requirements for many rare chaparral species, and many rare species may be more vulnerable to stressors than common species. Although restoring chaparral for many purposes, including habitat restoration, may focus on shrub species, approximately 70% of the plant taxa in chaparral that are considered rare, threatened, or endangered (i.e., California Native Plant Society (CNPS) ranks 1A and 1B) are either annual or perennial herbs (CNPS 2016).

Herbaceous species that require specific conditions to germinate and/or bloom may remain unapparent for years, making assessment of restoration success difficult.

Among the shrub species of chaparral, those that are slow-maturing, fire-dependent, obligate seeders are increasingly threatened by short fire-return intervals (Zedler 1995) and are, therefore, of greater conservation concern than most resprouters. For example, the rare morro manzanita (*Arctostaphylos moroensis*) requires especially long fire-free periods for successful recruitment (Odion and Tyler 2002) and therefore would not be a good restoration candidate in areas with short fire-return intervals. Wet-season burns reduced recruitment of several rare species (Parker 1987). Species that are locally or globally rare because of their high vulnerability to stressors (e.g., altered fire regime, competition from non-native annuals, and predation by domestic animals) may be poor candidates for restoration unless (or until) those stressors can be eliminated. For instance, the endangered shrub Nevin's Barberry (*Berberis nevinii*) is challenging to establish in wildlands because of human and grazing/burrowing animal impacts, but is available commercially and thrives in protected gardens (CNPS n.d.). Rare species may be particularly sensitive to changes in fire regime and efforts should be taken to consider whether recurrent unnatural disturbances are likely to impede restoration success.

13.3.4 Plant Traits for a Resilient Chaparral Community

Successful decadal-scale restoration requires persistence and resilience of the restored chaparral to future perturbations. Thus resilience to future fire or drought should be an important goal in chaparral restoration. Resilience of chaparral is due in large part to species traits and life-forms that enable post-disturbance regeneration (see Chaps. 1 and 2). Species may be selected for restoration based on these life-forms and traits to provide stability and ecosystem services to sites with novel disturbances. Resprouting species have the ability to recover from shoot loss, rendering them relatively resilient to above-ground disturbances produced by fire, grazing, and manual cutting of stems (Mooney and Hobbs 1986). Combining traditionally recognized responses to fire with considerations of life-form, seedbank dynamics, and time-to-maturation has proven useful in describing the rate at which different species spread and persist on burned sites (Keeley et al. 2006). Applying such concepts to chaparral restoration suggests that species that mature quickly, and produce seeds that do not require fire to germinate, may spread most rapidly on a restoration site. These traits will also contribute to the persistence of their populations after disturbance (Mooney and Hobbs 1986). Many of these species, however, are sage scrub species, suffrutescent species or subshrubs, rather than dominant species of mature chaparral (Keeley et al. 2006). Therefore, their role in ecosystem function and chaparral development on a restoration site should be considered as part of goal setting, in addition to their contribution to resilience (see Sect. 13.5.2.4).

13.3.5 Restoration under Changing Climate and Landscape Conditions

Although restoration goals are generally linked to a reference condition that depicts a less disturbed system, changing climate or landscape conditions may limit the degree to which full restoration of a chaparral community is feasible or even desirable on a site (Harris et al. 2006; Hobbs et al. 2009, 2014). Landscape conditions include fragmentation, urbanization, frequent fire, invasion, or vegetation type-conversion, or other changes in the adjacent landscape. Certain species may no longer persist in their previous location due to changing climate, air pollution, or altered disturbance regimes (see Sect. 13.4). Some species may not be welcome near human habitation (e.g., large predators and venomous animals), and some characteristics of the vegetation may be considered too hazardous to create at the wildland-urban interface (e.g., dangerous fuel structure or flammability). Where climate has changed to such an extent that chaparral restoration is likely to fail (see Sect. 13.4.1.2 and Chap. 14), options may be limited to deciding not to restore or accepting a different community, such as coastal sage scrub, as an acceptable "restoration" endpoint. Where the human landscape has rendered some characteristics of chaparral undesirable, novel ecosystems that include some historical elements of chaparral and some new elements that are compatible with current and future conditions may add to the success of restoration projects and help practitioners attain restoration goals (Wiens and Hobbs 2015; Miller and Bestelmeyer 2016 and references therein).

Restoration within the wildland-urban interface may often involve use of novel vegetation combinations because fire behavior from typical chaparral (characterized by high fuel loads) can threaten human communities. Early searches for low-volume shrubs that could replace non-native annual grasses on fuelbreaks (Nord and Green 1977) or stabilize slopes near human habitation (McKell et al. 1966; Westman 1976) included many non-native species. In general, species that have relatively thick leaves, a low tendency to hold dead tissue high in the canopy, and low production of volatile and resinous compounds tend to ignite less readily and burn more slowly than other species (Montgomery and Cheo 1971; Green 1981; Schwilk 2003). Chaparral shrubs with these characteristics, mostly resprouters of mixed chaparral, appear on lists of plants approved for fuel modification zones by various organizations and municipalities in California (e.g., County of Riverside [n.d.]; Santa Monica Mountains Fire Safe Alliance 2010). Although these species may provide a number of ecosystem services attributed to chaparral, if the species did not originally exist on the site, their establishment may not be viewed as traditional ecological restoration.

13.4 Limitations that Affect Project Design and Restoration Success

Assessing site conditions and alleviating stressors is an important early step in any restoration plan (Whisenant 1999; Galatowitsch 2012). Stressors may be abiotic (frequent fire or fire suppression, drought, soil disturbance, erosion, air pollution) or biotic (competition from non-native species, animal impacts, seedbank loss, lack of germination cues for refractory seeds). Limitations may determine the extent to which restoration goals can be achieved and shape planning and implementation. Understanding the degree to which these stressors and limitations can (or cannot) be overcome may help inform decisions on whether successful restoration is feasible on a site.

13.4.1 Stressors

13.4.1.1 Drought and Summer Moisture Availability

The summer dry season experienced by chaparral has prompted studies that collectively provide an understanding of drought stress physiology and adaptations of mature plants (Parker et al. 2016, see Chap. 1). Adaptations to drought include deep-rooting, resistance to xylem cavitation, and sclerophyllous leaves. Drought tolerance in woody plants is generally high but varies among shrub species. Recent prolonged drought has caused shrub mortality in areas that have had stable chaparral stands for many decades (Davis et al. 1999; Paddock et al. 2013; Pratt et al. 2014), and some species of resprouters suffered high postfire mortality during a single drought year (Pratt et al. 2008). Mesic coastal populations of chaparral shrubs are more sensitive to drought than inland populations (Vasey et al. 2012). General strategies of drought tolerant (reseeding) and drought avoiding (resprouting) chaparral species as well as adapted microsites (e.g., north- and south-facing slopes) are recognized (see Chaps. 2 and 15). However, drought tolerance changes with shrub age, and less is known about moisture requirements and the seedling establishment niche, yet these are critical for successful restoration. During postfire succession, even in average precipitation years, 90% or more of seedlings succumb during the dry summer (Kummerow et al. 1985; Moreno and Oechel 1992). Oechel (1988) identified minimum water potentials for postfire seedling establishment, a concept that can be applied to restoration practice (e.g., providing irrigation). Obligate resprouters have higher recruitment from seed in shade than in full sunlight, related to their higher moisture requirement (Pratt et al. 2008). An understanding of moisture requirements for establishing seedlings is key to active restoration of severely disturbed areas where there is little cover and potentially poor water infiltration. Seeding or planting may need to be done in multiple years if plants suffer severe drought mortality in the first years of a project.

13.4.1.2 Climate Change

There is evidence that climate change is shifting chaparral ecotones and causing increased mortality of some species under drought (see Chap. 14). For instance, several chaparral shrub species moved upslope an average of 65 m (213 feet) in response to climate warming over 30 years in desert chaparral (Kelly and Goulden 2008), and three of seven resprouting chaparral shrubs experienced the highest ever recorded mortality during a severe drought following a fire (Pratt et al. 2014). One climate change modeling effort showed that many chaparral species would be reduced to an average of 50% of their current area in conservation reserve lands with 2.2 °C warming in the next 50 years (Principe et al. 2013). In addition, a warming climate has been cited as one cause of increased fire frequency (see Chap. 1). Given the potential for unsuitable conditions and permanent vegetation shifts, it suggests that consideration should be given to planting species adapted to projected climate conditions. For instance, as the climate warms, species common to sage scrub may be better adapted to future conditions than chaparral shrubs. Alternatively, sage scrub shrubs may be used as nurse plants to facilitate establishment of chaparral shrubs (see Sect. 13.5.2.4) A recommended approach for shifting climate envelopes is to plant species adapted to the new conditions of the area, which in southern California would include selecting more drought-adapted species (Harris et al. 2006).

13.4.1.3 Nutrient Availability, Nitrogen Deposition, and Invasion

Mature chaparral soils tend to be relatively low in mineral nitrogen (N) and phosphorus, as these are immobilized both in living plant tissue and in slowly decomposing, high C:N litter (Parker et al. 2016, see Chap. 1). Following fire, nutrients are mineralized and initially taken up by fast-growing native herbs (Hanan et al. 2016). However, where seeds of non-native grasses and forbs have dispersed into chaparral, they have a postfire advantage of earlier germination phenology, higher densities, and higher growth rates under high soil N (Chiariello 1989; Padgett and Allen 1999; Cox and Allen 2008), so they are better able to take up the initial flush of mineral nutrients postfire than native herbs and shrubs. These non-native grasses and forbs would be expected to have a high reproductive output of high quality seeds in such settings, thus setting the stage for suppression of desired native species.

Anthropogenic N deposition increases shrub growth in unburned sage scrub stands (Vourlitis 2012) while also increasing productivity of non-native annuals (Fenn et al. 2010; Valliere et al. 2017). Invasion is more likely after a fire, as most annuals will not invade the understory of a closed canopy shrubland. Shrub mortality under elevated N or drought that creates canopy openings and promotes invasion has been documented in sage scrub (Vourlitis 2017; Valliere et al. 2017) but not, to date, in chaparral. As with sage scrub, N deposition may operate in conjunction with frequent fire and drought to cause conversion of chaparral towards a grassy shrub savanna. The relationship between N deposition and type-conversion in sage scrub shows a high probability of conversion above a critical load of 11 kg N ha^{-1} $year^{-1}$ (Cox et al. 2014). If similar analyses are conducted for chaparral ecosystems these could be important inputs for developing restoration site plans. Of the total land area of chaparral, 14.6% receives more than 10 kg N ha^{-1} $year^{-1}$ deposition (~2 kg N ha^{-1} $year^{-1}$ is considered background for clean air), suggesting vulnerability to elevated N (Fenn et al. 2010). Legislative efforts have been enacted to reduce N deposition, and restoration may be most effective in regions that fall below critical N loads (Fenn et al. 2010).

13.4.1.4 Altered Fire Frequency

Fire frequency, whether shorter or longer intervals than natural fire regimes, may alter chaparral plant composition and affect restoration potential. In southern California fire frequency has increased adjacent to developed areas because of human-caused ignitions (Keeley et al. 2005b; Syphard et al. 2007, see Chap. 1). Fire-return intervals of less than 5–10 years will kill obligate seeders before they mature, as well as potentially killing resprouters that have not recovered fully from the previous fire (Zedler et al. 1983; Keeley et al. 2005b; Keeley and Brennan 2012; Lippitt et al. 2013). A single short interval fire may not be enough to cause complete loss of chaparral in most areas (S. Ma unpublished data; Meng et al. 2014), but species composition may be changed if obligate seeders are reduced or eliminated (Zedler et al. 1983; Keeley and Brennan 2012). Chaparral stands are frequently colonized by non-native grasses and forbs 2–5 years after fire if they are located near a seed source (Keeley et al. 2005a). If a fire occurs again during this period, the lower fire intensity allows more non-native seeds to survive, and non-native species cover is higher during the next recovery period (Beyers et al. 1998; Keeley and Brennan 2012). High N deposition may exacerbate this situation (Allen et al. 1998; Padgett and Allen 1999; Fenn et al. 2010, see Sect. 13.4.1.3), increasing the potential for a grass-fire cycle.

Non-native grasses are strong competitors with seedlings of chaparral shrubs, as was demonstrated in early studies by the inability of obligate seeder shrubs to establish in burn sites heavily seeded with non-native annual grass (Schultz et al. 1955). For decades, non-native grasses were seeded postfire in an effort to reduce immediate postfire erosion (Robichaud et al. 2000, see Chap. 7). These seeding projects were seldom successful in reducing erosion, but high grass densities displaced native herbaceous fire-followers and reduced the density of native shrub seedlings (Beyers 2004). Postfire grass seeding is seldom done in southern California today (Wohlgemuth et al. 2009), but a lesson from those studies is that non-native grasses can interfere with successful restoration (Roy 2009; Engel et al. in prep; VinZant in prep-b).

In contrast to non-native grass invasion, tree invasion occurs at the chaparral-woodland ecotone in the mesic ranges of chaparral under long-term fire suppression (Vasey et al. 2012). Coast live oak (*Quercus agrifolia*) invades coastal chaparral unless there has been a fire within the past 70 years (Callaway and Davis 1993; Van Dyke et al. 2001). Encroachment of Douglas fir (*Pseudotsuga menziesii*) into chaparral has been noted in Santa Clara County (Greenlee et al. 1983) and Marin County (Sparling 1994; Dunne and Parker 1999; Horton et al. 1999). Fire suppression will continue where chaparral abuts human development, so if a restoration goal is the maintenance of diverse chaparral, prescribed fire may be needed to reverse tree encroachment. Conversely, invasion by coast live oak (*Q. agrifolia*) may have a negative feedback on fire regimes because oaks are less flammable than shrubs, and this may be a desired outcome near homes.

Fig. 13.2 Left photo: heavily deer-grazed scrub oak (*Quercus berberidifolia*) regenerating from salvaged root material in revegetation area. Right photo: oak sprouts 5 months after fencing (with meter stick in foreground). Green plant to right of meter stick is deerweed (*Acmispon glaber*) that was ungrazed. Photos by C. D'Antonio

13.4.1.5 Human Activities and Domestic Animals

Some activities of humans and their domestic animals can degrade chaparral and pose challenges to restoration. Humans, vehicles, and livestock can directly destroy plantings, disturb soil, and spread seed of non-native annuals into restoration sites. Domestic and feral cats are well known for decimating populations of native animals in many parts of the world (e.g., Marzluff and Ewing 2001; Medina et al. 2011). Studies in southern California suggest that restoration of native birds in shrubland near urban areas may only succeed where restored patches attract top predators like coyotes or otherwise discourage the presence of the meso-predators (domestic cats, raccoons and opossums) that reduce bird populations (Soulé et al. 1988; Crooks and Soulé 1999; Galatowitsch 2012). Human welfare, however, may conflict with attempts to encourage substantial populations of some predators near human habitation.

13.4.1.6 Native Animals

Although native animals are integral parts of chaparral ecosystems, their actions can benefit or slow restoration efforts. Granivory and herbivory can deplete seeds and damage seedlings and resprouting shrubs (Fig. 13.2). The magnitude of these processes varies with site degradation, fire frequency, animal composition, fluctuations in animal populations (van Mantgem et al. 2015), distance to cover for herbivores (site size), and the balance between direct negative impacts of these animals on native plants and indirect positive effects of herbivory on competing, non-native vegetation.

Grazing and browsing by native animals can create bare zones at the edges of shrub stands (Bartholomew 1970) and reduce herbaceous vegetation in the understory (Swank and Oechel 1991; Tyler 1995). Such herbivory may benefit restoration by reducing competition from non-native annual species and by preventing encroachment of perennial non-native species (Lambrinos 2006). Direct damage to chaparral species, however, may hinder restoration. Flowering of chaparral geophytes is reduced by leaf loss (Borchert and Tyler 2009; Williams and Burck unpublished data). Shrub establishment and survival may be reduced by herbivory, but assessments of its importance in chaparral dynamics, such as recovery after fire, yield conflicting results (e.g., Davis 1967; Bullock 1991; Tyler 1995; Potts et al. 2010; Ramirez et al. 2012). The rapidity with which seedlings become resilient to herbivory varies among species. Seedlings of some species (e.g., scrub oaks [*Quercus berberidifolia*]) can resprout following shoot loss their first year. Others can resprout when they are a bit older, and even seedlings of obligate seeders can survive some level of browsing (e.g., 2 or 3 years postfire, Mills 1986; Tyler and D'Antonio 1995). Even in resprouters, root:shoot allocation shifts that depress root growth following herbivory may render seedlings more susceptible to drought-induced mortality (e.g., McPherson 1993). Thus, the use of native browsers and grazers to control non-native annual vegetation is likely to benefit restoration efforts only if native species are protected from herbivory.

Below-ground browsing (root damage) can be more serious than shoot loss. Burrowing animals, such as Botta's pocket gopher (*Thomomys bottae*), damage roots and bury seed, presenting obstacles to woody plant establishment in forestry, agriculture, and restoration projects in California (e.g., Crouch 1982; Borchert et al. 1989; Witmer and Engeman 2007; Tyler et al. 2008). Adams et al. (1997), studying native oak plantings in California rangelands, noted that "pocket gophers can present a prolonged threat" to restoration efforts, because saplings remained vulnerable until they were fairly large in diameter. Thus, the presence of pocket gophers and the distribution of site characteristics that discourage them (e.g., rocky soil) should be considered when evaluating the likely success of chaparral restoration efforts.

13.4.2 Seed Limitation

Seed limitation is broadly acknowledged to be highly important in influencing community development and critical to passive habitat restoration, however little work has been done on seeding as an active restoration strategy in chaparral. Seed limitation can be extreme when no propagules are available or moderate when safe sites (sensu Harper 1977) outnumber available seed. Here we focus on broad considerations related to the seedbank, seed predation, seed availability, and germination cues.

Obligate seeders have a persistent seedbank composed of refractory (with high dormancy) seeds and are stimulated to germinate based on a range of cues typically related to fire (Parker and Kelly 1989; Keeley 1991, see Chap. 2). Because the adult plants are destroyed by fire, these species are dependent upon a robust seedbank for persistence (Zammit and Zedler 1988; Keeley 1991). Seed longevity in these species is considered to be great (Parker and Kelly 1989), yet finding commercially available seed or collecting it for restoration is difficult because of the generally low seed output and specific germination requirements (Emery 1988). Obligate resprouters often have a transient, or non-refractory, seedbank composed of seeds that are generally killed by fire and depend on animal or wind dispersal during fire-free intervals. Thus seed availability, ease of germination, seeding rate necessary for establishment of adults, and safe sites for establishment may be very different for these broad functional shrub groups, and restoration methods will vary as a result.

Seedbanks of obligate seeders can be depleted through a number of means. Repeated short-interval fires can stimulate germination but not replenish the seedbank if a second fire occurs before plants have reached reproductive maturity (Keeley and Brennan 2012). While individual plants may mature in less than 10 years (Zammit and Zedler 1993), some communities may require multiple decades to regenerate a seedbank large enough to maintain the population (Odion and Tyler 2002). Intense seed predation by some rodents (Keeley and Hayes 1976; Parker and Kelly 1989; Deveny and Fox 2006) can also deplete the seedbank. Non-native annual grasses and forbs can support granivores (Orrock and Witter 2010), and so as non-native annuals invade or dominate degraded chaparral, granivores may reduce the remaining seedbank, further restricting the likelihood of passive recovery. During active restoration, the control of seed predators may be important in the initial stages.

Finally, seed limitation in refractory seeds can arise if seeds are present in the seedbank but are in secondary dormancy and appropriate cues are not experienced (e.g., heat, charate or smoke compounds, period of stratification). Seed limitation for non-refractory species could arise if the degraded area is large and seed sources for re-colonization or seed collection are lacking. Active restoration might then focus on direct seeding with exposure to appropriate germination cues, planting seedlings, or planting nurse plants that might attract animal dispersers of the desired seeds.

Even if seed is available, getting the right germination cues can be challenging. Many chaparral species germinate following a fire cue, generally heat or smoke (Keeley 1991). For this reason, seed treatments that mimic fire are successful in inducing germination in the absence of fire (Wilkin et al. 2013). Heat treatment by placing seeds in boiling water or an oven have successfully increased germination of obligate seeding species of *Ceanothus* (Quick 1935, 1959; Radwan and Crouch 1977) and facultative seeders such as sugar bush (*Rhus ovata*) (Stone and Juhren 1951) and chamise (*Adenostoma fasciculatum*) (Christensen and Muller 1975). Other sources suggest additional dormancy-breaking techniques, such as acid treatment for *A. fasciculatum* (Emery 1988), illustrating the complexity of selecting germination treatments. Perhaps not surprisingly, use on a scale large enough for restoration has been limited. Charate (charred wood) can also induce germination in obligate seeders, such as some *Arctostaphylos* species, and resprouters, such as *Adenostoma fasciculatum* (Keeley 1984, 1991), and could be artificially applied, or charred wood of cleared shrubs could be raked into the soil of small sites to aid germination. Smoke can also induce germination of obligate seeding shrub and herbaceous species (Keeley and Bond 1997; Keeley and Fotheringham 1998). Scarification is another useful technique to induce seed germination. Hadley (1961) found scarifying bigpod ceanothus (*C. megacarpus*, obligate seeder) seeds increased germination almost fourfold compared to a heat treatment. Stone and Juhren (1951) found *R. ovata* (resprouter) had the highest germination rate when the seed coat was removed. Germination rates can further be improved upon by combining treatments, for example scarification and heat. However, best practices seem to be species-specific (Keeley 1991) and can even vary within a species (Stone and Juhren 1951). The application of these laboratory studies to field restoration, particularly on a large scale, has rarely been done (but see Sect. 13.5.2.2).

13.5 How to Restore Chaparral

There are a limited number of available studies on chaparral that document both passive and active forms of restoration. Passive and active restoration fall along a gradient of treatments depending on the extent of damage and residual propagule bank. Passive restoration yields a community of species that remain on site post-disturbance and species that colonize naturally. Choosing elements to restore is largely a process involved in active restoration. A mixed approach that involves passive restoration of some species and active restoration of others (those that do not readily establish or colonize naturally) may be appropriate for achieving specific restoration goals. Passive restoration techniques emphasize recovery that does not include replanting or other intervention, while active restoration emphasizes fire management, revegetation, and soil treatments.

13.5.1 Passive Restoration

Passive recovery from the seedbank has been evaluated after various disturbances to chaparral, including grazing and physical disturbance to soils, and studies of these disturbances indicate that the rate and pathway of natural recovery of chaparral is generally slow compared to patterns of postfire recovery (Keeley et al. 2006, see Chap. 1). For example, succession on former chaparral land in abandoned agricultural and urban construction soils resulted in alternative trajectories dominated by sage scrub species, such as *Eriogonum fasciculatum* and broom baccharis (*Baccharis sarothroides*) with an understory of non-native annuals in sites up to 70 years post-disturbance (Stylinski and Allen 1999; Stratton 2005, 2009). Deliberate type-conversions in the San Gabriel Mountains in the 1960s were

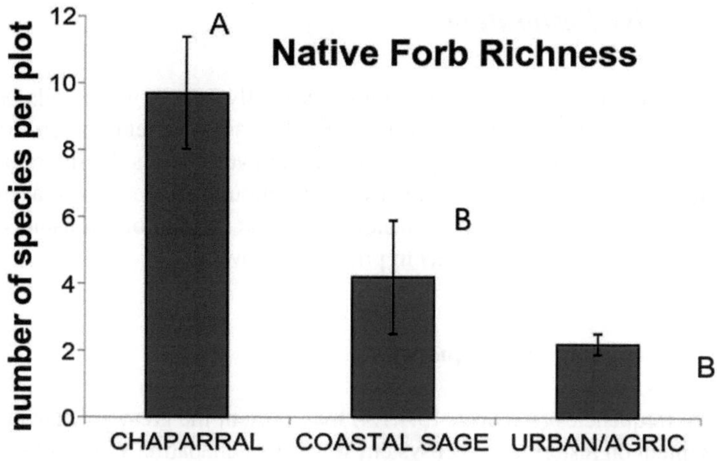

Fig. 13.3 Richness of naturally colonizing (not seeded) native forbs was highest on revegetated roadsides of Interstate 15, San Diego County, when the adjacent vegetation was chaparral rather than coastal sage scrub or urban/agricultural lands (Allen and Heindl 1993, unpublished). The cover of colonizing forbs ranged from 2% to 7% and the cover of planted shrubs averaged 47% across all sites, but no shrub species colonized that were not in the seed mix

invaded by similar species over the following decades. After a fire re-burned the sites in 2002, some chaparral species colonized but at lower abundance than in adjacent undisturbed chaparral (Hubbert et al. 2012b; Corcoran et al. in prep). Removal of domestic grazing animals enabled a degree of natural recovery of chaparral and sage scrub on Santa Catalina and Santa Cruz Islands, but active restoration was needed in some areas to control weeds, reduce erosion, and re-introduce rare species (O'Malley 1991; Halvorson 1994; Beltran et al. 2014). Abandoned roads are also challenging to restore as they have depauperate seedbanks and would need active restoration to re-introduce missing species (Holl et al. 2000). Roadsides actively revegetated with native shrubs were naturally colonized by as many as 10 species of unseeded native forbs, but no species of unseeded shrubs (Allen and Heindl 1993). Along roadsides the highest rate of re-colonization was adjacent to undisturbed chaparral compared to weedy sage scrub or urban development (Fig. 13.3, Allen and Heindl 1993). Where topsoil has been removed or altered, active restoration is usually needed to re-introduce most of the dominant species, otherwise the result could be a novel ecosystem with unusual combinations of native and non-native species (Stylinski and Allen 1999). While such a community may not meet biological diversity goals, it might suffice for other ecosystem services such as soil stability.

13.5.2 Active Restoration

The decision to engage in active restoration typically occurs when soils have been disturbed sufficiently to cause erosion and deplete soil nutrients, microorganisms, and plant propagules, or when vegetation type-conversion results in an alternative stable state. Such sites will not recover quickly enough to meet societal or agency goals for ecosystem services, or will not recover at all without assistance. In this section we review active treatments to promote recovery.

13.5.2.1 Managing Fire Frequency

Altered fire frequencies (see Sect. 13.4.1.4) may present the greatest challenge to land managers intent on restoring the ecosystem services of chaparral. Where shortened fire-return intervals have led to non-native grass invasion, a return to normal fire regimes would allow chaparral to develop a dense and less invasible canopy and also burn sufficiently hot to kill a higher proportion of the non-native seedbank (Keeley and Brennan 2012). Human-related ignitions are responsible for most of the fires in southern California, which usually threaten human developments (see Chap. 12). Use of deliberate type-conversions, fuelbreaks, and fuel modification zones (e.g., mastication of most shrubs with a scattered few left standing) between homes and wildlands may help to protect public safety and may also reduce fire spread from homes to nearby chaparral (Conard and Weise 1998). However, mastication treatments provide only temporary fuel reduction (Keeley et al. 2014; Brennan and Keeley 2015), and non-native herbaceous species may colonize shrub inter-spaces and provide flashy fuel cover (Potts and Stephens 2009). Fuelbreaks require maintenance by animal (grazing) or mechanical means. These vegetation sacrifice areas obviously do not provide the ecosystem services of chaparral but may allow chaparral to persist beyond them. Where chaparral has already been degraded by too-frequent fire, revegetation of areas beyond fuelbreaks may be needed if the return of chaparral ecosystem services is desired (see Sect. 13.5.2.3).

Periodic prescribed fire would prevent ingrowth of trees into fire-suppressed chaparral stands, but air quality concerns (Ahuja 2006) and caution regarding the chance for escapes may limit the use of fire as a vegetation management tool at the wildland-urban interface (Husari et al. 2006). When prescribed fire can be used, the operational window is usually well outside the natural fire season (e.g., during cooler and wetter periods when fire can be more easily be controlled, Green 1981). The resulting fire intensity and effects may be very different from those during the natural summer fire season: burning could be less complete and soils are often wet rather than dry, resulting in less soil heating. As a result, there might be low germination of heat-requiring seeds, such as many *Ceanothus* species (Kelly and Parker 1984; Parker 1990). In southern California these effects may not be extensive (Beyers and Wakeman 2000), but studies in northern California chaparral have generated greater concern (Parker 1987). Implementing prescribed fire in small stands of rare chaparral species, such as that described by Van Dyke et al. (2001), could be difficult. Trees can be removed mechanically to reduce shading in critical situations, but other beneficial effects of time-appropriate fire are hard to simulate.

13.5.2.2 Soil Treatments

Erosion Control

Stabilizing steep hillsides is challenging once shrubs have been completely removed by disturbance and may require an ecological engineering approach that combines surface soil stabilization treatments (erosion control mats, erosion barriers, channel stabilization, re-contouring) with seeding and planting. However, most of the information about the success of these treatments is anecdotal. Erosion barriers or mulch may be applied on small scales to protect urban areas or sensitive species habitat. For instance, mulch increased acorn germination for *Quercus berberidifolia* establishment in the Channel Islands (Stratton 2005). Anecdotal observations after broadcast seeding in utility right of way corridors in the Angeles National Forest suggested that hydromulching was comparable to imprinting (machine-made patterned depressions) for increasing shrub establishment from seed, although both produced only 5%–15% vegetation cover after 1 year (VinZant in prep-b). Effects of hydromulch on erosion and runoff were not measured in these corridors. First year hillslope sediment movement was substantially lower on hydromulched plots than control plots at two burned chaparral sites near Santa Barbara, California (Wohlgemuth et al. 2011). In another study, benefits of hydromulch on erosion control were limited to the first 2–4 months postfire in San Diego County (Hubbert et al. 2012a).

Soil Inoculum

Other soil treatments include use of mycorrhizal inoculum, topsoil, and fertilizer. Although most chaparral species are arbuscular mycorrhizal (AM), the below-ground mycorrhizal community is complex because it includes ectomycorrhizal (EM) oaks and pines, arbutoid mycorrhizal *Arctostaphylos* and other ericaceous species, and *Adenostoma fasciculatum* that can switch between EM or AM under wet or dry conditions, respectively (Allen 1991; Allen et al. 1999). Duff and soil from native oaks improved germination and establishment of acorns because it introduced EM inoculum, while uninoculated nursery transplants of arbutoid Santa Catalina island manzanita (*Arctostaphylos catalinae*) had low survival in former cropland (Stratton 2005). Alternatively, use of fresh chaparral soil added to planting holes for nursery stock of AM shrubs had no effect on their growth (Stratton 2005; VinZant in prep-b). AM shrubs are generalists in terms of their AM fungal associates, and when planted into soils dominated by AM grasses appropriate inoculum is likely to occur in the soil (Nelson and Allen 1993). By contrast, EM oaks and arbutoid manzanitas require their own host-specific inoculum.

Studies from other ecosystems have shown that non-native plant species may cause major shifts in soil microbial species composition and in nutrient cycling (Ehrenfeld 2003; Owen et al. 2013), but the non-native grasses of chaparral had weak feedbacks as there were few differences in microbial communities or nutrient cycling between chaparral and adjacent invaded soils (Dickens and Allen 2014). This suggests that the soil microbial community of Mediterranean-type climate annual grass invaded chaparral will seldom be limiting to plant growth, except in the case of EM or arbutoid fungi that are absent from grassland. In a meta-analysis of 22 restoration studies (none from chaparral), local AM or EM inoculum promoted greater biomass, while commercial inoculum did not (Maltz and Treseder 2015). Some AM plants are highly responsive to inoculation and perform better with inoculation throughout their lifespan. However, mycorrhizal responsiveness of chaparral plants is sparsely studied. Mycorrhizal plants of seasonally dry climates are better able to access water and survive drought (Querejeta et al. 2007), so inoculation is recommended before they are transplanted to the field.

Topsoiling

Where topsoil has been removed for construction purposes, replacing topsoil is recommended, as was tested in a utility corridors in Santa Barbara County (D'Antonio and Howald 1990) and in the Angeles National Forest (VinZant in prep-b). In the latter study, observations indicate that seeded plants had improved growth on topsoil compared to subsoil. Topsoil is seldom available for spreading across large areas, so VinZant (in prep-b) recommends that construction plans include salvaging and re-spreading topsoil. D'Antonio and Howald (1990) documented abundant regeneration of maritime chaparral forbs and early successional shrubs in topsoil that was replaced or conserved (Fig. 13.4). However, soils with a dense non-native seedbank should be avoided for salvaging or solarized (heat-treated by covering with clear or black plastic) during storage to minimize the need for post-restoration weed control. Studies of topsoil handling in *Banksia* woodland restoration in the Mediterranean-type climate region of Australia have revealed that dry-season handling of topsoil yielded much better results than wet-season handling, and that stockpiling soil for even 1 year reduced effectiveness of top-soil replacement. The most effective treatments in the study involved moving topsoil from newly disturbed areas to areas undergoing restoration (Rokich et al. 2000). Similar principles may apply to topsoil handling in chaparral restoration.

Fertilizer

If topsoil is not available to restore severely disturbed sites, soil amendments such as fertilizers and mulch may be beneficial, in addition to mycorrhizal inoculation. An important caveat is that chaparral soils tend to be fairly low in nutrients and

Fig. 13.4 Vegetative cover in second growing season of pipeline revegetation project. Topsoil was conserved during construction and replaced and many unseeded natives recovered from the seedbank. Most species visible here are short-lived perennials or sage scrub species. Surrounding vegetation is 50–60-year old chaparral on sandy soils (D'Antonio and Howald 1990). Burton Mesa region, Santa Barbara County. Photo by Carla D'Antonio

organic matter (Parker et al. 2016), and there is a danger of over-fertilizing and promoting weeds, if careful prescriptions for fertilizer rates are not followed. Non-native grasses have higher growth rates in response to nitrogen fertilizer and any level that is added may promote non-native over native species (Allen et al. 1998). Even in the absence of invasion, elevated soil nutrients will shift native species composition to dominance by species with plastic responses to fertilization, and gradually exclude slower-growing chaparral shrubs (Pasquini and Vourlitis 2010). This suggests that sites with natural topsoil are unlikely to need fertilization, and sites where topsoil is absent may only benefit from fertilizer addition to levels no higher than native soils based on an understanding of plant nutrient needs.

Charate

A soil treatment that could be considered in some settings to benefit plant establishment is burning piles of woody material over the restoration site, especially where a native seedbank is believed to persist. The resulting charred material is then raked over the disturbed area to replace charate that would be released from a natural fire. Such a treatment was used in the Purissima Hills in Santa Barbara County in a pipeline corridor through chaparral and Bishop pine (*Pinus muricata*) stands (D'Antonio and Howald 1990). Species that had not been seeded re-generated in the charate-influenced areas (Fig. 13.5).

Fig. 13.5 Thickleaf yerba santa (*Eriodictyon crassifolium*) seedling growing out of soil in experimental burn pile/charate spread area during second growing season in a pipeline corridor. Photo by Carla D'Antonio

13.5.2.3 Revegetation Treatments

Weeding

Control of non-native species may be sufficient to enable chaparral shrub recovery in invaded areas that have a native propagule bank. Grass-specific herbicide was used to control bulbous canarygrass (*Phalaris aquatica*) on Santa Catalina Island and enabled establishment of chaparral species from the seedbank (Stratton 2005). Weeding accompanied by seeding and planting is the more frequent approach, as heavily grass-invaded chaparral often has a compromised seedbank. For instance, no native shrub seedlings emerged at all from a weeded, formerly grass dominated site in Riverside County (Engel et al. in prep). A summary of weed control activities in invaded utility corridors of the Angeles National Forest included hand weeding and mechanical and herbicide treatments. Although data were not collected from these sites and no unweeded controls were maintained, managers interviewed concluded that weeding was essential for the establishment of seeded shrub species (VinZant in prep-b). Controlled studies showed that seeded black sage (*Salvia mellifera*) and *Eriogonum fasciculatum* (common to both sage scrub and chaparral) had high germination rates and required no seed dormancy-breaking treatment, but only survived in plots hand-weeded of annual grasses, otherwise there was

100% mortality during summer (Schultz 1996). A review of studies on controlling non-native annuals common to chaparral and sage scrub also demonstrated the importance of reduced competition for sage scrub establishment (Allen et al. in prep).

Seeding

Seeding is the preferred method of active restoration because it is less expensive than nursery transplants, but the practice has had limited success in chaparral. Germination rates are notoriously low for many chaparral species, in part because of their specialized dormancy-breaking requirements (see Sect. 13.4.2) and often low percentages of viable seed from wild collections (Wall and Macdonald 2009). Seed application of *Adenostoma fasciculatum, Quercus berberidifolia, Rhus ovata,* and skunk bush (*R. aromatica*) was unsuccessful in spite of appropriate seed treatment (liquid smoke, acid scarification, stratification) and non-native grass control by herbicide prior to seeding, most likely due to severe drought during the study (Engel et al. in prep). Chaparral seeding trials on Santa Catalina Island were less successful than native species emerging from the seedbank of abandoned agricultural fields (Stratton 2005), but there was no mention of seed pre-treatment in these studies. Sage scrub species considered to be early successional in chaparral, such as *Eriogonum fasciculatum*, deerweed (*Acmispon glaber*), and *Salvia mellifera*, established successfully from seed in a pipeline corridor revegetation effort in Santa Barbara County, while seeded chaparral species were largely unsuccessful (D'Antonio and Howald 1990). In a successful chaparral seeding effort, Roy (2009) planted smoke-treated *Adenostoma fasciculatum* seeds in an abandoned rock quarry soil, and observed high germination rates and adequate survival in weeded plots, but poor survival in plots with competition from native and non-native annuals. Summer irrigation reduced dry-season mortality, but establishment occurred in the absence of irrigation when plots were weeded. Taken together, these studies show that drought and competition from non-native annuals are the greatest limiting factors to establishment of chaparral shrubs from seed. Although seed dormancy-breaking requirements have been tested in the laboratory for a wide range of chaparral species (Keeley and Keeley 1987; Keeley and Fotheringham 1998, see Sect. 13.4.2), we could find few references to their use in restoration projects in California in the scientific literature. By contrast, smoke is routinely used as a pre-treatment for seed germination in restoring areas impacted by mines in Australian shrublands (Roche et al. 1997).

Fig. 13.6 Restored abandoned corridor in chamise chaparral on Del Sur Ridge, Angeles National Forest. Container transplants of California sagebrush (*Artemisia californica*), California buckwheat (*Eriogonum fasciculatum*), black sage (*Salvia mellifera*), and Eastwood's manzanita (*Arctostaphylos glandulosa*) were initially irrigated and protected from deer browsing by netting. Photo is taken 2 years after planting. Photo by Jan Beyers

Seed Collection

The preferred seed collection area should be as local as possible to avoid genetic concerns (Vander Mijnsbrugge et al. 2010), although this may be difficult if the degraded area is large and seed sources are distant. US Forest Service guidelines recommend preferably within 8 km (~5 miles) (at most 16 km [~10 miles]) distance and within 150–300 m (492–984 ft) in elevation (VinZant in prep-a). However, these guidelines may need to be relaxed under warming climate conditions. Adapted seed collection zones could be experimentally tested using transplant gardens. Guidelines for southern California chaparral seed processing after collection are summarized in Wall and Macdonald (2009) and Emery (1988) that lists pre-planting seed treatments for many chaparral species.

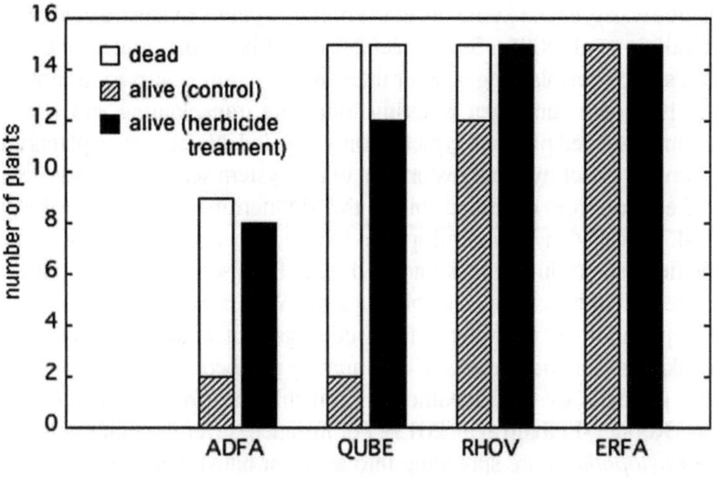

Fig. 13.7 First-year establishment of containerized shrubs was improved for 3 of 4 species (ADFA and QUBE, slight increase for RHOV) by herbicide treatment to control non-native annual grasses and forbs. ADFA = *Adenostoma fasciculatum* (chamise), QUBE = *Quercus berberidifolia* (California scrub oak), RHOV = *Rhus ovata* (sugar bush), ERFA = *Eriogonum fasciculatum* (California buckwheat) (redrawn from Engel et al. in prep)

Nursery Transplants

Chaparral restoration has been most successful from nursery transplants, although this is admittedly a more costly approach than seeding and has only been practiced in small areas (see Chap. 15). Anecdotal reports from utility corridor restoration in the Angeles National Forest indicated that shrub transplants had up to 40% cover the first year, while seeding only resulted in up to 20% (VinZant in prep-b, Fig. 13.6). Some of the successful transplants were hand-watered using deep pipe irrigation that delivers water to the lower rooting zone through a pipe installed at planting (Bainbridge 2007). Deep-rooted (34 cm) transplants of lemonade berry (*Rhus integrifolia*) and laurel sumac (*Malosma laurina*) with consistent irrigation had greater survival than shallow-rooted (10 cm) transplants of the same species (Burkhart 2006). Transplants of 35 shrub species on Santa Catalina Island had marginally greater survival of some species with hand-watering. Interestingly, a water-absorbing polymer decreased survival of some species (Stratton 2005). Stratton (2005) concluded that watering was not required for survival of most of these 35 species in years with average precipitation, but hypothesized that watering would increase the return on investment of transplants of more sensitive species in a drought year. In addition to hand-watering, herbicide treatments reduced annual weed competition and further increased the transplant survival of some shrub species (Engel et al. in prep.), (Fig. 13.7).

In the Mediterranean Basin, transplantation of shrubs was successful if they were pre-conditioned to avoid transplant shock (Vallejo and Alloza 2012), but survival was highly variable depending on the site (Maestre et al. 2006). Polymer gels increased transplant survival in one study (Clemente et al. 2004), and the success of many shrub species was greatest in the presence of nurse plants (Maestre et al. 2001) or tree shelters (Vallejo et al. 2006). Such practices could be implemented in California.

Overall, shrub transplanting is labor intensive. It requires nursery establishment and care for 3–12 months and then watering following transplanting and weeding in the field, in contrast seeded plots are typically only weeded. Nursery transplants are justified where the cost is offset by rapid restoration of ecosystem services, as was demonstrated by reduced erosion from degraded soils in the Mediterranean Basin (Le Houérou 2000; Vallejo et al. 2006). Many chaparral species have not been successfully established from seed in the field, so producing containerized stock from seed or cuttings may be the only option for certain dominant species of chaparral. We are not aware of restoration-scale practical application of transplants from cuttings, but cuttings of some species are routinely taken in botanic gardens. Clean nursery protocols are critically important for restoration projects to avoid pathogen transmission from contaminated nursery transplants (Rooney-Latham et al. 2015). For instance, several species of the pathogenic oomycete *Phytophthora* are spreading into adjacent native chaparral populations from contaminated nursery material (Sims et al. n.d.), a situation that can be avoided by proper sterilization of nursery soils.

13.5.2.4 Establishing Early Successional Species to Initiate Succession

Successional patterns provide information that is useful in ecological restoration (Walker et al. 2007), but this approach to restoring chaparral has not been well explored. Although some obligate resprouters may require the shade of larger shrubs for seedling establishment (Keeley 1992; Pratt et al. 2008), such facilitation plays little role in early stages of postfire chaparral recovery, the context in which chaparral succession is most frequently studied (e.g., Hanes 1971; Kummerow et al. 1985). In restoring chaparral on disturbed or type-converted sites, however, facilitation may play a role. Woody plants that colonize severely disturbed chaparral are often soft-leaved semi-drought deciduous subshrubs of coastal sage (Callaway and Davis 1998; Stylinski and Allen 1999). Where chaparral shrubs have been observed to establish on sites formerly dominated by non-native annuals, they have done so after colonization by soft-leaved shrubs (Callaway and Davis 1993; Brennan 2015). Drought deciduous sage scrub dominants are less likely than evergreen plants to ameliorate the environment of chaparral shrub seedlings during the hot summer drought, and the limited information available suggests that evergreen species may be more effective than drought deciduous species in serving as nurse plants for other woody species (Callaway and D'Antonio 1991). The potential value of using faster-growing evergreen plants to facilitate the establishment of chaparral on disturbed sites requires investigation.

Drought deciduous subshrubs of sage scrub and postburn chaparral (e.g., *Eriogonum fasciculatum*, *Artemisia californica*, and *Salvia mellifera*) are often included in plant palettes used for restoration of disturbed sites (VinZant in prep-a). These species establish more readily and grow more quickly than deeply rooted, slower-growing dominants of chaparral, thereby covering the soil and providing some

Table 13.1 Suggested monitoring criteria over time for chaparral restoration

Criterion	Categories and approaches
Community, measurements:	
Native plant cover (total)	Using transects/plots/cover classes
Native woody plant cover	By life-form, e.g., obligate seeders versus resprouters or sage scrub versus chaparral species
Native woody plant density	By species
Native plant richness	Include all life-forms: forbs, grasses, subshrubs, shrubs
Non-native plant cover (total)	All life-forms in plots or transects
Bird, mammal, herptile usage	Richness, point counts/abundance, nesting, etc.
Invertebrate abundance	Pollinator richness, sweep net counts, pitfall traps
Ecosystem process measurements:	
Net primary production, plant C storage	Biomass accumulation over time, litter accumulation, decomposition rate, % bare soil, NDVI/EVI over time or compared to reference
Soil compaction, nutrients, C storage, microbial functioning	Organic matter accumulation, bulk density, C in profile, microbial biomass
Sediment movement, soil erosion	Soil slips, mudflows, sediment yield
Water yield	Streamflow before/after restoration actions or captured runoff
Nitrate in runoff	In-stream flow before and after restoration actions

degree of soil stabilization more rapidly. Seed mixes for restoration may focus on early seral species that lack the complex germination requirements of many chaparral species (VinZant in prep-a). Whether these species facilitate or retard succession to mature chaparral in restoration settings, however, is not yet known.

13.6 Assessing Restoration Success

The assessment of success in a restoration project depends on the specific goals set during the planning process. These goals should be specific enough to have led to a list of measurable criteria upon which to base decisions for further follow-up actions. Despite the recognized importance of ecosystem processes in most restoration projects, specific measurable criteria for most restoration projects are species based (Ruiz-Jaen and Aide 2005, Table 13.1). They typically include plant cover, richness, and density. Because sites targeted for restoration are likely to be denuded of native vegetation, broad measures of native cover and woody plant density are appropriate. Due to the importance of chaparral for animal diversity, including some estimates of animal presence could be another useful measure of success. Chaparral restoration is frequently oriented to restoration of ecosystem processes such as reducing erosion, so success criteria might also include ecosystem processes such as erosion or nutrient loss (Table 13.1). Plant cover and presence of bare soil can be useful indicators of erosion potential if direct measurements are not made (Robichaud et al. 2000).

A typical restoration project has a reference site against which the measured criteria are compared. Yet in chaparral restoration, the difficulty of restoration toward diverse evergreen sclerophyllous shrubland communities, suggests that success criteria are more likely to be based on broad measurements of native shrub cover and ecosystem functioning rather than comparison to diversity of a reference community. For monitoring ecosystem function, measured outcomes may be compared against both intact reference sites and pre-restoration values using a before/after type statistical design. If restoration sites are not converging toward reference sites or showing directional improvement over time, a need for further restoration planning and actions would be indicated.

13.6.1 Prospects for Long-Term Restoration Success

Active chaparral restoration is too recent and limited in extent to make definitive conclusions about prospects for success on scales larger than road beds or construction pads. The easiest species to establish from seed are herbaceous plants or those shrubs that are considered early successional or are common within sage scrub, such as *Eriogonum fasciculatum*, *Salvia* spp., *Acmispon glaber*, or *Artemisia californica*, whereas true chaparral shrubs are more easily established from container-grown plants (VinZant in prep-b). Most chaparral shrub seeds require pretreatment of some kind to germinate, which is easiest to do in small batches used for propagating nursery stock. In the absence of heat or smoke treatment, only species without a fire-related cue for germination are easy to seed in large quantities or likely to passively colonize restoration sites. In small-scale projects, however, plant cover was most quickly established on restoration sites using chaparral container stock and consistent irrigation (Stratton 2005; VinZant in prep-b).

Sage scrub species have been found to establish from seed without irrigation in years of average precipitation in sage scrub restoration sites (Padgett and Allen 1999), a precipitation regime that is below that for the more mesic regime of chaparral. This suggests they would be suited to chaparral restoration sites in drought years, when true chaparral species would be more difficult to establish, or at the lower elevation limits of chaparral. The decision to replace one vegetation type (chaparral) with another (sage scrub) cannot be made lightly and has become a major debate among conservation and restoration ecologists struggling to preserve species under climate change (Seddon 2010). However, sage scrub species colonize naturally into severely disturbed soil, ecotones with chaparral, and deliberately type-converted areas. They are faster-growing than chaparral shrubs, more drought tolerant, and thus more resilient under frequent fire. The transition from chaparral to sage scrub already occurs in some areas and provides land managers with a suite of species that may perform some desired ecosystem services.

It is unlikely, however, that soft-leaved, deciduous, shallower-rooted species will perform all of the same ecosystem services as hard-leaved, evergreen, deep-rooted chaparral species in the long-term. For instance, sage scrub roots that are 1.5 m (~5 ft) deep should help stabilize surface soils, but they likely will not be as effective at slope stabilization nor sequester as much carbon in the long-term as deep-rooted chaparral shrubs. The comparative roles of sage scrub and chaparral in carbon sequestration is a topic in need of further study. Sage scrub vegetation also does not provide the same habitat structure for animals as true chaparral. For conservation reserves there may be aesthetic issues and diversity shifts in selecting a suite of sage scrub species where chaparral once occurred, but these species may facilitate the establishment of chaparral species (see Sect. 13.5.2.4). Where rapid and more reliable establishment of chaparral is a project goal, the use of nursery transplants, protection from herbivory, weeding, irrigation and soil treatments may be needed in the most severely disturbed sites. Across large public landscapes, trade-offs between the costs of restoration and ecosystem services will need to be assessed. Clearly, additional research on incorporating large scale seed dormancy-breaking treatments, facilitation, and understanding long-term successional processes for chaparral restoration is critically needed.

References

Adams, Jr, T. E., P. B. Sands, W. H. Weitkamp, and M. E. Stanley. 1997. Oak seedling establishment by artificial regeneration on California rangelands. Pages 213–222 *in* N. H. Pillsbury, J. Verner, and W. D. Tietje, editors. Proceedings of a Symposium on Oak Wood- lands: Ecology, Management, and Urban Interface Issues, March 19–22 1996, San Luis Obispo, California, USA. General Technical Report PSW-GTR-160. USDA Forest Service, Pacific Southwest Research Station, Albany, California, USA.

Ahuja, S. 2006. Fire and air resources. Pages 481-498 *in* N. G. Sugihara, J. W. Van Wagtendonk, K. E. Shaffer, J. Fites-Kaufman, and A. E. Thode, editors. Fire in California's ecosystems. University of California Press, Berkeley, California, USA.

Allen, E. B., J. S. Brown, and M. F. Allen. 2001. Restoration of plant, animal and microbial diversity. Pages 185-202 *in* S. Levin, editor. Encyclopedia of biodiversity, Vol. 5. Academic Press, San Diego, California, USA.

Allen, E. B., and B. Heindl. 1993. Roadside vegetation of I15. Final Report to California Department of Transportation District 11, San Diego, California, USA.

Allen, E. B., C. McDonald, and B. E. Hilbig. in prep. Long-term prospects for restoration of coastal sage scrub: invasive species, nitrogen deposition, and novel ecosystems. *In* M. Narog, editor. Proceedings of the Chaparral Restoration Workshop, Arcadia, California, June 17-20, 2013. General Technical Report. USDA Forest Service, Pacific Southwest Research Station, Albany, California, USA.

Allen, E. B., P. E. Padgett, A. Bytnerowicz, and R. Minnich. 1998. Nitrogen deposition effects on coastal sage vegetation of southern California. Pages 131-140 *in* A. Bytnerowicz, editor. Proceedings of the International Symposium on Air Pollution and Climate Change Effects on Forest Ecosystems, Riverside, California, USA. February 5-9, 1996. General Technical Report PSW-GTR-166, Albany, California, USA, USDA Forest Service, Pacific Southwest Research Station.

Allen, M. F. 1991. The ecology of mycorrhizae. Cambridge University Press, Cambridge, UK.

Allen, M. F., L. M. Egerton-Warburton, E. B. Allen, and O. Karen. 1999. Mycorrhizae in *Adenostoma fasciculatum*: a combination of unusual ecto- and endo-forms. Mycorrhiza 8:225-228.

Bainbridge, D. A. 2007. A guide for desert and dryland restoration. New hope for arid lands. Island Press, Washington, D.C., USA.

Bartholomew, B. 1970. Bare zone between California shrub and grassland communities: the role of animals. Science 170:1210-1212.

Beltran, R. S., N. Kreidler, D. H. Van Vuren, S. A. Morrison, E. S. Zavaleta, K. Newton, B. R. Tershy, and D. A. Croll. 2014. Passive recovery of vegetation after herbivore eradication on Santa Cruz Island, California. Restoration Ecology 22:790-797.

Bentley, J. R. 1967. Conversion of chaparral areas to grassland: techniques used in California. USDA Forest Service Agriculture Handbook 328:1-35.

Beyers, J. L. 2004. Postfire seeding for erosion control: effectiveness and impacts on native plant communities. Conservation Biology 18:947-956.

Beyers, J. L., C. D. Wakeman, P. M. Wohlgemuth, and S. G. Conard. 1998. Effects of post-fire grass seeding on native vegetation in southern California chaparral. Pages 52-64 *in* Proceedings, Nineteenth Annual Forest Vegetation Management Conference: wildfire rehabilitation. Forest Vegetation Management Conference, Redding, California, USA, January 20-22, 1998. https://www.fs.fed.us/psw/publications/4403/EffectsofPostfire.pdf

Beyers, J. L., and C. D. Wakeman. 2000. Season of burn effects in southern California chaparral. Pages 45-55 *in* J. E. Keeley, M. Baer-Keeley, and C. J. Fotheringham, editors. 2nd Interface between Ecology and Land Development in California. US Geological Survey Open-Fire Report 00-61.

Biswell, H. H. 1954. The brush control problem in California. Journal of Range Management 7:57-62.

Borchert, M. I., F. W. Davis, J. Michaelsen, and L. D. Oyler. 1989. Interactions of factors affecting seedling recruitment of blue oak (*Quercus douglasii*) in California. Ecology 70:389-404.

Borchert, M., and C. M. Tyler. 2009. Patterns of post-fire flowering and fruiting in *Chlorogalum pomeridianum* var. *pomeridianum* in southern California chaparral. International Journal of Wildland Fire 18:623-630.

Brennan, S. G. 2015. The long-term dynamics of coyote brush invasion in a type-converted landscape of southern California. Thesis. California State University, Long Beach, California, USA.

Brennan, T. J., and J. E. Keeley. 2015. Effect of mastication and other mechanical treatments on fuel structure in chaparral. International Journal of Wildland Fire 24:949-963.

Bullock, S. H. 1991. Herbivory and the demography of the chaparral shrub *Ceanothus greggii* (Rhamnaceae). Madroño 38:63-72.

Burkhart, B. 2006. Selecting the right container for revegetation success with tap-rooted and deep-rooted chaparral and oak species. Ecological Restoration 24:87-92.

Callaway, R. M., and C. M. D'Antonio. 1991. Shrub facilitation of coast live oak establishment in central California. Madroño 38:158-169.

Callaway, R. M., and F. W. Davis. 1993. Vegetation dynamics, fire, and the physical environment in coastal central California. Ecology 74:1567-1578.

Callaway, R. M., and F. W. Davis. 1998. Recruitment of *Quercus agrifolia* in central California: the importance of shrub-dominated patches. Journal of Vegetation Science 9:647-656.

CalTrans. 2008. Roadside Erosion Control and Management Studies, CTSW-RT-08-067-01-1. California Department of Transportation. www.dot.ca.gov/hq/LandArch/16_la_design/research/docs/recm.pdf

Chiariello, N. R. 1989. Phenology of California grasslands. Pages 47-58 *in* L. F. Huenneke and H. A. Mooney, editors. Grassland structure and function: California annual grassland. Kluwer Academic Publishers, Dordrecht, Netherlands.

Christensen, N. L., and C. H. Muller. 1975. Relative importance of factors controlling germination and seedling survival in *Adenostoma* chaparral. American Midland Naturalist 93:71-78.

Clemente, A. S., C. Werner, C. Maguas, M. S. Cabral, M. A. Martins-Loucao, and O. Correia. 2004. Restoration of a limestone quarry: effect of soil amendments on the establishment of native Mediterranean sclerophyllous shrubs. Restoration Ecology 12:20-28.

CNPS. 2016. Inventory of rare and endangered plants (online edition, v8-02). California Native Plant Society, Sacramento, California, USA. http://www.rareplants.cnps.org

CNPS. n.d. Nevin's barberry. http://calscape.org/Berberis-nevinii

Conard, S. G. and D. R. Weise. 1998. Management of fire regime, fuels and fire effects in southern California chaparral: lessons from the past and thoughts for the future. Pages 342-350 *in* T. L. Pruden and L. A. Brennan, editors. Fire in ecosystem management: shifting the paradigm from suppression to prescription. Tall Timbers Fire Ecology Conference Proceedings, No. 20. Tall Timbers Research Station, Tallahassee, Florida, USA.

Corbett, E. S., and R. M. Rice. 1966. Soil slippage increased by brush conversion. Research Note PSW-RN-128. USDA Forest Service, Pacific Southwest Forest and Range Experiment Station, Berkeley, California, USA.

Corcoran, B. C., M. G. Narog, and P. M. Wohlgemuth. in press. Shrub recruitment 10 years following fire on type-converted and native chaparral watersheds of San Dimas Experimental Forest, California. *In* M. Narog, editor. Proceedings of the Chaparral Restoration Workshop, Arcadia, California, June 17-20, 2013. General Technical Report. USDA Forest Service.

County of Riverside. No date. County of Riverside, California friendly plant list. http://rctlma.org/Portals/7/documents/landscaping_guidelines/comprehensive_plant_list.pdf

Cox, R. D., and E. B. Allen. 2008. Composition of soil seed banks in southern California coastal sage scrub and adjacent exotic grassland. Plant Ecology 198:37-46.

Cox, R. D., K. L. Preston, R. F. Johnson, R. A. Minnich, and E. B. Allen. 2014. Influence of landscape-scale variables on vegetation conversion in southern California, USA. Global Ecology and Conservation 2:190-203.

Crooks, K. R., and M. E. Soulé. 1999. Mesopredator release and avifaunal extinctions in a fragmented system. Nature 400:563-566.

Crouch, G. L. 1982. Pocket gophers and reforestation on western forests. Journal of Forestry 80:662-664.

D'Antonio, C. M. and A. Howald. 1990. Evaluating the effectiveness of hydroseed mixes, topsoil conservation and other revegetation techniques: a case study in Santa Barbara County, California, USA. Pages 338-348 *in* H. G. Hughes and T. M. Bonnicksen, editors. Restoration '89: the new management challenge. Proceedings 1st Annual Meeting of the Society for Ecological Restoration, Oakland, California, January 16-20, 1989, University of Wisconsin Arboretum, Madison, Wisconsin, USA.

D'Antonio, C. M., and P. M. Vitousek. 1992. Biological invasions by exotic grasses, the grass/fire cycle, and global change. Annual Review of Ecology and Systematics 23:63-87.

Davis, J. 1967. Some effects of deer browsing on chamise sprouts after fire. American Midland Naturalist 77:234-238.

Davis, S. D., F. W. Ewers, J. Wood, J. J. Reeves, and K. J. Kolb. 1999. Differential susceptibility to xylem cavitation among three pairs of *Ceanothus* species in the Transverse Mountain Ranges of southern California. Ecoscience 6:180-186.

Day, K., J. Berg, H. Brown, T. Crow, J. Morrison, G. Nowacki, D. Puckett, R. Sallee, T. Schenck, and B. Wood. 2006. Ecosystem restoration: a framework for restoring and maintaining the national forests and grasslands. USDA Forest Service. http://www.fs.fed.us/restoration/documents/RestFramework_final_010606.pdf

Deveny, A. J., and L. R. Fox. 2006. Indirect interactions between browsers and seed predators affect the seed bank dynamics of a chaparral shrub. Oecologia 150:69-77.

Dickens, S. J. M., and E. B. Allen. 2014. Exotic plant invasion alters chaparral ecosystem resistance and resilience pre- and post-wildfire. Biological Invasions 16:1119-1130.

Dunne, J. A., and V. T. Parker. 1999. Species-mediated soil moisture availability and patchy establishment of *Pseudotsuga menziesii* in chaparral. Oecologia. 119:36-45.

Ehrenfeld, J. G. 2003. Effects of exotic plant invasions on soil nutrient cycling processes. Ecosystems 6:503-523.

Emery, D. E. 1988. Seed propagation of native California plants. Santa Barbara Botanic Garden, Santa Barbara, California, USA.

Engel, M. D., K. Williams, C. J. McDonald, and J. L. Beyers. in prep. The feasibility of chaparral restoration on type-converted slopes. *In* M. Narog, technical coordinator. Chaparral restoration: a paradigm shift. Proceedings of the Chaparral Workshop, Arcadia, California June17–20, 2013. General Technical Report. USDA Forest Service, Pacific Southwest Research Station, Albany, California, USA.

Fenn, M. E., E. B. Allen, S. B. Weiss, S. Jovan, L. H. Geiser, G. S. Tonnesen, R. F. Johnson, L. E. Rao, B. S. Gimeno, F. Yuan, T. Meixner, and A. Bytnerowicz. 2010. Nitrogen critical loads and management alternatives for N-impacted ecosystems in California. Journal of Environmental Management 91:2404-2423.

Force, D. C. 1990. Ecology of insects in California chaparral. Research Paper PSW-RP-201. USDA Forest Service, Pacific Southwest Research Station, Albany, California, USA.

Galatowitsch, S. M. 2012. Ecological restoration. Sinauer Associates, Sunderland, Massachusetts, USA.

Green, L. R. 1981. Burning by prescription in chaparral. General Technical Report PSW-GTR-51. USDA Forest Service, Pacific Southwest Forest and Range Experiment Station, Berkeley, California, USA.

Greenlee, J. M., J. H. Langenheim, and D A. Benson. 1983. Vegetation, fire history, and fire potential of Big Basin Redwoods State Park, California. Final Report to California Department of Parks and Recreation, Sacramento. 108 p. Cited in Greenlee, J. M., and J. H. Langenheim. 1990. Historic fire regimes and their relation to vegetation patterns in the Monterey Bay area of California. American Midland Naturalist 124:239-253.

Greipsson, S. 2011. Restoration ecology. Jones & Bartlett, Sudbury, Massachusetts, USA.

Hadley, E. B. 1961. Influence of temperature and other factors on *Ceanothus megacarpus* seed germination. Madroño 16:132-138.

Halvorson, W. L. 1994. Ecosystem restoration on the California Channel Islands. Pages 485-490 *in* W. L. Halvorson and G. J. Maender, editors. The Fourth California Islands Symposium: Update on the Status of Resources. Santa Barbara Natural History Museum, Santa Barbara, California, USA.

Hanan, E. J., C. M. D'Antonio, D. A. Roberts, and J. P. Schimel. 2016. Factors regulating nitrogen retention during the early stages of recovery from fire in coastal chaparral ecosystems. Ecosystems 19:910-926.

Hanes, T. L. 1971. Succession after fire in the chaparral of southern California. Ecological Monographs 41:27-52.

Harper, J. L. 1977. Population biology of plants. Academic Press, London, UK.

Harris, J. A., R. J. Hobbs, E. Higgs, and J. Aronson. 2006. Ecological restoration and global climate change. Restoration Ecology 14:170-176.

Hellmers, H., J. S. Horton, G. Juhren, and J. O'Keefe. 1955. Root systems of some chaparral plants in southern California. Ecology 36:667-678.

Hobbs, R. J., E. Higgs, and J. A. Harris. 2009. Novel ecosystems: implications for conservation and restoration. Trends in Ecology & Evolution 24:599-605.

Hobbs, R. J., E. Higgs, C. M. Hall, P. Bridgewater, F. S. Chapin, E. C. Ellis, J. J. Ewel, L. M. Hallett, J. Harris, K. B. Hulvey, and S. T. Jackson. 2014. Managing the whole landscape: historical, hybrid, and novel ecosystems. Frontiers in Ecology and the Environment 12:557-564.

Holl, K. D., H. N. Steele, M. H. Fusari, and L. R. Fox. 2000. Seed banks of maritime chaparral and abandoned roads: potential for vegetation recovery. Journal of the Torrey Botanical Society 127:207-220.

Horton, T. R., T. D. Bruns, and V. T. Parker. 1999. Ectomycorrhizal fungi associated with *Arctostaphylos* contribute to *Pseudotsuga menziesii* establishment. Canadian Journal of Botany 77:93-102.

Hubbert, K. R., P. M. Wohlgemuth, and J. L. Beyers. 2012a. Effects of hydromulch on post-fire erosion and plant recovery in chaparral shrublands of southern California. International Journal of Wildland Fire 21:155-167.

Hubbert, K. R., P. M. Wohlgemuth, J. L. Beyers, M. G. Narog, and R. Gerard. 2012b. Post-fire soil water repellency, hydrologic response, and sediment yield compared between grass-converted and chaparral watersheds. Fire Ecology 8:143-162.

Husari, S., H. T. Nichols, N. G. Sugihara, and S. L. Stephens. 2006. Fire and fuel management. Pages 444-465 *in* N. G. Sugihara, J. W. Van Wagtendonk, K. E. Shaffer, J. Fites-Kaufman, and A. E. Thode, editors. Fire in California's ecosystems. University of California Press, Berkeley, California, USA.

Keeley, J. E. 1984. Factors affecting germination of chaparral seeds. Bulletin of Southern California Academic Science 83:113-120.
Keeley, J. E. 1991. Seed germination and life history syndromes in the California chaparral. The Botanical Review 57:81-116.
Keeley, J. E. 1992. Recruitment of seedlings and vegetative sprouts in unburned chaparral. Ecology 73:1194-1208.
Keeley, J. E., M. Baer-Keeley, and C. J. Fotheringham. 2005a. Alien plant dynamics in Mediterranean-climate California shrublands. Ecological Applications 15:2109-2125.
Keeley, J. E., and W. J. Bond. 1997. Convergent seed germination in South African fynbos and California chaparral. Plant Ecology 133:153-167.
Keeley, J. E., and T. J. Brennan 2012. Fire-driven alien invasion in a fire-adapted ecosystem. Oecologia 169:1043-1052.
Keeley, J. E., and C. J. Fotheringham. 1998. Smoke-induced seed germination in California chaparral. Ecology 79:2320-2336.
Keeley, J. E., C. J. Fotheringham, and M. Baer-Keeley. 2005b. Determinants of postfire recovery and succession in Mediterranean-climate shrublands of California. Ecological Applications 15:1515-1534.
Keeley, J. E., C. J. Fotheringham, and M. Baer-Keeley. 2006. Demographic patterns of postfire regeneration in Mediterranean-climate shrublands of California. Ecological Monographs 76:235-255.
Keeley, J. E., and R. L. Hayes. 1976. Differential seed predation on two species of *Arctostaphylos* (Ericaceae). Oecologia 24:71-81.
Keeley, J. E., and S. C. Keeley. 1987. Role of fire in the germination of California chaparral and coastal sage scrub. American Journal of Botany 74:653-653.
Keeley, J. E., and C. C. Swift. 1995. Biodiversity and ecosystem functioning in Mediterranean-climate California. Pages 121-193 *in* G. W. Davis and D. M. Richardson, editors. Mediterranean-type ecosystems. Springer, Berlin, Germany.
Keeley, J. E., D. R. Weise, and T. Brennan. 2014. Effectiveness and effects of mastication fuel treatments in non-forested vegetation of southern California. Final Report JFSP Project Number 10-1-01-23. http://www.firescience.gov/projects/10-1-01-23/project/10-1-01-23_final_report.pdf
Kelly, A. E., and M. L. Goulden. 2008. Rapid shifts in species distribution with recent climate change. Proceedings of the National Academy of Sciences 105:11823-11826.
Kelly, V. R., and V. T. Parker. 1984. The effects of wet season fires on chaparral vegetation in Marin County, California. Unpublished report submitted to the Marin Municipal Water District.
Kummerow, J., B. A. Ellis, and J. N. Mills. 1985. Post-fire seedling establishment of *Adenostoma fasciculatum* and *Ceanothus greggii* in southern California chaparral. Madroño 32:158-157.
Lambrinos, J. G. 2006. Spatially variable propagule pressure and herbivory influence invasion of chaparral shrubland by an exotic grass. Oecologia 147:327-334.
Le Houerou, H. N. 2000. Restoration and rehabilitation of arid and semiarid Mediterranean ecosystems in north Africa and west Asia: a review. Arid Soil Research and Rehabilitation 14:3-14.
Lippitt, C. L., D. A. Stow, J. F. O'Leary, and J. Franklin. 2013. Influence of short-interval fire occurrence on post-fire recovery of fire-prone shrublands in California, USA. International Journal of Wildland Fire 22:184-193.
Maestre, F. T., S. Bautista, J. Cortina, and J. Bellot. 2001. Potential for using facilitation by grasses to establish shrubs on a semiarid degraded steppe. Ecological Applications 11:1641-1655.
Maestre, F. T., J. Cortina, and R. Vallejo. 2006. Are ecosystem composition, structure, and functional status related to restoration success? A test from semiarid Mediterranean steppes. Restoration Ecology 14:258-266.
Maltz, M. R., and K. K. Treseder. 2015. Sources of inocula influence mycorrhizal colonization of plants in restoration projects: a meta-analysis. Restoration Ecology 23:625-634.
Marzluff, J. M., and K. Ewing. 2001. Restoration of fragmented landscapes for the conservation of birds: a general framework and specific recommendations for urbanizing landscapes. Restoration Ecology 9:280-292.

McKell, C. M., V. Scoutemyer, C. Perry, L. Pyeatt, and J. R. Goodin. 1966. Hillside clearing and revegetation of fire hazard areas. California Agriculture 20:9-11.

McPherson, G. R. 1993. Effects of herbivory and herb interference on oak establishment in a semi-arid temperate savanna. Journal of Vegetation Science 4:687-692.

Medina, F. M., E. Bonnaud, E. Vidal, B. R. Tershy, E. S. Zavaleta, C. Josh Donlan, B. S. Keitt, M. Corre, S. V. Horwath, and M. Nogales. 2011. A global review of the impacts of invasive cats on island endangered vertebrates. Global Change Biology 17:3503-3510.

Meng, R., P. E. Dennison, C. M. D'Antonio, and M. A. Moritz. 2014. Remote sensing analysis of vegetation recovery following short-interval fires in southern California shrublands. PLoS ONE 9:e110637.

Miller, J. R., and B. T. Bestelmeyer. 2016. What's wrong with novel ecosystems, really? Restoration Ecology 24:577-582.

Mills, J. N. 1986. Herbivores and early postfire succession in southern California chaparral. Ecology 67:1637-1649.

Minnich, R. A., and C. J. Bahre. 1995. Wildland fire and chaparral succession along the California Baja-California boundary. International Journal of Wildland Fire 5:13-24.

Moldenke, A. R. 1976. California pollination ecology and vegetation types. Phytologia 34:305-361.

Montgomery, K. R., and P. C. Cheo. 1971. Effect of leaf thickness on ignitibility. Forest Science 17:475-478.

Mooney, H. A., and E. Zavaleta, editors. 2016. Ecosystems of California. University of California Press, Berkeley, California, USA.

Mooney, H. A., and R. J. Hobbs. 1986. Resilience at the individual plant level. Pages 65-82 *in* B. Dell, A. J. M. Hopkins, and B. B. Lamont, editors. Resilience in Mediterranean-type ecosystems. Dr. Junk Publishers, Dordrecht, Netherlands.

Moreno, J. M., and W. C. Oechel. 1992. Factors controlling postfire seedling establishment in southern California chaparral. Oecologia 90:50-60.

Murphy, A. H., and O. A. Leonard. 1974. Chaparral shrub control: grazing, herbicides. California Agriculture January:10-13.

Nelson, L. L., and E. B. Allen. 1993. Restoration of *Stipa pulchra* grasslands: effects of mycorrhizae and competition from *Avena barbata*. Restoration Ecology 1:40-50.

Nord, E. A., and L. R. Green. 1977. Low-volume and slow-burning vegetation for planting on clearings in southern California chaparral. Research Paper PSW-RP-124. USDA Forest Service, Pacific Southwest Forest and Range Experiment Station, Berkeley, California, USA.

Odion, D., and C. Tyler. 2002. Are long fire-free periods needed to maintain the endangered, fire-recruiting shrub *Arctostaphylos morroensis* (Ericaceae)? Conservation Ecology 6:4.

Oechel, W. C. 1988. Seedling establishment and water relations after fire in a Mediterranean ecosystem. Pages 34-45 *in* E. B. Allen, editor. The reconstruction of disturbed arid lands. Westview Press, Boulder, Colorado, USA.

O'Malley, P. G. 1991. Large-scale restoration on Santa Catalina Island, California. Restoration and Management Notes 9:7-15.

Orrock, J. L., and M. S. Witter. 2010. Multiple drivers of apparent competition reduce re-establishment of a native plant in invaded habitats. Oikos 119:101-108.

Owen, S. M., C. H. Sieg, N. C. Johnson, and C. A. Gehring. 2013. Exotic cheatgrass and loss of soil biota decrease the performance of a native grass. Biological Invasions 15:2503-2517.

Paddock, W. A. S. III, S. D. Davis, R. B. Pratt, A. L. Jacobsen, M. F. Tobin, J. Lopez-Portillo, and F. W. Ewers. 2013. Factors determining mortality of adult chaparral shrubs in an extreme drought year in California. Aliso 31:49-57.

Padgett, P. E., and E. B. Allen. 1999. Differential responses to nitrogen fertilization in native shrubs and exotic annuals common to Mediterranean coastal sage scrub of California. Plant Ecology 144:93-101.

Parker, V. T. 1987. Effects of wet-season management burns on chaparral vegetation: implications for rare species. Pages 233-237 *in* T. S. Elias, editor. Conservation and management of rare and endangered plants. California Native Plant Society, Sacramento, California, USA.

Parker, V. T. 1990. Problems encountered while mimicking nature in vegetation management: an example from a fire-prone vegetation. Pages 231-234 *in* R. S. Mitchell, C. J. Sheviak, and D. J. Leopold, editors. Ecosystem management: rare species and significant habitats. New York State Museum Bulletin No. 471. The State Education Department, Albany, New York, USA.

Parker, V. T., and V. R. Kelly. 1989. Seed banks in California chaparral and other Mediterranean climate shrublands. Pages 231-256 *in* M. A. Leck, V. T. Parker and R. L. Simpson, editors. Ecology of soil seed banks. Academic Press, San Diego, California, USA.

Parker, V. T., R. B. Pratt, and J. E. Keeley. 2016. Chaparral ecosystems. Pages 479-508 *in* H. A. Mooney and E. Zavaleta, editors. Ecosystems of California. University of California Press, Berkeley, California, USA.

Pasquini, S. C., and G. L. Vourlitis. 2010. Post-fire primary production and plant community dynamics in chaparral stands exposed to varying levels of nitrogen deposition. Journal of Arid Environments 74:310-314.

Potts, J. B., E. Marino, and S. L. Stephens. 2010. Chaparral shrub recovery after fuel reduction: a comparison of prescribed fire and mastication techniques. Plant Ecology 210:303-315.

Potts, J. B., and S. L. Stephens. 2009. Invasive and native plant responses to shrubland fuel reduction: comparing prescribed fire, mastication, and treatment season. Biological Conservation 142:1654-1657.

Pratt, R. B., A. L. Jacobsen, R. Mohla, F. W. Ewers, and S. D. Davis. 2008. Linkage between water stress tolerance and life history type in seedlings of nine chaparral species (Rhamnaceae). Journal of Ecology 96:1252-1265.

Pratt, R. B., A. L. Jacobsen, A. R. Ramirez, A. M. Helms, C. A. Traugh, M. F. Tobin, M. S. Heffner, and S. D. Davis. 2014. Mortality of resprouting chaparral shrubs after a fire and during a record drought: physiological mechanisms and demographic consequences. Global Change Biology 20:893-907.

Principe, Z., J. B. MacKenzie, B. Cohen, J. M. Randall, W. Tippets, T. Smith, and S. A. Morrison. 2013. 50-year climate scenarios and plant species distribution forecasts for setting conservation priorities in southwestern California v.1. The Nature Conservancy of California, San Francisco, California, USA.

Querejeta, J. I., L. M. Egerton-Warburton, and M. F. Allen. 2007. Hydraulic lift may buffer rhizosphere hyphae against the negative effects of severe soil drying in a California oak savanna. Soil Biology & Biochemistry 39:409-417.

Quick, C. R. 1935. Notes on the germination of *Ceanothus* seeds. California Botanical Society 3:135-140.

Quick, C. R. 1959. *Ceanothus* seeds and seedlings on burns. Madroño 15:79-81.

Radwan, M. A., and G. L. Crouch. 1977. Seed germination and seedling establishment of redstem ceanothus. The Journal of Wildlife Management 41:760-766.

Ramirez, A. R., R. B. Pratt, A. L. Jacobsen, and S. D. Davis. 2012. Exotic deer diminish post-fire resilience of native shrub communities on Santa Catalina Island, southern California. Plant Ecology 213:1037-1047.

Robichaud, P. R., J. L. Beyers, and D. G. Neary. 2000. Evaluating the effectiveness of postfire rehabilitation treatments. General Technical Report RMRS-GTR-63. USDA Forest Service, Rocky Mountain Research Station, Fort Collins, Colorado, USA.

Roche, S., J. M. Koch, and K. W. Dixon. 1997. Smoke enhanced seed germination for mine rehabilitation in the southwest of Western Australia. Restoration Ecology 5:191-203.

Rokich, D. P., K. W. Dixon, K. Sivasithamparam, and K. A. Meney. 2000. Topsoil handling and storage effects on woodland restoration in Western Australia. Restoration Ecology 8:196-208.

Rooney-Latham, S., C. L. Blomquist, T. Swiecki, E. Bernhardt, and S. J. Frankel. 2015. First detection in the USA: new plant pathogen, *Phytophthora tentaculata*, in native plant nurseries and restoration sites in California. Native Plants Journal 16:23-25.

Rosario, J. A., and E. W. Lathrop. 1974. Comparison of vegetation structure in modified and natural chaparral. Journal of Range Management 27:310-312.

Roy, C. L. 2009. Restoration of *A. fasciculatum* at Rocky Canyon Granite Quarry, San Luis Obispo, California. Thesis. California Polytechnic State University, San Luis Obispo, California, USA.

Ruiz-Jaen, M. C., and T. M. Aide. 2005. Restoration success: how is it being measured? Restoration Ecology 13:569-577.
Santa Monica Mountains Fire Safe Alliance. 2010. A road map to fire safety: how to create defensible space in the Santa Monica Mountains. Los Angeles County, Califorenia, USA. http://www.rcdsmm.org/sites/default/files/userfiles/file/Fire%20Safety/RoadMaptoFireSafety.pdf
Sauvajot, R. M., M. Buechner, D. A. Kamradt, and C. M. Schonewald. 1998. Patterns of human disturbance and response by small mammals and birds in chaparral near urban development. Urban Ecosystems 2:279-297.
Schultz, G. P. 1996. Seedling establishment and competition in coastal sage scrub and annual grassland. Thesis. University of California Riverside, Riverside, California, USA.
Schultz, A. M., and H. H. Biswell. 1952. Competition between grasses reseeded on burned brushlands in California. Journal of Range Management 5:338-345.
Schultz, A. M., J. L. Launchbaugh, and H. H. Biswell. 1955. Relationship between grass density and brush seedling survival. Ecology 36:226-238.
Schwilk, D. W. 2003. Flammability is a niche construction trait: canopy architecture affects fire intensity. American Naturalist 162:725-733.
Seddon, P. J. 2010. From reintroduction to assisted colonization: moving along the conservation translocation spectrum. Restoration Ecology 18:796-802.
SER [Society for Ecological Restoration]. 2004. International Science & Policy Working Group. The SER International Primer on Ecological Restoration. Society for Ecological Restoration International, Tucson, Arizona, USA. www.ser.org
Sims, L., M. Garbelotto, S. Tjosvold, D. Chambers, S. Koike, C. Conforti, A. Shor, A. Forrestel and L. Stringer. n.d. *Phytophthora* in wildlands, native plant nurseries, and management. http://ucnfa.ucanr.edu/files/250502.pdf
Soulé, M. E., D. T. Bolger, A. C. Alberts, J. Wrights, M. Sorice, and S. Hill. 1988. Reconstructed dynamics of rapid extinctions of chaparral-requiring birds in urban habitat islands. Conservation Biology 2:75-92.
Sparling, P. M. 1994. Invasion of *Pseudotsuga menziesii* into chaparral: analysis of spatial and temporal patterns. Thesis. San Francisco State University, San Francisco, California, USA.
Stone, E. C., and G. Juhren. 1951. The effect of fire on the germination of the seed of *Rhus ovata* Wats. American Journal of Botany 38:368-372.
Stratton, L. C. 2005. Santa Catalina Island experimental restoration trials: developing appropriate techniques. Pages 213-227 *in* D. K. Garcelon and C. A. Schwemm, editors. Proceedings of the 6th California Islands Symposium, Ventura, California, USA, December 1-3, 2003. National Park Service Technical Publication CHIS-05-01. Institute for Wildlife Studies, Arcata, California, USA.
Stratton, L. C. 2009. Restoration strategies for overcoming limitations to scrub oak regeneration on Catalina Island. Pages 185-200 *in* C. C. Damiani and D. K. Garcelon, editors. Proceedings of the 7th California Islands Symposium, Oxnard, California, USA, February 5-8, 2003. Institute for Wildlife Studies, Arcata, California, USA.
Stylinski, C. D., and E. B. Allen. 1999. Lack of native species recovery following severe exotic disturbance in southern Californian shrublands. Journal of Applied Ecology 36:544-554.
Swank, S. E., and W. C. Oechel. 1991. Interactions among the effects of herbivory, competition, and resource limitation on chaparral herbs. Ecology 72:104-115.
Syphard, A. D., V. C. Radeloff, J. E. Keeley, T. J. Hawbaker, M. C. Clayton, S. I. Stewart, and R. B. Hammer. 2007. Human influence on California fire regimes. Ecological Applications 17:1388-1402.
Tyler, C. M. 1995. Factors contributing to postfire seedling establishment in chaparral: direct and indirect effects. Journal of Ecology 83:1009-1020.
Tyler, C. M., and C. M. D'Antonio. 1995. The effects of neighbors on the growth and survival of shrub seedlings following fire. Oecologia 102:255-264.
Tyler, C. M., F. W. Davis, and B. E. Mahall. 2008. The relative importance of factors affecting age-specific seedling survival of two co-occurring oak species in southern California. Forest Ecology and Management 255:3063-3074.

Vallejo, V. R., and J. A. Alloza. 2012. Post-fire management in the Mediterranean basin. Israel Journal of Ecology & Evolution 58:251-264.

Vallejo, R., J. Aronson, J. G. Pausas, and J. Cortina. 2006. Restoration of Mediterranean woodlands. Pages 193-207 *in* J. Van Andel and J. Aronson, editors. Restoration ecology, Blackwell Publishing. Malden, Massachusetts, USA.

Valliere, J. M., I. C. Irvine, L. S. Santiago, and E. B. Allen. 2017. High N, dry: experimental nitrogen deposition exacerbates native shrub loss and invasion during extreme drought. Global Change Biology 23:4333–4345.

Van Dyke, E., K. D. Holl, and J. R. Griffin. 2001. Maritime chaparral community transition in the absence of fire. Madroño 48:221-229.

van Mantgem, E. F., J. E. Keeley, and M. Witter. 2015. Faunal responses to fire in chaparral and sage scrub in California, USA. Fire Ecology 11:128-148.

Vander Mijnsbrugge, K., A. Bischoff, and B. Smith. 2010. A question of origin: where and how to collect seed for ecological restoration. Basic and Applied Ecology 11:300-311.

Vasey, M. C., M. E. Loik, and V. T. Parker. 2012. Influence of summer marine fog and low cloud stratus on water relations of evergreen woody shrubs (*Arctostaphylos*: Ericaceae) in the chaparral of central California. Oecologia 170:325-337.

VinZant, K. in prep-a. Guidelines for choosing a restoration plant palette and collecting native plant materials. *In* M. Narog, technical coordinator. Chaparral restoration: a paradigm shift. Proceedings of the Chaparral Workshop, Arcadia, California June17-20, 2013. General Technical Report. USDA Forest Service, Pacific Southwest Research Station, Albany, California, USA.

VinZant, K. in prep-b. Restoration in type converted and heavily disturbed chaparral: lessons learned. In M. Narog, technical coordinator. Chaparral restoration: a paradigm shift. Proceedings of the Chaparral Workshop, Arcadia, California June17-20, 2013. General Technical Report. USDA Forest Service, Pacific Southwest Research Station, Albany, California, USA.

Vourlitis G. L. 2012. Aboveground net primary production response of semi-arid shrublands to chronic experimental dry-season N input. Ecosphere 3:1-9.

Vourlitis, G. L. 2017. Chronic N enrichment and drought alter plant species abundance in a Mediterranean-type semi-arid woodland. Oecologia. 184: 267–277.

Walker L. R., J. Walker, R. J. Hobbs. 2007. Linking restoration and ecological succession. Springer, London, UK.

Wall, M., and J. Macdonald. 2009. Processing seeds of California native plants for conservation, storage, and restoration. Rancho Santa Ana Botanic Garden Occasional Publication, Number 10, Claremont, California, USA.

Westman, W. E. 1976. Vegetation conversion for fire control in Los Angeles. Urban Ecology 2:119-137.

Whisenant, S. G. 1999. Repairing damaged wildlands. Cambridge University Press, Cambridge, UK.

Wiens, J. A., and R. J. Hobbs. 2015. Integrating conservation and restoration in a changing world. BioScience 65:302-312.

Wilkin, K. M. 2009. Mimicking fire for post-mining restoration success at Rocky Canyon Quarry. Thesis. California Polytechnic State University, San Luis Obispo, California, USA.

Wilkin, K., V. L. Holland, D. Keil, and A. Shaffner. 2013. Mimicking fire for successful chaparral restoration. Madroño 60:165-172.

Witmer, G. W., and R. M. Engeman. 2007. Subterranean rodents as pests: The case of the pocket gopher. Pages 287-299 *in* S. Begall, H. Burda, and C. E. Schleich, editors. Subterranean rodents. Springer, Berlin, Germany.

Wohlgemuth, P. M., J. L. Beyers, and K. R. Hubbert. 2009. Rehabilitation strategies after fire: the California, USA experience. Pages 511-536 *in* A. Cerdá and P. R. Robichaud, editors. Fire effects on soils and restoration strategies. Volume 5 of series: land reconstruction and management. Science Publishers, Enfield, New Hampshire, USA.

Wohlgemuth, P. M., J. L. Beyers, and P. R. Robichaud. 2011. The effectiveness of aerial hydromulch as an erosion control treatment in burned chaparral watersheds in southern California. Pages 162-167 *in* C. N. Medley, G. Patterson, and M. J. Parker, editors. Proceedings of the Fourth Interagency Conference on Research in the Watersheds. Fairbanks, Alaska, USA, September

26-30, 2011. US Department of the Interior, Geological Survey, Scientific Investigations Report 2011-5169.
Zammit, C. A., and P. H. Zedler. 1988. The influence of dominant shrubs, fire, and time since fire on soil seed banks in mixed chaparral. Vegetatio 75:175-187.
Zammit, C. A., and P. H. Zedler. 1993. Size structure and seed production in even-aged populations of *Ceanothus greggii* in mixed chaparral. Journal of Ecology 81:499-511.
Zedler, P. H. 1995. Fire frequency in southern California shrublands: Biological effects and management options. Pages 101-112 *in* J. E. Keeley and T. Scott, editors. Brushfires in California wildlands: ecology and resource management. International Association of Wildland Fire, Fairfield, Washington, USA.
Zedler, P. H., C. R. Gautier, and G. S. McMaster. 1983. Vegetation change in response to extreme events: The effect of a short interval between fires in California chaparral and coastal scrub. Ecology 64:809-818.

Chapter 14
Climate Change Trends for Chaparral

Nicole A. Molinari, Emma C. Underwood, John B. Kim, and Hugh D. Safford

Abstract Chaparral vegetation is a dominant and unique feature of California's Mediterranean-type climate. The evergreen shrubs that characterize chaparral are well adapted to long, hot, dry summers and extreme fluctuations in inter-annual precipitation. Despite the ability of chaparral species to tolerate climatic extremes, the integrity of the chaparral ecosystem is currently being challenged by rising temperatures, increased variability in precipitation, and longer and more persistent droughts. Climate scenarios for California project continued warming through the century leading to increased physiological stress, canopy thinning, and mortality of chaparral vegetation across portions of the state. In some instances, however, chaparral vegetation may expand into forested landscapes. Climate change forecasts suggest enhanced fire activity, including an extended fire season and more frequent large fires. In this already stressed system, non-climate stressors, like increased fire frequencies, can lead to decreased shrub biomass, loss of species diversity, and conversion to other vegetation types. Chaparral in southern California is already trending toward conversion to dominance by non-native annual grasses, and climate projections suggest that this trend will continue in the future. In this chapter, we evaluate historical and projected climate trends in California and explain how they might directly and indirectly affect the integrity and persistence of chaparral on the landscape. We show that the interaction of climate and non-climate stressors can drive landscape level conversion of shrublands to non-native annual grasses leading to the loss of social and ecological benefits provided by the ecosystem. We provide

N. A. Molinari (✉)
USDA Forest Service, Los Padres National Forest, Goleta, CA, USA
e-mail: nmolinari@fs.fed.us

E. C. Underwood
University of California, Davis, CA, USA

University of Southampton, UK

J. B. Kim
USDA Forest Service, Pacific Northwest Research Station, Corvallis, OR, USA

H. D. Safford
USDA Forest Service, Pacific Southwest Region, Vallejo, CA, USA

University of California, Davis, CA, USA

E. C. Underwood et al. (eds.), *Valuing Chaparral*, Springer Series on Environmental Management, https://doi.org/10.1007/978-3-319-68303-4_14

a detailed review of projected changes in carbon storage for one of the (under-valued) ecosystem services provided by chaparral. We conclude by highlighting key management lessons from our review, and point to a few high priority information gaps that must be filled by future research.

Keywords Carbon storage · Drought · Dynamic Global Vegetation Model · Fire regimes · MC1 vegetation shifts · Vulnerability

14.1 Introduction

The effect of climate on chaparral vegetation in California is expected to manifest directly through climate exposure and indirectly through interactions between climate and other stressors, like altered fire regimes, non-native species, urban encroachment and nitrogen deposition. Many of the non-climate stressors are considered in other chapters (see Chaps. 12 and 15), so here we focus on the effects of drought, precipitation, and temperature as they relate to fire regime, composition, distribution, and productivity of chaparral ecosystems.

Chaparral vegetation is widespread throughout California and the structure and composition of these shrublands vary with latitude, elevation, and biophysical properties. As a consequence, the exposure and sensitivity of chaparral to direct and indirect climate effects will vary geographically. The first step to assessing potential climate impacts is to identify areas that may be at high risk to climatic change and climate-induced changes in fire regime, since fire is the principal disturbance factor in California's chaparral landscapes (Keeley and Safford 2016). In this chapter, we use recent downscaled climate projections to describe statewide patterns in expected climate change. We then identify ecoregions with a high cover of chaparral shrubland and evaluate future projections in temperature and precipitation by vegetation type. Finally, we summarize lessons for managers and identify key information gaps that need to be filled for proper climate change planning in chaparral ecosystems.

14.2 Climate Change Trends

The distributions of species and vegetation types are sensitive to a variety of climatic changes including the direction, magnitude, seasonality, and the periodicity of temperature and precipitation events (Field et al. 1999). The first step to evaluating the effects of climate on chaparral vegetation is to establish historical and projected trends in precipitation and temperature at the finest scale possible. While the direction of precipitation change across California is uncertain, there is unanimity that warming will continue into the future (Hayhoe et al. 2004; Collins et al. 2013). The effect of projected warming on chaparral ecosystems is most likely to be realized through increased drought conditions. Drought is most simply defined by Mann and Gleick (2015) as the "mismatch between the amounts of water nature provides and the amount of water the environment demands" and is quantified in numerous ways (e.g., Palmer Drought Index, climatic water deficit). Furthermore, it can be represented as seasonal changes (difference in available water between winter and summer) to which chaparral species are well adapted, or long-term trends that may push chaparral species to the edge of their physiological tolerance. Below we discuss three elements of climate change—drought, precipitation, and temperature—using long-term historical and future averages.

14.2.1 Drought and Precipitation

Between 2012 and 2015, California experienced a multi-year record setting drought that included the lowest yearly precipitation and highest temperatures on record (Diffenbaugh et al. 2015). Tree ring data suggest that the 2012–2014 drought was the most severe in the past 1200 years (Griffin and Anchukaitis 2014). While the precipitation deficits beginning in 2012 are not unprecedented in the paleoclimate, when coupled with rising temperatures, the current drought stands out as the most severe since the ninth century A.D. (Griffin and Anchukaitis 2014). Despite there being a higher frequency of low precipitation years in the past few decades, it is unclear whether these precipitation trends will continue into the future (Diffenbaugh et al. 2015; Mann and Gleick 2015).

Historical analyses do not provide a clear picture for generalized trends in precipitation for chaparral ecosystems across California. Over the past century (1900–1939 compared to 1970–2009) average annual precipitation in the state of California has increased by 26 mm (Rapacciuolo et al. 2014). While the state average exhibited an increase in precipitation, smaller scale ecoregional analyses show substantial variability in the direction and magnitude of precipitation. For example, the Southwestern California Region, which encompasses the most extensive cover of chaparral (Holland and Keil 1995), experienced the largest decline in rainfall in the state (19.5 mm between 1900–1939 and 1970–2009) (Rapacciuolo et al. 2014). It is important to note that southern California supports the highest inter-annual variability in precipitation in the United States (Dettinger et al. 2011), with recent trends showing continued increases in variability between years (Sawyer et al. 2014; Molinari et al. 2016). As an extreme example of this tendency, 2 water-years after the historic 2012–2014 drought we experienced one of the wettest years in California history.

Future projections of precipitation are also spatially variable and uncertain. We use three downscaled climate change projections to illustrate the uncertainty in future precipitation across California (Fig. 14.1). We selected projections by three general circulation models (GCM's) under the RCP8.5 climate change scenario, which represents a future with no coordinated greenhouse gas mitigation (Riahi et al. 2011). The three selected projections represent a range of possible future

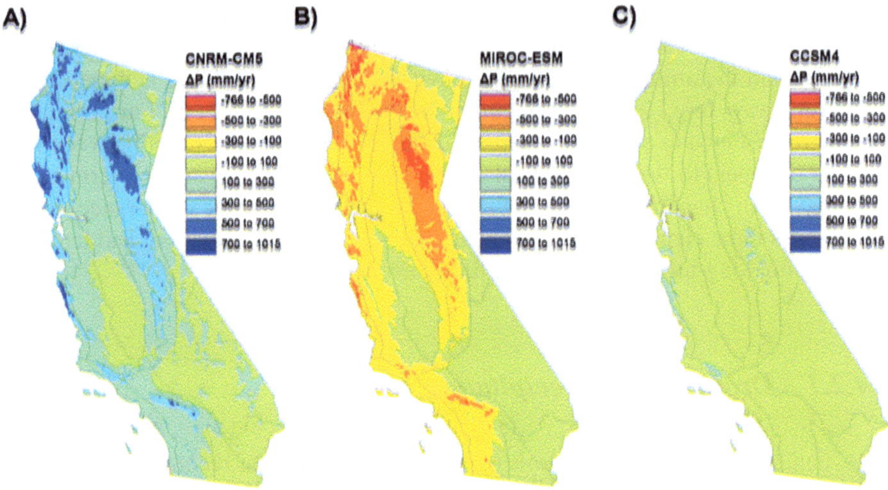

Fig. 14.1 Change in precipitation across the state from current (1951–1980) to end of century (2070–2099). The data reflect a range of climate projections: (**a**) warmer-wetter (CNRM-CM5), (**b**) hotter-drier (MIROC-ESM), and (**c**) an approximation for the ensemble mean (CCSM4). Future precipitation for California were projected using the NASA NEX-DCP30 climate dataset, which comprise 30+ Coupled Model Inter-comparison Project, Phase 5 (CMIP5) GCM outputs downscaled to 30 arc second (~800 m) resolution at monthly time steps (Thrasher et al. 2013)

climates under RCP8.5: CNRM-CM5 (warmer-wetter), MIROC-ESM (hotter-drier), and CCSM4 (closest to the ensemble mean of 18 projections assessed by Flint and Flint 2014). While the amount of change varies markedly between the warmer-wetter (CNRM-CM5) and hotter-drier (MIROC-ESM) projections, the spatial areas that are most impacted by these changes are similar (Fig. 14.1). It is the higher elevation areas of the state which experience the greatest change in precipitation between current and end of century conditions, these include the northern Sierra Nevada, the northern Coast Ranges and Klamath mountains, and the Transverse Ranges in southern California. In CNRM-CM5, these areas experience more rainfall compared to historic and in MIROC-ESM they experience less precipitation (mm/year) relative to other areas of the state including the central Valley, the Modoc Plateau in the north, and the Mojave Desert in southeastern California.

This disparity with respect to the direction and magnitude of future precipitation projections reflect earlier climate modelling studies (Hayhoe et al. 2004; Dettinger 2005; Cayan, Maurer, Dettinger, Tyree, & Hayhoe, 2008). Some suggest there will be small changes in overall precipitation (Cayan et al. 2008), while Hayhoe et al. (2004) project a decrease in precipitation at the end of the century under the highest emission scenarios. Seasonality of rainfall is likely to remain relatively constant (Cayan et al. 2008), however rarer, more intense rainfall events have already become more common across the US and are expected to increase over the next century (Melillo et al. 2014).

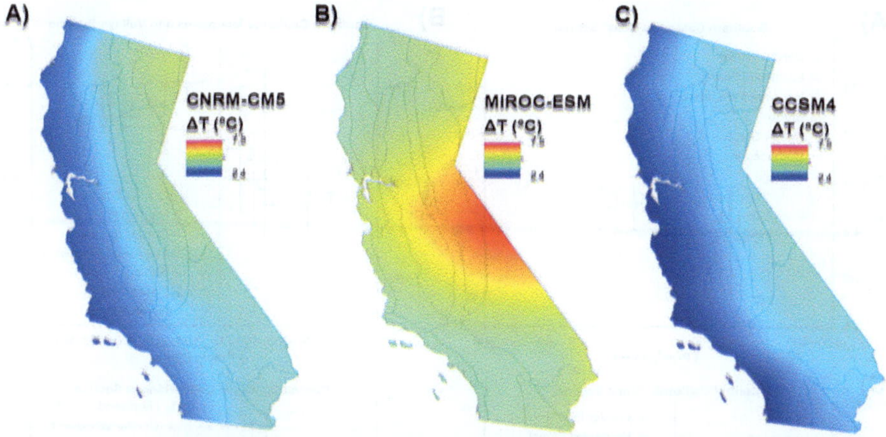

Fig. 14.2 Change in temperature across the state from current (1951–1980) to end of century (2070–2099). The data reflect a range of CMIP5 climate projections: (**a**) warmer, wetter (CNRM-CM5), (**b**) hotter drier (MIROC-ESM), and (**c**) an approximation for the ensemble mean (CCSM4). Data are downscaled from the NASA NEX-DCP30 climate dataset using the RCP8.5 climate change scenario

14.2.2 Temperature

Unlike precipitation trends, there is a broad consensus that temperatures in California have increased and will continue on this trajectory into the future. The historical climate analysis conducted by Rapacciuolo et al. (2014) shows an overall increase in mean annual temperature of 0.45 °C across California between historical (1900–1939) and modern (1970–2009) times, with southern parts of the state nearly doubling in comparison to the state average (Southwestern California Region = 0.70 °C and Sonoran Desert Region = 0.95 °C).

Future projections indicate that warming trends will continue across a broad range of general circulation models and emission scenarios and the greatest increase in temperature will likely occur in the summer (Hayhoe et al. 2004). As with precipitation, we used the NASA NEX-DCP30 climate data and focused on the same three GCM's under the RCP8.5 climate change scenario: warmer-wetter CNRM-CM5 (warmer-wetter), MIROC-ESM (hotter-drier), and CCSM4 (closest to the ensemble mean) and used them to project changes in temperature across California (Fig. 14.2). The three climate projections show broadly similar spatial patterns of temperature change between current and end of century. The warmer-wetter and ensemble-mean projections (CNRM-CM5 and CCSM4) show the smallest changes in temperature along the west coast that gradually increase inland towards the Sierra Nevada, with the greatest changes occurring in the eastern, continental part of the state (Fig. 14.2a, c). The hotter-drier projection (MIROC-ESM) also shows eastern parts of the state warming more, but with a strong center of

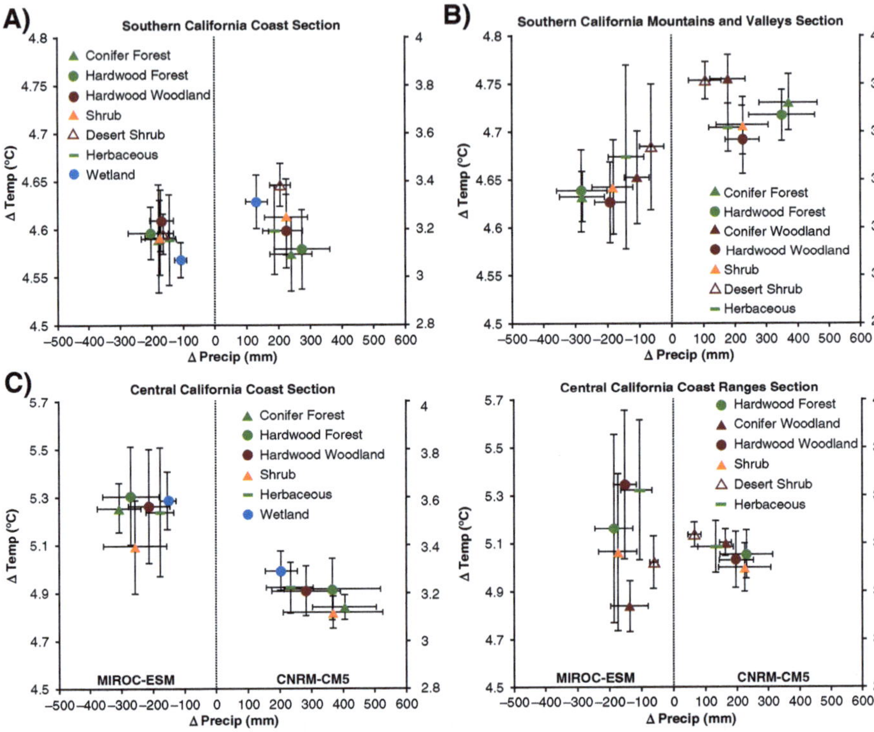

Fig. 14.3 Change in precipitation and temperature for different vegetation types by Bailey Ecoregion sections where shrubland cover exceeds 20%. The hotter-drier projection (MIROC-ESM) is on the left side of the graphs and the warmer-wetter projection (CNRM-CM5) on the right. Vegetation data are based on FRAP (2015)

extreme warming focused in the central and southern Sierra Nevada and Basin and Range Province (Fig. 14.2b).

14.2.3 Understanding Climate Change Trends for Different Vegetation Types

Given the spatial variability associated with climate projections, it is challenging to anticipate how these state-wide trends will affect vegetation at finer spatial scales. To evaluate future climate exposure for various vegetation types across distinct geographic areas, we summarized the change in end of century (2070–2099) temperature and precipitation projections for seven vegetation types as defined by the

California Wildlife Habitat Relations program (FRAP 2015): shrub, conifer forest, hardwood forest, hardwood woodland, desert scrub, herbaceous, and wetland (Fig. 14.3). The analysis was conducted using two climate change projections included in the NASA NEX-DCP30 dataset: warmer-wetter (CNRM-CM5) and hotter-drier (MIROC-ESM) projections as described above (RCP8.5). These two climate projections were summarized by vegetation type and Bailey's Ecoregion sections (Miles and Goudey 1997). Only ecoregion sections with greater than 20% cover of shrubland vegetation were included in the analysis: the Central California Coast, Southern California Coast, Central California Coast Ranges, and Southern California Mountains and Valleys. This analysis provides the opportunity to consider climate effects in more detail and develop hypothesizes about the future status of chaparral across these diverse geographic areas.

Shrublands across the four ecoregions included in the analysis vary in their exposure to changing climate. Independent of the climate change projection, shrublands in the Southern California Coast and Southern California Mountains and Valleys ecoregions are projected to be no more or less exposed to changes in temperature and precipitation than other vegetation types (Fig. 14.3a, b). In contrast, forests (conifer and hardwood) are projected to experience the most pronounced changes in precipitation in the Southern California Mountains and Valleys Ecoregion (Fig. 14.3b). In the Central California Coast Ecoregion, shrublands are projected to experience the smallest change in temperature relative to other vegetation types within the ecoregion, indicating that they may be less vulnerable to climate-induced stress (Fig. 14.3c).

Under the **hotter-drier climate** projection, forested vegetation types in the Southern California Mountain and Valleys and Central California Coast ecoregions are projected to experience a pronounced decline in precipitation relative to other vegetation types. These forests could experience heightened mortality from physiological stress, bark beetle attack, or increased fire severity (Allen et al. 2010; van Mantgem et al. 2013) that could result in conversion to more drought and fire tolerant neighboring vegetation types like montane or lower elevation chaparral. This climate projection also predicts shrublands in the Central California Coast Ecoregion will experience a large decline in precipitation and increase in temperature compared with shrublands in other ecoregions, resulting in higher vulnerability to climate effects and potential conversion to more xeric vegetation types like grassland. Under the **warmer-wetter climate** projection, shrublands in the Central California Coast Ecoregion are also expected to experience the greatest increases in precipitation change relative to shrublands in other ecoregions, making them potentially susceptible to invasion by mesic tree species, while the Southern California Mountains and Valleys are projected to experience great increases in temperature at the end of the century.

14.3 Effects of Changing Climate on Fire Regime

14.3.1 Overview of the Climate-Fire Relationship

Climatic variability affects fire at a variety of temporal scales (Restaino and Safford 2018). In the short-term, climate influences fine fuel moisture, ignition frequencies, and fire spread rates. In the medium term (annual to interannual), climate influences the abundance of fuels as well as the continuity of fine fuels and the moisture content of coarser fuels. Over the long-term (decadal to centennial) climate has a major influence on the species pool that persists in a given location. Interactions between these species and their physical characteristics and more direct influences of climate on fire lead to the distinctive fire regime and vegetation structure that are characteristic of a given place and time on a landscape, and changes in these variables and their interactions can have major ecological implications.

A fundamental trade-off exists between climate effects on the spatial structure of vegetation versus fuel moisture, and conceptually this has been represented as a continuum. Vegetation in more productive regions (e.g., maritime forests) are often considered "climate-limited" and are characterized as having high fuel continuity and sufficient biomass capable of carrying fire, but flammability is usually limited by high fuel moisture (Krawchuk and Moritz 2011; Pausas and Paula 2012; Steel et al. 2015). At the opposite end of the continuum are "fuel-limited" ecosystems (e.g., deserts, dry pine and oak forests) that are sufficiently flammable during the fire season, yet relatively slow fuel accumulation (relative to the frequency of potential fire starts) inhibits ignition and fire spread (Krawchuk et al. 2009). Chaparral has been considered intermediate to these extremes, as fuels are perennially abundant and fuel moisture typically declines to combustible levels with the summer drought. Given these characteristics, chaparral landscapes within southern California are considered "ignition-limited" (Steel et al. 2015; Keeley and Syphard 2016) and therefore may be less affected by climate-induced changes in fire and more sensitive to anthropogenic factors that affect ignitions (Keeley and Syphard 2015, 2016, 2017).

While climate can contribute to fire regimes (described below), it is important to recognize that non-climatic factors, such as fauna, non-native plant species, and socio-economic factors, also contribute to deviations in fire regime within shrublands (Pausas and Keeley 2014a). In southern California, contemporary fires are largely caused by human ignitions rather than lightning events, with the probability of lightning ignition increasing to the east and with elevation. In fact, Syphard et al. (2007) found that human-related variables like population density and distance to developed communities are the best predictors of fire frequency in southern California (see Chap. 12). In particular, exurban or intermediate density development is linked to the highest level of ignitions (Syphard et al. 2007). Areas that are especially vulnerable to this type of development in southern California are the wildlife-urban interface (WUI) zones, transition zones where human communities and structure meet or mix with undeveloped wildland at risk of wildfire. Furthermore, the spread of non-native herbaceous vegetation around communities adjacent to the WUI exacerbates fire risk due to the presence of readily available fine fuels. With its dense (637 people per mi^2 [260 ha]) and growing (11% between 1990 and 2000) human population (Underwood et al. 2009), southern California is at particular risk for increasing fire frequencies at lower elevation sites.

14.3.2 Chaparral and the Climate-Fire Relationship

Fire is an important structuring force in chaparral ecosystems and perturbations to the fire regime, especially those that enhance fire frequency, can alter the composition and abundance of native chaparral species (see Chap. 2). Climate-driven increases in fire activity is already apparent in the western United States (Littell et al. 2009; Dennison et al. 2014; Abatzoglou and Williams 2016; Westerling 2016) and future projections suggest fire activity will persist and possibly accelerate in the future due to increased growth of fuels under higher CO_2 (and in some cases increased precipitation), decreased fuel moisture from warmer temperatures, and possibly increased thundercell activity (Price and Rind 1994; Lenihan et al. 2008; Westerling and Bryant 2008). These climate-fire analyses, however, encompass broad geographies that incorporate a diversity of climate and vegetation types, each with a unique relationship with fire (Littell et al. 2009; Keeley and Syphard 2016). Therefore, we provide below a more focused analysis to uncover the subtleties of the climate-fire relationship and the role it plays within chaparral ecosystems.

Unlike forested ecosystems in the Sierra Nevada, where burned area is partially related to spring and summer temperatures, area burned in lower elevation non-forested landscapes is largely unresponsive to seasonal changes in temperature. In part, this may be reflective of shrublands, especially in the South Coast Ecoregion, already exceeding a climate threshold that is conducive to large fires and the previously described importance of ignition limitations within this ecosystem (Keeley and Syphard 2017). Fire activity in southern California shrublands is not entirely immune to climate-fire interactions however, and there is evidence pointing towards antecedent (both seasonal and prior year) rainfall amounts influencing fire activity (Davis and Michaelsen 1995; Littell et al. 2009; Dennison and Moritz 2009; Keeley and Syphard 2017). Using historical rainfall and burn area data, Keeley and Syphard (2017) highlight the positive relationship between prior year rainfall and burned area in chaparral dominated ecoregions (Central Coast and South Coast). They note that this relationship has become more pronounced over the past 50 years, likely coinciding with the rise in abundance of non-native annual grasses within shrublands. Annual grass production is sensitive to precipitation, with years of high rainfall having the potential to generate a dense and continuous flashy fuel bed that can contribute to increased fire activity in the subsequent year.

Future climate-fire projections for southern and central California align with historical analyses (see Keeley and Syphard 2017) and predict reductions in fire frequency in low-elevation shrublands, presumably driven by a dampening of fine fuel production under future drier climates (Westerling and Bryant 2008). Lenihan et al. (2008) used the MC1 dynamic vegetation model to forecast the effects of climate on vegetation distribution, ecosystem productivity, and fire across California. They found that the future expansion of grasslands enhanced the rate of fire spread leading to more area burned. This was especially true in the Central California Coast and South Coast ecoregions where increased fire activity further promoted grassland vegetation at the expense of woody life-forms. The effect of future climate projections on productivity strongly influenced fire regime such that fire frequency was enhanced under the cooler and less dry climate projection (PCM-A2) due to an increase in fine flammable grasses. Together, these studies highlight the interaction between precipitation and non-native annual grass production as a crux of the climate-fire relationship within chaparral ecosystems in southern California. The effects of climate on fire regime may be most pronounced in degraded chaparral landscapes with a high cover of non-native annuals (e.g., "chaparral savanna") that can rapidly respond to inter-annual changes in moisture availability.

Over longer time scales, drought has been associated with mega-fires in southern California (Keeley and Zedler 2009). Long duration, extreme drought can result in branch dieback and in more extreme cases mortality of chaparral shrubs (Keeley 1975; Parsons et al. 1981; Schlesinger et al. 1982; Mahall and Wilson 1986; Davis et al. 2002; Paddock et al. 2013). Keeley and Zedler (2009) propose that the low moisture content associated with dead material can increase the spread of fire by enhancing the likelihood of spot fire ignitions. Furthermore, the accumulation of dead vegetation associated with extended drought will leave a legacy of dead fuels on the landscape, potentially making chaparral more fire prone across a diversity of climate years.

14.3.3 Climate and Seasonality of Fire

Climate warming and changes in precipitation patterns can also impact the seasonality and duration of the wildfire season. In forests across the western USA, the length of the wildfire season has grown in relation to fuel aridity associated with spring and summer temperatures and timing of snowmelt (Westerling et al. 2006; Westerling 2016). In southern California, there are historically two fire seasons, a spring-summer (May to September) and fall (October to December) season, the latter often characterized by extreme "Santa Ana" wind events (see Chap. 15). Brown et al. (2012) predict that drier spring and warmer temperatures will elevate fire danger in April and May leading to an increase in fires during this historically low season. A study in Los Angeles County, California found reduced springtime precipitation drives live fuel moisture to critically low levels that coincide with the largest conflagrations (Dennison and Moritz 2009). Therefore, future climatic conditions that advance the onset of springtime drought will result in an extension of the burn season and influence the size of fires within chaparral ecosystems. The expansion of the fire season is particularly concerning since early season fire coincides with peak growth and reduced carbohydrate storage for resprouting species (Jones and Laude 1960; Mooney and Chu 1974). Changes in climate that promote an early onset to the fire season may enhance postfire mortality, particularly for species with a resprouting life history that rely on carbohydrate reserves for survival (Pratt et al. 2012). Therefore, a higher occurrence of early season fire may result in compositional changes that shift dominance away from resprouting species and toward functional types that rely on regeneration from seeds after fire (Beyers and Wakeman 2000).

Changes in the magnitude and frequency of extreme events that drive fire weather may also affect chaparral communities. In southern California, fire is often fanned and spread by strong winds which originate in the interior and blow downslope towards the coast generally in the fall (known as Santa Ana winds). As the air is forced to descend it compresses and warms with a corresponding decrease in humidity. Weather conditions like these Santa Ana winds cause spikes in daily temperature and are important predictors of fire occurrence, size, and rate of spread, especially in years with low spring precipitation (Davis and Michaelsen 1995). The largest, most damaging fires in southern California occur during the fall Santa Ana wind season (Keeley and Fotheringham 2001; Jin et al. 2014), yet there is uncertainty surrounding the effects of climate on the seasonality and frequency of Santa Ana wind events. On the one hand, the projected increase in minimum winter temperatures in the desert could cause a decrease in the temperature gradient between the desert and the ocean, leading to a reduction in the duration and frequency of Santa Ana events (Hughes et al. 2011). Also the seasonality of the peak in dry offshore winds is predicted to shift from September–October to November–December (Miller and Schlegel 2006). With this projection, by mid-century, Santa Ana wind events are expected to decrease in frequency (Miller and Schlegel 2006; Hughes et al. 2011; Jin et al. 2015). In contrast, reductions in relative humidity with higher temperature and increased wind speed may exacerbate fire conditions under Santa Ana wind events in the future (Hughes et al. 2011; Jin et al. 2015). Under these assumptions, Jin et al. (2015) project that area burned in Santa Ana wind-driven fires will increase by more than 60% by the 2040–2060 period (and that the area of fuel-driven burning will increase by 77%).

14.4 Understanding the Effects of Changing Climate on Chaparral Vegetation

Vegetation patterns on the landscape are dynamic. In southern California, climate impacts may manifest as broad scale state changes where shrublands transition to other vegetation types (e.g., grassland or woodland) or they may be finer scale, more subtle alterations in vegetation structure and community composition. There is a diversity of climate and non-climate factors that drive vegetation change. In chaparral shrublands, these factors include grazing, fire, non-native species, land use change, nitrogen deposition and, of course, climate (Haidinger and Keeley 1993; Callaway and Davis 1998; Tyler et al. 2007). It is important to note that these factors do not operate independently, making it challenging to untangle the effects of climate on chaparral vegetation.

The vulnerability of chaparral to global change factors, including climate, is a function of exposure, sensitivity of the species to perturbations, and the ability for species to adapt (IPCC 2007). There have been several approaches to understanding the vulnerability of California ecosystems to changes in climate and related factors (Chornesky et al. 2015), and chaparral-specific assessments have been completed through collaborative efforts that couple expert opinion with literature reviews. One such project led by the USDA Forest Service, the California Landscape Conservation Cooperative (LLC), and EcoAdapt identified factors underlying chaparral vulnerability, outlined adaptation opportunities to offset the effect of a changing climate, and identified research gaps (see Box 14.1).

14.4.1 Spatial Vegetation Modeling: Landscape Scale Patterns

Over the past 20 years there have been multiple modeling approaches that evaluate vegetation shifts using various climate scenarios. One of the most comprehensive techniques is the use of MC1 Dynamic Global Vegetation Model (DGVM), which simulates vegetation biogeography, biogeochemistry, and fire dynamics on a gridded landscape (Daly et al. 2000; Bachelet 2001). MC1-DGVM is a dynamic model (as opposed to an equilibrium model, which simulates conditions in a stable climate): MC1 runs on a monthly time step, and is driven by monthly climate data from 1895 to 2100. MC1 produces annual maps of plant functional types (e.g., grassland, shrubland, conifer forest) and MC1 simulations of California have been updated periodically as new climate projections based on a suite of general circulation models and emission scenarios are published (see Lenihan et al. 2003; Hayhoe et al. 2004; Lenihan et al. 2008; Shaw et al. 2011).

There are some general patterns that emerge from MC1 DGVM for California shrublands. First, the overall area occupied by shrublands across the state declines in all climate change projections considered (Lenihan et al. 2003; Hayhoe et al. 2004; Lenihan et al. 2008), except in Shaw et al. (2011), which we discuss below. Lenihan et al. (2008) project a −30% to −55% change in shrubland coverage at the end of the century across California. Second, with warmer-drier climates, shrublands are lost to grasslands due to greater fire activity (this process being reinforced by the highly combustible nature of grass fuels), and as the result of grasses being better suited competitively for drier conditions (Lenihan et al. 2003; Hayhoe et al. 2004). With a warmer-drier climate, grassland coverage is projected by Lenihan et al. (2008) to increase by 70–80% across California by the end of the century. Third, under warmer-wetter conditions, shrubland losses can occur in a number of ways, including replacement by forest types where higher precipitation and fuel

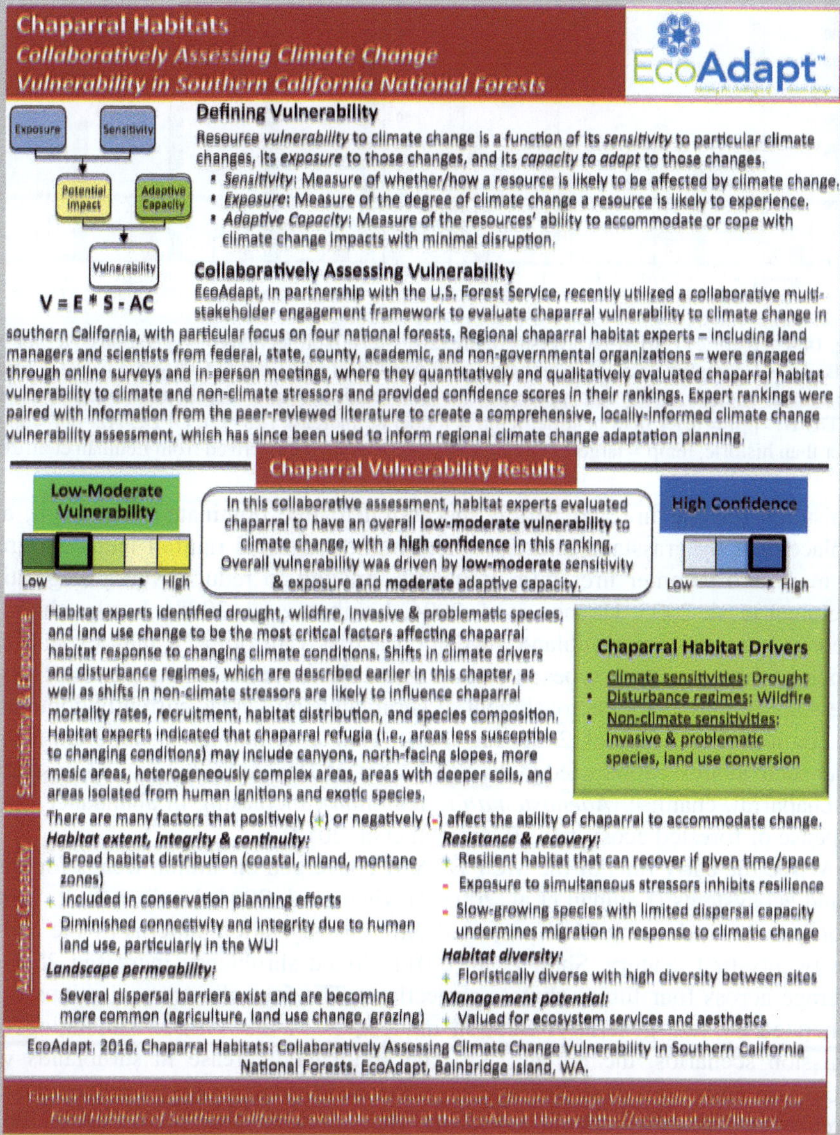

Chaparral Habitats
Collaboratively Assessing Climate Change Vulnerability in Southern California National Forests

EcoAdapt

V = E * S - AC

Defining Vulnerability

Resource *vulnerability* to climate change is a function of its *sensitivity* to particular climate changes, its *exposure* to those changes, and its *capacity to adapt* to those changes.

- *Sensitivity*: Measure of whether/how a resource is likely to be affected by climate change.
- *Exposure*: Measure of the degree of climate change a resource is likely to experience.
- *Adaptive Capacity*: Measure of the resources' ability to accommodate or cope with climate change impacts with minimal disruption.

Collaboratively Assessing Vulnerability

EcoAdapt, in partnership with the U.S. Forest Service, recently utilized a collaborative multi-stakeholder engagement framework to evaluate chaparral vulnerability to climate change in southern California, with particular focus on four national forests. Regional chaparral habitat experts – including land managers and scientists from federal, state, county, academic, and non-governmental organizations – were engaged through online surveys and in-person meetings, where they quantitatively and qualitatively evaluated chaparral habitat vulnerability to climate and non-climate stressors and provided confidence scores in their rankings. Expert rankings were paired with information from the peer-reviewed literature to create a comprehensive, locally-informed climate change vulnerability assessment, which has since been used to inform regional climate change adaptation planning.

Chaparral Vulnerability Results

Low-Moderate Vulnerability
Low → High

In this collaborative assessment, habitat experts evaluated chaparral to have an overall **low-moderate vulnerability** to climate change, with a **high confidence** in this ranking. Overall vulnerability was driven by **low-moderate** sensitivity & exposure and **moderate** adaptive capacity.

High Confidence
Low → High

Sensitivity & Exposure

Habitat experts identified drought, wildfire, invasive & problematic species, and land use change to be the most critical factors affecting chaparral habitat response to changing climate conditions. Shifts in climate drivers and disturbance regimes, which are described earlier in this chapter, as well as shifts in non-climate stressors are likely to influence chaparral mortality rates, recruitment, habitat distribution, and species composition. Habitat experts indicated that chaparral refugia (i.e., areas less susceptible to changing conditions) may include canyons, north-facing slopes, more mesic areas, heterogeneously complex areas, areas with deeper soils, and areas isolated from human ignitions and exotic species.

Chaparral Habitat Drivers

- Climate sensitivities: Drought
- Disturbance regimes: Wildfire
- Non-climate sensitivities: Invasive & problematic species, land use conversion

Adaptive Capacity

There are many factors that positively (+) or negatively (-) affect the ability of chaparral to accommodate change.

Habitat extent, integrity & continuity:

\+ Broad habitat distribution (coastal, inland, montane zones)
\+ Included in conservation planning efforts
\- Diminished connectivity and integrity due to human land use, particularly in the WUI

Landscape permeability:

\- Several dispersal barriers exist and are becoming more common (agriculture, land use change, grazing)

Resistance & recovery:

\+ Resilient habitat that can recover if given time/space
\- Exposure to simultaneous stressors inhibits resilience
\- Slow-growing species with limited dispersal capacity undermines migration in response to climatic change

Habitat diversity:

\+ Floristically diverse with high diversity between sites

Management potential:

\+ Valued for ecosystem services and aesthetics

EcoAdapt. 2016. Chaparral Habitats: Collaboratively Assessing Climate Change Vulnerability in Southern California National Forests. EcoAdapt, Bainbridge Island, WA.

Further information and citations can be found in the source report, *Climate Change Vulnerability Assessment for Focal Habitats of Southern California*, available online at the EcoAdapt Library: http://ecoadapt.org/library.

Box 14.1 Collaborative approach to evaluate chaparral vulnerability to climate change in southern California. Assessment lead by EcoAdapt non-profit organization

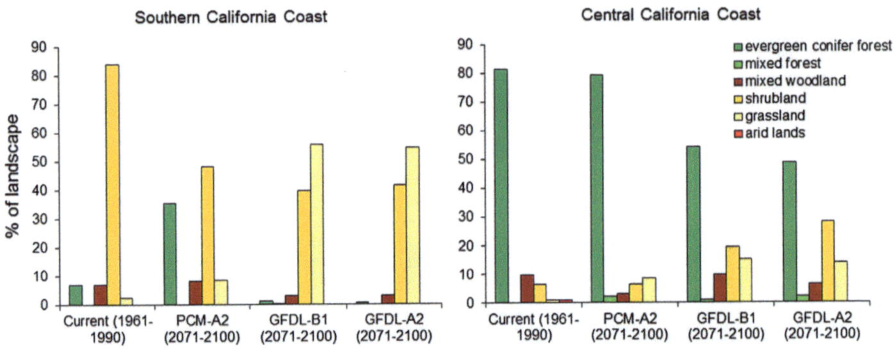

Fig. 14.4 Historic (1961–1990) versus future (2071–2100) projections of vegetation cover for two of Bailey's Ecoregion sections: Southern California Coast and Central California Coast. Climate projections include: PCM-A2: ppt = similar to historic, temp = moderate increase (<3 °C), GFDL-B1: ppt = drier than historic, temp = moderate increase (<3 °C), GFDL-A2: ppt = much drier than historic, temp = largest increase (>4 °C). Original data derived from Lenihan et al. 2008

moistures impede fire and permit succession to tree dominated vegetation, and replacement by grassland in areas where higher seasonal rainfall increases grass biomass and summer fires burn frequently enough to reduce woody vegetation (Lenihan et al. 2003; Hayhoe et al. 2004; Lenihan et al. 2008). Lastly, although there is a net loss in shrubland vegetation across California, it may expand and replace other vegetation types in certain areas. For instance, drought and increased fire activity may favor encroachment of shrublands into areas currently dominated by conifers (Lenihan et al. 2008, Fig. 14.4). This latter finding is supported from a finer scale analysis conducted in the San Francisco Bay Area showing a key species of chaparral, chamise (*Adenostoma fasciculatum*), increasing in abundance at the expense of forested ecosystems (Cornwell et al. 2012).

While the early runs of the MC1 DGVM converged on similar trends for shrubland ecosystems (Lenihan et al. 2003; Hayhoe et al. 2004; Lenihan et al. 2008), more recent work using MC1 projected different vegetation patterns by the end of the twenty-first century. Shaw et al. (2011) found shrublands increased 25% on average across four future climate projections. The four climate projections were generated by two general circulation models, PCM and GFDL, simulating two emission scenarios, the B1 and A2. The simulated increase in shrublands was accompanied by an overall decrease in grassland vegetation. They attribute the differences between their results and earlier MC1 projections to more accurate representations of land cover types, enhanced shrub survival from more efficient water use associated with enhanced CO_2, and the use of an "advanced fire model" (Shaw et al. 2011). According to Shaw et al. (2011) their fire model simulated frequent low intensity fire that was ineffective at killing woody lifeforms (Shaw et al. 2011), however chaparral burns nearly exclusively in stand-replacing crown fires (Keeley and Fotheringham 2001) and therefore this model may artificially enhance the persistence of shrubland ecosystems into the future.

While MC1 outputs are useful at informing large scale patterns in vegetation change, the lumping of various shrubland types (e.g., montane chaparral, mixed chaparral, coastal sage scrub, sagebrush steppe) into a single shrubland plant functional type may mask species-specific vegetation dynamics, such as dispersal and regeneration mechanisms. Indeed, even though MC1 DGVM simulates vegetation response to a constantly changing climate, it does not explicitly simulate seed dispersal and germination, resilience to fire, and regeneration dynamics, which may be key processes that govern chaparral loss and expansion at fine spatial scales. Another related limitation of MC1 is the relatively coarse spatial resolution of the simulations. MC1 simulations are performed typically at 800 m to 10 km grid resolutions, as dictated by the availability of climate change data. Simulating shrubland cover across large geographic areas, like the state of California at those scales, may mask fine spatial scale patterns. To show geographic variation more quantitatively, we dissected the Lenihan et al. (2008) statewide MC1 outputs by Bailey's Ecoregion sections, to help bring clarity to differences in projected end of the century shrubland cover at finer geographic scales. Our analysis showed that shrubland cover along the South Coast Ecoregion tracks statewide trends and shows a decline in shrublands at the end of the century (Fig. 14.4). In contrast, under the warmest-driest climate projections (GFDL-B1 and GFDL-A2) the Central Coast Ecoregion exhibits shrubland expansion, mostly at the expense of conifer forest (Fig. 14.4).

14.4.2 Finer Scale Analyses: Movement of Communities and Drought Resistance of Populations

Mediterranean-type climate regions exhibit high levels of plant richness and endemism (Cowling et al. 1996; Underwood et al. 2009), yet they are also among those most sensitive to climate and land use change (Sala et al. 2000; Hoekstra et al. 2005). The fragmented landscapes of California, and southern California in particular, will limit the ability of chaparral species to shift their ranges in response to climate change (Underwood et al. 2009; Lawson et al. 2010; Riordan and Rundel 2014) and therefore reduce their adaptive capacity (see Box 14.1). When species dispersal is unrestricted, the distribution of vegetation in California is expected to move upslope and poleward, and in some cases toward the coast in response to climate change (Lenihan et al. 2003; Hayhoe et al. 2004; Lenihan et al. 2008; Loarie et al. 2008; Riordan and Rundel 2014). While such long-term shifts are difficult to observe, observations of short-term shifts in response to drought and climate variability can offer insight into the trends we are likely to see under warmer and drier climates. Kelly and Goulden (2008) compared a 2006–2007 plant survey to a 1977 survey along a 2314 m (7592 ft) gradient transect through desert scrub, pinyon-juniper woodland, chaparral shrubland, and conifer forest in the Santa Rosa Mountains in southern California. They found the average elevation of the dominant 10 plant species rose approximately 65 m (213 ft) between historic and modern surveys. Dominant chaparral species sugar bush (*Rhus ovata*), *perplexing ceanothus* (*Ceanothus perplexans*), and Muller's oak (*Quercus cornelius-mulleri*'s) moved upslope by an average of 56 m (184 ft). Fellows and Goulden (2012) measured the distribution of live and dead trees in the San Jacinto Mountains of southern California to further explore these patterns. They recorded a 37 ± 33 m (121 ± 108 ft) upslope shift in mid-montane species' distributions following the 2002–2004 prolonged drought period. Both studies attribute this upward 'lean' to small scale climate variability (Kelly and Goulden 2008; Fellows and Goulden 2012), although other authors suggest that for at least some of the observed species, the shift may be better explained by self-thinning associated with different disturbance histories (Schwilk and Keeley 2012). Given that warming in the southwestern United States is expected to intensify evapotranspiration and drought (Fellows and Goulden 2012), it is likely that the montane vegetation shifts observed in these studies will be observed on larger spatial and temporal scales in the coming years.

Making predictions about climate-induced shifts in community structure within chaparral requires an understanding of the sensitivity of chaparral species to climate extremes. Ecophysiological studies have provided mechanistic evidence for the ability of chaparral species to cope with climatic extremes and aid in better understanding the thresholds that lead to mortality. Physiological capacity and morphology help define the climatic conditions and thresholds that lead to the decline of some shrub species and rise in abundance of others. Some mature chaparral species are able to tolerate extreme levels of dehydration (measured as water stress induced cavitation, Ψ_{50}) and this tolerance can exceed that of neighboring communities, like coastal sage scrub and Mojave Desert scrub (Kolb and Davis 1994; Jacobsen et al. 2007a).

Within chaparral, however, there is a diversity of adaptations for mature shrubs to deal with drought stress, including potential trade-offs between dehydration avoidance (e.g., high root:shoot, stomatal closure, growth in mesic sites) and dehydration tolerance (e.g., physiological adaptations such as resistance to water stress induced cavitation) (see Chap. 1) and these trade-offs have been linked to postfire functional type (e.g., "resprouters": fire tolerant species with the capacity to resprout following fire versus "*obligate seeders*": species that are killed by fire and only recruit from seedbanks in the soil) (Pausas et al. 2016). Despite mature obligate seeding species possessing superior physiological adaptations for tolerating drought stress (Jacobsen et al. 2007b; Pratt et al. 2007), resprouting species with extensive root systems may be better equipped for avoiding extreme drought conditions when compared to non-resprouting species (Parsons et al. 1981; Paddock et al. 2013). Old-growth stands of *Adenostoma fasciculatum*, a facultative seeder, can experience seasonal evapotranspiration that exceeds precipitation, suggesting that deeply rooted resprouters can avoid drought by accessing more perennial water sources (Luo et al. 2007). In general, resprouting functional types may be favored under more intense drought conditions that manifest with warmer-drier climate futures at the expense of non-resprouting functional types (Paddock et al. 2013; Pausas et al. 2016). Severe drought can result in branch dieback of mature individuals and in some instances death (Schlesinger et al. 1982; Mahall and Wilson 1986; Davis et al.

2002; Paddock et al. 2013). In turn, this dieback of mature chaparral will result in canopy gaps that expose the soil to increased solar radiation. Under such conditions, soil heating may result in "novel germination", high seedling mortality, and depletion of the seedbank for species that usually require fire cues for germination (Stephen Davis, pers. comm.).

As drought and fire become more prevalent in the western United States (Westerling and Bryant 2008), it is important to consider the response of chaparral vegetation to these interacting stressors. Following fire, survival of resprouts and seedlings may be particularly vulnerable to drought conditions. For instance, resprouted stems can be more susceptible to water stress than undisturbed chaparral individuals (Saruwatari and Davis 1989; Ramirez et al. 2012; Pratt et al. 2014). Pratt et al. (2014) found increased mortality of resprouted vegetation in *Adenostoma fasciculatum*, greenbark ceanothus (*Ceanothus spinosus*) and toyon (*Heteromeles arbutifolia*) under drought conditions relative to resprouts in average rainfall years. When compared to prefire shrub density, fire and drought together led to population declines in all resprouting species at a site in the Santa Monica Mountains, California, yet hoaryleaf ceanothus (*Ceanothus crassifolius*), the only obligate seeder included in the study, increased in density (Pratt et al. 2014). Other studies have also shown promise for non-resprouters to withstand drought conditions following fire (Frazer and Davis 1988; Thomas and Davis 1989; Pratt et al. 2008) and their dehydration tolerance is hypothesized to be an adaptation for recruiting into dry and open environments following fire (Pausas and Keeley 2014b). In addition, the distribution of non-resprouters on xeric microsites suggest they are able to recruit and tolerate these conditions (Schlesinger et al. 1982; Frazer and Davis 1988), which may result in an expansion of this postfire functional type with warmer-drier climate projections. Obligate resprouters are most abundant in mesic microsites (Meentemeyer et al. 2001; Cornwell et al. 2012) and climate sensitivity of vegetation on north-facing, low insolation slopes is high (Ackerly et al. 2015). Therefore obligate resprouters may retract in their distribution with increased drought, but could expand with warmer-wetter climates (Cornwell et al. 2012). It is important to note, however, that climate stressors are not acting independently of other global change factors, and obligate seeding species are the most susceptible to short-interval fires, which may challenge their ability to reproduce and expand into drier microsites in the future.

14.5 Consequences of Climate-Induced Changes on Ecosystem Services: Carbon Storage

The turnover of vegetation types (e.g., chaparral to grassland or forest to chaparral), as predicted with changing climate and fire regime in California, can have significant impacts on both conservation and human derived benefits from natural ecosystems (Chan et al. 2006). The effects of climate-induced vegetation change will impact a diverse spectrum of ecosystem services, ranging from cultural services like landscape aesthetics; to provisioning services like water supply; to supporting services like pollination; to regulating services that contribute to climate stabilization through carbon storage. Here we focus on climate change impacts on carbon storage.

Greenhouse gases have increased dramatically in the atmosphere since the industrial revolution, and annual CO_2 emissions have nearly doubled since the 1970s (IPCC 2007). One way that increases in atmospheric CO_2 are offset is through carbon sequestration by photosynthetic organisms, which then store the C in their tissue. Sequestered carbon will remain in plant tissues until abscission, death, or a disturbance like fire. Climate change may alter the capacity of chaparral to store carbon in one of two ways. First, enriched atmospheric CO_2 could increase photosynthesis, thereby promoting more efficient carbon fixation that allows for greater water use efficiency. The relative response of ecosystems to enriched CO_2 is hypothesized to be greatest in drought stressed systems, like chaparral (Mooney et al. 1991). Under more extreme drought conditions however, reductions in photosynthetic area from branch or leaf dieback or degraded physiological function can reduce the capacity for mature old-growth chaparral to sequester carbon (Luo et al. 2007). Second, climate-driven shifts in vegetation type, especially shifts that result in turnover of the dominant life-form (e.g., forest to shrubland or shrubland to grassland) can alter the total amount of carbon stored.

Changes in the above-ground live carbon pool are already evident across California (excluding agriculture and urban lands), with a 69 Tg decrease from 2001 to 2010 (Gonzalez et al. 2015). Wildfire and other disturbances that affect vegetation were the primary drivers for carbon loss during this 9 year period. Wildfire in central and southern California shrublands accounted for one-third of this carbon stock decrease, while vegetation type-conversion in shrublands led to a 15 Tg decline in above-ground carbon (Gonzalez et al. 2015).

Evaluating the effects of large scale vegetation shifts on carbon can be achieved using the MC1 dynamic vegetation model (described above). Shaw et al. (2011) most thoroughly explored the effects of climate change on carbon sequestration and coupled the MC1 monthly carbon outputs with an economic analysis to evaluate the potential impacts of climate change on human-derived benefits. They found that warmer-wetter futures (PCM-B1) will enhance carbon storage across the state with the potential for economic gains as estimated using the social cost of carbon up to $22 billion annually by 2070. In contrast, hotter-drier conditions generated by the highest emission scenario (CCSM3-A2) will diminish carbon storage with an annual economic loss of $62 billion by 2070 (Shaw et al. 2011). Earlier MC1 modeling efforts (Lenihan et al. 2008) also found higher carbon storage at the end of the century for cooler-less dry projections and declines in total ecosystem carbon over the same time period for warmer-drier climates. Losses in carbon were largely attributed to reduction of carbon in live vegetation driven by the decline of woody species and the increase in grasses (Lenihan et al. 2008). Furthermore, type-conversion from native shrubland to non-native grassland has the potential to change type-converted areas from carbon sinks to carbon sources (Bradley et al. 2006).

14.6 Conclusions and Future Directions

Given the spatial complexity and uncertainty of future climate conditions, it is difficult to provide resource managers with guidance for developing climate change adaptation and mitigation strategies (although see Box 14.1). To better direct this effort, we highlight a few primary management considerations as they pertain to chaparral and climate change and then describe research needs and important next steps to evaluate the long- and short-term effects of climate change on chaparral ecosystems.

14.6.1 Management Considerations

- Managers need to focus on preventing wildfire at all costs in chaparral shrublands in California. Despite numerous accounts of chaparral burning too frequently, we know of no area of chaparral that ecologically requires fire at a higher frequency than the current frequency. The warming climate, higher incidence of drought, and increasing pressure from other anthropogenic stressors (e.g., non-native species, nitrogen input) exacerbate the problem and result in situations where increasingly frequent fire is leading to quasi-permanent type-conversions to annual non-native grassland.
- Restoration efforts in burned chaparral should consider the warmer and seasonally drier conditions that will likely characterize most of the chaparral landscapes in the future. Vegetation resilience and restoration success may be enhanced by planning for a "worst case" climate scenario. Selecting species mixes and genotypes based on climatic parameters, as well as on tolerance for fire, will be important considerations for restoration planning. Non-resprouting species tend to be more tolerant of drought conditions, while resprouting species tend to recover more rapidly and reliably after fires that are occurring at relatively high frequencies.
- The warming climate is likely to accelerate seasonal dry down of fuel moisture and over the long-term, warmer and drier conditions and periodic drought may contribute to shrub die-off and the build-up of dead fuel. These patterns will be apparent in both mature and young stands. When considering the use of prescribed fire in chaparral, it will be important to acknowledge the negative interaction between postfire recovery and drought. Furthermore, young chaparral stands that have experienced mortality due to postfire drought may be less effective at dampening fire behavior during fuel-driven fire events.
- Non-native annual grass production will be most responsive to annual and seasonal changes in precipitation. Therefore the development of fuel management plans that incorporate the control of annual flashy fuels is of critical importance to limiting ignitions.

- At lower elevations and on southerly and westerly exposures, it may be difficult in the future to maintain shrub cover where fire frequencies cannot be moderated. In such sites, some thought should be given to novel ecosystems that may be introduced and managed to provide for continued ecosystem services.

14.6.2 Outstanding Research Needs

- The recent severe drought across California has heightened interest in the physiological ability of chaparral species to withstand and regenerate under extreme drought conditions. Continued research in this arena will be valuable to determining the thresholds of drought tolerance, yet this effort needs to be coupled with long-term monitoring of dieback and mortality of various shrub species and age classes. Together these data will inform which species are most vulnerable to drought, and help identify which geographic areas are most susceptible to compositional change and type-conversion. The application of this knowledge includes identification of target species for restoration under warmer-drier climates, protection of climate refugia that may function as important seed sources, and prioritization of lands for restoration activities.
- Despite recent trends towards warmer-drier conditions across much of California, it is important to acknowledge the uncertainty in predicting future precipitation and consider the potential effects of warmer-wetter conditions on chaparral. While dynamic vegetation and fire models have incorporated various climate projections into their outputs, empirical studies have been slow to integrate and evaluate the impacts of warmer-wetter climate conditions on chaparral ecosystems. Recent water enhancement experiments in the Santa Monica Mountains, California, support the idea that increased rainfall will enhance shrub survivorship after fire (Pausas et al. 2016), suggesting a positive influence of warmer-wetter futures on lowland chaparral stands. Yet, in more productive regions, warmer-wetter futures may promote encroachment of mixed evergreen species leading to the loss of shrub dominance (Lenihan et al. 2003). Given these patterns and predictions, studies directed at evaluating the effects of enhanced precipitation across species and geographies will help tease apart the finer-scale effects of warmer-wetter futures across the range of chaparral.
- Drought induced dieback of shrubs will interact with fire intensity, seasonality, and ignitability. The influence of climate and shrub dieback on fire regime is critical to understand as it will influence management decisions, especially as they pertain to prefire planning at the wildland-urban interface. In particular, research that aims to quantify the extent of mortality and its influence on fire behavior will be of utmost importance.
- Restoration of degraded chaparral is appealing to land managers as a means of enhancing biodiversity and maintaining ecosystem services (see Chap. 15). Re-establishing native shrub species under current climate conditions has proven challenging (see Chap. 13) and this may be exacerbated by hotter-drier conditions

in the future. One approach to improving restoration success is to select species that are best suited for future climatic and disturbance conditions. To achieve this, a deeper understanding of the ability of chaparral species to adapt to changing conditions is needed. This includes exploring the genetic variability within species, and comparing physiological and regeneration potential across species.

References

Abatzoglou, J. T., and A. P. Williams. 2016. Impact of anthropogenic climate change on wildfire across western US forests. Proceedings of the National Academy of Sciences 113:11770-11775.

Ackerly, D. A., W. K. Cornwell, S. B. Weiss, L. E. Flint, and A. L. Flint. 2015. A geographic mosaic of climate change impacts on terrestrial vegetation: which areas are most at risk? PLoS ONE 10:e0130629.

Allen, C. D., A. K. Macalady, H. Chenchouni, D. Bachelet, N. McDowell, M. Vennetier, T. Kitzberger, A. Rigling, D. D. Breshears, and E. T. Hogg. 2010. A global overview of drought and heat-induced tree mortality reveals emerging climate change risks for forests. Forest Ecology and Management 259:660-684.

Bachelet, D. 2001. MC1, a dynamic vegetation model for estimating the distribution of vegetation and associated ecosystem fluxes of carbon, nutrients, and water. General Technical Report PNW-GTR-508. Corvallis, Oregon, USA, USDA Forest Service, Pacific Northwest Research Station.

Beyers, J. L., and C. D. Wakeman. 2000. Season of burn effects in southern California chaparral. Pages 45-56 *in* J. E. Keeley, M. Baer-Keeley, and C. J. Fotheringham, editors. Second interface between ecology and land development in California. US Geological Survey Open-File Report 00-62. US Geological Survey, Sacramento, California, USA.

Bradley, B. A., R. Houghton, J. F. Mustard, and S. P. Hamburg. 2006. Invasive grass reduces aboveground carbon stocks in shrublands of the western US. Global Change Biology 12:1815-1822.

Brown, T. J., C. A. Kolden, and J. T. Abatzoglou. 2012. Assessing fuels treatments in southern California National Forests in the context of climate change. Joint Fire Science Program Research Report. http://digitalcommons.unl.edu/jfspresearch/30/

Callaway, R., and F. Davis. 1998. Recruitment of *Quercus agrifolia* in central California: the importance of shrub-dominated patches. Journal of Vegetation Science 9:647-656.

Cayan, D. R., E. P. Maurer, M. D. Dettinger, M. Tyree, and K. Hayhoe. 2008. Climate change scenarios for the California region. Climatic Change 87:21-42.

Chan, K. M., M. R. Shaw, D. R. Cameron, E. C. Underwood, and G. C. Daily. 2006. Conservation planning for ecosystem services. PLoS Biology 4:2138-2152.

Chornesky, E. A., D. D. Ackerly, P. Beier, F. W. Davis, L. E. Flint, J. J. Lawler, P. B. Moyle, M. A. Moritz, M. Scoonover, and K. Byrd. 2015. Adapting California's ecosystems to a changing climate. BioScience 65:247-262.

Collins, M., R. Knutti, J. Arblaster, J. L. Dufresne, T. Fichefet, P. Friedlingstein, X. Gao, W. J. Gutowski, T. Johns, G. Krinner, M. Shongwe, C. Tebaldi, A. J. Weaver, and M. Wehner. 2013. Long-term climate change: projections, commitments and irreversibility. Pages 1029-1136 *in* T. G. Stocker, D. Qin, G. K. Plattner, M. Tignor, S. K. Allen, J. Boschung, A. Nauels, Y. Xia, V. Bex, and P. M. Midgley, editors. Climate change 2013: the physical science basis, Cambridge University Press, Cambridge, UK.

Cornwell, W. K., S. A. Stuart, A. Ramirez, C. R. Dolanc, J. H. Thorne, and D. D. Ackerly. 2012. Climate change impacts on California vegetation: physiology, life history, and ecosystem change. California Energy Commission CEC-500-2002-023. Sacramento, California, USA.

Cowling, R. M., P. W. Rundel, B. B. Lamont, M. K. Arroyo, and M. Arianoutsou. 1996. Plant diversity in Mediterranean-climate regions. Trends in Ecology & Evolution 11:362-366.

Daly, C., D. Bachelet, J. M. Lenihan, R. P. Neilson, W. Parton, and D. Ojima. 2000. Dynamic simulation of tree-grass interactions for global change studies. Ecological Applications 10:449-469.

Davis, S. D., F. W. Ewers, J. S. Sperry, K. A. Portwood, M. C. Crocker, and G. C. Adams. 2002. Shoot dieback during prolonged drought in *Ceanothus* (Rhamnaceae) chaparral of California: a possible case of hydraulic failure. American Journal of Botany 89:820-828.

Davis, F. W., and J. Michaelsen. 1995. Sensitivity of fire regime in chaparral ecosystems to climate change. Pages 435-456 *in* J. M. Moreno and W. C. Oechel, editors. Global change and Mediterranean-type ecosystems. Springer, New York, New York, USA.

Dennison, P. E., S. C. Brewer, J. D. Arnold, and M. A. Moritz. 2014. Large wildfire trends in the western United States, 1984-2011. Geophysical Research Letters 41:2928-2933.

Dennison, P. E., and M. A. Moritz. 2009. Critical live fuel moisture in chaparral ecosystems: a threshold for fire activity and its relationship to antecedent precipitation. International Journal of Wildland Fire 18:1021-1027.

Dettinger, M. D. 2005. From climate-change spaghetti to climate-change distributions for the 21st century California. San Francisco Estuary and Watershed Science 3:Article 4.

Dettinger, M. D., F. M. Ralph, T. Das, P. J. Neiman, and D. R. Cayan. 2011. Atmospheric rivers, floods and the water resources of California. Water 3:445-478.

Diffenbaugh, N. S., D. L. Swain, and D. Touma. 2015. Anthropogenic warming has increased drought risk in California. Proceedings of the National Academy of Sciences 112:3931-3936.

Fellows, A. W., and M. L. Goulden. 2012. Rapid vegetation redistribution in southern California during the early 2000s drought. Journal of Geophysical Research 117:G03025.

Field, C. B., G. C. Daily, F. W. Davis, S. Gaines, P. A. Matson, J. Melack, and N. L. Miller. 1999. Confronting climate change in California: ecological impacts on the golden state. A report of the Union of Concerned Scientists, Cambridge, Massachusetts and the Ecological Society of America, Washington D.C., USA.

FRAP [Fire Resource and Assessment Program], California Department of Forestry and Fire Protection. 2015. FRAP vegetation geodatabase. http://frap.fire.ca.gov/data/frapgisdata-sw-fveg_download

Frazer, J., and S. Davis. 1988. Differential survival of chaparral seedlings during the first summer drought after wildfire. Oecologia 76:215-221.

Gonzalez, P., J. J. Battles, B. M. Collins, T. Robards, and D. S. Saah. 2015. Aboveground live carbon stock changes of California wildland ecosystems, 2001–2010. Forest Ecology and Management 348:68-77.

Griffin, D., and K. J. Anchukaitis. 2014. How unusual is the 2012–2014 California drought? Geophysical Research Letters 41:9017-9023.

Haidinger, T. L., and J. E. Keeley. 1993. Role of high fire frequency in destruction of mixed chaparral. Madroño 40:141-147.

Hayhoe, K., D. Cayan, C. B. Field, P. C. Frumhoff, E. P. Maurer, N. L. Miller, S. C. Moser, S. H. Schneider, K. N. Cahill, and E. E. Cleland. 2004. Emissions pathways, climate change, and impacts on California. Proceedings of the National Academy of Sciences 101:12422-12427.

Hoekstra, J. M., T. M. Boucher, T. H. Ricketts, and C. Roberts. 2005. Confronting a biome crisis: global disparities of habitat loss and protection. Ecology Letters 8:23-29.

Holland, V., and D. J. Keil. 1995. California vegetation. Kendall/Hunt Publishing Company, Dubuque, Iowa, USA.

Hughes, M., A. Hall, and J. Kim. 2011. Human-induced changes in wind, temperature and relative humidity during Santa Ana events. Climatic Change 109:119-132.

IPCC [Intergovernmental Panel on Climate Change]. 2007. Climate change 2007: synthesis report. Contribution of working groups I, II and III to the fourth assessment report of the intergovernmental panel on climate change, Intergovernmental Panel on Climate Change, Geneva, Switzerland.

Jacobsen, A. L., R. B. Pratt, S. D. Davis, and F. W. Ewers. 2007a. Cavitation resistance and seasonal hydraulics differ among three arid Californian plant communities. Plant, Cell & Environment 30:1599-1609.

Jacobsen, A. L., R. B. Pratt, F. W. Ewers, and S. D. Davis. 2007b. Cavitation resistance among 26 chaparral species of southern California. Ecological Monographs 77:99-115.

Jin, Y., M. L. Goulden, N. Faivre, S. Veraverbeke, F. Sun, A. Hall, M. S. Hand, S. Hook, and J. T. Randerson. 2015. Identification of two distinct fire regimes in Southern California: implications for economic impact and future change. Environmental Research Letters 10:094005.

Jin, Y., J. T. Randerson, N. Faivre, S. Capps, A. Hall, and M. L. Goulden. 2014. Contrasting controls on wildland fires in Southern California during periods with and without Santa Ana winds. Journal of Geophysical Research: Biogeosciences 119:432-450.

Jones, M. B., and H. M. Laude. 1960. Relationships between sprouting in chamise and the physiological condition of the plant. Journal of Range Management 13:210-214.

Keeley, J. E. 1975. Longevity of nonsprouting *Ceanothus*. American Midland Naturalist 93:504-507.

Keeley, J. E., and C. Fotheringham. 2001. Historic fire regime in southern California shrublands. Conservation Biology 15:1536-1548.

Keeley, J. E., and H. D. Safford. 2016. Fire as an ecosystem process. Pages 27-45 *in* H. A. Mooney and E. Zavaleta, editors. Ecosystems of California. University of California Press, Berkeley, California, USA.

Keeley, J. E., and A. D. Syphard. 2015. Different fire-climate relationships on forested and non-forested landscapes in the Sierra Nevada ecoregion. International Journal of Wildland Fire 24:27-36.

Keeley, J. E., and A. D. Syphard. 2016. Climate change and future fire regimes: examples from California. Geosciences 6:37.

Keeley, J., and A. D. Syphard. 2017. Different historical fire-climate patterns in California. International Journal of Wildland Fire 26:253-268.

Keeley, J. E., and P. H. Zedler. 2009. Large, high-intensity fire events in southern California shrublands: debunking the fine-grain age patch model. Ecological Applications 19:69-94.

Kelly, A. E., and M. L. Goulden. 2008. Rapid shifts in plant distribution with recent climate change. Proceedings of the National Academy of Sciences 105:11823-11826.

Kolb, K. J., and S. D. Davis. 1994. Drought tolerance and xylem embolism in co-occurring species of coastal sage and chaparral. Ecology 74:648-659.

Krawchuk, M. A., M. A. Moritz, M. A. Parisien, J. Van Dorn, and K. Hayhoe. 2009. Global pyrogeography: the current and future distribution of wildfire. PloS ONE 4:e5102.

Krawchuk, M. A., and M. A. Moritz. 2011. Constraints on global fire activity vary across a resource gradient. Ecology 92:121-132.

Lawson, D. M., H. M. Regan, P. H. Zedler, and J. Franklin. 2010. Cumulative effects of land use, altered fire regime and climate change on persistence of *Ceanothus verrucosus*, a rare, fire-dependent plant species. Global Change Biology 16:2518-2529.

Lenihan, J. M., D. Bachelet, R. P. Neilson, and R. Drapek. 2008. Response of vegetation distribution, ecosystem productivity, and fire to climate change scenarios for California. Climatic Change 87:215-230.

Lenihan, J. M., R. Drapek, D. Bachelet, and R. P. Neilson. 2003. Climate change effects on vegetation distribution, carbon, and fire in California. Ecological Applications 13:1667-1681.

Littell, J. S., D. McKenzie, D. L. Peterson, and A. L. Westerling. 2009. Climate and wildfire area burned in western US ecoprovinces, 1916–2003. Ecological Applications 19:1003-1021.

Loarie, S. R., B. E. Carter, K. Hayhoe, S. McMahon, R. Moe, C. A. Knight, and D. D. Ackerly. 2008. Climate change and the future of California's endemic flora. PLoS ONE 3:e2502.

Luo, H., W. C. Oechel, S. J. Hastings, R. Zulueta, Y. Qian, and H. Kwon. 2007. Mature semiarid chaparral ecosystems can be a significant sink for atmospheric carbon dioxide. Global Change Biology 13:386-396.

Mahall, B., and C. Wilson. 1986. Environmental induction and physiological consequences of natural pruning in the chaparral shrub *Ceanothus megacarpus*. Botanical Gazette 147:102-109.

Mann, M. E., and P. H. Gleick. 2015. Climate change and California drought in the 21st century. Proceedings of the National Academy of Sciences 112:3858-3859.

Meentemeyer, R. K., A. Moody, and J. Franklin. 2001. Landscape-scale patterns of shrub-species abundance in California chaparral-the role of topographically mediated resource gradients. Plant Ecology 156:19-41.

Melillo, J. M., T. T. Richmond, and G. Yohe. 2014. Climate change impacts in the United States: the third national climate assessment, US Global Change Research Program, Washington D.C., USA. http://nca2014.globalchange.gov/downloads

Miles, S. R., and C. B. Goudey. 1997. Ecological subregions of California: section and subsection descriptions. R5-EM-TP-005. San Fransisco, California, USA, USDA Forest Service, Pacific Southwest Region.

Miller, N. L., and N. J. Schlegel. 2006. Climate change projected fire weather sensitivity: California Santa Ana wind occurrence. Geophysical Research Letters 33:15.

Molinari, N. A., S. C. Sawyer, and H. D. Safford. 2016. A summary of current trends and probable future trends in climate and climate-driven processes in the Cleveland National Forest and neighboring lands. USDA Forest Service, Pacific Southwest Region, Vallejo, California, USA. https://www.fs.usda.gov/Internet/FSE_DOCUMENTS/fseprd497636.pdf

Mooney, H. A., and C. Chu. 1974. Seasonal carbon allocation in *Heteromeles arbutifolia*, a California evergreen shrub. Oecologia 14:295-306.

Mooney, H., B. Drake, R. Luxmoore, W. Oechel, and L. Pitelka. 1991. Predicting ecosystem responses to elevated CO_2 concentrations. BioScience 41:96-104.

Paddock III, W. A., S. D. Davis, R. B. Pratt, A. L. Jacobsen, M. F. Tobin, J. López-Portillo, and F. W. Ewers. 2013. Factors determining mortality of adult chaparral shrubs in an extreme drought year in California. Aliso: A Journal of Systematic and Evolutionary Botany 31:49-57.

Parsons, D. J., P. W. Rundel, R. P. Hedlund, and G. A. Baker. 1981. Survival of severe drought by a non-sprouting chaparral shrub. American Journal of Botany 68:973-979.

Pausas, J. G., and J. E. Keeley. 2014a. Abrupt climate-independent fire regime changes. Ecosystems 17:1109-1120.

Pausas, J. G., and J. E. Keeley. 2014b. Evolutionary ecology of resprouting and seeding in fire-prone ecosystems. New Phytologist 204:55-65.

Pausas, J. G., and S. Paula. 2012. Fuel shapes the fire-climate relationship: evidence from Mediterranean ecosystems. Global Ecology and Biogeography 21:1074-1082.

Pausas, J. G., R. B. Pratt, J. E. Keeley, A. L. Jacobsen, A. R. Ramirez, A. Vilagrosa, S. Paula, I. N. Kaneakua-Pia, and S. D. Davis. 2016. Towards understanding resprouting at the global scale. New Phytologist 209:945-954.

Pratt, R., A. Jacobsen, K. Golgotiu, J. Sperry, F. Ewers, and S. Davis. 2007. Life history type and water stress tolerance in nine California chaparral species (Rhamnaceae). Ecological Monographs 77:239-253.

Pratt, R., A. Jacobsen, R. Mohla, F. Ewers, and S. Davis. 2008. Linkage between water stress tolerance and life history type in seedlings of nine chaparral species (Rhamnaceae). Journal of Ecology 96:1252-1265.

Pratt, R. B., A. L. Jacobsen, J. Hernandez, F. W. Ewers, G. B. North, and S. D. Davis. 2012. Allocation tradeoffs among chaparral shrub seedlings with different life history types (Rhamnaceae). American Journal of Botany 99:1464-1476.

Pratt, R. B., A. L. Jacobsen, A. R. Ramirez, A. M. Helms, C. A. Traugh, M. F. Tobin, M. S. Heffner, and S. D. Davis. 2014. Mortality of resprouting chaparral shrubs after a fire and during a record drought: physiological mechanisms and demographic consequences. Global Change Biology 20:893-907.

Price, C., and D. Rind. 1994. The impact of a 2× CO_2 climate on lightning-caused fires. Journal of Climate 7:1484-1494.

Ramirez, A., R. Pratt, A. Jacobsen, and S. Davis. 2012. Exotic deer diminish post-fire resilience of native shrub communities on Santa Catalina Island, southern California. Plant Ecology 213:1037-1047.

Rapacciuolo, G., S. P. Maher, A. C. Schneider, T. T. Hammond, M. D. Jabis, R. E. Walsh, K. J. Iknayan, G. K. Walden, M. F. Oldfather, and D. D. Ackerly. 2014. Beyond a warming fingerprint: individualistic biogeographic responses to heterogeneous climate change in California. Global Change Biology 20:2841-2855.

Restaino, C. R., and H. D. Safford. 2018. Fire and climate change. Pages 493-505 *in* J. W. van Wagtendonk, N. G. Sugihara, S. L. Stephens, A. E. Thode, K. E. Shaffer, and J. A. Fites-Kaufman, editors. Fire in California's ecosystems. Second edition. University of California Press, Berkeley, California, USA.

Riahi, K., S. Rao, V. Krey, C. Cho, V. Chirkov, G. Fischer, G. Kindermann, N. Nakicenovic, and P. Rafaj. 2011. RCP 8.5-A scenario of comparatively high greenhouse gas emissions. Climatic Change 109:33-57.

Riordan, E. C., and P. W. Rundel. 2014. Land use compounds habitat losses under projected climate change in a threatened California ecosystem. PloS ONE 9:e86487.

Sala, O. E., F. S. Chapin, J. J. Armesto, E. Berlow, J. Bloomfield, R. Dirzo, E. Huber-Sanwald, L. F. Huenneke, R. B. Jackson, and A. Kinzig. 2000. Global biodiversity scenarios for the year 2100. Science 287:1770-1774.

Saruwatari, M., and S. Davis. 1989. Tissue water relations of three chaparral shrub species after wildfire. Oecologia 80:303-308.

Sawyer, S. C., J. Hooper, and H. D. Safford. 2014. A summary of current trends and probable future trends in climate and climate-driven processes for the Angeles and San Bernardino National Forests. USDA Forest Service, Pacific Southwest Region, Vallejo, California, USA. https://www.fs.usda.gov/Internet/FSE_DOCUMENTS/stelprdb5445379.pdf

Schlesinger, W. H., J. T. Gray, D. S. Gill, and B. E. Mahall. 1982. *Ceanothus megacarpus* chaparral: a synthesis of ecosystem processes during development and annual growth. Botanical Review 48:71-117.

Schwilk, D. W., and J. E. Keeley. 2012. A plant distribution shift: temperature, drought or past disturbance. PLoS ONE 7:e31173.

Shaw, M. R., L. Pendleton, D. R. Cameron, B. Morris, D. Bachelet, K. Klausmeyer, J. MacKenzie, D. R. Conklin, G. N. Bratman, and J. Lenihan. 2011. The impact of climate change on California's ecosystem services. Climatic Change 109:465-484.

Steel, Z. L., H. D. Safford, and J. H. Viers. 2015. The fire frequency-severity relationship and the legacy of fire suppression in California forests. Ecosphere 6:1-23.

Syphard, A. D., V. C. Radeloff, J. E. Keeley, T. J. Hawbaker, M. K. Clayton, S. I. Stewart, and R. B. Hammer. 2007. Human influence on California fire regimes. Ecological Applications 17:1388-1402.

Thomas, C., and S. Davis. 1989. Recovery patterns of three chaparral shrub species after wildfire. Oecologia 80:309-320.

Thrasher, B., J. Xiong, W. Wang, F. Melton, A. Michaelis, and R. Nemani. 2013. Downscaled climate projections suitable for resource management. Eos, Transactions American Geophysical Union 94:321-323.

Tyler, C. M., D. C. Odion, and R. M. Callaway. 2007. Dynamics of woody species in the California grassland. Pages 169-179 *in* M. Stromberg, J. Corbin, and C. D'Antonio, editors. California Grasslands Ecology and Management. University of California Press, Berkeley, California, USA.

Underwood, E. C., J. H. Viers, K. R. Klausmeyer, R. L. Cox, and M. R. Shaw. 2009. Threats and biodiversity in the mediterranean biome. Diversity and Distributions 15:188-197.

van Mantgem, P. J., J. C. Nesmith, M. Keifer, E. E. Knapp, A. Flint, and L. Flint. 2013. Climatic stress increases forest fire severity across the western United States. Ecology Letters 16:1151-1156.

Westerling, A. L. 2016. Increasing western US forest wildfire activity: sensitivity to changes in the timing of spring. Philosophical Transactions of the Royal Society of London B: Biological Sciences 371:20150178.

Westerling, A., and B. Bryant. 2008. Climate change and wildfire in California. Climatic Change 87:231-249.

Westerling, A. L., H. G. Hidalgo, D. R. Cayan, and T. W. Swetnam. 2006. Warming and earlier spring increase western US forest wildfire activity. Science 313:940-943.

Chapter 15
Managing Chaparral Resources on Public Lands

Hugh D. Safford, Emma C. Underwood, and Nicole A. Molinari

Abstract Southern California supports some of the highest biodiversity in the United States, but it also suffers from very heavy visitor use, a large influx of non-native species, high levels of air pollution, steep and erosive slopes, and the most unpredictable precipitation regime in the nation. Wildland vegetation in southern California is dominated by highly flammable shrublands like chaparral. As a result, public lands in southern California are exceptionally fire-prone. Annually they experience more economic and environmental damage from wildfire than any other part of the US. Management in southern California shrubland ecosystems has traditionally focused heavily on fire and fuels, but degraded terrestrial and aquatic ecosystems and hundreds of rare, threatened, and endangered species require a more holistic approach, especially with growing human populations and their needs for ecosystem services, and the developing threat of climate change. In this chapter we categorize the major management priorities on public lands in southern California and explore their inter-relationships. We also identify a suite of ecosystem services provided by chaparral landscapes, and we assess how current management priorities interact with and impact these services. Major tensions exist between certain management focus areas, especially recreation and fuel management, and other management priorities and the ecosystem services we assessed. We show how an ecosystem service-based approach to chaparral management can help to better elucidate and resolve conflicts in chaparral management.

H. D. Safford (✉)
USDA Forest Service, Pacific Southwest Region, Vallejo, CA, USA

University of California, Davis, CA, USA
e-mail: hughsafford@fs.fed.us

E. C. Underwood
University of California, Davis, CA, USA

University of Southampton, UK

N. A. Molinari
USDA Forest Service, Los Padres National Forest, Goleta, CA, USA

E. C. Underwood et al. (eds.), *Valuing Chaparral*, Springer Series on Environmental Management, https://doi.org/10.1007/978-3-319-68303-4_15

Keywords Conservation · Ecosystem services · Fire · Fuel management · Recreation · Restoration

15.1 Introduction

Chaparral is one of most widely distributed terrestrial ecosystems in California, and it is found in some form on nearly every major federal conservation unit in the state (Fig. 1.1). Fully 60% (840,000 of 1.4 million ha) of the four southern California national forests support chaparral vegetation, and even the Mendocino National Forest, 650 km (400 miles) north of Santa Barbara, is almost one-quarter chaparral. Millions of Californians live in communities that are carved like islands from the chaparral sea, and tens of thousands reside in rural homes surrounded by chaparral on all sides.

Every year we are reminded that these are dangerous places to live. Chaparral vegetation is the principal fuel in the most fire-prone region in the United States (southern California), and the average year sees hundreds or thousands of homes and businesses burned, people killed or injured, and hundreds of millions of dollars in fire suppression costs and fire-related economic losses (Safford 2007; Halsey 2008; Rahn 2009). Yet people continue to pour into chaparral dominated landscapes and poor planning decisions continue to be made at the community and county level, decisions that have major implications for human economies and human lives (Halsey 2008). Ecological consequences of these planning decisions include direct ramifications, such as the loss of habitat, but also indirect ramifications, most consequentially related to fuel and fire management decisions designed to protect the community after it is built and inhabited.

Fire management has become a major industry in southern California at all jurisdictional levels. Within the US Forest Service, fire management now consumes >70% of the budget in southern California, compared to >50% nationally (USDA 2015a). Many, if not most, major management decisions made by US Forest Service units in southern California—which manage the vast bulk of chaparral habitat in the region—revolve around fire or fuels at some level. For example, suppressing fires to protect human lives and infrastructure; managing vegetation to reduce fuel; carrying out postfire rehabilitation and restoration work; or controlling non-native species in frequently burned areas. Other federal and state lands like the Santa Monica Mountains National Recreation Area (National Park Service), Camp Pendleton (Department of Defense) or Mount San Jacinto State Park (California State Parks) are experiencing similar situations (Table 15.1).

This myopic focus on the chaparral fire problem clouds and impoverishes people's understanding of the ecosystem values of chaparral. Indeed, given the major exposure Californians have to chaparral, it is extraordinary the extent to which it has been misunderstood and even reviled (see Chaps. 1, 5, and 11). This antipathy for chaparral has roots that go way back (see Chap. 5). Early settlers in southern California saw a troublesome brushland with trifling economic potential or value. It

Table 15.1 Major land management units (>5000 ha or 12,355 acres) in southern California in 2016 with large areas of chaparral[a]

Unit	Level	Agency	Size (ha)	Management plan and citation	Management focus areas[b]
Los Padres National Forest	Federal	US Forest Service	708,502	Southern California National Forests Land Management Plan, 2005 (USDA 2005)	Watershed protection, fire protection, conservation, recreation, flood protection, education
Angeles National Forest	Federal	US Forest Service	283,401	Ibid	Ibid
San Bernardino National Forest	Federal	US Forest Service	275,304	Ibid	Ibid
Cleveland National Forest	Federal	US Forest Service	186,235	Ibid	Ibid
Santa Rosa & San Jacinto Mtns. National Monument	Federal	Bureau of Land Management and Forest Service	113,360[c]	SRSJNM Final Management Plan, 2004 (USDI and USDA 2004)	Conservation, recreation
Channel Islands National Park	Federal	National Park Service	101,215	CINP Final General Management Plan, 2015 (USDI 2015)	Conservation, restoration, recreation, education
Santa Monica Mtns. National Recreation Area	Federal	National Park Service	64,777	SMMNRA General Management Plan, 2002 (USDI 2002)	Conservation, recreation, education
Camp Pendleton	Federal	US Marine Corps	50,607	CP Integrated Natural Resources Management Plan (USMC 2012)	Conservation, restoration, recreation
Vandenberg Air Force Base	Federal	US Air Force	40,486	VAFB Integrated Natural Resources Management Plan, 2011–2015 (USAF [US Air Force]. 2011)	Conservation, restoration, recreation
Cuyamaca Rancho State Park	State	CA Dept. of Parks and Recreation	10,121	CRSP General Plan, 2015 (State of California 2015)	Conservation, education, recreation
Hungry Valley State Vehicular Recreation Area	State	CA Dept. of Parks and Recreation	7692	HVSVRA General Plan, 1981 (State of California 1981)	Recreation, conservation, restoration

(continued)

Table 15.1 (continued)

Unit	Level	Agency	Size (ha)	Management plan and citation	Management focus areas[b]
Mount San Jacinto State Park	State	CA Dept. of Parks and Recreation	5668	MSJSP General Plan, 2002 (State of California 2002)	Conservation, recreation

[a]Since early 2016, two new national monuments (NM) have been proclaimed that contain large areas of chaparral (and US Forest Service and BLM lands): San Gabriel Mountains NM and Sand to Snow NM. These two National Monuments are too new to have approved management plans, thus they are not included here
[b]Management focus areas are ordered according to their emphasis in the plan
[c]Includes portions of San Bernardino National Forest

was a common misconception that chaparral represented a degraded, deforested condition and this belief led to widespread conifer planting in the late nineteenth and early twentieth centuries (Burns and Sauer 1992). Fire was used to clear land for home sites and crops, to open access for miners and livestock herders, and to improve forage (Plummer 1911; Patric and Hanes 1964; Keeley and Fotheringham 2003). By the end of the nineteenth century, such was the destruction of native vegetation in the San Gabriel and San Bernardino mountains that great vacillations in water supply—lower water delivery overall, punctuated by periodic devastating floods—and destructive landslides generated real economic harm in the valleys below, particularly to the croplands and large fruit orchards that characterized the greater Los Angeles Basin at the time (Plummer 1911). This galvanized public opinion and led to the establishment of the nation's second forest reserve (after Yellowstone) in the San Gabriel Mountains in 1892, and soon after other forest reserves in the neighboring mountains (e.g., Trabuco Canyon [Santa Anas] in 1893, San Jacinto in 1897). It is noteworthy that unlike most national forest lands, these landscapes were primarily protected to prevent watershed degradation (rather than to guard timber resources), principally through fire suppression (Plummer 1911).

Early conservationists and foresters like John Muir and Abbot Kinney (Kinney 1888; Muir 1918) were respectful of, if not enamored by chaparral, but as the southern California population grew, the respect turned largely to disdain. By the 1960s and 1970s, even scientific opinion had turned against chaparral, and the literature of the period is full of publications expounding the multiple benefits of clearing chaparral by whatever means necessary (e.g., Macey and Gilligan 1961; Bentley 1967; Corbett and Crouse 1968; Cable 1975; Riggan and Dunn 1982). These putative benefits included increasing runoff, generating better forage for livestock and game, making recreational and hunting use of the landscape easier, harvesting biomass for energy, and reducing fire-fighting costs. Even the pioneering and otherwise progressive fire ecologist Howard Biswell prefaced his 1974 study of chaparral fire ecology by suggesting that understanding the relationship between chaparral and fire would better serve the purposes of people who wanted to control and remove it (Biswell 1974).

Today we understand that chaparral has its own value. It is an important repository of biodiversity, and chaparral dominated counties like San Diego and Ventura are among the most biodiverse political jurisdictions in the nation (see Chaps. 2 and 3). Intact chaparral cover is an important component of the habitat required by hundreds of rare and endangered species, and the important but often forgotten ecosystem service of pollination is highly dependent on the proximity of healthy chaparral and other native vegetation to cropland and orchards (Kremen et al. 2004; Klein et al. 2012). Intact chaparral cover also protects soil from the direct impact of heavy rain, and live chaparral roots prevent soil erosion and overland flow (see Chaps. 7 and 8, Wohlgemuth et al. 2009) during both the dry and wet seasons. Chaparral sequesters much more carbon than the grass ecosystems that replace it when it is degraded (see Chaps. 6 and 12, Bohlman et al. in press) and intact chaparral may reduce fire ignitions during the high fire season due to high live fuel moisture and lower combustibility relative to the dried foliage of annual species. Intact chaparral with its characteristic dense cover and closed canopy is also relatively resistance to plant invasions.

Management of chaparral habitats is fraught with controversy (see Chaps. 5 and 11, Keeley 2002a; Halsey 2008). Many of these controversies find their roots in the different perspectives people and organizations have of the value of chaparral. Today, the extraordinary threat of fire and too many people, hundreds of threatened and endangered species, degraded watersheds, climate change, and water issues are re-focusing and intensifying the chaparral debate. More people are interested in conserving chaparral, at the same time that more people are exposed to the threat of fire; more people require water collected from chaparral dominated watersheds, at the same time that droughts are becoming more common and more pronounced; threats to rare species' survival are mounting at the same time that the housing market has recovered from the 2008 crash and subdivision growth is on the rise.

Sustainable management of chaparral will require that we balance these competing needs. Ideally, this is best done through a transparent process that considers proposed actions in the light of their impacts on a broad suite of ecosystem services. It is our thesis that much of the polemic in chaparral management can be resolved, or at least made less acute, by taking a more holistic view, and considering and integrating information on the range of ecosystem services provided by chaparral landscapes that are detailed in the chapters of this book. In this chapter we identify five major areas of public agency management focus in chaparral ecosystems, and summarize the inter-linked issues and controversies surrounding them. Along the way we highlight, through a series of case studies, progress that has been made in integrating multiple natural resources and ecosystem services in prioritizing, planning, and implementing management actions.

15.2 Areas of Management and Ecosystem Service Focus in Chaparral Ecosystems

To assess the overlap between areas of management focus and ecosystem services we perused the General Plans, or Land and Resource Management Plans, for the major federal and state conservation units and other designations in southern

California that included large proportions of chaparral, and some level of resource protection as a goal or mission (Table 15.1). We identified the overarching management focus areas for each unit, as outlined in each planning document (Table 15.1), and also the principal ecosystem services that each planning document identified as a good or service that could be enhanced or provided through management.

Generalized across the management plans, we found the principal management focus areas fall into six broad categories. In approximate order of their importance (i.e., either as directly identified in the management plans or based on their relative emphasis in the plan text [Table 15.1]) they are:

- Conservation
- Recreation management
- Fuel management
- Fire management
- Restoration
- Education

The principal ecosystem services acknowledged (either directly or indirectly) in the planning documents, also in approximate order of their importance (definition as above), are:

- Water provision
- Reduction of erosion and flooding
- Facilitation of recreation opportunities
- Protection of biodiversity
- Provision of aesthetic landscapes
- Carbon sequestration

Below, we describe the inter-relationships between the management focus areas highlighted in the general management plans and their connections to the six ecosystem services, and we provide a snapshot of the allocation of funding. These relationships (management priorities, ecosystem services, and budget) are highlighted within each of the management focus area subsections below to provide guidance for the reader. Note that we do not address education here, as it is discussed in depth in Chap. 11. Three case studies relating to these management priorities are presented in Boxes 15.1–15.3.

15.3 Major Chaparral Management Priorities, their Inter-relationships, and their Influences on Ecosystem Services

15.3.1 Conservation

By conservation, we refer both to biological conservation, which includes those human actions that seek to maintain genes, species, communities, and ecosystems that make up the earth's biological diversity (Soulé 1985), and natural resource

Box 15.1 Thinking Big in Chaparral Conservation

California is a national leader in collaborative efforts to protect large, interconnected landscapes for multiple species and ecosystems. The California Natural Community Conservation Act (1993, revised 2003) led to the Natural Community Conservation Planning (NCCP) Program. NCCP is managed by the California Department of Fish and Wildlife and, where there are federally listed species in question, the US Fish and Wildlife Service. NCCPs are focused on preserving intact ecosystems rather than single species, and became necessary in places like southern California where the single species focus and reactionary implementation of the national and California Endangered Species Acts were proving insufficient and controversial (Pollak 2001a). Although they are largely voluntary agreements, NCCPs bring stakeholders to the table by providing heightened predictability about the future regulatory environment, and more certitude about what can be developed and what must be conserved. Pollak (2001b) lists four major achievements of the NCCP program:

- A planning outlook that is more regional and longer term
- Development of large, interconnected networks of conservation reserves and corridors
- A setting for collaborative efforts by sometimes diametrically opposed groups
- Streamlining of regulatory processes and enhanced certainty regarding conservation and development actions

The original NCCP Act arose in response to conservation versus development controversies related to the potential listing of the California gnatcatcher in the late 1980s and early 1990s (Pollak 2001a). The primary focus of early NCCPs was the identification and protection, often through acquisition, of sage scrub habitat—a close relative of chaparral—which is key to gnatcatcher survival in southern California. Today NCCPs exist for species and habitats across California, centered in areas with extreme development pressures and multiple listed or potentially listed species, like southern California, the San Francisco Bay Area, and the central and southern Central Valley. Chaparral habitats are a focus of at least 16 NCCPs in California, 11 of them south of the Tehachapi Mountains (see map link on https://www.wildlife.ca.gov/Conservation/Planning/NCCP).

The NCCP process has provided major impetus to agencies, governments, businesses, private landowners, and conservation Non-Governmental Organizations (NGOs) to cooperate in developing workable compromises that attempt to balance development and conservation across large landscapes. For example, the Santa Rosa Plateau and lands draining to the Santa Margarita River were identified by animal tracking and corridor mapping (Luke et al. 2004; Beier et al. 2006) as key to maintaining ecological connectivity between the coast, the Santa Ana Mountains, and the Palomar Mountains (Fig. 15.1).

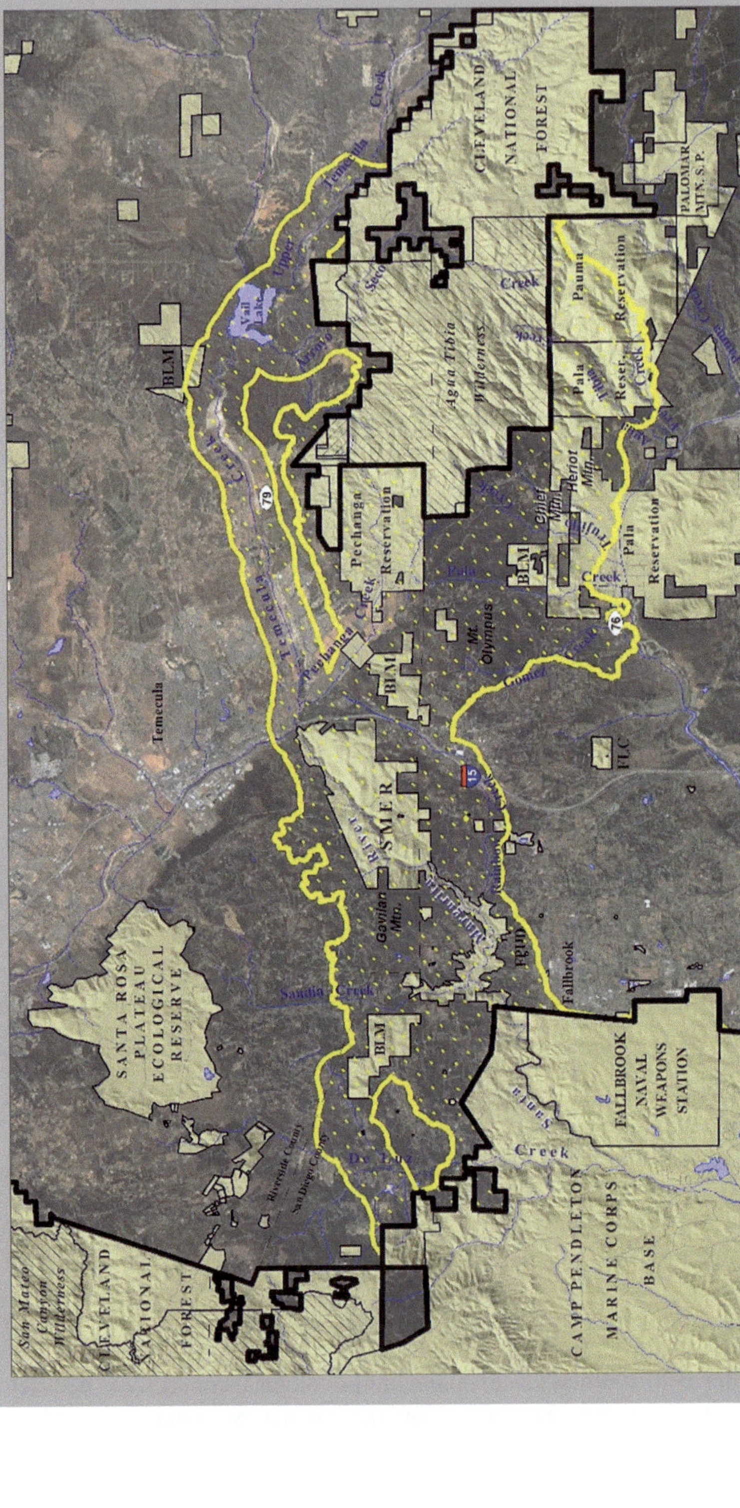

Fig. 15.1 Proposed Santa Ana Mountains-Palomar Mountain habitat corridor. Yellow perimeter indicates linkage design. From Luke et al. 2004

However, the logistical and political hurdles to saving this linkage corridor are daunting, which is further compounded by development pressures and increasing land prices. Two NCCPs in the area, the Western Riverside County Multiple Species Habitat Conservation Plan (County 2003), and San Diego North County Multiple Species Conservation Plan, have mapped proposed reserves, however the checkerboard of land ownership makes implementation extremely challenging. Large properties in the area belong to the US Forest Service, the US Marine Corps, the US Navy, a number of tribal reservations, Riverside County Parks, San Diego State University, the US Bureau of Land Management, The Audubon Society, and California State Parks, but there are thousands of smaller parcels as well, most privately owned. The I-15 freeway also represents a major barrier to connectivity. The Nature Conservancy and other NGOs have prioritized land acquisitions in the corridor (and other corridors on and adjacent to the Santa Rosa Plateau), but land prices and unwillingness to sell at the current time are proving to be major obstacles. The Santa Ana-Palomar linkage will be a test case of the NCCP process and its ability to conserve truly critical ecological landscapes in the face of overwhelming economic and development pressures (Morrison and Boyce 2008).

conservation, which is related but focuses more on the sustainability of resources necessary for human survival and well-being. Based on our management plan review, conservation is the principal management focus of most major public management units in southern California (Table 15.1). Southern California is overall the most biodiverse part of the United States, and it also supports the highest number of threatened, endangered, and sensitive (TES) species in the contiguous 48 states (Dobson et al. 1997; Stein et al. 2000).

High levels of habitat fragmentation constitute one of the paramount threats to biodiversity and TES species on chaparral lands throughout California. Chaparral and related shrubland types like sage scrub used to cover virtually uninterrupted swathes of millions of hectares along much of the California coast, but the rapid growth of human populations and associated urban and suburban areas have restricted natural habitat to numerous terrestrial islands that have lost much or all of their connectivity with nearby wildlands (Stephenson and Calcarone 1999). This has resulted in chaparral (and, even more so, sage scrub) dominated regions in California being major hotspots of species endangerment for multiple taxonomic groups (Dobson et al. 1997). As a consequence, the need for regional strategic conservation planning efforts has become paramount (Zedler 1996; Stephenson and Calcarone 1999), and over the last 25 years, southern California has become a national model for large scale, multiparty habitat and species conservation planning (see Box 15.1).

With respect to the other management priorities, conservation has an important influence on recreation (the second ranked management focus area), positively through the preservation of natural areas and open spaces that permit recreational use and educational opportunities, and in some cases negatively through the prohibition of certain kinds of recreational use (e.g., in wilderness areas or the Sespe Condor Sanctuary). Conservation has both positive and negative influences on fire and fuel management. Conserved areas are typically farther from urban areas and support low densities of access routes so human ignition densities tend to be lower, but the large areas of contiguous fuels in conserved areas can lead to greater fire spread and such areas are sometimes off-limits legally or logistically to fuel management activities, as well as certain fire management activities (USDA 2005; Syphard et al. 2008). Strategic land acquisition and conservation in areas of high fire hazard may be a valuable tool to reduce fire risk to homes while protecting biodiversity and other ecosystem services (Butsic et al. 2016). Conservation has a primarily positive influence on restoration, as conserved areas typically require less restoration, freeing restoration funds and efforts for more degraded habitats. Interestingly, although conservation is the primary focus of most chaparral dominated management units in southern California, it is not necessarily the top priority in terms of greatest spatial area in publicly owned chaparral. This is because the four national forests belong to the National Forest System, which is managed under a different set of laws and regulations than national, state, and county parks and is compelled to consider a wider swath of public uses (Wilcove 1989; Thomas 1996, Table 15.1).

Given the large and growing population in southern California and other chaparral areas, and the increasing occurrence of droughts and warmer, drier fire seasons, there are inevitably strong tensions between conservation and recreation and conservation and fuel management (see next section). At the present time, these tensions represent two of the most important resource management issues in California chaparral landscapes.

Of the general management priorities we identified in our review, conservation is the only one to have a primarily positive influence on all of the ecosystem service categories. For example, conserving native shrubland secures sediment on steep slopes, provides habitat for plant and animal species, increases the aesthetic value of the landscape, and stores carbon both above- and below-ground (the latter of which can be considerable for certain chaparral shrub species, especially postfire resprouters). Conservation also provides open space for many types of recreation, but strict conservation policies can have negative influences on opportunities for recreation, especially high-impact activities such as motorized recreation and suction mining.

It is difficult to determine exact budget numbers related to support of conservation on public lands in southern California or other areas dominated by chaparral. On some federal lands (e.g., US Forest Service), actual management of wildlife species falls to the California Department of Fish and Wildlife (DFW). In these cases, most actions carried out by the federal agency other than DFW are related to habitat management rather than direct species management (this is not the case with the National Park Service, which manages its own wildlife populations). Also, funds that are directly earmarked for conservation are a relatively small part of most agencies' budgets, and much conservation-related work is accomplished as part of projects funded by other areas of the budget or from trust-funds or partner funding. For example, federal agencies can purchase important private properties for conservation purposes with funds from the Land and Water Conservation Fund, which is derived by fees collected from offshore oil and gas drilling. State agencies carry out a large component of their conservation work through allocation of State bond funds. The long list of habitat, multi-species, and natural community conservation plans in California involves many public and private partners, with the bulk of funding often coming from bonds, foundations, or private donations. There is also a large area of overlap between restoration and conservation, which is the primary goal of most restoration projects, and most large scale restoration projects are funded through multi-party collaboratives.

15.3.2 Recreation

Recreation is the second most cited management focus area, which reflects the mandates of the public land management agencies, the proximity of the management units to the ~23 million residents of southern California, and the general lack of publicly accessible open space. The draft management plan for the new San Gabriel National Monument (USDA 2015b) is a good example of the difficult balancing act experienced by management agencies tasked with conserving species, ecosystems, and resources, while at the same time serving a huge clientele of recreational users whose cumulative actions comprise a major conservation threat. The plan notes that more than 15 million people live within 90 min of this national monument, but only 2% of the monument is suitable for recreational use, and within that 2% most use is concentrated at a handful of sites with water access (USDA 2015b) (see Chap. 10 for more information on recreational use and photo of water site use on the San Gabriel National Monument).

Other recreation issues for the public land management agencies in southern California include the high diversity of types of recreational use, some of which conflict (e.g., hikers versus mountain bikers versus horse riders; bird-watchers versus motorcycles; shooters versus hikers; fishermen versus suction miners, Moore 1994); ethnic differences in type of use (see Chap. 10, Baas et al. 1993); illegal recreational activities (e.g., unpermitted Off Highway Vehicle (OHV) use, unpermitted trail and campsite construction, illegal shooting) (Hartley 1986; McIntyre and Weeks 2002), and generally low budgets for recreation management.

With respect to management priorities, recreation has both direct and indirect effects associated with conservation (see Chap. 10 for a detailed study of recreation use on public lands). There is a direct negative relationship between conservation and recreation and this ranges from slightly to moderately negative in cases of low-impact recreation (non-motorized, low densities of users) to highly negative in cases of high-impact recreation, such as popular OHV trails (Havlick 2002; Reed and Merenlender 2008). However, there is also an indirect positive relationship, in that recreation is the principal connection for many users with natural ecosystems and such users are more likely to consider conservation an important goal (Lee 2011).

Recreationists can also have an important impact on fire management and fuel management, as areas with high recreational use are often major sources of ignitions and are more likely to require fuel reduction, not to mention enforcement of fire safety regulations (use of spark arresters, campfire restrictions, parking and road closures, etc.). Recreationist and recreational facility presence on the landscape can also greatly complicate fire operations (Bricker et al. 2008).

With respect to the ecosystem services, legal recreation itself has little relationship to water provision, however, excessive recreational use of streams can have major negative impacts on water quality. For example, short reaches of the East Fork of the San Gabriel River may see as many as 8000 people per day during the summer. The picnics and barbecues associated with this use generate over four hundred 122 L (32 gallon) bags of trash per day, according to US Forest Service estimates. Trash discarded in the streambed and river terrace area severely impairs the river and as a consequence a total maximum daily load (TMDL) has been established. In addition, roads and trails supporting recreation can be major sources of erosion and can impact soil retention. In terms of biodiversity, recreational use at the high levels common in southern California has negative influences, especially for larger animals like ungulates and predators (Knight and Gutzwiller 1995; Czech et al. 2000) and aquatic biota in heavily-used streams. The aesthetics of public lands can experience minor to moderate impacts from legal recreation, such as increased traffic and trash, but these escalate in high-impact recreation sites such as vehicular recreation areas or natural water features, or when recreationists are the ignition sources for wildfires (not an uncommon occurrence [Prestemon et al. 2013]).

In contrast, illegal recreational use has a variety of deleterious impacts on ecosystem services. These range from unpermitted OHV and other trail impacts on erosion, water quality, biodiversity, and aesthetics (Havlick 2002), to illegal shooting impacts on wildlife (direct mortality and indirect, through ingestion of lead ammunition), increased fire risk, and even human health (LPFW [Los Padres Forest Watch] 2016). Some of these impacts, such as OHV use, are often concentrated in areas where fuel management or fire clears dense vegetation from portions of the landscape.

Although recreation is a management priority—and public lands in southern California are among the most visited public lands in the US—the actual annual budgets allocated to recreation by most agencies are small (Kaczynski and Crompton 2006). For example, in 2016 the US Forest Service budget for "recreation, heritage and wilderness" was approximately 5% of the total budget (https://www.fs.fed.us/sites/default/files/media/2015/07/fy2016-budget-overview-update.pdf). For this reason, US Forest Service units in southern California have implemented (controversial) user fees to help pay the cost of, among other things, maintaining recreational facilities and dealing with resource damage by recreationists.

15.3.3 Fuel Management

Because chaparral landscapes are highly flammable and because human communities are so dispersed within these flammable landscapes, chaparral fuel management is a major focus for all public land managers in southern California. On southern California public lands, most ecosystem-disturbing activities carried out by management agencies relate in some fashion to fuel management. These can be either strategic prefire activities, such as prescribed fires or the removal of vegetation by bulldozers or masticators in fuelbreaks across the landscape, or tactical activities carried out in the heat of active fire control (which fall under fire management, below).

For decades, California land managers performed large scale fuel manipulations on chaparral landscapes with little public attention. However, advances in our understanding of the value and vulnerabilities of chaparral landscapes and the trade-offs associated with such work have led to increased public interest in the nature of and philosophy behind fuel manipulations. These advances include: (1) scientific information documenting the deleterious effects of reduced chaparral cover on a number of ecosystem properties and their function, (2) the development of a better understanding of the natural fire regime of chaparral vegetation, and (3) a growing recognition that reduced woody fuels do not act as a reliable barrier to fire spread under severe fire weather conditions, e.g., fires driven by Santa Ana winds.

In the first case, we now have a much better idea of the ecological trade-offs associated with the quasi-permanent removal of native vegetation cover that the fuelbreak system entails. Fuelbreaks do not cover a large portion of the southern California landscape, for example, the Cleveland National Forest fuelbreak network comprises less than 2.5% of the forest area (estimate based on draft Strategic Fuelbreak Analysis: 514 km (319 miles) of fuelbreaks at a maximum of 90 m (295 ft) width, T. Metzger, US Forest Service, pers. comm.), but they are very obvious to observers, and their environmental costs are high. These costs include increased non-native species invasion, higher soil loss rates, more variable runoff, and less groundwater recharge, enhanced access for unpermitted motorized vehicle entry, aesthetic impacts, and under certain circumstances higher local fire hazard (due to herbaceous fuels dominating where woody fuels have been reduced) (Corbett and Rice 1966; Merriam et al. 2006; Halsey 2008; Wohlgemuth et al. 2009).

In the second case, there is now broad agreement that the natural fire regime of most low- and mid-elevation chaparral ecosystems is characterized by infrequent (fire-return intervals between 30 and 100+ years), high severity fires (Minnich 2001; Van de Water and Safford 2011). Although chaparral is resilient to variability in fire frequency, we now know that persistent fires at return intervals of less than about 15 years reduce woody vegetation and enhance non-native grass invasion (especially in locations with high atmospheric nitrogen input, which is widespread in southern California) to the point that such sites eventually transform to quasi-permanent weedy grasslands (Zedler et al. 1983; Keeley 2006; Keeley and Brennan 2012). Today large areas of the mountain foothills of southern California been transformed from chaparral and sage scrub vegetation to non-native grasslands and weed fields. In turn, these sites are the major sources of non-native plant seeds that dominate the contemporary soil seed pool in southern California, and are also the major sources of human fire ignitions that define the modern southern California experience (see Chap. 12, Syphard et al. 2007a). It is also worth noting that application of prescribed fire on many chaparral landscapes represents a further increase in fire frequency on lands where the major source of degradation is already an overabundance of fire.

Table 15.2 Fuel and fire characteristics from 1970 to 2013 for two fire seasons on southern California's US Forest Service units (fire statistics from chaparral burning only)

Fire season	Name	Mean live fuel moisture (%)[a]	Mean monthly number of fires[b,c]	Mean fire size (ha)[b]	Median fire size (ha)[b]	Mean annual max fire size (ha)[b]	Absolute max fire size (ha)[b]	Mean annual burned area (ha)[b]	Mean monthly burned area (ha)[b]
May to Sept[a]	Spring–summer	93 (65–135)	2.9	2327	131	17,763	97,311	33,593	6719
Oct to Dec	Santa Ana	78 (65–90)	1.2	4858	335.4	8619	109,589	17,687	5896

[a](County of Los Angeles 2016). Mean of mid-month shrub live fuel moisture measurements, 1981–2016

[b]Fire perimeter data (FRAP 2014) only include fires reaching at least 4 ha (10 acres) in size on US Forest Service lands. The number of fires <4 ha is much higher than this number. Fire-return interval departure data (FRID) are from the USDA (2015c)

cThe Santa Ana season is tied to the development of cooler desert temperatures than coastal temperatures, and such a situation usually begins in October. Some years this occurs in late September however, so the boundary between September and October is necessarily artificial

In the third case, the historic fires of 2003 and 2007 made clear the limitations of southern California fuel reduction treatments in stopping fires under severe weather conditions (i.e., high winds, high temperatures, and conditions of drought, Keeley et al. 2004; Keeley et al. 2009). It should be noted that these limitations were understood in land management circles well before the 2000s (see, for example, Pillsbury 1963; Tyrrel 1982, and discussion in Cermak 2005), but these catastrophic events brought the issue into full focus.

A major debate has been ongoing for decades regarding the relative roles of fuels versus extreme weather in driving chaparral fire behavior (Minnich 1983, 2001; Conard and Weise 1998; Keeley et al. 1999; Keeley 2002a; Moritz et al. 2004; Keeley et al. 2009; Jin et al. 2014). In part, this is because of the different natures of the two fire seasons in southern California. The chaparral fire season in southern California stretches through much of the year, but it is best described as two distinct "subseasons", which occur back-to-back and are then separated by the (ever-shortening) rainy season in the winter and early spring. General characteristics of the two fire seasons are given in Table 15.2. Most fires in southern California occur in the late spring to late summer (Keeley and Fotheringham 2003), when air humidity and fuel moistures are higher, and winds are mostly maritime and moderate. In the second fire season, which lasts from early fall to early winter, there are fewer fires, but fuel moistures are lower than in the summer (the rainy season often does not arrive till December or January), and periods of high atmospheric pressure in the interior deserts bring strong winds from the east (föhn-type winds known as "Santa Anas") that carry hot, dry air into southwestern California (Keeley and Fotheringham 2003). The two fire seasons support similar monthly burned areas (although the longer length of the spring-summer season results in more overall burned area in this season [Jin et al. 2014]), but by far the most economic loss, destruction of homes, and fatalities are during the Santa Ana season (Safford 2007; Halsey 2008; Keeley et al. 2009). In the spring-summer fire season the fuel structure of chaparral is more likely to have important effects on fire spread and control. In the Santa Ana season, the windy conditions that make fire control difficult to impossible to control are more common and fuel structure is not a reliable barrier to fire spread (Keeley 2002a; Keeley et al. 2009; Syphard et al. 2011).

Recent work has made it clear that the gradual shift from the spring-summer season to the fall-winter season is characterized by changes in the relative balance between fuels and extreme weather driving fire behavior in chaparral (Keeley 2002a; Keeley et al. 2009; Jin et al. 2014). Of course there are fires, and portions of fires, that violate this rule in both fire seasons every year, but the overall pattern is well-understood by most managers and scientists. During fuel-driven fires in chaparral, fuel manipulations are valuable from a tactical standpoint, as long as they are strategically placed in areas safe enough to deploy firefighters (Syphard et al. 2011). Under extreme weather conditions, these fuel manipulations are less effective (and in many cases simply ineffective) because even very young chaparral vegetation is flammable under dry, hot conditions; winds carry embers far from the flaming front to jump over areas of fuel treatment; aerial attack is nearly impossible; and many firefighters are deployed in and adjacent to areas of human infrastructure to protect structures and lives (Keeley 2002a; Moritz et al. 2004). The understanding that fuel manipulations tend to be differentially effective in the two fire seasons has led to a focus on the characteristics of spring-summer fires for designing fuelbreaks (e.g., the ongoing US Forest Service strategic analysis of the southern California fuelbreak network), and a focus on fire prevention, community planning, and structure hardening (adapting homes to reduce vulnerability to fire) for the inevitable Santa Ana fires in the fall and winter (see Box 15.2).

Concerning the influence of the fuel management priority on the other management priorities: fuel management has the most strongly negative direct relationship with conservation, but for many spring-summer fires and some fall (Santa-Ana) fires fuel management is an important contributor to fire management success, so there is a very important spatial relationship to consider, namely that areas that are "sacrificed" for fuel reduction can act to conserve adjacent areas (Keeley 2002a; Keeley and Safford 2016). Effects of fuel management on legal recreation are probably minimal, except from the standpoint of aesthetics. Obviously, fuel reduction has important influences on fire management in conditions where fuels limit fire spread. As noted above, this is common in the spring-summer fire season, and less common in the fall fire season when most human infrastructure and lives are lost. Unlike fuel reduction in many semi-arid conifer forests in California (e.g., ponderosa pine or mixed conifer forest, Winford et al. 2015), fuel treatment implementation on chaparral landscapes is not a restorative treatment but rather a local resource sacrifice executed for the purpose of realizing benefits on adjacent lands. As such, its direct relationship to restoration—as with conservation—is negative. Indeed, abandoned fuelbreaks may require substantial restoration. Even so, fuelbreak construction may be necessary in some situations to provide at least a modicum of protection for adjacent chaparral (or other vulnerable vegetation) stands that have been restored.

Box 15.2 Reducing Fire Impacts on Natural Resources and Communities through Science-Based Collaboration

The Santa Monica Mountains National Recreation Area (SMMNRA) is embedded within the Los Angeles metropolitan area and provides recreation opportunities to over 12 million nearby residents (Fig. 15.2). The SMMNRA, like the majority of chaparral dominated landscapes in southern California, has deviated from historical fire-return intervals with some areas burning 5–11 times in the last 90 years. Repeated fires rank among the greatest threats to the park's natural resources, equal in importance to urbanization and fragmentation (NPS 2015). When coupled with concerns over community protection, reducing fire frequency has become a key issue for SMMNRA.

Most catastrophic wildland fires in the SMMNRA occur under extreme wind events (usually "Santa Ana" winds) that increase the rate of fire spread and impede control efforts. Between 1990 and 2009, 80% of economic losses occurred under Santa Ana winds (Jin et al. 2014). Under these conditions, landscape level fuel treatments are less effective at limiting the rate of fire spread. Consequently, the SMMNRA has refocused its efforts on two areas: (1) limiting fire frequency by reducing the number of ignitions during extreme weather conditions, and (2) encouraging the creation of defensible space around houses and retrofitting homes to reduce vulnerabilities to fire.

To accomplish these goals, the SMMNRA developed the Santa Monica Mountains Wildland Fire Resilient Landscape Collaborative (SMMWFRLC).

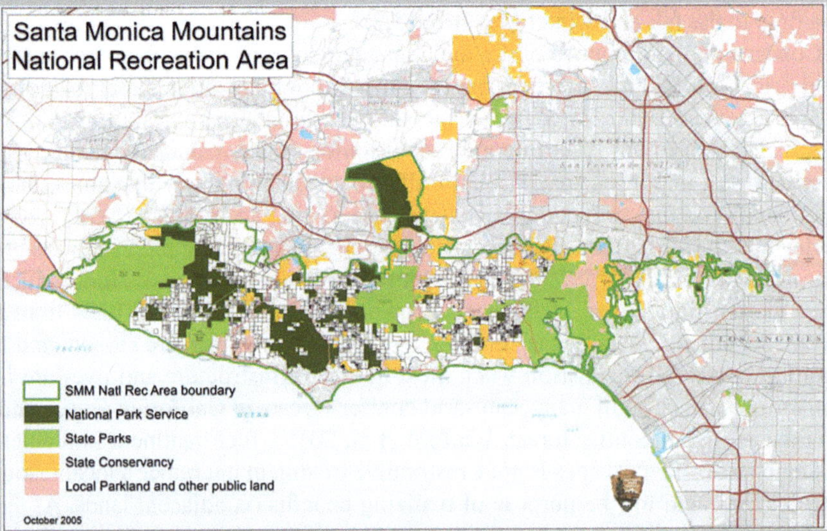

Fig. 15.2 SMMNRA comprises a diversity of ownership and an extensive wildland-urban interface. Gray and uncolored land is privately owned (image compliments of the National Park Service)

This collaborative brings together land management agencies, local fire departments, city and county governments, NGOs, fire safe councils, business and homeowner groups with a shared vision of "promoting a healthy, diverse, productive native landscape where local communities are safer and exposed to fewer major wildfires" (NPS 2015).

The SMMWFRLC has defined desired outcomes and measureable objectives for achieving their goals through coordination, communication, risk-based prioritization of treatments, and homeowner incentives across the SMMNRA. Examples of desired outcomes include:

- Improve resilience through fewer wildland fire ignitions and increased time between fires. A number of strategies will be employed to achieve this goal, including installation of concrete barriers to keep vehicles on the roadway within key ignition corridors, enhanced arson watch patrols during extreme weather events, closing parks and banning power tools during red flag weather, and relocation of powerlines under-ground in high wind corridors. Success will be measured as an increase in the average fire-return interval and a shift in the time-since-last-fire distribution for the SMMNRA.
- Improve ecosystem function through promoting native shrub cover and diversity. Actions include the control of non-native invading species, active restoration to enhance shrub recovery in disturbed or degraded areas, and reduction of fire ignitions to prevent type-conversion. Success will be measured through field monitoring and remotely sensed imagery.
- These actions will be guided by the emerging understanding that landscape resilience and maximization of ecosystem services in chaparral depends on reducing human fire ignitions and encouraging the establishment and persistence of native vegetation.

In relation to ecosystem services, chaparral fuel treatment has a necessarily negative direct effect on most of them. It removes native vegetation, reduces groundwater recharge and increases local runoff, increases erosion, and reduces landscape scenic quality. These negative effects are magnified as fuel treatments become larger. However, it is important to reiterate that the usefulness and importance of fuel treatments are not realized on the lands actually treated, but in those neighboring lands that are protected from fire. In addition, if a landscape has a high probability of burning in a large, severe event, strategically located and implemented fuel treatments and prescribed fires may increase erosion, runoff, and non-native plant invasion, but they may do so at a lower magnitude and in a more controlled fashion than a wildfire (or a series of wildfires) that burns a large area under severe weather conditions (Riggan et al. 1994; Regelbrugge 2001). It is also important to note that current-day fuel treatment practices are strongly constrained by the NEPA process and more mitigations are constantly being incorporated into these practices.

In terms of the budget devoted to fuel management, over the last 3 years (2014–2016) the southern California national forests have been allocated an average annual total of $7.26 million for wildland fuels management, which is approximately one-ninth of the budget (11.4%) allocated to wildland fire suppression (internal USFS budget numbers, G. Macias, US Forest Service, pers. comm.). Each national forest allocates these funds across a broad spectrum of activities which include salary, projects, travel, and training in addition to fuel management. This level of funding is far short of what is necessary to maintain the current fuelbreak network, and ongoing Forest Service assessments in southern California are considering how to consolidate the fuelbreak network to a set of strategically effective and defendable locations that can be adequately maintained under realistic budget scenarios (T. Metzger and S. Fillmore, US Forest Service, pers. comm.).

15.3.4 Fire Management

Fire management in chaparral vegetation is focused on immediate suppression of unplanned fires across all parts of the landscape. The number of human-caused ignitions in southern California, and other chaparral areas near urban areas in California, is extraordinary and the area of fire remains consistently high even as fire readiness and suppression budgets soar and associated staff dedicated to this task expand (Safford 2007). Although early researchers and managers believed that chaparral "needs" high frequencies of fire (e.g., Hanes 1971; Biswell 1974), the weight of evidence suggests that pre-EuroAmerican settlement fire-return intervals (FRIs) were in the range of 30–100 years (mean around 50–70 years, Minnich 2001; Van de Water and Safford 2011). Considerations of the natural history and regeneration strategies of chaparral shrub species also suggest that FRIs <15 years or so are sufficient to reduce most chaparral stands to grassland (Keeley 2006). Most chaparral dominated landscapes in California (but not all of them) occur in areas that support low densities of lightning ignitions (Keeley and Fotheringham 2001; Safford and Van de Water 2014), and the majority of pre-EuroAmerican fires would have been set by Native Americans, which probably had major influence on the distribution of grasslands and shrublands in coastal California (see Chap. 4, Keeley 2002b).

Today, many areas of chaparral in southern California, and some areas elsewhere in the State, support FRIs that are near or below 20 years (Safford and Van de Water 2014). These areas are all found adjacent to human communities, where human ignitions and flashy, herbaceous fuels interact to create zones of very high fire hazard (Syphard et al. 2007b) (Fig. 15.3).

Fire management has major influences on all of the other management focus areas and all of the ecosystem services. With respect to the management priorities: for much of southern California, fire management plays a major role in promoting conservation of unburned and infrequently burned areas. Although fire suppression has had a deleterious effect on many semi-arid conifer forests in the western US by removing a formerly important ecological process, full fire suppression is a man-

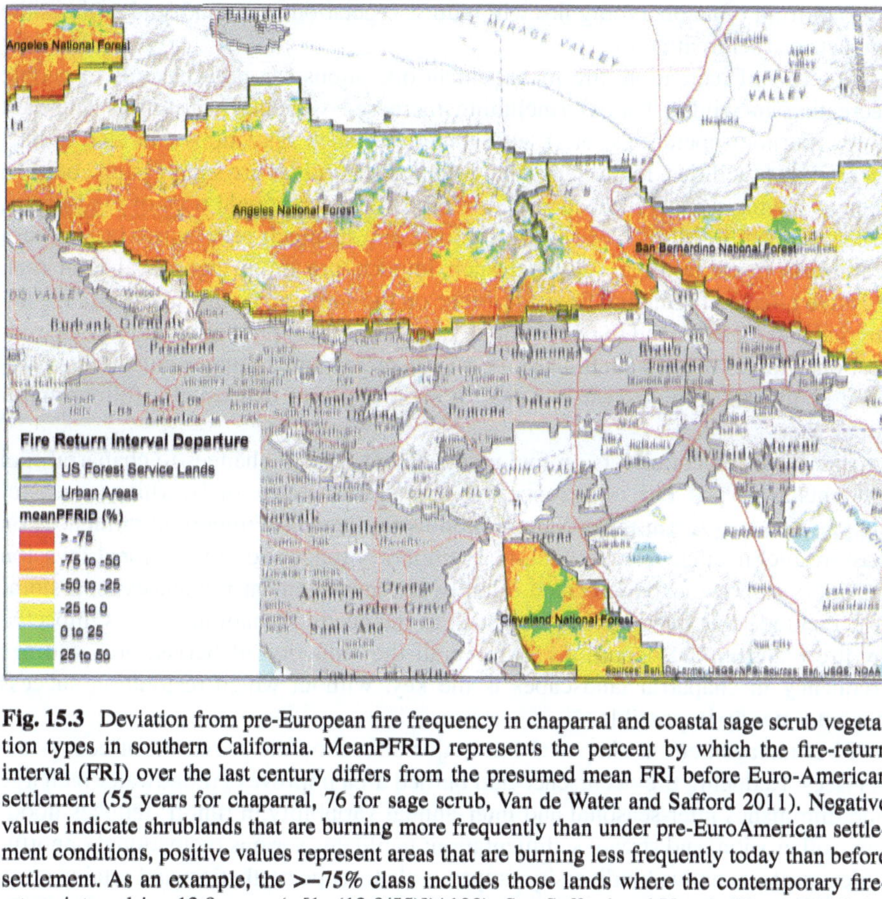

Fig. 15.3 Deviation from pre-European fire frequency in chaparral and coastal sage scrub vegetation types in southern California. MeanPFRID represents the percent by which the fire-return interval (FRI) over the last century differs from the presumed mean FRI before Euro-American settlement (55 years for chaparral, 76 for sage scrub, Van de Water and Safford 2011). Negative values indicate shrublands that are burning more frequently than under pre-EuroAmerican settlement conditions, positive values represent areas that are burning less frequently today than before settlement. As an example, the >–75% class includes those lands where the contemporary fire-return interval is <13.8 years (–[1–(13.8/55)])*100). See Safford and Van de Water (2014) for details

agement necessity in southern California chaparral, where fire frequencies are well above the natural range of variation on much of the landscape (Safford and Van de Water 2014; Keeley and Safford 2016, Fig. 15.3). Although the concept used to be derided, there is such a thing as chaparral old-growth, and it is rare and becoming rarer by the year. Such older stands support more intact soils, less stream sedimentation, more regular streamflows, fewer non-native species, and important recruitment sites for shade tolerant resprouting shrubs whose seeds are dispersed by animals (Keeley 1992).

There are some short- to medium-term negative influences of fire management on conservation, mostly at a local scale. These include soil disturbance caused by firelines created during firefighting, especially when bulldozers are used, creation of temporary roads and helispots, subsequent non-native species invasion, and illegal OHV use within these features. Fire retardant drops by aircraft also have impacts, particularly in riparian and aquatic habitats, and the burning of vegetation to reduce fuels in front of an oncoming fire will reduce vegetation cover and lead to the same impacts as fuel treatments.

In terms of recreation, fire management operations can pose hazards to people recreating in wildland areas (including increased vehicular traffic, air retardant drops, burnout operations, etc.), and sometimes such operations can also impact recreational facilities. In the sense that it impedes burning, fire suppression could be seen as having negative effects on certain postfire recreational uses of landscapes, such as postfire wildflower viewing. However, so much area burns in southern California in an average year that such impacts are more theoretical than real. Positive effects of reducing fire occurrence on the southern California landscape are myriad and these extend to recreation. Examples include protection of physical recreational infrastructure such as trails, roads, picnic and campgrounds, and avoidance of the often multi-year closures of such facilities that usually follow major fires.

Fire management effects on fuel management involve changes to chaparral fuel amounts and types. Fire occurrence will temporarily reduce fuels, while successful suppression will retain fuels. Suppression tactics can have important influences on postfire fuel management as well, i.e., temporary roads and unintentional invasion corridors as previously described. Finally, fire management influences on restoration include some negative impacts, such as the need to rehabilitate areas impacted by fire management operations, but overall the reduction of burned area and fire frequency in chaparral landscapes is the key, without which restoration success becomes nearly impossible.

Fire management in chaparral has major effects on the provision of ecosystem services. Reducing fire occurrence and burned area improves groundwater recharge and diminishes inter-seasonal and inter-annual variability in runoff; reduces flooding, soil erosion, and stream sedimentation; facilitates various recreational opportunities; protects native biodiversity; improves landscape aesthetics; and retains more carbon. At local spatial scales, and short- to medium-term temporal scales, fire management operations can also have negative effects on these services, primarily through direct impacts to vegetation, soil, and water through heavy machinery operation, fire retardant drops, and burnout operations (see Backer et al. 2004 for a more detailed discussion).

Of all of the management focus areas referenced in the management plans we reviewed, fire management is by far the most expensive priority. For example, the four southern California national forests all spend more than two-thirds of their annual budgets on fire-related expenses, and this does not include the funds that come from the national office to support the suppression of large fires that become regional or national emergencies. On the Los Padres National Forest, the percentage of the fiscal year 2015 allocated budget spent on fire relative to recreation and restoration was 74:6:2 (S. Shaw, US Forest Service, pers. comm.). Even so, in recent years fire management costs have been so high that the US Forest Service has had to routinely "borrow" funds from other management functions, a short-term fix that is creating medium- and long-term problems by diluting the agency's capacities and accomplishments in areas other than fire suppression (USDA 2015b). In landscapes dominated by frequent-fire forest types (such as yellow pine and mixed conifer forests in the Sierra Nevada or the southern California mountains), shifting fire suppression funds to prefire fuel management and more wildland fire use is likely to pay off in lower overall cost, better ecological and ecosystem service outcomes, and less environmental damage due to severe fire (North et al. 2015). In chaparral landscapes, funding for fuel management needs to be sufficient to maintain strategically useful fuelbreaks (which it currently is not), but the very different relationship between fire and chaparral and the high level of fire risk to human lives and infrastructure means that fire suppression will necessarily remain a major focus. Enhanced focus on fire prevention and education, structure hardening and home retrofitting, and changes in community land planning are likely to have major payoffs in chaparral dominated landscapes as well (Keeley 2002a; Safford 2007; Halsey 2008).

15.3.5 Restoration

Although restoration is the management priority with the fewest mentions in the management plans we reviewed, its importance has been elevated in recent years. Current US Forest Service Region 5 guidance is to integrate principles of ecological restoration into all relevant facets of US Forest Service business (USDA 2011). In addition, California Proposition 84 (2006) provides hundreds of millions of dollars for restoration projects on state and federal lands, and restoration is an important focus area for the National Park Service, as well as on the two military bases included in our study (Table 15.1).

The growing focus on restoration is based on the understanding that ecologically intact chaparral landscapes are best for providing for and balancing the various ecosystems services that southern Californians desire from the wildlands that surround their communities. Although chaparral has historically been considered a resilient vegetation type able to quickly recover following disturbance, post-disturbance recovery of chaparral is increasingly compromised in contemporary southern California. Chaparral degradation—and loss of associated ecosystem services—is typically driven by repeat disturbance events like fire or grazing (see Chap. 12), and may be exacerbated by post-disturbance drought (Pratt et al. 2014) and atmospheric nitrogen deposition (Pasquini and Vourlitis 2010), among other things.

To illustrate the impact of some of these factors for managing chaparral resources, we combined data on climatic water deficit (Flint et al. 2013), nitrogen deposition (Fenn et al. 2010), and fire frequency (Safford et al. 2011) across three fires in the Santa Clara River watershed: the Copper (occurred in 2002), Piru (2003) and Ranch (2007) fires. In each of the input data layers (Fig. 15.4a–c) the higher the value, the more negative the conditions for chaparral. For example, higher values of climatic water deficit indicate greater stress on the vegetation, higher levels of nitrogen deposition are associated with non-native annual grass invasion and persistence, and

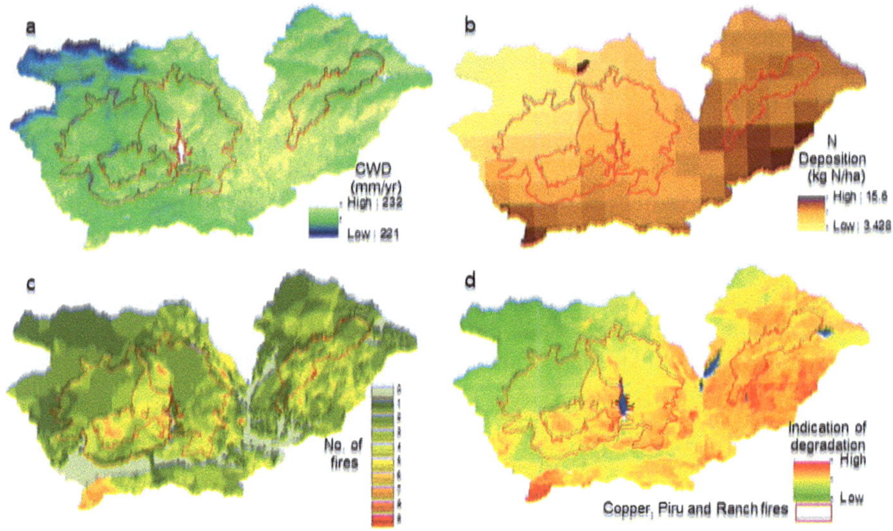

Fig. 15.4 Integration of (**a**) climatic water deficit (Flint et al. 2013), (**b**) Nitrogen deposition (Fenn et al. 2010), and (**c**) the number of fires since 1910 (Safford et al. 2011), to provide an indication of chaparral degradation (**d**) across three fires in the Santa Clara River watershed

high fire frequency has been linked with type-conversion of chaparral to non-native grasslands. The values in each of the input data layers were normalized by transforming from original units into deciles, and then summed to identify a range of degradation levels across the landscape (Fig. 15.4d). Higher degradation is indicated in the eastern part of the study area and within the southern, lower elevation regions of each of the fire perimeters, while areas in the northwest of the study area have relatively low levels of degradation.

The widespread nature of chaparral degradation and the importance of ecosystem services provided by chaparral landscapes to humans in southern California have prompted interest in chaparral restoration. While restoration of other California shrubland types, like coastal sage scrub or sagebrush steppe, has been well studied and broadly implemented, the viability of restoring chaparral is still in question (see Chap. 13). To this point, a few projects have demonstrated success in small, highly managed locations (e.g., Engel 2014), but success on the landscape scale is yet to come (see Chap. 13). To a great extent, the uncertainty in chaparral restoration comes from numerous difficult-to-control threats that thwart recovery. These include overly frequent fire, nitrogen deposition, recreational use, invasion by non-native species, and prolonged drought. Reducing the stressors that inhibit chaparral recovery is no small task, and support for this cause will require a broad understanding of the ecological, social, and economic value of chaparral ecosystems. This book is an effort in that direction.

Originally, restoration in the western United States tended to be narrowly seen as the restitution of ecosystem compositions and structures from before the period of

Euro-American settlement (Safford et al. 2012). In highly degraded landscapes like modern southern California however, such goals are unrealistic and unattainable for large portions of the landscape. Here, a landscape framework will need to be developed that permits identification of those places where traditional restorations are feasible, those places where novel conditions will need to be accommodated, and those places where a hybrid approach is most likely to succeed (Hobbs et al. 2014). In our opinion, moving forward with chaparral restoration will have more likelihood of success and gaining social support if it is approached through the lens of re-establishing ecosystem services (see Box 15.3).

With respect to the other management priorities, conservation is strongly positively influenced by restoration, except for local, largely ephemeral impacts that may result from ground-disturbing activities. Restoration has a largely positive medium- to long-term influence on recreation, since it restores aesthetic landscapes, but in the short-term it may require closure of recreation sites. The relationship between restoration and fuel and fire management is an important but complicated one. In many western US landscapes, ecological restoration is synonymous with the re-introduction of fire, but the reverse is the case in most of southern California, where fire frequencies are higher than the natural range of variation even under strict fire suppression policies (Safford and Van de Water 2014). Restoration in southern California thus requires the reduction of fire on the landscape, but this is much easier said than done. Probably the major conundrum in chaparral restoration is that most areas in major need of restoration are near human communities, which are the source of almost all fire ignitions as well as non-native, highly flammable annual grasses. In addition, restoration of tall, dense chaparral near areas of human habitation can increase fire risk, depending on the topography, weather patterns, and flammability of the local human environment. Restoration of older chaparral stands has unavoidable consequences for fire and fuel management in the form of changed fuel loading and structure.

Like conservation, restoration has primarily strongly positive influences on all of the ecosystem services referred to in the management plans we reviewed (see details of these services under Sect. 15.3.1). This is largely because the provision of many ecosystem services is generally maximized where there is intact native vegetation. These positive influences are the primary drivers behind the growing interest in chaparral restoration in southern California.

That said, like conservation and recreation, restoration receives only a small portion of the annual allocated budget in any of the management units we reviewed. The largest funding sources come from outside agency budgets, for example from California Proposition 84 funding, or funds acquired by the US Forest Service via negotiated settlements or litigation by the US Department of Justice against corporations or other entities who ignite costly wildfires. The latter source has resulted in some large allocations to select southern California watersheds, but to this point very little of the money has been spent in actually restoring chaparral, and the very short life-span of the funds (current US Forest Service policy is to use them within 3 years) means that real investment in restoration, which in highly degraded landscapes is a decades-long proposition, is nearly impossible.

15.4 Integrating Ecosystem Services into Chaparral Management

An understanding of ecosystem services, their quantity and value (see Chap. 9), can provide a framework to underpin the management of chaparral shrubland. This information can be provided by ecosystem service maps (see Box 15.3), detailing the pattern and quantity of ecosystem services across a landscape. Such maps can help spatially prioritize work, especially when there is overlap between multiple high value services and the management strategies needed to sustain them (Chan et al. 2006; Schroter and Remme 2016). This allows greater efficiencies to be achieved with limited resources, e.g., an area managed for carbon storage might also provide benefits for biodiversity and recreational use. Alternatively, maps of ecosystem services can illuminate trade-offs in natural resource management and policy decisions, for example in quantifying how vegetation removal during fuel management can impact carbon storage, biodiversity, or sediment erosion retention for a particular location.

Spatial maps of ecosystem services also provide a foundation for assessing the persistence of services under future conditions, for example, quantifying changing water runoff with a warmer, drier climate. Alternately, changes can be viewed from a demand perspective, for example the increased demand for water provision associated with growing populations and urbanized areas (Balvanera et al. 2001). Based on these findings, management decisions can be made as to what strategies will ensure the continued or increased provision of ecosystem services in the long-term.

15.4.1 Ecosystem Services and Restoration

From a broad perspective, the theory and principles of re-establishing key ecosystem components are relatively well-established, however an increasing focus on ecosystem services presents a shift in objectives in ecological restoration. Some fear an ecosystem services focus will undermine efforts to restore and conserve biodiversity, while others believe such a focus is necessary in order to guarantee human support for restoration endeavors, as well as to enable restoration efforts that occur at dimensions that are likely to actually make a difference at regional and global scales (Bullock et al. 2011; Safford et al. 2012; Alexander et al. 2016). In addition, strategies such as Payment for Ecosystem Services (PES) may present opportunities for funding conservation or restoration efforts. Despite the uncertainty and polemic in adopting a more ecosystem service-centric approach to restoration, evidence suggests an alignment between traditional principles of restoration and enhancement of ecosystem services. For example, a review of 89 studies across a range of ecosystem types showed that restoration increased biodiversity by an average of 44% while also increasing the provision of a suite of key ecosystem services (Benayas et al. 2009).

To this point, there has been minimal integration of information on ecosystem services in restoration planning and prioritization in chaparral landscapes. In general, focal areas for chaparral restoration have tended to be opportunistic and tied to habitat mitigation (see Chap. 13), stakeholder or academic/research interests, or specific funding sources, e.g., fire restoration funds allocated to a specific fire. However, integrating data on ecosystem services into decision making can contribute substantially to focusing, directing, and justifying restoration efforts.

The immediate impact of fire on chaparral and the ecosystem services it provides is substantial (see, for example, prefire compared to postfire sediment erosion in Box 15.3). However, relative to other ecosystems, healthy chaparral stands tend to be resilient to fire and recover relatively quickly. Some services, such as reduction of erosion, are greatly impacted in the first year postfire, but erosional loss of sediment from chaparral hillslopes can be down to prefire levels within 2 years of burning if vegetation recovery is normal (Wohlgemuth 2015). Postfire studies also show that chaparral shrublands generally recover their prefire biomass within about a decade (Black 1985; Bohlman et al. in press), much faster than forests burned at similar severity. With some ecosystem services, such as recreation, the rate of postfire recovery is less clear. Wildfires affect aesthetic values and recreational activities, with people tending to prefer vegetated, unburned landscapes (Hesseln et al. 1984). At the same time, the spectacular flush of wildflowers in the spring after a chaparral fire is a major attraction for botanically-inclined visitors.

In summary, when chaparral fires burn within the natural range of variation (NRV, 30–100+ years between fires), we can conclude that most ecosystem services provided by chaparral recover in a decade or less. The challenge is that much of the southern California landscape is experiencing fire at considerably higher frequencies than the NRV, and there are interacting stressors—such as nitrogen deposition and non-native plants—that further complicate the picture. Areas of degraded chaparral will necessarily provide attenuated ecosystem services. In Box 15.3 we use a recent fire from the Angeles National Forest to illustrate how data on prefire ecosystem services can be combined with information on postfire vulnerability and degradation to assist in the selection of restoration sites.

15.5 Conclusion

In this chapter we reviewed the relationships among five management focus areas on chaparral dominated public lands in southern California, as well as the relationships between these management focus areas and a suite of ecosystem services. Positive relationships among the focus areas and ecosystem services characterize situations where management actions can have multiple salutary outcomes, whereas negative relationships can help to identify situations where creative solutions are required.

Although some management focus areas have generally positive direct relationships with most other focus areas and ecosystem services (e.g., conservation and restoration), and some have generally negative direct relationships (e.g., fuel management and recreation), simple relationships were hard to find. Conservation and restoration do not benefit all segments of society, for example, and they certainly run afoul of certain economic interests.

Fuel management is a particularly complicated situation, and source of much of the tension that arises between managers and certain segments of the public in southern California and other chaparral areas. Management of chaparral fuels has negative direct effects on all of the ecosystem services we analyzed, and it runs counter to many of the other management focus areas. However, fuel management is absolutely necessary in chaparral landscapes that support human habitation, especially where such habitation has been spread across the landscape with little regard for human safety, and where the source of most fires is human ignitions. The real impact of chaparral fuel management needs to be assessed across landscapes and not in single localities, and through time and not at a given instant. From the ecological viewpoint, chaparral fuel management needs to be understood as a local resource sacrifice made in order to gain a benefit at the landscape scale. Because of its environmental impacts, such work must be carried out carefully and after comprehensive strategic analysis of the short- and long-term, local and regional impacts (Syphard et al. 2011). Careful, environmentally conscious fuel treatment planning and implementation is becoming progressively more common in chaparral landscapes. We describe one of these recent analyses in and around the Santa Monica Mountains National Recreation Area in Box 15.2.

Another, similar analysis—the so-called Strategic Fuelbreak Assessment—is underway on the four national forests in southern California. The overall objective of the analysis is to identify the fuelbreak system with "the highest probability of assisting fire suppression operations and maximizing the potential for long term maintenance." The desired outcome is a fuelbreak network that is as much meat and as little fat as possible, one that supports fuel treatments that are strategic and necessary for fire control and also maintainable in the long run. Current fuel treatments that are neither necessary nor maintainable will be allowed to revert to natural conditions. Although the analysis does not explicitly incorporate environmental variables, the process will provide an opportunity to employ the ecosystem service viewpoint in prioritizing and implementing restoration actions on those lands that will be removed from the network.

Recreation is the other area of major tension in chaparral management. Recreational use of chaparral lands, especially in southern California and the San Francisco Bay Area, is beginning to overwhelm both management agencies and ecosystems. Federal and state budgets for recreation management are small, and compounding the problem is the lack of philosophical coherence within the recreation community, where notable feuds exist between various subpopulations. The recent naming of two national monuments on lands of mixed US Forest Service and Bureau of Land Management jurisdiction in southern California (San Gabriel Mountains NM and Sand to Snow NM) was driven to a great extent by popular dismay at the inability of current recreational and educational opportunities to meet demand. Given that these lands are already heavily impacted by human use and

Box 15.3 Prioritizing Areas for Restoration in the Copper Fire
Emma C. Underwood and Allan D. Hollander

When the Copper Fire burned 7284 ha (18,000 acres) of chaparral in 2002 it had numerous impacts: it damaged recreation facilities, destroyed habitat for threatened and endangered plant species, and increased erosion from sub-catchments which impacted roads and aquatic habitats (Fig. 15.5). This case study illustrates how ecosystem service data can help prioritize sites for post-fire restoration efforts with the intent of maximizing the re-establishment of native chaparral while simultaneously enhancing the long-term provision of ecosystem services. We used a two-step process which first assessed the pre-fire quantity and pattern of four ecosystem services (water runoff, groundwater recharge, biodiversity, and carbon storage). Second, we assessed the postfire vulnerability of the landscape and the suitability for restoration by identifying: (1) locations where restoration efforts could help prevent high levels of sediment erosion, and (2) locations offering more suitable conditions for regeneration based on their fire history.

Identifying areas with high provision of ecosystem services

Data on each of the four ecosystem services were extracted from spatial data generated for a broader, southern California study area (see Chap. 9 for details of methods). Prefire patterns of water runoff were relatively high in the

Fig. 15.5 Location of the Copper Fire in the Angeles National Forest (upper map) and topographic features of the upper Copper Fire perimeter referred to in the case study

upper northwestern corner of the Copper and lowest at the southern end (Fig. 15.6a). Prefire patterns of groundwater recharge were relatively high along the base of Bee Canyon and the perimeter on the western side of San Francisquito Canyon (Fig. 15.6b). The mean annual recharge within the fire perimeter was 66 mm/year and runoff was 54 mm/year. Patterns of biodiversity (Fig. 15.6c) represent an irreplaceability index with higher values indicating areas of greater importance for meeting conservation goals for each conservation target. These targets included sensitive species, natural vegetation types, landscape connectivity, Watershed Condition Class, and streams for the federally endangered, southern California steelhead trout. Locations with higher irreplaceability values were in the lower portion of the Copper Fire on the western side of San Francisquito Canyon. The prefire estimate of carbon storage for the Copper Fire used the Enhanced Vegetation Index (EVI)

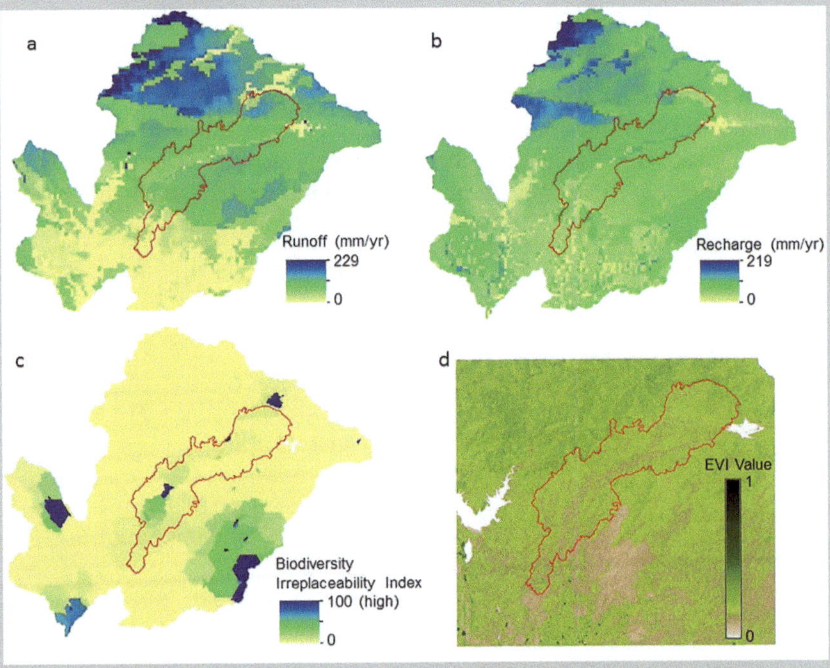

Fig. 15.6 Prefire patterns of ecosystem services in the hydrological units (HUC12) that intersect with the Copper Fire (red perimeter) shown in their original mapping units; (**a**) water runoff (270 m resolution [18 acres]), (**b**) groundwater recharge (270 m), (**c**) biodiversity (4.04-6475 ha [10–16,000 acre] polygons [minimum mapping unit 200 m]), and (**d**) the Enhanced Vegetation Index (30 m or 0.2 acres) from May 2002 used as a proxy for carbon storage (see Chap. 9 for details)

from Landsat TM imagery as a proxy. EVI values ranging from 0 to 1 prefire (May 2002) showed relatively uniform values within the fire perimeter, with lower values indicating less vegetation along Bee Canyon and San Francisquito Canyon Road, and slightly higher values in the northern tip of the fire perimeter associated with the higher elevations of Jupiter Mountain (1219 m or 4000 ft) (Fig. 15.6d).

We normalized the values in each of these services by converting from their original units to deciles, and then summed the four layers to identify prefire priorities for the provision of ecosystem services (Fig. 15.7). For the purposes of this case study, we assumed that areas with highest values of runoff, recharge, biodiversity, and biomass (carbon storage) contribute most to the provisioning of services and therefore should be priorities for restoration. High value areas of these four services combined are found in the northern, higher elevation end of the fire perimeter, around Jupiter Mountain, and along the western edge of the San Francisquito Canyon. The (summed) ecosystem service values decrease in the southern, lower elevation portion of the analysis area toward the city of Valencia.

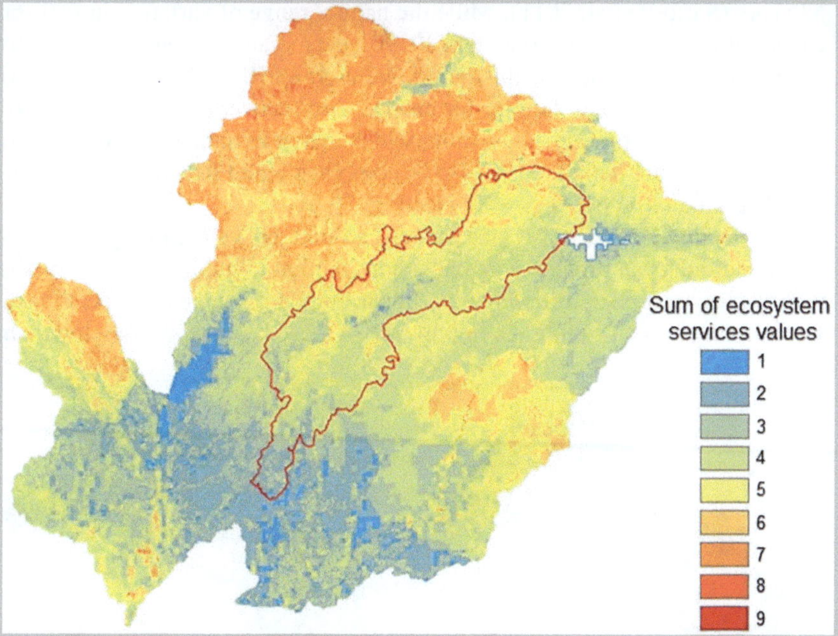

Fig. 15.7 Summation of the values of four ecosystem services across the Copper Fire (red perimeter)

Assessing Postfire Vulnerability

Evaluating the effects of fire on sediment erosion

To identify areas of high erosion vulnerability postfire we determined which subcatchments had the greatest change in sediment yield between prefire and immediately postfire. Sediment erosion was modeled using GeoWEPP software (Renschler 2003) under prefire and immediate postfire conditions. Prefire erosion was generally less than 5 tons/ha/year. (Fig. 15.8), compared to postfire where the majority of subcatchments ranged from 5 to 20 tons/ha/year. By visually inspecting the percent change between prefire and postfire sediment erosion we identified a threshold of 300% to provide a reasonably small area of the fire for the purposes of this case study. In practice, this threshold could be identified with resource managers familiar with the landscape. These areas of high erosion vulnerability were overlaid onto the map of ecosystem services provision (Fig. 15.9).

Assessing postfire regeneration suitability

Fire history was used to indicate sites within the Copper Fire that may be susceptible to degradation and type-conversion to annual non-native species. The current Fire-Return Interval (FRI) within the perimeter ranged from 8 to 53 years (Safford et al. 2011), while the natural range of variation in FRI for chaparral and serotinous conifer is 30–90 years (Van de Water and Safford 2011). For the purposes of illustration, we used a FRI of 21 years or less to indicate areas of low suitability for chaparral to recover postfire (Fig. 15.9). Other layers, like nitrogen deposition (which increases annual grass invasion) and climatic water deficit, could also be combined with FRI to help determine the likelihood of recovery postfire (see Fig. 15.4).

Combining ecosystem service values and data on vulnerability and suitable conditions can provide a valuable contribution to guiding the prioritization of chaparral restoration postfire, given limited resources and personnel. Determining how to weight the importance of these layers and assessing their

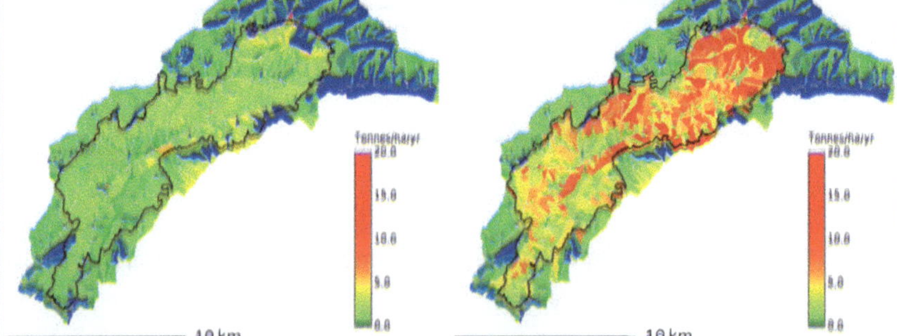

Fig. 15.8 Sediment erosion yield (tons/ha/year) by subcatchment within the Copper Fire under prefire and immediate postfire conditions

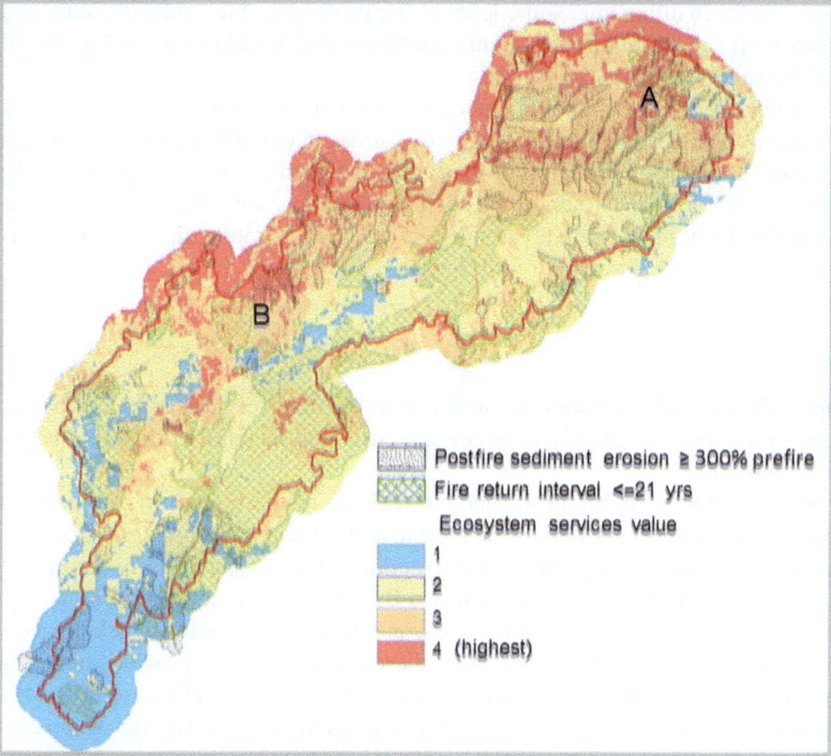

Fig. 15.9 Summation of the values of four ecosystem services condensed into four classes using natural breaks for ease of interpretation, overlain with areas where sediment erosion postfire is modeled to be ≥300% greater than prefire erosion, and areas where the fire-return interval is <=21 years. Label A (high service values and high erosion postfire) and Label B (high service values and high fire-return intervals) indicate possible locations for restoration activities that are discussed in the text

combined impact on the landscape depends on the restoration goals at the site and direction from the land management plan for the national forest, among other things.

In this illustration using the Copper Fire, we identified two locations with similar ecosystem service values but different erosion and fire characteristics. The first focused on the area around Jupiter Mountain (labelled 'A' in Fig. 15.9), where high levels of the four ecosystem services and high postfire erosion suggest a high potential payoff for restoration actions. The fire-return interval in this area is longer than the 21-year threshold, which suggests a greater probability of restored chaparral vegetation surviving long enough to develop into a resilient state. Area B is another area of high ecosystem service provision (Fig. 15.9), with some areas of high erosion, but successful restoration here could be threatened by higher fire frequencies given the current FRI

<21 years. In either site, some level of fire protection, fuels management, and non-native plant treatment is likely necessary if restoration is to be successful in the medium- to long-term.

Other data that could be included to prioritize areas for restoration include recreation values (e.g., trails, campsites), areas of cultural importance, or aesthetics. In addition, integrating maps of ecosystem services under future climate scenarios, or climatic data such as climatic water deficit, can indicate whether focal restoration areas selected today will support similar conditions for restored species in the long-term.

other anthropogenic stressors, an objective, ecosystem service-based approach will be best positioned to clearly compare the costs and benefits of different management alternatives.

Our case studies highlight the need to scale up restoration efforts from the local to the landscape level, as well as the need to better integrate ecosystem services into planning and prioritizing management actions. This is particularly important in southern California and the San Francisco Bay Area, where tens of millions of people live in close proximity to public lands, and there is huge (often unacknowledged) demand for the ecosystem benefits that these natural habitats provide, such as recreation and educational opportunities, water provision, protection from floods and debris flows, aesthetic landscapes, and the remarkable biodiversity for which coastal California is renowned. One of the key benefits of integrating ecosystem services into planning and management decision making is that it necessitates a broader perspective over longer timeframes and larger areas. For example, although fuel management negatively affects many ecosystem services at the location of the fuel treatment, under many conditions these features can help to stop undesired wildfires, protecting human lives and infrastructure but also ensuring continued provision of important ecosystem services in off-site habitats that are sensitive to frequent burning. The negative ecological impacts of a strategic fuelbreak network that is the product of a careful, holistic planning process are also likely to be much less in sum than the impacts of a large wildfire under severe weather conditions. The ecosystem service viewpoint also provides an impetus to develop multi-partner collaborations, and to integrate ecological and socio-economic aspects into chaparral management. This viewpoint will be particularly valuable in cases where management focus areas conflict—such as between conservation and recreation or conservation and fuel management—and budgets and resources are limited.

References

Alexander, S., J. Aronson, O. Whaley, and D. Lamb. 2016. The relationship between ecological restoration and the ecosystem services concept. Ecology and Society 21:34.

Baas, J. M., A. Ewert, and D. J. Chavez. 1993. Influence of ethnicity on recreation and natural environment use patterns: managing recreation sites for ethnic and racial diversity. Environmental Management 17:523-539.

Backer, D. M., S. E. Jensen, and G. R. McPherson. 2004. Impacts of fire suppression activities on natural communities. Conservation Biology 18:937-946.

Balvanera, P., G. C. Daily, P. R. Ehrlich, T. H. Ricketts, S. Bailey, S. Kark, C. Kremen, and H. Prereira. 2001. Conserving biodiversity and ecosystem services. Science 291:2047-2048.

Beier, P., K. L. Penrod, C. Lulke, W. D. Spencer, and C. Cabañero. 2006. South Coast missing linkages: restoring connectivity to wildlands in the largest metropolitan area in the USA. Pages 555-586 *in* K. R. Crooks and M. Sanjayan, editors. Connectivity conservation. Cambridge University Press, Cambridge, UK.

Benayas, J. M. R., A. C. Newton, A. Diaz, and J. M. Bullock. 2009. Enhancement of biodiversity and ecosystem services by ecological restoration: a meta-analysis. Science 325:1121-1124.

Bentley, J. R. 1967. Conversion of chaparral to grassland: techniques used in California. Agriculture Handbook 328. Washington, D.C., USA, USDA Forest Service.

Biswell, H. H. 1974. Effects of fire on chaparral. Pages 321-364 *in* T. T. Kozlowski, editor. Fire and ecosystems. Academic Press, New York, New York, USA.

Black, C. H. 1985. Biomass, nitrogen, and phosphorus accumulation over a southern California fire cycle chronosequence. Pages 445-458 *in* J. D. Tenhunen, F. M. Catarino, O. L. Lange, and W. C. Oechel, editors. Plant response to stress functional analysis in Mediterranean ecosystems. Springer-Verlag, Berlin, Germany.

Bohlman, G. N., E. C. Underwood, and H. D. Safford. Estimating biomass in California's chaparral and coastal sage scrub shrublands. Madroño, in press.

Bricker, K. S., D. J. Chavez, and W. W. Hendricks. 2008. Recreation and fire management in the urban National Forests: a study of manager perspectives. Pages 69-86 *in* D. J. Chavez, J. D. Absher, and P. L. Winter, editors. Fire social science research from the Pacific Southwest Research Station: studies supported by National Fire Plan funds. General Technical Report PSW-GTR-2009. USDA Forest Service, Pacific Southwest Research Station, Albany, California, USA.

Bullock, J. M., J. Aronson, A. C. Newton, R. F. Pywell, and J. M. R. Benayas. 2011. Restoration of ecosystem services and biodiversity: conflicts and opportunities. Trends in Ecology and Evolution 26:541-549.

Burns, C., and J. Sauer. 1992. Resistance by natural vegetation in the San Gabriel Mountains of California to invasion by introduced conifers. Global Ecology and Biogeography Letters 2:46-51.

Butsic, V., A. D. Syphard, A. Bar-Massada, and J. E. Keeley. 2016. Can private land conservation reduce wildfire risk to homes? A case study in San Diego County, California, USA. Landscape and Urban Planning 157:161-169.

Cable, D. R. 1975. Range management in the chaparral type and its ecological basis: the status of our knowledge. Research Paper RM-RP-155. Fort Collins, Colorado, USA, USDA Forest Service, Rocky Mountain Forest and Range Experiment Station.

Cermak, R. W. 2005. Fire in the forest. A history of fire control on the National Forests in California, 1898-1956. Publication R5-FR-003. USDA Forest Service, Pacific Southwest Region, Vallejo, California, USA.

Chan, K. M. A., M. R. Shaw, D. R. Cameron, E. C. Underwood, and G. C. Daily. 2006. Conservation planning for ecosystem services. PLoS Biology 4:1238-2152.

Conard, S. G., and D. R. Weise. 1998. Management of fire regime, fuels, and fire effects in southern California chaparral: lessons from the past and thoughts for the future. Pages 342-350 *in*

T. L. Pruden, and L. A. Brennan, editors. Tall timbers fire ecology conference proceedings, no. 20. Tall Timbers Research Station, Tallahassee, Florida, USA.

Corbett, E. S., and R. P. Crouse. 1968. Rainfall interception by annual grass and chaparral. Research Paper PSW-RP-48. Berkeley, California, USA, USDA Forest Service, Pacific Southwest Forest and Range Experiment Station.

Corbett, E. S., and R. M. Rice. 1966. Soil slippage increased by brush conversion. Research Note PSW-128. Berkeley, California, USA, USDA Forest Service, Pacific Southwest Forest and Range Experiment Station.

County of Los Angeles. 2016. Vegetation Management Program live fuel moisture 1981-2017, all areas, all fuels - chamise, ceanothus, sages, sagebrush. County of Los Angeles Fire Department, Forestry Division, Vegetation Management Unit, Pacoima, California, USA. http://www.fire.lacounty.gov/lfm/LFMAllArea1.pdf

Czech, B., P. Krausman, and P. Devers. 2000. Economic associations among causes of species endangerment in the United States. BioScience 50:593-601.

Dobson, A. P., J. P. Rodriguez, W. M. Roberts, and D. S. Wilcove. 1997. Geographic distribution of endangered species in the United States. Science 275:550-553.

Engel, M. D. 2014. The feasibility of chaparral restoration on type-converted slopes. Thesis. California State University, San Bernardino, California, USA.

Fenn, M. E., E. B. Allen, S. B. Weiss, S. Jovan, L. H. Geiser, G. S. Tonnesen, R. F. Johnson, L. E. Rao, B. S. Gimeno, F. Yuan, T. Meixner, A. Bytnerowicz. 2010. Nitrogen critical loads and management alternatives for N-impacted ecosystems in California. Journal of Environmental Management 91:2404-2423.

Flint, L. E., A. L. Flint, J. H. Thorne, and R. Boynton. 2013. Fine-scale hydrologic modeling for regional applications: the California Basin Characterization Model development and performance. Ecological Processes 2:25-46.

FRAP [Fire Resouce and Assesment Program], California Department of Forestry and Fire Protection. 2014. California fire perimeter geodatabase, calendar year 2013. http://frap.fire.ca.gov/data/frapgisdata-sw-fireperimeters_download

Halsey, R. W. 2008. Fire, chaparral, and survival in southern California. Second edition. Sunbelt Publications, San Diego, California, USA.

Hanes, T. L. 1971. Succession after fire in the chaparral of southern California. Ecological Monographs 41:27-52.

Hartley, M. S. 1986. An analysis of recreation management on southern California National Forests. USDA Forest Service, Pacific Southwest Region, California, USA. https://www.fs.fed.us/psw/publications/hartley/hartley_papers.pdf

Havlick, D. 2002. No place distant: roads and motorized recreation on America's public lands. Island Press, Washington, D.C., USA.

Hesseln, H., J. B. Loomis, and A. González-Cabán. 1984. Comparing the economic effects of fire on hiking demand in Montana and Colorado. Journal of Forest Economics 10:21-35.

Hobbs, R. J, E. Higgs, C. M. Hall, P. Bridgewater, F. S. Chapin III, E. C. Ellis, J. J. Ewel, L. M. Hallett, J. Harris, K. B. Hulvey, S. T. Jackson, P. L. Kennedy, C. Kueffer, L. Lach, T. C. Lantz, A. E. Lugo, J. Mascaro, S. D. Murphy, C. R. Nelson, M. P. Perring, D. M. Richardson, T. R. Seastedt, R. J. Standish, B. M. Starzomski, K. N. Suding, P. M. Tognetti, L. Yakob, and L. Yung. 2014. Managing the whole landscape: historical, hybrid, and novel ecosystems. Frontiers in Ecology and the Environment 12:557-564.

Jin, Y., J. T. Randerson, N. Faivre, S. Capps, A. Hall, and M. L. Goulden. 2014. Contrasting controls on wildland fires in southern California during periods with and without Santa Ana winds. Journal of Geophysical Research: Biogeosciences 119:432-450.

Kaczynski, A. T., and J. L. Crompton. 2006. Financing priorities in local governments: where do park and recreation services rank? Journal of Park and Recreation Administration 24:84-103.

Keeley, J. E. 1992. Recruitment of seedlings and vegetative sprouts in unburned chaparral. Ecology 73:1194-1208.

Keeley, J. E. 2002a. Fire management of California shrubland landscapes. Environmental Management 29:395-408.

Keeley, J. E. 2002b. Native American impacts on fire regimes of the California coast ranges. Journal of Biogeography 29:303-320.
Keeley, J. E. 2006. South Coast Bioregion. Pages 350-390 *in* N. G. Sugihara, J. W. van Wagtendonk, K. E. Shaffer, J. Fites-Kaufman, and A. E. Thode, editors. Fire in California's ecosystems. University of California Press, Berkeley, California, USA.
Keeley, J. E., and T. J. Brennan. 2012. Fire driven alien invasion in a fire-adapted ecosystem. Oecologia 69:1043-1052.
Keeley, J. E., C. J. Fotheringham, and M. Morais. 1999. Reexamining fire suppression impacts on brushland fire regimes. Science 284:1829-1832.
Keeley, J. E., and C. J. Fotheringham. 2001. Historic fire regime in southern California shrublands. Conservation Biology 15:1536-1548.
Keeley, J. E., and C. J. Fotheringham. 2003. Impact of past, present, and future fire regimes on North American Mediterranean shrublands. Pages 218-262 in T. T. Veblen, W. L. Baker, G. Montenegro, and T. W. Swetnam, editors. Fire and climatic change in temperate ecosystems of the western Americas. Springer, New York, New York.
Keeley, J. E., C. J. Fotheringham, and M. Moritz. 2004. Lessons from the October 2003 wildfires in southern California. Journal of Forestry 102:26-31.
Keeley, J. E., H. D. Safford, C. J. Fotheringham, J. Franklin, and M. A. Moritz. 2009. The 2007 Southern California wildfires: lessons in complexity. Journal of Forestry 107:287-296.
Keeley, J. E., and H. D. Safford. 2016. Fire as an ecosystem process. Pages 27-45 *in* H. A. Mooney and E. Zavaleta, editors. Ecosystems of California. University of California Press, Berkeley, California, USA.
Kinney, A. 1888. The forests of Los Angeles, San Diego, and San Bernardino counties, California. Report on the forest conditions in the Rocky Mountains and other papers. Government Printing Office, Washington, D.C., USA.
Klein, A. M., C. Brittain, S. D. Hendrix, R. Thorp, N. Williams, and C. Kremen. 2012. Wild pollination services to California almond rely on semi-natural habitat. Journal of Applied Ecology 49:723-732.
Knight, R. L., and K. J. Gutzwiller. 1995. Wildlife and recreationists. Coexistence through management and research. Island Press, Washington, D.C., USA.
Kremen, C., N. M. Williams, R. L. Bugg, J. P. Fay, and R. W. Thorp. 2004. The area requirements of an ecosystem service: crop pollination by native bee communities in California. Ecology Letters 7:1109-1119.
Lee, T. H. 2011. How recreation involvement, place attachment and conservation commitment affect environmentally responsible behavior. Journal of Sustainable Tourism 19:895-915.
LPFW [Los Padres Forest Watch]. 2016. Forest in the crosshairs: the environmental and health impacts of target shooting in the Los Padres National Forest. Los Padres Forest Watch, Santa Barbara, California, USA. http://lpfw.org/wp-content/uploads/2016/03/Forest-in-the-Crosshairs-Final-for-web.pdf
Luke, C., K. Penrod, C. R. Cabañero, P. Beier, W. Spencer, and S. Shapiro. 2004. A linkage design for the Santa Ana-Palomar Mountains connection. San Diego State University Field Stations Program, San Diego, California and South Coast Wildlands, Idyllwild, California, USA. http://www.scwildlands.org/reports/SCML_SantaAna_Palomar.pdf
Macey, A., and J. Gilligan. 1961. Man, fire and chaparral: a conference on southern California wildland research problems. University of California Conference Center at Lake Arrowhead, February 10-11, 1961. Agricultural Publication, Berkeley, California, USA.
McIntyre, D. L., and J. R. Weeks. 2002. Environmental impacts of illegal immigration on the Cleveland National Forest in California. The Professional Geographer 54:392-405.
Merriam, K. E., J. E. Keeley, and J. L. Beyers. 2006. Fuel breaks affect nonnative species abundance in Californian plant communities. Ecological Applications 16:515-527.
Minnich, R. A. 1983. Fire mosaics in southern California and northern Baja California. Science 219:1287-1294.
Minnich, R. A. 2001. An integrated model of two fire regimes. Conservation Biology 15:1549-1553.
Moore, R. L. 1994. Conflicts on multiple-use trails: synthesis of the literature and state of the practice. US Department of Transportation, Federal Highways Administration and National

Recreational Trails Advisory Committee, Washington, D.C., USA. http://ntl.bts.gov/lib/6000/6300/6341/Conflicts.pdf

Moritz, M. A., J. E. Keeley, E. A. Johnson, and A. A. Schaffner. 2004. Testing a basic assumption of shrubland fire management: how important is fuel age? Frontiers in Ecology and the Environment 2:67-72.

Morrison, S. A., and W. M. Boyce. 2008. Conserving connectivity: some lessons from mountain lions in southern California. Conservation Biology 23:275-285.

Muir, J. 1918. Steep trails. Houghton-Mifflin, Boston, Massachusetts, USA.

North, M. P., S. L. Stephens, B. M. Collins, J. K. Agee, G. Aplet, J. F. Franklin, and P. Z. Fulé. 2015. Reform forest fire management. Science 349:1280-1281.

NPS [National Park Service]. 2015. Foundation document Santa Monica Mountains National Recreation Area. May 2015. US Department of Interior, National Park Service, Santa Monica Mountains, Thousand Oaks, California, USA.

Pasquini, S. C., and G. L. Vourlitis. 2010. Post-fire primary production and plant community dynamics in chaparral stands exposed to varying levels of nitrogen deposition. Journal of Arid Environments 74:310-314.

Patric, J. H, and T. L. Hanes. 1964. Chaparral succession in a San Gabriel Mountain area of California. Ecology 45:353-360.

Pillsbury, A. F. 1963. Principles for the management of steep chaparral lands undergoing urbanization in the city of Los Angeles, City of Los Angeles Office of Civil Defense, Los Angeles, California, USA.

Plummer, F. G. 1911. Chaparral. Studies in the dwarf forests, or elfin-wood, of southern California. Bulletin 85, US Department of Agriculture, Forest Service. Washington, DC, USA.

Pollak, D. 2001a. Natural Community Conservation Planning. The origins of an ambitious experiment. California Research Bureau, Sacramento, California, USA.

Pollak, D. 2001b. The future of habitat conservation? The NCCP experience in southern California. California Research Bureau, Sacramento, California, USA.

Pratt, R. B., A. L. Jacobsen, A. R. Ramirez, A. M. Helms, C. A. Traugh, M. F. Tobin, M. S. Heffner, and S. D. Davis. 2014. Mortality of resprouting chaparral shrubs after a fire and during a record drought: physiological mechanisms and demographic consequences. Global Change Biology 20:893-907.

Prestemon, J. P., T. J. Hawbaker, M. Bowden, J. Carpenter, M. T. Brooks, K. L. Abt, R. Sutphen, and S. Scranton. 2013. Wildfire ignitions: a review of the science and recommendations for empirical modeling. General Technical Report SRS-GTR-171. Ashville, North Carolina, USA, USDA Forest Service, Southern Research Station.

Rahn, M. 2009. Report. Wildfire impact analysis. San Diego State University, San Diego, California, USA. http://newscenter.sdsu.edu/sdsu_newscenter/images/rahn2009fireanalysis.pdf

Reed, S. E. and A. M. Merenlender. 2008. Quiet, nonconsumptive recreation reduces protected area effectiveness. Conservation Letters 1:146-154.

Regelbrugge, J. C. 2001. Role of prescribed burning in the management of chaparral ecosystems in southern California. Pages 19-26 *in* J. E. Keeley, M. B. Keeley, and C. J. Fotheringham, editors. Second interface between ecology and land development in California. Open-File Report 00–62, US Geological Survey, Sacramento, California, USA.

Riggan, P. J. and P. H. Dunn. 1982 Harvesting chaparral biomass for energy- an environmental assessment. General Technical Report PSW-GTR-58. USDA Forest Service, Pacific Southwest Forest and Range Experimental Station, Berkeley, California, USA.

Riggan, P. J., S. E. Franklin, J. A. Brass, and F. E. Brooks. 1994. Perspectives on fire management in Mediterranean ecosystems of southern California. Pages 140–162 *in* J. M. Moreno and W. C. Oechel, editors. The role of fire in Mediterranean type ecosystems. Springer-Verlag, New York, New York, USA.

Renschler, C. S. 2003. Designing geo-spatial interfaces to scale process models: the GeoWEPP approach. Hydrological Processes 17:1005-1017.

Riverside County. 2003. Western Riverside County multiple species habitat conservation plan. http://www.rctlma.org/Portals/0/mshcp/volume1/sec1.html#1.1

Safford, H. D. 2007. Man and fire in Southern California: doing the math. Fremontia 35:25-29.

Safford, H. D., G. Hayward, N. Heller, and J. A. Wiens. 2012. Climate change and historical ecology: can the past still inform the future? Pages 46-62 *in* J. A. Wiens, G. Hayward, H. D. Safford, and C.M. Giffen, editors. Historical environmental variation in conservation and natural resource management. John Wiley & Sons, New York, New York, USA.

Safford, H. D., K. Van de Water, and D. Schmidt. 2011. California Fire Return Interval Departure (FRID) map, 2010 version. USDA Forest Service, Pacific Southwest Region and The Nature Conservancy California. http://www.fs.fed.us/r5/rsl/clearinghouse/r5gis/frid/

Safford, H. D., and K. M. Van de Water. 2014. Using Fire Return Interval Departure (FRID) analysis to map spatial and temporal changes in fire frequency on National Forest lands in California. Research Paper PSW-RP-266, USDA Forest Service, Pacific Southwest Research Station, Albany, California, USA.

Schroter, M., and R. P. Remme. 2016. Spatial prioritisation for conserving ecosystem services: comparing hotspots with heuristic optimisation. Landscape Ecology 31:431-450.

Soulé, M. E. 1985. What is conservation biology? BioScience 35:727-274.

State of California. 1981. Hungry Valley State Vehicular Recreation Area general plan, Sacramento, California, USA, California Department of Parks and Recreation. http://www.parks.ca.gov/?page_id=21299

State of California. 2002. Mount San Jacinto State Park general plan. California Department of Parks and Recreation Southern Service Center, San Diego, California, USA. http://www.parks.ca.gov/?page_id=21299

State of California. 2015. Cuyamaca Rancho State Park general plan and environmental impact report. California Department of Parks and Recreation Southern Service Center, San Diego, California, USA. http://www.parks.ca.gov/?page_id=21299

Stein, B. A., L. S. Kutner, and J. S. Adams. 2000. Precious heritage: the status of biodiversity in the United States. Oxford University Press, New York, New York, USA.

Stephenson, J. R., and G. M. Calcarone. 1999. Southern California mountains and foothills assessment. General Technical Report PSW-GTR-175. Albany, California, USA, USDA Forest Service, Pacific Southwest Research Station.

Syphard, A. D., K. C. Clarke, and J. Franklin. 2007a. Simulating fire frequency and urban growth in southern California coastal shrublands, USA. Landscape Ecology 22:431-455.

Syphard, A. D., J. E. Keeley, and T. J. Brennan. 2011. Comparing the role of fuel breaks across southern California national forests. Forest Ecology and Management 261:2038-2048.

Syphard, A. D., V. C. Radeloff, J. E. Keeley, T. J. Hawbaker, M. K. Clayton, S. I. Stewart, and R. B. Hammer. 2007b. Human influence on California fire regimes. Ecological Applications 17:1388-1402.

Syphard, A. D., V. C. Radeloff, N. S. Keuler, R. S. Taylor, T. J. Hawbaker, S. I. Stewart, and M. K. Clayton. 2008. Predicting spatial patterns of fire on a southern California landscape. International Journal of Wildland Fire 17:602-613.

Thomas, J. W. 1996. Forest Service perspective on ecosystem management. Ecological Applications 6:703-705.

Tyrrel, R. R. 1982. Chaparral in southern California. Pages 22-26 *in* C. E. Conrad, and W. C. Oechel, technical coordinators. Proceedings of the symposium on dynamics and management of Mediterranean-type ecosystems. General Technical Report PSW-GTR-58. USDA Forest Service, Pacific Southwest Forest and Range Experiment Station, Berkeley, California, USA.

USAF [US Air Force]. 2011. Integrated natural resources management plan, plan period 2011-2015. US Air Force, Vandenberg Air Force Base, California, USA. www.dtic.mil/cgi-bin/GetTRDoc?AD=ADA550681

USDA [US Department of Agriculture]. 2005. Southern California National Forests land management plan. USDA Forest Service, Vallejo, California, USA. http://www.fs.usda.gov/Internet/FSE_DOCUMENTS/fsbdev7_007721.pdf

USDA [US Department of Agriculture]. 2011. Region 5 ecological restoration leadership intent. USDA Forest Service, Pacific Southwest Region, Vallejo, California, USA. https://www.fs.usda.gov/Internet/FSE_DOCUMENTS/stelprdb5351674.pdf

USDA [US Department of Agriculture]. 2015a. The rising cost of wildfire operations: effects on the Forest Service's nonfire work. USDA Forest Service, Washington, D.C., USA. https://www.fs.fed.us/sites/default/files/2015-Rising-Cost-Wildfire-Operations.pdf

USDA [US Department of Agriculture]. 2015b. San Gabriel National Monument draft land and resource management plan. USDA Forest Service, Angeles National Forest, Arcadia, California, USA.

USDA [US Department of Agriculture]. 2015c. Fire-return interval departure (FRID) geodatabase, calendar year 2014. http://www.fs.usda.gov/detail/r5/landmanagement/gis/?cid=STELPRDB5327836

USDI [US Department of the Interior]. 2002. Santa Monica Mountains National Recreation Area general management plan and environmental impact statement. US Department of the Interior, National Park Service, Calabasas, California, USA. https://www.nps.gov/samo/learn/management/gmp-general-management-plan-documents.htm

USDI and USDA [US Department of the Interior and US Department of Agriculture]. 2004. Santa Rosa and San Jacinto Mountains National Monument final management plan and record of decision, North Palm Springs, California, USA, US Department of the Interior, Bureau of Land Management, and USDA Forest Service. http://www.blm.gov/ca/st/en/prog/nlcs/SantaRosa_SanJacintoMtns_NM/management_plan.html

USDI [US Department of the Interior]. 2015. Channel Islands National Park final general management plan/wilderness study/environmental impact statement. US Department of the Interior, National Park Service, Ventura, California, USA. https://www.nps.gov/chis/learn/management/general-management-plan.htm

USMC [US Marine Corps]. 2012. Camp Pendleton integrated natural resources management plan. US Marine Corps, Camp Pendleton, California, USA. http://www.pendleton.marines.mil/Staff-Agencies/Environmental-Security/Natural-Resources-Management-Plan/

Van de Water, K. M., and H. D. Safford. 2011. A summary of fire frequency estimates for California vegetation before Euroamerican settlement. Fire Ecology 7:26-58.

Wilcove, D. S. 1989. Protecting biodiversity in multiple-use lands: lessons from the US Forest Service. Trends in Ecology and Evolution 4:385-388.

Winford, E. M., J. T. Stevens, and H. D. Safford. 2015. Fuel treatment impacts on the ecology of yellow pine and mixed conifer forests: a synthesis. California Agriculture 69:150-156.

Wohlgemuth, P. M. 2015. Hillslope erosion and small watershed sediment yield before and after fire in southern California. Pages 977-986 *in* Proceedings of the 3rd Joint Federal Interagency Conference (10th Federal Interagency Sedimentation Conference and 5th Federal Interagency Hydrologic Modeling Conference), Reno, Nevada, April 19-23, 2015. Washington, D.C., USA.

Wohlgemuth, P. M., K. R. Hubbert, J. L. Beyers, and M. G. Narog. 2009. Post-fire watershed response at the wildland/urban interface, southern California. Pages 137-142 *in* R. M. T. Webb and D. J. Semmens, editors. Planning for an uncertain future - monitoring, integration, and adaptation. Proceedings of the Third Interagency Conference on Research in the Watersheds, Estes Park, Colorado, September 8-11, 2008. Scientific Investigations Report 2009-5049, US Geological Survey, Estes Park, Colorado, USA.

Zedler, J. B. 1996. Coastal mitigation in southern California: the need for a regional restoration strategy. Ecological Applications 6:84-93.

Zedler, P. H., C. R. Gautier, and G. S. McMaster. 1983. Vegetation change in response to extreme events: the effect of short interval between fires in California chaparral and coastal scrub. Ecology 64:809-818.

Chapter 16
Summary: The Past, Present, and Future of California Chaparral

Hugh D. Safford, Emma C. Underwood, Nicole A. Molinari, and Jon E. Keeley

Chaparral ecosystems have covered expansive swaths of low- and mid-elevation California for millions of years. Like the world's four other Mediterranean-type climate (MTC) regions, the California landscape is biologically diverse, with plentiful resources. Consequently, when human immigrants arrived from Asia in the late Pleistocene they found ample sustenance to support their needs and in turn, learned to manage these ecosystems and leave their own mark on the landscape (see Chaps. 1–3). By the time of the arrival of Europeans in California in the late eighteenth century, Native American populations near the coast were perhaps the largest of any indigenous peoples in North America. Native American management of chaparral habitats was extensive and locally intensive, and the variegated landscape that Spanish explorers and missionaries encountered near the coast and at lower elevations was largely the product of indigenous management, with fire being the central management tool (see Chap. 4).

H. D. Safford
USDA Forest Service, Pacific Southwest Region, Vallejo, CA, USA

University of California, Davis, CA, USA

E. C. Underwood
University of California, Davis, CA, USA

University of Southampton, UK

N. A. Molinari
USDA Forest Service, Los Padres National Forest, Goleta, CA, USA

J. E. Keeley
U.S. Geological Survey, Three Rivers, CA, USA

University of California, Los Angeles, CA, USA

E. C. Underwood et al. (eds.), *Valuing Chaparral*, Springer Series on Environmental Management, https://doi.org/10.1007/978-3-319-68303-4_16

Euro-American settlement in California brought a new set of cultural norms and management practices. Fire was feared, indigenous peoples were erased, and settlement and large scale resource extraction became the economic driving forces (see Chap. 12). Because chaparral did not provide many commodities necessary to support Euro-American populations, it was viewed as an impediment, and its reduction or removal an imperative. Ironically, maltreatment of the mountainsides and watersheds surrounding southern California settlements led ultimately to the establishment of some of the first federal public lands in the West, as it was clear to these early land stewards the important role chaparral played in protecting watersheds (see Chaps. 6 and 7). However, competing with these values was an ever-growing belief that chaparral landscapes could better provide other services, such as grazing, timber, or housing (see Chaps. 5, 11, and 15).

The paucity of lightning ignitions along the California coast led to long fire-free periods before humans arrived, however over the last 12,000 years human ignitions have dominated the chaparral fire regime, and as human populations have increased, so have ignitions. Today millions of Californians live embedded in landscapes that were, or are still, dominated by chaparral and other related shrubland types, and the number of annual fire ignitions is far higher than the historical range of variation. The dense nature of chaparral fuels leads inevitably to high intensity crown fires that spread rapidly across the landscape, posing a serious threat to human populations that have chosen to live within these watersheds. Unlike many western forests where fire suppression has altered natural fire regimes, in California chaparral fire suppression is necessary to offset the huge increase in anthropogenic ignitions. This need is so critical that it has resulted in an inordinate focus of management dollars on fire protection, with little time, money, or interest remaining to adequately grapple with the myriad other problems that afflict chaparral landscapes. These include non-native plants and animals, air pollution and atmospheric nutrient inputs, threatened and endangered species, erosion and sedimentation, debris flows and floods, climate change, heavy recreational use, suburban expansion, and so on (see Chaps. 6–9, 11, 12, 14, and 15).

However, there is an optimistic future for chaparral. Despite chaparral habitats being bulldozed, burned-off or degraded, people have finally begun to take notice. Once unimaginable, a chaparral conservation ethic is developing, and it is infusing decisions and opinions in government agencies, academia, environmental organizations, and the general public. In science, a measure of this is the exponential increase in Google Scholar-cited studies involving both "chaparral" and" conservation", rising from 88 between 1951 and 1955, to 6510 between 2011 and 2015 (5-year increments from 1951 to 2015; $y = (7E\text{-}65)e^{0.0777x}$; $R^2 = 0.985$). In the management sphere it is reflected in changes in the US Forest Service Forest Plans between the original plans from the mid-1980s and the most recent 2001 edition, in an uptick in restoration projects in chaparral habitats, and in recent progressive fuel treatment planning processes occurring on both national park and national forest lands (see Chaps. 12 and 15). In the political realm, three large national monuments (San Gabriel Mountains, Sand to Snow, Berryessa-Snow Mountain; ranging in size from 62,000 ha to 140,000 ha (153,205 to 345,946 acres) have been created in chaparral dominated landscapes since 2014. In the public sphere, the evidence is in the rising number of chaparral-related environmental education facilities and curricula (see Chap. 11), and in the number of recent symposia and workshops focusing on chaparral conservation and management issues, including the 2011 MEDECOS conference (the International Mediterranean Ecosystems Conference) at the University of California, Los Angeles, the 2013 and 2015 US Forest Service Southern California Chaparral Symposia, and the chaparral sessions at the 2016 Natural Areas Conference at the University of California, Davis.

This book builds on these efforts to amplify the public's understanding of the value of chaparral. Its pages are filled with rich descriptions and images of the intrinsic and intangible value of these landscapes, appreciated by Native people, early settlers, and present populations alike. As John Muir noted, chaparral is a "quickly available retreat"...where within a "few hours lowlanders can get well up into the sky and find refuge ... while breathing reviving ozone[sic], they may absorb the beauty about them" (Muir 1901). The book has also presented in depth a more formal description of "ecosystem services," one which articulates, quantifies, and values the stocks and flows of services and identifies specific beneficiaries. It exemplifies the range of ecosystem services provided by chaparral, from regulating services such as carbon storage (see Chap. 6), water provision and water quality (see Chap. 8), to provisioning services such as sediment erosion retention and flood mitigation (see Chap. 7), habitat and supporting services (see Chaps. 2 and 3), to cultural services such as recreation (see Chap. 10) which also encompasses chaparral's intrinsic value. An appreciation by society of both these formal ecosystem service values as well as the informal appreciation of chaparral's raw and natural beauty are needed in concert.

However, as is typically the case, the value of such services only becomes apparent as the services become scarce. If current trends continue, further diminishment in the supply of ecosystem services provided by chaparral seems inevitable. Future climate projections foresee temperatures rising another 2 °C to 5 °C by the end of the century (see Chap. 14), demographic projections suggest chaparral dominated counties in the San Francisco Bay Area and southern California may see human population increases of 20–50% by 2040 (CEF 2015), while high fire frequency and subsequent invasion of non-native annuals has the potential to change many of California's chaparral landscapes forever (see Chap. 12). These trends do not bode well for the sustainability of chaparral habitats and the services they provide.

In drawing this book to a close, we would like to finish by highlighting some of the key issues and challenges currently facing chaparral landscapes and the people who manage and live in them:

- **Ecosystem services:** Unlike in forested ecosystems, the recognition that there are valuable ecosystem services associated with chaparral landscapes is a recent development. For example, California carbon storage assessments have mapped chaparral areas as blanks for years, but now we know that rates of carbon sequestration and sustained carbon storage in chaparral can compare to some forest types. The quantification, spatial mapping, and economic valuation of these services and incorporation of this information into project planning and prioritization activities can revolutionize how we manage chaparral (see Chap. 9 and Box 15.3).

- **Fire suppression:** Given the continued expansion of communities and infrastructure into wildland areas and the need to protect human lives and homes, funding for fire suppression will likely continue to dominate the budgets of public agencies tasked with managing chaparral landscapes. However, steps can be taken to reduce fire risks while minimizing damage to chaparral habitats, such as pre-emptive approaches focused on ignition reduction, savvy land use planning, and structure hardening. Implementing these ideas is achievable through collaborations between city and county governments, NGOs, land management agencies, local fire departments, fire safe councils, and business and homeowner groups. In addition, strategically placed and maintained fuelbreaks will continue to aid in fire suppression, but success will be maximized when fuelbreak treatments are coupled with community preparedness that reduces structure vulnerabilities.
- **Land use planning:** For many decades, land use planning decisions have been made by cities and counties in chaparral landscapes with little consideration of the impacts on human safety and fire protection, logistics, and cost, or the reciprocal impacts on chaparral landscapes from the increased ignitions that derive from housing subdivisions. The bloated fire management budgets, and dwindling resource and recreation budgets, in the southern California public land management agencies are due directly to choices made in planning, development, and zoning departments across the region. An average year sees hundreds to thousands of homes lost to fire, human fatalities and injuries, and millions to billions of dollars in economic losses. Development pressures, and potential profits, are huge in many chaparral landscapes. Perhaps an enhanced understanding of the issues and values at risk can lead to a more reasonable balance between economic growth, human safety, and environmental protection. We hope our book will inform the debate.
- **Restoration:** Chaparral was once thought to be a resilient and stable vegetation type on the landscape, however with repeat disturbance, drought-induced dieback, and large scale type-conversion, there is a nascent movement to restore areas of degraded chaparral. Although there is much to be learned about how to accomplish chaparral restoration across broad scales, small scale projects indicate that topsoil preservation, the use of container stock, and non-native species suppression are fundamental to success, as is protection of the site from excessive fire. Land managers can also draw from assessments of ecosystem services to guide restoration priorities and focus efforts on sites where re-establishing chaparral aligns with the long-term provision of high value ecosystem services (see Box 15.3).
- **Old-growth chaparral:** Chaparral communities over 75 years old are top priorities for protection, not only because of the remarkable biodiversity they support and the ecosystem services they provide, but also because of impending threats from urban development, fragmentation, increased fire frequencies, invasive non-native plants, and warming temperatures. The fact is that very few areas remain in southern California chaparral that have escaped fire for three-quarters of a century. Such areas are ecologically unique, and they are highly threatened by all of the trends that we have highlighted throughout this book. We believe that the preservation of old-growth chaparral for its own sake should be a priority for management agencies.

- **Future climates and human demography:** Projected changes in the climate and human population will challenge the long-term integrity of chaparral landscapes. Temperature and precipitation directly influence the structure and function of chaparral communities as well as disturbances like fire that impact them. Increased densities of humans and their infrastructure will also heat the air, use more water, and ignite more fires. Where chaparral will continue to grow in 100 years is challenging to predict, as it depends on interactions between the climate, fire, and human actions. Predictions are that human populations in chaparral areas will grow by up to 50% by the middle of the century, and we are confident that the climate will be warmer. But we do not know how much warmer, and we do not know whether the warmth will be accompanied by more or less precipitation. Most published future climate studies project drier summers and more drought, but recent modeling is suggesting the possibility of wetter conditions under some scenarios. With such divergent futures possible, it seems important for the resource management agencies to undertake scenario planning. This sort of effort is underway, or is in development, in at least some agencies. For example, a climate change vulnerability and adaptation assessment was recently completed for the southern California national forests (see Box 15.1).
- **Relieving tension between conservation and recreation:** Chaparral landscapes are the natural backyard for tens-of-millions of urban residents in southern California, the San Francisco Bay Area, and elsewhere. At the same time, these landscapes support essential ecosystem services and hundreds of threatened, endangered, and sensitive species. Some areas in these landscapes are suffering from overuse, but other areas could realistically support more use. One of the chief challenges for management agencies in chaparral landscapes will be to find the proper balance between human use and conservation. Three recently named national monuments in chaparral dominated geographies (see Chap. 15) attest to the value that people place in these landscapes. The heightened profile that such designations provide will hopefully result in more public investment in modern, low-impact recreation infrastructure, enhanced enforcement, and perhaps most importantly, increased educational outreach.
- **Relieving tension between conservation and fuel management:** Another major challenge for public land management agencies is balancing conservation and fuel management priorities. Land management agencies have the unenviable task of protecting human communities that were typically built with nary a thought of the hazards posed by the surrounding environment. Some level of fuel reduction is clearly necessary, but the vegetation loss and diminishment of ecosystem services associated with such work is an important regulating factor. Looking forward, fuel management planners and the public need to acknowledge that the environmental risks of overly frequent burning in chaparral landscapes are real and significant, and that the success of fuel reduction is largely contingent on firefighter access and weather conditions. At the same time, the public needs to better understand that the real value of fuel treatment in chaparral landscapes is not usually realized in the here and now, but rather in long-term risk reduction, and broad scale protection of environmental values and ecosystem

services, usually in other, untreated areas. The major threat to both chaparral sustainability and human safety is overly frequent fire, and strategic fuel management is a necessary part of the response. In the not-too-distant future, fuel management planning (and other management actions) will benefit from the availability of spatial information on ecosystem services and how they are likely to be changed by potential management actions and disturbances. Cost-benefit assessments based on this kind of information should help to clear some of the smoke that clouds this issue currently.

It seems obvious to us that the optimal path for resolving these challenges and enhancing the conservation and sustainability of chaparral lies in uncovering and promoting the multiple benefits chaparral provides. Our book approaches the chaparral ecosystem from various angles—cultural, biological, historical, environmental, ecological, sociological, managerial—but a common theme to all of its chapters is the importance of considering the whole ecosystem and the different perspectives people hold of it. We firmly believe that an ecosystem service-based approach, coupled with a greater appreciation of its non-quantifiable components, can provide a broader and more practical understanding of the value of chaparral. In the end, we hope our book provides a common starting point for incorporating these varying perspectives of chaparral value into more robust, strategic, and creative solutions for protecting chaparral landscapes today and in the future.

Reference

CEF (California Economic Forecast). 2015. California county-level economic forecast 2015–2040. California Department of Transportation, Sacramento, California, USA. http://www.dot.ca.gov/hq/tpp/offices/eab/docs/Full%20Report%202015.pdf.

Index

E. C. Underwood et al. (eds.), *Valuing Chaparral*, Springer Series on Environmental Management, https://doi.org/10.1007/978-3-319-68303-4

B

C

MIX
Papier aus verantwortungsvollen Quellen
Paper from responsible sources
FSC® C105338

If you have any concerns about our products,
you can contact us on
ProductSafety@springernature.com

In case Publisher is established outside the EU,
the EU authorized representative is:
Springer Nature Customer Service Center GmbH
Europaplatz 3, 69115 Heidelberg, Germany

Printed by Libri Plureos GmbH
in Hamburg, Germany